Spatial Capture-Recapture

Spatial Capture-Recapture

J. Andrew Royle

Richard B. Chandler

Rahel Sollmann

Beth Gardner

USGS Patuxent Wildlife Research Center
North Carolina State University, USA

AMSTERDAM • BOSTON • HEIDELBERG • LONDON
NEW YORK • OXFORD • PARIS • SAN DIEGO
SAN FRANCISCO • SINGAPORE • SYDNEY • TOKYO

Academic Press is an imprint of Elsevier

Academic Press is an imprint of Elsevier
225, Wyman Street, Waltham, MA 02451, USA
The Boulevard, Langford Lane, Kidlington, Oxford OX5 1GB, UK
Radarweg 29, PO Box 211, 1000 AE Amsterdam, The Netherlands

Library of Congress Cataloging-in-Publication Data
A catalog record for this book is available from the Library of Congress

British Library Cataloguing-in-Publication Data
A catalogue record for this book is available from the British Library

ISBN: 978-0-12-405939-9

For information on all Academic Press publications
visit our website at http://store.elsevier.com

Printed in the United States of America
14 15 16 17   9 8 7 6 5 4 3 2 1

Contents

PART III ADVANCED SCR MODELS

Foreword

In the early 1990s, Ullas Karanth asked my advice on estimating tiger density from camera trap data. Historic uses of camera traps had been restricted to wildlife photography and the documentation of species presence. Ullas had the innovative idea to extend these uses to inference about tiger population size, density, and even survival and movement by exploiting the individual markings of tigers. I had worked on development and application of capture-recapture models, so we began a collaboration that focused on population inferences based on detection histories of marked tigers. Early on in this work, we had to consider how to deal with two problems associated with the spatial distributions of both animals and traps.

The first problem was that of heterogeneous capture probabilities among animals resulting from the positions of their ranges relative to trap locations. Animals with ranges centered in the middle of a trapping array are much more likely to encounter traps and be captured than animals with range centers just outside the trapping array. Ad hoc abundance estimators were available to deal with such heterogeneity, and we initially resolved to rely primarily on such estimators for our work.

Ullas was more interested in tiger density (defined loosely as animals per unit area) than in abundance, and the second problem resulted from our need to translate abundance estimates into estimates of density. This translation required inference about the total area sampled, that is the area containing animals exposed to sampling efforts. In the case of fixed sampling devices such as traps and cameras, the area sampled is certainly greater than the area covered by the devices themselves (e.g., as defined by the area of the convex hull around the array of devices), but how do we estimate this area? This problem had been recognized and considered since the 1930s, and ad hoc approaches to solving it included nested grids, assessment lines, trapping webs, and use of movement information from either animal recaptures or radiotelemetry data. We selected an approach using distances between captures of animals.

We thus recognized these two problems caused by spatial distribution of animals and traps, and we selected approaches to deal with them as best we could. We were well aware of the ad hoc nature of our pragmatic solutions. In particular, we viewed the use of movement information based on recaptures to translate our abundance estimates into density estimates as the weak link in our approach to inference about density.

In the early 2000s, Murray Efford developed a novel approach to inference about animal density based on capture-recapture data. The manuscript on this work was rejected initially by a top ecological journal without review (an interesting comment on the response of our peer-review system to innovation), but was published in Oikos in 2004. The approach was anchored in a conceptual model of the trapping process in which an animal's probability of being captured in any particular trap was a decreasing function of the distance between the animal's home range center and the trap. This assumed relationship was very similar to the key relationship on which distance sampling methods are based. Efford viewed the distribution of animal range centers as being governed by a spatial point process, and the target of estimation was the intensity of this process, equivalent to animal density in the study area. Efford (2004) initially used an ad hoc approach to inference based on inverse prediction. He later teamed with David Borchers to develop a formal likelihood approach to estimation (Borchers and Efford, 2008 and subsequent papers).

At about the same time that Efford was formalizing his approach, Andy Royle developed a similar approach for the related problem of density estimation based on locations of captures of animals obtained during active searches of prescribed areas (as opposed to captures in traps with fixed locations). Andy approached the inference problem using explicit hierarchical models with both a process component (the spatial distribution of animal range centers and a probability distribution reflecting movement about those centers) and an observation component (non-zero capture probability for locations within the surveyed area and zero outside this area). He used the data augmentation approach that he had just developed (Royle et al., 2007) to deal with animals in the population that are never captured, and he implemented the model using Markov chain Monte Carlo sampling (Royle and Young, 2008). Ullas and I asked Andy for help (Figure 1) with inference about tiger densities, and he extended his approach to deal with fixed trap locations by modeling detection probability as a function of the distance between range center and trap, thus solving our two fundamental problems emanating from spatial distributions of animals and traps (Royle et al., 2009a,b).

FIGURE 1

Jim Nichols (left) discussing capture-recapture with K. Ullas Karanth (right) and Andy Royle (middle) at Patuxent Wildlife Research Center, October 15, 2007.

The preceding narrative about the solution of two inference problems faced by Ullas Karanth and me was presented to motivate interest in the models that are the subject of *Spatial Capture-Recapture*. SCR models provide a formal solution to the problem of heterogeneous capture probabilities associated with locations of animal ranges relative to trap locations. They also provide a formal and direct (as opposed to ad hoc and indirect) means of estimating density, naturally defined for SCR models as the number of range centers per unit area. This motivation is perhaps adequate, but it is certainly incomplete. As noted in this book's Introduction, SCR models should not be viewed simply as extensions of standard capture-recapture models designed to solve specific spatial problems. Rather, SCR models represent a much more profound development, dealing explicitly with ecological processes associated with animal locations and movement as well as with the spatial aspects of sampling natural populations. They provide improvements over standard capture-recapture models in our abilities to address questions about demographic state variables (density, abundance) and processes (survival, recruitment), and they provide new possibilities for addressing questions about spatial organization and space use by animals.

As the promise of SCR models has become recognized, work on them has proliferated over the last 5 years, with substantive new developments led in part by the authors of this book, Andy Royle, Richard Chandler, Rahel Sollmann, and Beth Gardner. Because of this explosive development, it is no longer possible to consult one or two key papers in order to learn about SCR. Royle and colleagues recognized the need for a synthetic treatment to integrate this work and place it within a common framework. They wrote *Spatial Capture-Recapture* in order to fill this need.

The history of methodological development in quantitative ecology contains numerous examples of synthetic books and monographs that have been extremely influential in advancing the use of improved inference procedures. *Spatial Capture-Recapture* will become a part of this history, serving as a catalyst for use and further development of SCR methods. The writing style is geared to a biological readership such that this book will provide a single source for biologists interested in learning about SCR models. The statistical development is sufficiently rigorous and complete that this synthesis of existing work should serve as a springboard for statisticians interested in extensions and new developments. I believe that *Spatial Capture-Recapture* will be an extremely important book.

Spatial Capture-Recapture is organized around four major sections (plus appendices). The first, "Background and Concepts," provides motivation for SCRs and a history of relevant concepts and modeling. Two chapters are devoted to statistical background, one including material introducing random variables, common probability distributions, and hierarchical models. The second chapter on statistical background develops the concept of SCRs as generalized linear mixed models, with some emphasis on Bayesian inference methods for such models. Also included in this section is a chapter on standard (non-spatial) capture-recapture models for closed populations. This chapter helps motivate SCRs and introduces the idea of data augmentation as an approach to dealing with zero-inflated models for inference about abundance. The authors develop a primitive SCR model in this chapter by noting that location data for captured animals can be viewed as individual covariates.

The second major section, "Basic SCR Models," begins with a complete development of SCRs as hierarchical models with observation and spatial point process components. Included is a clear discussion of space use by animals, important because any model of the detection process implies a model for space use. A chapter is devoted to likelihood analysis of SCR models including both model development and an introduction to software available for fitting models. Another chapter is devoted to various approaches to modeling variation in encounter probability. A variety of basic models are introduced, as well as approaches to modeling covariates associated with traps, time, individual capture history, and individual animals (e.g., sex, body mass, random effects models). The chapter on model selection and assessment does not provide an omnibus, one-size-fits-all statistic. Rather, it describes useful approaches including AIC for likelihood analyses and both DIC and the Kuo and Mallick (1998) indicator variable approach for Bayesian analyses. For assessing model adequacy, they use the Bayesian *p*-value approach (Gelman et al., 1996) applied to different components of model fit. Another chapter is devoted to the encounter process which requires attention to the nature of the detection device (e.g., can an animal be caught only once or multiple times during an occasion, do traps permit catches of multiple or only single individuals, can an individual be detected multiple times by the same device) and the kinds of data produced by these devices. The final chapter in this section deals with the important topic of study design. A fundamental design tradeoff involves the competing needs to capture enough animals (sample size) and to attain a reasonably high average capture probability, and the authors emphasize the need for designs that represent a good compromise rather than those that emphasize one component to the exclusion of the other. General recommendations about trap spacing and clustering, and use of ancillary data (telemetry) are discussed as well. The material in this section is extremely important in conveying the basic principles underlying SCR modeling and, as such, will be the section of primary interest to many readers.

The next section, "Advanced SCR Models," will be of great interest to ecologists, not just because of the advanced model structures presented, but because of the ecological questions that become accessible using these methods. For example, the authors show how spatial variation in density can be modeled as a function of spatial covariates associated with all locations in the state space. Similarly, the authors relax the assumption of basic SCR models that encounter probability is a function of Euclidean distance between range center and trap, and focus instead on the "least cost path" between the range center and trap. The least cost path concept is modeled by including resistance parameters related to habitat covariates, and is relevant to the ecological concepts of connectivity and variable space use. The authors note ecological interest in resource selection functions, which focus on animal use of space as a function of specific resource or habitat covariates and which are typically informed by radiotelemetry data. They present a framework for development of joint models that combine SCR and resource selection function telemetry data. In some situations, sampling is done via a search encounter process rather than using detection devices with fixed locations, and SCR models are extended to deal with these. Models are developed for combining data from sampling at multiple sites or across multiple occasions. The extension of the SCR framework to models for open populations permits inference about the processes of survival, recruitment, and movement. Inference about time-specific changes in space use is also directly accessible using this approach, and I anticipate a great many advances in the development and application of open population SCR models.

The final section, "Super-Advanced SCR Models," includes a technical chapter on development of MCMC samplers for the primary purpose of providing increased flexibility in SCR modeling. A chapter of huge potential importance introduces SCR models for unmarked populations, relying on the spatial correlation structure of resulting count data to draw inferences about animal distribution and density. These models will see widespread use in studies employing remote detection devices (camera traps, acoustic detectors) to sample animals that do not happen to have individually recognizable visual patterns or acoustic signatures. In many sampling situations, some animals will be individually identifiable and many will not, and the authors develop mark-resight models to combine detection data from these two classes of animals. The final chapter provides a glimpse of the future by pointing to a sample of neat developments that should be possible using the conceptual framework provided by SCR models.

I very much like the writing style of the authors and found the book relatively easy to read (there were exceptions), with clear presentations of important ideas. Most models are illustrated nicely with actual examples and corresponding sample computer code (frequently WinBUGS).

In summary, I repeat my claim that *Spatial Capture-Recapture* is an extremely important and useful book. A thorough read of the section on basic SCR models provides a good understanding of exactly how these models are constructed and how they "work" in terms of underlying rationale. The two sections on advanced SCR models present a thorough account of the current state of the art written by those who have largely defined this state. As an ecologist, I found myself thinking of one potential application of these models after another. These methods will free ecologists to begin to think more clearly about interesting questions concerning the statics and dynamics of space use by animals. The ability to draw inferences about distribution and density of animals based on counts of unmarked individuals using remote detection devices has the potential to revolutionize conservation monitoring programs.

So does *Spatial Capture-Recapture* solve the inference problems encountered by Ullas Karanth and me two decades ago? You bet. But it does so much more than that. Andy, Richard, Rahel, and Beth, thanks for an exceptional contribution.

James D. Nichols
Patuxent Wildlife Research Center

Preface

Capture-recapture (CR) models have been around for well over a century, and in that time they have served as the primary means of estimating population size and demographic parameters in ecological research. The development of these methods has never ceased, and each year new and useful extensions are presented in ecological and statistical journals. The seemingly steady clip of development was recently punctuated with the introduction of spatial capture-recapture (SCR; a.k.a. spatially explicit capture-recapture, or SECR) models, which in our view stand to revolutionize the study of animal populations. The importance of this new class of models is rooted in the fact that they acknowledge that both ecological processes and observation processes are inherently spatial. The purpose of this book is to explain this statement, and to bring together all of the developments over the last few years while offering researchers practical options for analyzing their own data using the large and growing class of SCR models.

CR and SCR have been thought of mostly as ways to "estimate density" with not so much of a direct link to understanding ecological processes. So one of the things that motivated us in writing this book was to elaborate on, and develop, some ideas related to modeling ecological processes (movement, space usage, landscape connectivity) in the context of SCR models. The incorporation of spatial ecological processes is where SCR models present an important improvement over traditional, non-spatial CR models. SCR models explicitly describe exposure of individuals to sampling that results from the juxtaposition of sampling devices or traps with individuals, as well as the ecologically intuitive link between abundance and area, both of which are unaccounted for by traditional CR models. By including spatial processes, these models can be adapted and expanded to directly address many questions related to animal population and landscape ecology, wildlife management and conservation. With such advanced tools at hand, we believe that, but for some specific situations, traditional closed population models are largely obsolete, except as a conceptual device.

So, while we do have a lot of material on density estimation in this book—this is problem #1 in applied ecology—we worked hard to cover a lot more of the spatial aspect of population analysis as relevant to SCR. There are many books out there that cover spatial analysis of population structure that are more theoretical or mathematical, and there are many books out there that cover sampling and estimation, but that are *not* spatial. Our book bridges these two major ideas as much as is possible as of, roughly, mid-late 2012.

Themes of this book

In this book, we try to achieve a broad conceptual and methodological scope from basic closed population models for inference about population density, movement, space usage, and resource selection, and open population models for inference about vital rates such as survival and recruitment. Much of the material is a synthesis of recent research but we also expand SCR models in a number of useful directions, including the development of explicit models of landscape connectivity based on ecological or least-cost distance (Chapter 12), use of telemetry information to model resource selection with SCR (Chapter 13), and to accommodate unmarked individuals (Chapter 18), and many other new topics that have only recently, or not yet at all, appeared in the literature. Our intent is to provide a comprehensive resource for ecologists interested in understanding and applying SCR models to solve common problems faced in the study of populations. To do so, we make use of hierarchical models (Royle and Dorazio, 2008), which allow great flexibility in accommodating many types of capture-recapture data. We present many example analyses of real and simulated data using likelihood-based and Bayesian methods—examples that readers can replicate using the code presented in the text and the resources made available online and in our accompanying **R** package `scrbook`.

The conceptual and methodological themes of this book can be summarized as follows:

1. *Spatial ecology:* Much of ecology is about spatial variation in processes (e.g., density) and the mechanisms (e.g., habitat selection, movement) that determine this variation. Temporal variation is also commonly of interest and we cover this as well, but in less depth.
2. *Spatial observation error:* Observation error is ubiquitous in ecology, especially in the study of free-ranging vertebrates, and in fact the entire 100+ year history of capture-recapture studies has been devoted to estimating key demographic parameters in the presence of observation error because we simply cannot observe all the individuals that are present, and we can't know their fates even if we mark them all. What has been missing in most of the capture-recapture methods is an acknowledgment of the spatial context of sampling and the fact that capture (or detection) probability will virtually always be a function of the distance between traps and animals (or their home ranges).
3. *Hierarchical modeling:* Hierarchical models (HM) are the perfect tool for modeling spatial processes, especially those of the type covered in this book, where one process (the observation process) is conditionally related to another (the ecological process). We make use of HMs throughout this book, and we do so using both Bayesian and classical (frequentist, likelihood-based) modes of inference. These tools allow us to mold our hypotheses into probability models, which can be used for description, testing, and prediction.
4. *Model implementation:* We consider proper implementation of the models to be very important throughout the book. We explore likelihood methods using existing software such as the **R** package `secr` (Efford, 2011a), as well as development of custom solutions along the way. In Bayesian analyses of SCR models, we emphasize the use of the **BUGS** language for describing models. We also show readers how to devise their own MCMC algorithms for Bayesian analysis of SCR models, which can be convenient (even necessary) in some practical situations.

Altogether, these elements provide for a formulation of SCR models that will allow the reader to learn the fundamentals of standard modeling concepts and ultimately implement complex hierarchical models. We also believe that while the focus of the book is spatial capture-recapture, the reader will be able to apply the general principles that we cover in the introductory material (e.g., principles of Bayesian analysis) and even the advanced material (e.g., building your own MCMC algorithm) to a broad array of topics in general ecology and wildlife science. Although we aim to reach a broad audience, at times we go into details that may only be of interest to advanced practitioners who need to extend capture-recapture models to unique situations. We hope that these advanced topics will not discourage those new to these methods, but instead will allow readers to advance their own understanding and become less reliant on restrictive tools and software.

Computing[1]

We rely heavily on data processing and analysis in the **R** programming language, which by now is something that many ecologists not only know about, but use frequently. We adopt **R** because it is free, has a large community that constantly develops code for new applications, and it gives the user flexibility in data processing and analyses. There are some great books on **R** out there, including Venables and Ripley (2002), Bolker (2008), and Zuur et al. (2009), and we encourage those new to **R** to read through the manuals that come with the software. We use a number of **R** packages in our analyses, which are described in Appendix 1, and moreover, we provide an **R** package containing the scripts and functions for all of our analyses (see below).

We also rely on the various implementations of the **BUGS** language including **WinBUGS** (Lunn et al., 2000) and **JAGS** (Plummer, 2003). Because **WinBUGS** is not in active development any more, we are transitioning to mainly using **JAGS**. Sometimes models run better or mix better in one or the other. As a side note, we don't have much experience with **OpenBUGS** (Thomas et al., 2006), but our code for **WinBUGS** should run just the same in **OpenBUGS**. The **BUGS** language provides not only a computational device for fitting models but it also emphasizes understanding of what the model is and fosters understanding of how to construct models. As our good colleague Marc Kéry wrote (Kéry, 2010, p. 30) "**BUGS** frees the modeler in you." While we mostly use **BUGS** implementations, we do a limited amount of developing our own custom MCMC algorithms (see Chapter 17) which we find very helpful for certain problems where **BUGS/JAGS** fail or prove to be inefficient.

You will find a fair amount of likelihood analysis throughout the book, and we have a chapter that provides the conceptual and technical background for how to do this, and several chapters use likelihood methods exclusively. We use the **R** package

[1]Use of product names does not imply endorsement by the federal government.

secr (Efford et al., 2009a) for many analyses, and we think people should use this tool because it is polished, easy to use, fairly general, has the usual **R** summary methods, and has considerable capability for doing analysis from start to finish. In some chapters we discuss models that we have to use likelihood methods for, but which are not implemented (at the time when we wrote this book) in secr (e.g., Chapters 12 and 13). These provide good examples of why it is useful to understand the principles and to be able to implement these methods yourself.

The R package scrbook

As we were developing content for the book it became clear that it would be useful if the tools and data were available for readers to reproduce the analyses and also to modify so that they can do their own analysis. Almost every analysis we did is included as an **R** script in the scrbook package. The **R** package will be very dynamic, as we plan to continue to update and expand it.

The scrbook package can be downloaded by following links on the website: https://sites.google.com/site/spatialcapturerecapture/. Support for the scrbook package can also be found there.

The package is not meant to be general-purpose, flexible software for doing SCR models but, rather, a set of examples and templates illustrating how specific things are done. Code can be used by the reader to develop methods tailored to his/her situation, or possibly even more general methods. Because we use so many different software packages and computing platforms, we think it's impossible to put all of what is covered in this book into a single integrated package. The scrbook package is for educational purposes and not for production or consulting work.

Organization of this book

We expect that readers have a basic understanding of statistical models and classical inference (What is frequentist inference? What is a likelihood? Generalized linear model? Generalized linear mixed model?), Bayesian analysis (What is a prior distribution? And a posterior distribution?), and have used the **R** programming environment and maybe even the **BUGS** language. The ideal candidate for reading this book has basic knowledge of these topics; however, we do provide introductory chapters on the necessary components, which we hope can serve as a brief and cursory tutorial for those who might have only limited technical knowledge, e.g., many biologists who implement field sampling programs but do not have extensive experience analyzing data.

To that extent, we introduce Bayesian inference in some detail because we think readers are less likely to have had a class in that and we also wanted to produce a standalone product. Because we do likelihood analysis of many models, there is an introduction to the relevant elements of likelihood analysis in Chapter 6, and the implementation of SCR models in the package `secr` (Efford, 2011a). Our intent was to provide all of the material you need in one place, but naturally this led to one of the deficiencies with the book: it's a little bit long-winded, especially in the first, introductory part. This should not discourage you, and if you already have extensive background in the basics of statistical inference, you can skip straight ahead to the specifics of SCR modeling, starting with Chapter 5.

In the following chapters we develop a comprehensive synthesis and extension of spatial capture-recapture models. Roughly the first third of the book is introductory material. In Chapter 3 we provide the basic analysis tools to understand and analyze SCR models, namely generalized linear models (GLMs) with random effects, and demonstrate their analysis in **R** and **WinBUGS**. Because SCR models represent extensions of basic CR models, we cover ordinary closed population models in Chapter 4.

In the second section of the book, we extend capture-recapture to SCR models (Chapter 5), and discuss a number of different conceptual and technical topics including tools for likelihood inference (Chapter 6), analysis of model fit and model selection (Chapter 8), and sampling design (Chapter 10). Along with Chapters 7 and 9, this part of the book provides the basic introduction to spatial capture-recapture models and their analysis using Bayesian and likelihood methods.

The third section of the book covers more advanced SCR models. We have a number of chapters on spatial modeling topics related to SCR, including modeling spatial variation in density (Chapter 11), modeling landscape connectivity or "ecological distance" using SCR models (Chapter 12), and modeling space usage or resource selection (Chapter 13), which includes material on integrating telemetry data into SCR models. After this there are three chapters that involve some elements of modeling spatially or temporally stratified populations. We cover Bayesian multi-session models in Chapter 14, what we call "search-encounter" models in Chapter 15 and, finally, fully open models involving movement or population dynamics in Chapter 16. The reason we view the search-encounter models, Chapter 15, as a prelude to fully open models is that these models apply to situations where we observe the animal locations "unbiased by fixed sampling locations"—so we get to observe clean measurements of movement outcomes, which is a temporal process. When this is possible, we can resolve parameters of explicit movement models free of those that involve encounter probability. For example, one such model has two "scale" parameters: σ that determines the rate of decay in encounter probability from a sampling point or line, and τ which is the standard deviation of movements about an individuals activity center.

The final section of this book is what we call "Super-advanced SCR Models." We include a chapter on developing your own MCMC algorithms for SCR models because many advanced models require you to do this, or can be run more efficiently than in the **BUGS** language, and we thought some readers would appreciate a practical introduction to MCMC for ecologists. Following the MCMC chapter, we have a number of topics related to unmarked individuals (Chapter 18) or partially marked populations (Chapter 19). This last section of the book contains some research areas that we are currently developing but lays the foundation for further development of novel extensions and applications.

When this project was begun in 2008, the idea of producing a 550 page book would have been unimaginable—there wasn't that much material to work with. Optimistically, there was maybe a 250 page monograph that could have been squeezed out of the literature. But, during the project, great and new things appeared in the literature, and we developed new models and concepts ourselves, in the process of writing the book. There are at least 10 chapters in the book that we couldn't have thought about 5 years ago. We hope that the result is a timely summary and a lasting resource and inspiration for future developments.

Acknowledgments

The project owes a great intellectual debt to Jim Nichols, who has been an extraordinary mentor and colleague, and who generously shared his astounding insight into animal sampling and modeling problems, his knowledge of the literature and history of abundance and density estimation and, most importantly, his extremely valuable time. He has been an extremely helpful guy on all fronts. We are honored that Jim agreed to write the Foreword to introduce the book. We thank Marc Kéry for being a great friend and colleague, and for his creativity, energy, and enthusiasm in developing new ideas and presenting workshops on hierarchical modeling in ecology.

Special thanks to the following people: (1) Kimberly Gazenski whose support was invaluable. She worked on administrative, technical, and editorial aspects of the whole project. She maintained the BibTeX database, worked on the GitHub repository that housed the LaTeX and **R** package source trees, edited LaTeX files, tested **R** scripts, did GIS and **R** programming, analysis, debugging, and graphics; (2) Our WCS Tiger program colleagues K. Ullas Karanth and Arjun Gopalaswamy for continued support and collaboration on SCR problems; (3) Sarah Converse, our PWRC colleague, for her interest in SCR models and developing a number of methodological and application papers related to multi-session models, providing feedback on draft material, and friendship; (4) Murray Efford whose seminal 2004 Oikos paper first introduced spatial capture-recapture models. His **R** package `secr` is a powerful tool for analyzing spatial capture-recapture data used throughout the book. Murray also answered many questions regarding `secr` that were helpful in developing our applications and examples.

We thank the following people for providing data, photographs, or figures: Agustín Paviolo (jaguar data in Chapter 11, and the cover image). Michael Wegan, Paul Curtis, and Raymond Rainbolt (black bear data from Fort Drum, NY); Audrey Magoun (wolverine data and photographs); Cat Sun and Angela Fuller (black bear data in Chapter 13; and Chapter 20 photos); Joshua Raabe and Joseph Hightower (American shad photo and data in Chapter 16); Erin Zylstra (tortoise data in Chapter 4); Martha (Liz) Rutledge (Canada geese data and picture in Chapter 19); Craig Thompson (fisher data in Chapter 15 and photographs in Chapter 1); Jerrold Belant (black bear data in Chapter 10); Kevin and April Young (FTHL photograph, Chapter 15); Theodore Simons, Allan O'Connell, Arielle Parsons and Jessica Stocking (raccoon data in Chapter 19) Marty DeLong (weasel photograph, Chapter 20), and Bob Wiesner (mountain lion photograph, Chapter 15).

We thank the following people for reviewing one or more draft chapters and giving feedback along the way: David Borchers, Sarah Converse, Bob Dorazio, Angela Fuller, Tim Ginnett, Evan Grant, Tabitha Graves, Marc Kéry, Brett McClintock, Leslie New, Allan O'Connell, Krishna Pacifici, Agustín Paviolo, Brian Reich, Robin Russell, Sabrina Servanty, Cat Sun, Yifang Li, Earvin Balderama, and Chris Sutherland.

Additionally, many colleagues and friends, as well as our families, provided us with encouragement and feedback throughout this project and we thank them for their continued support.

PART

I

Background and Concepts

CHAPTER

Introduction

1

Space plays a vital role in virtually all ecological processes (Tilman and Kareiva, 1997; Hanski, 1999; Clobert et al., 2001). The spatial arrangement of habitat can influence movement patterns during dispersal, habitat selection, and survival. The distance between an organism and its competitors and prey can influence activity patterns and foraging behavior. Further, understanding distribution and spatial variation in abundance is necessary in the conservation and management of populations. The inherent spatial aspect of *sampling* populations also plays an important role in ecology as it strongly affects, and biases, how we observe population structure (Seber, 1982; Buckland et al., 2001; Borchers et al., 2002; Williams et al., 2002). However, despite the central role of space and spatial processes to both understanding population dynamics and how we observe or sample populations, a coherent framework that integrates these two aspects of ecological systems has not been fully realized either conceptually or methodologically.

Capture-recapture (CR) methods represent perhaps the most common technique for studying animal populations, and their use is growing in popularity due to recent technological advances that provide methods to study many taxa which before could not be studied efficiently, if at all. However, a major deficiency of classical capture-recapture methods is that they do not admit the spatial structure of the ecological processes that give rise to encounter history data, nor the spatial aspect of collecting these data. While many technical limitations of applying classical capture-recapture methods, related to their lack of spatial explicitness, have been recognized for decades (Dice, 1938; Hayne, 1950), it has only been very recently (Efford, 2004; Borchers, 2012) that spatially explicit capture-recapture methods—those which accommodate space—have been developed.

Spatial capture-recapture (SCR) methods resolve a host of technical problems that arise in applying capture-recapture methods to animal populations. However, SCR models are not merely an extension of technique. Rather, they represent a much more profound development in that they make ecological processes explicit in the model—processes of density, spatial organization, movement, and space usage by individuals. The practical importance of SCR models is that they allow ecologists to study elements of ecological theory using individual encounter data that exhibit various biases relating to the observation mechanisms employed. At the same time, SCR models can be used, and may be the only option, for obtaining demographic data on some of the rarest and most elusive species—information which is required for

Spatial Capture-Recapture. http://dx.doi.org/10.1016/B978-0-12-405939-9.00001-3

effective conservation. It is this potential for advancing both applied and theoretical research that motivated us to write this book.

1.1 The study of populations by capture-recapture

In the fields of conservation, management, and general applied ecology, information about abundance or density of populations and their vital rates is a basic requirement. To that end, a huge variety of statistical methods have been devised, and as we noted already, the most well developed are collectively known as capture-recapture (or capture-mark-recapture) methods. For example, the volumes by Otis et al. (1978), White et al. (1982), Seber (1982), Pollock et al. (1990), Borchers et al. (2002), Williams et al. (2002), and Amstrup et al. (2005) are largely synthetic treatments of such methods, and other contributions on modeling and estimation using capture-recapture are plentiful in the peer-reviewed ecology literature.

Capture-recapture techniques make use of individual *encounter history* data, by which we mean sequences of (usually) 0s and 1s denoting if an individual was encountered during sampling over a certain time period (occasion). For example, the encounter history "010" indicates that this individual was encountered only during the second of three trapping occasions. As we will see, these data contain information about encounter probability, and abundance, and other parameters of interest in the study of populations.

Capture-recapture has been important in studies of animal populations for many decades, and its importance is growing dramatically in response to technological advances that improve our ability and efficiency to obtain encounter history data. Historically, such information was only obtainable using methods requiring physical capture of individuals. However, new methods do not require physical capture or handling of individuals. A large number of detection devices and sampling methods produce individual encounter history data including camera traps (Karanth and Nichols, 1998; O'Connell et al., 2011), acoustic recording devices (Dawson and Efford, 2009), and methods that obtain DNA samples such as hair snares for bears, scent posts for many carnivores, and related methods which allow DNA to be extracted from scat, urine, or animal tissue in order to identify individuals. This book is concerned with how such individual encounter history data can be used to carry out inference about animal abundance or density, and other parameters such as survival, recruitment, resource selection, and movement using new classes of capture-recapture models, spatial capture-recapture,[1] which utilize auxiliary spatial information related to the encounter process.

As the name implies, the primary feature of SCR models that distinguishes them from traditional CR methods is that they make use of the spatial information inherent to capture-recapture studies. Encounter histories that are associated with information on the location of capture are *spatial encounter histories*. This auxiliary information is informative about spatial processes including the spatial organization of individuals,

[1] In the literature the term spatially explicit capture-recapture (SECR) is also used, but we prefer the more concise term.

variation in density, resource selection and space usage, and movement. As we will see, SCR models allow us to overcome critical deficiencies of non-spatial methods, and integrate ecological theory with encounter history data. As a result, this greatly expands the practical utility and scientific relevance of capture-recapture methods, and studies that produce encounter history data.

1.2 Lions and tigers and bears, oh my: genesis of spatial capture-recapture data

A diverse number of methods and devices exist for producing individual encounter history data with auxiliary spatial information about individual locations. Historically, physical "traps" have been widely used to sample animal populations. These include live traps, mist nets, pitfall traps, and many other types of devices. Such devices physically restrain animals until visited by a biologist, who removes the individual, marks it or otherwise molests it in some scientific fashion, and then releases it. Although these are still widely used, recent technological advances for obtaining encounter history data non-invasively have made it possible to study many species that were difficult if not impossible to study effectively just a few years ago. As a result, these methods have revolutionized the study of animal populations by capture-recapture methods, have inspired the development of spatially explicit extensions of capture-recapture, and will lead to their increasing relevance in the future. We briefly review some of these techniques here, which we consider in more detail in later chapters of this book.

1.2.1 Camera trapping

Considerable recent work has gone into the development of camera trapping methodologies (Figure 1.1). For a historical overview of this method see Kays et al. (2008) and Kucera and Barrett (2011). Several recent synthetic works have been published including Nichols and Karanth (2002), and an edited volume by O'Connell et al. (2011) devoted solely to camera trapping concepts and methods. As a method for estimating abundance, some of the earliest work that relates to the use of camera trapping data in capture-recapture models originate from Karanth (1995) and Karanth and Nichols (1998, 2000).

In camera trapping studies, cameras are often situated along trails or at baited stations, and individual animals are photographed and subsequently identified either manually by a person sitting behind a computer, or sometimes now using specialized identification software. Camera trapping methods are widely used for species that have unique stripe or spot patterns such as tigers (Karanth, 1995; Karanth and Nichols, 1998), ocelots (*Leopardus pardalis*; Trolle and Kéry, 2003, 2005), leopards (*Panthera pardus*; Balme et al., 2010), and many other cat species. Camera traps are also used for other species such as wolverines (*Gulo gulo*; Magoun et al., 2011; Royle et al., 2011b), and even species that are less easy to uniquely identify such as mountain lions (*Puma concolor*; Sollmann et al., 2013b) and coyotes (*Canis latrans*; Kelly et al., 2008).

FIGURE 1.1

Left: Wolverine being encounted by a camera trap (*Photo credit: Audrey Magoun*). Right: Tiger encountered by camera trap (*Photo credit: Ullas Karanth*).

We note that even for species that are not readily identified by pelage patterns, it might be efficient to use camera traps in conjunction with spatial capture-recapture models to estimate density (see Chapters 18 and 19).

1.2.2 DNA sampling

DNA obtained from hair, blood, or scat is now routinely used to obtain individual identity and encounter history information about individuals (Taberlet and Bouvet, 1992; Kohn et al., 1999; Woods et al., 1999; Mills et al., 2000; Schwartz and Monfort, 2008). A common method is based on the use of "hair snares" (Figure 1.2), which are widely used to study bear populations (Woods et al., 1999; Garshelis and Hristienko, 2006; Kendall et al., 2009; Gardner et al., 2010b). A sample of hair is obtained as individuals pass under or around barbed wire (or another physical mechanism) to

FIGURE 1.2

Left: Black bear in a hair snare (*Photo credit: M. Wegan*). Right: European wildcat loving on a scent stick (*Photo credit: Darius Weber*).

take bait. Hair snares and scent sticks have also been used to sample felid populations (García-Alaníz et al., 2010; Kéry et al., 2010) and other species. Research has even shown that DNA information can be extracted from urine deposited in the wild (e.g., in snow; see Valiere and Taberlet, 2000) and as a result this may prove another future data collection technique where SCR models are useful.

1.2.3 Acoustic sampling

Many studies of birds (Dawson and Efford, 2009), bats, and whales (Marques et al., 2009) now collect data using devices that record vocalizations. When vocalizations can be identified to individual at multiple recording devices, spatial encounter histories are produced that are amenable to the application of SCR models (Dawson and Efford, 2009; Efford et al., 2009b). Recently, these ideas have been applied to data on direction or distance to vocalizations by multiple simultaneous observers and related problems (D. L. Borchers, ISEC 2012 presentation).

1.2.4 Search-encounter methods

There are other methods which don't fall into a nice clean taxonomy of methods. Spatial encounter histories are commonly obtained by conducting manual searches of geographic sample units such as quadrats, transects or road or trail networks. For example, DNA-based encounter histories can be obtained from scat samples located along roads or trails or by specially trained dogs (MacKay et al., 2008) searching space (Figure 1.3). This method has been used in studies of martens, fishers

FIGURE 1.3

Left: A wildlife research technician for the USDA Forest Service holding a male fisher captured as part of the Kings River Fisher Project in the Sierra National Forest, California. Right: A dog handler surveying for fisher scat in the Sierra National Forest (*Photo credit: Craig Thompson*).

(Thompson et al., 2012), lynx, coyotes, birds, and many other species. A similar data structure arises from the use of standard territory or spot mapping of birds (Bibby et al., 1992) or area sampling in which space is searched by observers to physically capture individuals. This is common in surveys that involve reptiles and amphibians, e.g., we might walk transects picking up box turtles (Hall et al., 1999), or desert tortoises (Zylstra et al., 2010), or search space for lizards (Royle and Young, 2008).

These methods don't seem like normal capture-recapture in the sense that the encounter of individuals is not associated with specific trap locations, but SCR models are equally relevant for analysis of such data as we discuss in Chapter 15.

1.3 Capture-recapture for modeling encounter probability

We briefly introduced techniques used for the study of animal populations. These methods produce individual encounter history data, a record of where and when each individual was captured. We refer to this as a *spatial encounter history*. Historically, auxiliary spatial information has been ignored, and encounter history data have been *summarized* to simple "encounter or not" for the purpose of applying ordinary CR models. The basic problem with these ordinary (or "non-spatial") capture-recapture models is that they do not contain an explicit sense of space, the spatial information is summarized out of the data set, so we aren't able to use such models for studying movement, or resource selection, etc. Instead, ordinary capture-recapture models usually resort to a focus on models of "encounter probability," which is a nuisance parameter, seldom of any ecological relevance. We show an example here that is in keeping with the classical application of ordinary capture-recapture models.

1.3.1 Example: Fort Drum bear study

Here we confront the simplest possible capture-recapture problem—but one of great applied interest—estimating population size (N) and density from a standard capture-recapture study. We use this as a way to introduce some concepts and motivate the need for spatial capture-recapture models by confronting technical and conceptual problems that we encounter. The data come from a study to estimate black bear abundance on the U.S. Army's Fort Drum Military Installation in northern New York (Wegan (2008), see also Chapter 4 for more details). The specific data used here are encounter histories of 47 individuals obtained from an array of 38 baited "hair snares" during June and July 2006. The study area and locations of the 38 hair snares are shown in Figure 1.4. Barbed wire traps (see Figure 1.2) were baited and checked for hair samples each week for eight weeks. Analysis of these data appear in Gardner et al. (2009, 2010b), and we use the data in a number of analyses in later chapters.

Although each bear was captured, or not, in each of the 38 hair snares, we start by treating this data set as a standard capture-recapture data set and summarize to an

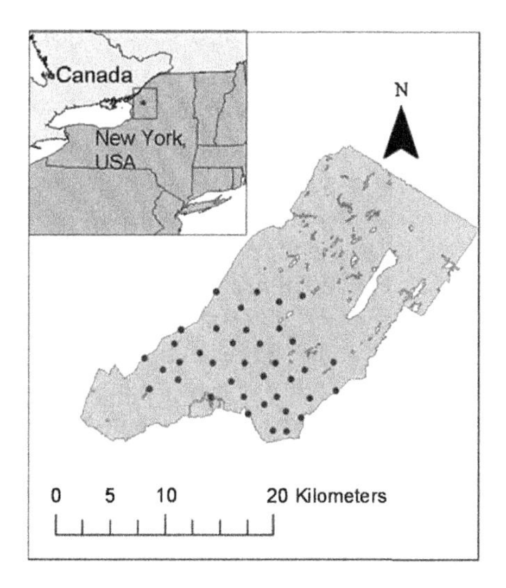

FIGURE 1.4

Locations of hair snares on Fort Drum, New York, operated during the summer of 2006 to sample black bears.

encounter history matrix with 47 rows and 8 columns with entries y_{ik}, where $y_{ik} = 1$ if individual i was captured, at any trap, in sample occasion k and $y_{ik} = 0$ otherwise. There is a standard closed population model for estimating N, colloquially referred to as "model M_0" (see Chapter 4), which assumes that encounter probability p is constant for all individuals and sample periods. We fitted model M_0 to the Fort Drum data using traditional likelihood methods, yielding the maximum likelihood estimate (MLE) of $\hat{N} = 49.19$ with an asymptotic standard error (SE) of 1.9.

The key issue in using such a closed population model regards how we should interpret this estimate of $\hat{N} = 49.19$ bears. Does it represent the entire population of Fort Drum? Certainly not—the trapping array covers less than half of Fort Drum as we see in Figure 1.4. So to estimate the total bear population size of Fort Drum, we would have to convert our $\hat{N}$ to an estimate of density (D) and extrapolate. To get at density, then, should we assert that $\hat{N}$ applies to the southern half of Fort Drum below some arbitrary line? Surely bears move on and off of Fort Drum without regard to hypothetical boundaries. Without additional information there is simply no way of converting this estimate of N to density, and hence it is really not meaningful biologically. To resolve this problem, we will adopt the customary approach of converting $\hat{N}$ to $\hat{D}$ by buffering the convex hull around the trap array. The convex hull has area 157.135 km^2. We follow Bales et al. (2005) in buffering the convex hull of the trap array by the radius of the mean female home range size.

The mean female home range radius was estimated for this study region to be 2.19 km (Wegan, 2008), and the area of the convex hull buffered by 2.19 km is

277.01 km^2. (**R** commands to compute the convex hull, buffer it, and compute the area are given in the **R** package `scrbook` which accompanies this book). Hence, the estimated density here is approximately 0.178 bears/km^2 using the estimated population size obtained by model M_0. We could assert that the problem has been solved, go home, and have a beer. But then, on the other hand, maybe we should question the use of the estimated home range radius—after all, this is only the female home range radius and home ranges very for many reasons. Instead, we may decide to rely on a buffer width based on one-half mean maximum distance moved (MMDM) estimated from the actual hair snare data as is more customary (Dice, 1938). In that case the buffer width is 1.19 km, and the resulting estimated density is increased to 0.225 bears/km^2, about 27% larger. But wait—some studies actually found the full MMDM (Parmenter et al., 2003) to be a more appropriate measure of movement (e.g., Soisalo and Cavalcanti, 2006). So maybe we should use the full MMDM which is 2.37 km, pretty close to the telemetry-based estimate and therefore providing a similar estimate of density (0.171 bears/km^2). So in trying to decide how to buffer our trap array we have already generated three density estimates. The crux of the matter is obvious: Although it is intuitive that N should scale with area—the number of bears should go up as area increases and go down as area decreases—in this ad hoc approach of accounting for animal movement N remains the same, no matter what area we assert was sampled. The number of bears and the area they live in are not formally tied together within the model. Estimating N and estimating the area N relates to are two completely independent analytical steps which are not related to one another by a formal model.

Unfortunately, our problems don't end here. In thinking about the use of model M_0, we might naturally question some of the basic assumptions that go into that model. The obvious one to question is that detection probability p is constant. One clear source of variation in p is variation *among individuals*. We expect that individuals may have more or less exposure to trapping due to their location relative to traps, and so we try to model this "heterogeneous" encounter probability phenomenon. To illustrate this phenomenon, here are the number of traps that each individual was encountered in:

```
# traps:  1   2   3  4  5  6  9
# bears:  23  13  6  2  1  1  1
```

meaning that 23 bears were captured in only 1 trap, and 1 bear was captured in 9 distinct traps, etc. The variation in trap-encounter frequencies suggests quite a range in traps exposed to by bears in the sampled population. Historically, researchers try to reduce spatial heterogeneity in capture probability by placing >1 trap per home range (Otis et al., 1978; Williams et al., 2002). This seems like a sensible idea but it is difficult to do in practice since we don't know where all the home ranges are and so we try to impose a density of traps that averages to something >1 per home range. An alternative solution is to fit models that allow for individual heterogeneity in p (Karanth, 1995). Such models have the colloquial name of "model M_h" (Otis et al., 1978). We fitted this model (see Chapter 4 for details) to the Fort Drum data using each of the three buffer widths previously described (telemetry, 1/2 MMDM and MMDM),

Table 1.1 Table on estimates of density (D, bears/km^2) for the Fort Drum data using models M_0 and M_h and different buffers. Model M_h here is a logit-normal mixture (Coull and Agresti, 1999). AIC is Akaike Information Criterion for model selection. Models with small values of AIC are favored (see Chapt. 8).

Model	Buffer	$\hat{D}$	SE	AIC
M_0	Telemetry	0.178	0.178	−72.58
M_0	MMDM	0.171	0.171	−72.58
M_0	1/2 MMDM	0.225	0.225	−72.58
M_h	Telemetry	0.341	0.144	−105.43
M_h	MMDM	0.327	0.138	−105.43
M_h	1/2 MMDM	0.432	0.183	−105.43

producing the estimates reported in Table 1.1. While we can tell by the models' AIC that M_h is clearly favored by more than 30 units, we might still not be entirely happy with our results. There is information in our data that could tell us something about the exposure of individual bears to the trap array—where they were captured, and how many times—but since space has no representation in our model, we can't make use of this information. Model M_h thus merely accounts for what we observe in our data (some bears were more frequently captured than others) rather than explicitly accounting for the processes that generated the data.

So what are we left with? Our density estimates span a range from 0.17 to 0.43 bears/km^2 depending on which estimator of N we use and what buffer strip we apply. Should we feel strongly about one or the other? Which buffer should we prefer? The Akaike Information Criterion (AIC) favors model M_h (Table 1.1), but does that model adequately account for the differences in exposure of individuals to the trap array? Are we happy with a purely phenomenological model for heterogeneity? It assumes that encounter probabilities for all individuals are independent and identically distributed (iid) random variables, and does not account for the explicit mechanism of induced heterogeneity. And, further, we have information about that (trap of capture) which model M_h ignores. Note also that choice of buffering method does not affect AIC. Therefore, if we choose one type of buffer, how do we compare our density estimates to those from other studies that may opt for a different kind of buffer? The fact that N does not scale with area, A, as part of the model, renders this choice arbitrary.

1.3.2 Inadequacy of non-spatial capture-recapture

The parameter N (population size) in an ordinary capture-recapture (CR) model is functionally unrelated to any notion of sample area, and so we are left taking arbitrary guesses at area, and matching it up with estimates of N from different models that do not have any explicit biological relevance. Clearly, there is not a compelling solution to be derived from this "estimate N and conjure up a buffer" approach and we are left not much wiser about bear density at Fort Drum than we were before we conducted this analysis, and certainly not confident in our assessments. Closed population models are not integrated with any ecological theory, so our N is not connected to the specific landscape in any explicit way.

The capture-recapture models that we used apply to truly closed populations—a population of goldfish in a fish bowl. Yet here we are applying them to a population of bears that inhabit a rich two-dimensional landscape of varied habitats, exposed to trapping by an irregular and sparse array of traps. It seems questionable that the same model that is completely sensible for a population of goldfish in a bowl should also be the right model for this population of bears distributed over a broad landscape. Ordinary capture-recapture methods are distinctly non-spatial. They don't admit spatial indexing of either sampling (the observation process) or of individuals (the ecological process). This leads immediately to a number of practical deficiencies with the application of ordinary capture-recapture models: (1) They do not provide a coherent basis for estimating density, a problem we struggled with in the black bear study. (2) Discarding information on the locations of capture so that ordinary CR models can be used *induces* a form of heterogeneity that can only at best be approximated by classical models of latent heterogeneity. (3) Ordinary CR models do not accommodate trap-level covariates which exist in a large proportion of real studies. (4) They do not accommodate formal consideration of any spatial process that gives rise to the observed data.

In subsequent chapters of this book, we resolve these specific technical problems related to density, model-based linkage of N and A, covariates, spatial variation, and related matters all within a coherent unified framework for spatial capture-recapture.

1.4 Historical context: a brief synopsis

Spatial capture-recapture is a relatively new methodological development, at least with regard to formal estimation and inference. However, the basic problems that motivate the need for formal spatially explicit models have been recognized for decades and quite a large number of ideas have been proposed to deal with these problems. We review some of these ideas here.

1.4.1 Buffering

The standard approach to estimating density even now is to estimate N using conventional closed population models (Otis et al., 1978) and then try to associate with this estimate some specific sampled area, say A, the area which is contributing individuals to the population for which N is being estimated. The strategy is to define A by placing a buffer of say W around the trap array or some polygon which encloses the trap array. The historical context is succinctly stated by O'Brien (2011) from which we draw this description:

> *"At its most simplistic, A may be described by a concave polygon defined by connecting the outermost trap locations (A_{tp}; Mohr (1947)). This assumes that animals do not move from outside the bounded area to inside the area or vice versa.*

Unless the study is conducted on a small island or a physical barrier is erected in the study area to limit movement of animals, this assumption is unlikely to be true. More often, a boundary area of width W (A_w) is added to the area defined by the polygon A_{tp} to reflect the area beyond the limit of the traps that potentially is contributing animals to the abundance estimate (Otis et al., 1978). The sampled area, also known as the effective area, is then $A(W) = A_{tp} + A_w$. Calculation of the buffer strip width (W) is critical to the estimation of density and is problematic because there is no agreed upon method of estimating W. Solutions to this problem all involve ad hoc methods that date back to early attempts to estimate abundance and home ranges based on trapping grids (see Hayne, 1949). Dice (1938) first drew attention to this problem in small mammal studies and recommended using one-half the diameter of an average home range. Other solutions have included use of inter-trap distances (Blair, 1940; Burt, 1943), mean movements among traps, maximum movements among traps (Holdenried, 1940; Hayne, 1949), nested grids (Otis et al., 1978), and assessment lines (Smith et al., 1971)."

The idea of using 1/2 mean maximum distance moved ("MMDM" Wilson and Anderson, 1985b) to create a buffer strip seems to be the standard approach even today, presumably justified by Dice's suggestion to use 1/2 the home range diameter, with the mean among individuals of the maximum distance moved being an estimator of home range diameter. Alternatively, some studies have used the full MMDM (e.g., Parmenter et al., 2003), because the trap array might not provide a full coverage of the home range (home ranges near the edge should be truncated) and so 1/2 MMDM should be biased smaller than the home range radius. And, sometimes home range size is estimated by telemetry (Karanth, 1995; Bales et al., 2005). Use of MMDM summaries to estimate home range radius is usually combined with an AIC-based selection from among the closed population models in Otis et al. (1978) which most often suggests heterogeneity in detection (model M_h). Almost all of these early methods were motivated by studies of small mammals using classical "trapping grids" but, more recently, their popularity in the study of wildlife populations has increased with the advent of new technologies, especially related to non-invasive sampling methods such as camera trapping. In particular, the series of papers by Karanth (1995) and Karanth and Nichols (1998, 2002) has led to fairly widespread adoption of these ideas.

1.4.2 Temporary emigration

Another intuitively appealing idea is that by White and Shenk (2000) who discuss "correcting bias of grid trapping estimates" by recognizing that the basic problem is like random temporary emigration (Kendall et al., 1997; Chandler et al., 2011; Ivan et al., 2013a,b) where individuals flip a coin with probability ϕ to determine if they are "available" to be sampled or not. White and Shenk's idea was to estimate ϕ from radio telemetry, as the proportion of time an individual spends in the study area. They obtain the estimated "super-population" size by using standard closed population models to estimate N and then obtain density by $\hat{D} = \hat{N}\hat{\phi}/A$ where A is the nominal area of the trapping array (e.g., minimum convex hull). A problem with this approach is that individuals that were radio collared represent a biased sample i.e., you fundamentally have to sample individuals randomly from the population *in proportion to their exposure to sampling* and that seems practically impossible to accomplish. In other words, "in the study area" has no precise meaning itself and is impossible to characterize in almost all capture-recapture studies. Deciding what is "in the study area" is effectively the same as choosing an arbitrary buffer which defines who is in the study area and who isn't. That said, the temporary emigration analogy is a good heuristic for understanding SCR models and has a precise technical relevance to certain models.

Another interesting idea is that of using some summary of "average location" as an individual covariate in standard capture-recapture models. Boulanger and McLellan (2001) use distance-to-edge (DTE) as a covariate in the Huggins-Alho (Huggins, 1989; Alho, 1990) type of model. Ivan (2012) uses this approach in conjunction with an adjustment to the estimated N obtained by estimating the proportion of time individuals are "on the area formally covered by the grid" using radio telemetry. We do not dwell too much on these different variations but we do note that the use of DTE as an individual covariate amounts to a type of intermediate model between simple closed population models and fully spatial capture-recapture models, which we address directly in Chapter 4.

While these procedures are all heuristically appealing, they are also essentially ad hoc in the sense that the underlying model remains unspecified or at least imprecisely characterized and so there is little or no basis for modifying, extending, or generalizing the methods. These methods are distinctly *not* model-based procedures. Despite this, there seems to be an enormous amount of literature developing, evaluating, and "validating" these literally dozens of heuristic ideas that solve specific problems, as well as various related tweeks and tunings of them haven't led to any substantive breakthroughs that are sufficiently general or theoretically rigorous.

1.5 Extension of closed population models

The deficiency with classical closed population models is that they have no spatial context. N is just an integer parameter that applies equally well to estimating the number of unique words in a book, the size of some population that exists in a computer, or a bucket full of goldfish. Surely it must matter whether the N items exist as words in a book, or goldfish in a bowl, or tigers in a patch of forest! That classical closed population models have no spatial context leads to a number of conceptual and methodological problems or limitations as we have encountered previously. More important, ecologists seldom care only about N—space is often central to objectives of many population studies—movement, space usage, resource selection, how individuals are distributed in space and in response to explicit factors related to land use or habitat. Because space is central to so many real problems, this is probably the number 1 reason that many ecologists don't bother with capture-recapture. They haven't seen capture-recapture methods as being able to solve their problems. Thus, the essential problem is that classical closed population models are too simple—they ignore the spatial attribution of traps and encounter events, movement and variability in exposure of individuals to trap proximity. These problems can be addressed formally by the development of more general capture-recapture models.

1.5.1 Toward spatial explicitness: Efford's formulation

The solution to the various issues that arise in the application of ordinary capture-recapture models is to extend the closed population model so that N becomes spatially explicit. Efford (2004) was the first to formalize an explicit model for spatial capture-recapture problems in the context of trapping arrays. He adopted a Poisson point process model to describe the distribution of individuals and essentially a distance sampling formulation of the observation model which describes the probability of detection as a function of individual location, regarded as a latent variable governed by the point process model. While earlier (and contemporary) methods of estimating density from trap arrays have been ad hoc in the sense of lacking a formal description of the spatial model, Efford achieved a formalization of the model, describing explicit mechanisms governing the spatial distribution of individuals and how they are encountered by traps, but adopted a more or less ad hoc framework for inference under that spatial model using a simulation-based method known as inverse prediction (Efford, 2004; Gopalaswamy, 2012).

Recently, there has been a flurry of effort devoted to formalizing inference under this model-based framework for the analysis of spatial capture-recapture data (Borchers and Efford, 2008; Royle and Gardner, 2011; Borchers, 2012; Gopalaswamy, 2012). There are two distinct lines of work which adopt the model-based formulation in terms of the underlying point process but differ primarily by the manner in which inference is achieved. One approach (Borchers and Efford, 2008) uses classical inference based on likelihood (see Chapter 6), and the other (Royle and Young, 2008) adopts a Bayesian framework for inference (Chapters 5 and 17).

1.5.2 Abundance as the aggregation of a point process

Spatial point process models represent a major methodological theme in spatial statistics (Cressie, 1991) and they are widely applied as models for many ecological phenomena (Stoyan and Penttinen, 2000; Illian et al., 2008). Point process models apply to situations in which the random variable in question represents the locations of events or objects: trees in a forest, weeds in a field, bird nests, etc. As such, it seems natural to describe the organization of individuals in space using point process models. SCR models represent the extension of ordinary capture-recapture by augmenting the model with a point process to describe individual locations.

Specifically, let $\mathbf{s}_i; i = 1, 2, \ldots, N$ be the locations of all individuals in the population. One of the key features of SCR models is that the point locations are latent, or unobserved, and we only obtain imperfect information about the point locations by observing individuals at trap or observation locations. Thus, the realized locations of individuals represent a type of "thinned" point process, where the thinning mechanism is not random but, rather, biased by the observation mechanism. It is also natural to think about the observed point process as some kind of a compound or aggregate point process with a set of "parent" nodes being the locations of individual home ranges or their centroids, and the observed locations as "offspring"—i.e., a Poisson cluster process (PCP). In that context, density estimation in SCR models is analogous to estimating the number of parents of a Poisson cluster process (Chandler and Royle, 2013).

Most of the recent developments in modeling and inference from spatial encounter history data, including most methods discussed in this book, are predicated on the view that individuals are organized in space according to a relatively simple point process model. More specifically, we often assume that individual activity centers are independent and identically distributed random variables, and they are distributed uniformly over some region. This is consistent with the assumption that the activity centers represent the realization of a Poisson point process or, if the total number of activity centers is fixed, then this is usually referred to as a binomial point process.

1.5.3 The activity center concept

In the context of SCR models, and because most animals we study by capture-recapture are not sessile, there is not a unique and precise biological definition of the point locations $\mathbf{s}$. Rather, we imagine these to be the centroid of individuals' home ranges, or the centroid of an individual's activities during the time of sampling, or even its average location measured with error (e.g., from a long series of telemetry measurements). In general, this point is unknown for any individual but, if we could track an individual over time and take many observations, then we could perhaps get a good idea of where that point is. We'll think of the collection of these points as defining the spatial distribution of individuals in the population.

We use the terms home range and activity center interchangeably. The term "home range center" suggests that models are only relevant to animals that exhibit behavior of establishing home ranges or territories, or central place foragers, and since not all species do that, perhaps the construction of SCR models based on this idea is flawed. However, the notion of a home range center is just a conceptual device and we don't view this concept as being strictly consistent with classical notions of animal territories. Rather our view is that a home range or territory is inherently dynamic, temporally, and thus it is a transient quantity—where the animal lived during the period of study, a concept that is analogous to the more conventional notion of utilization distributions. Therefore, whether or not individuals of a species establish home ranges is irrelevant because, once a precise time period is defined, this defines a distinct region of space that an individual must have occupied.

1.5.4 The state-space

Once we introduce the collection of activity centers, $\mathbf{s}_i; i = 1, 2, \ldots, N$, then the question "what are the possible values of $\mathbf{s}$?" needs to be addressed because the individual $\mathbf{s}_i$ is *unknown*. As a technical matter, we will regard them as random effects and, in order to apply standard methods of statistical inference, we need to provide a distribution for these random effects. In the context of the point process model, the possible values of the point locations are referred to as the "state-space" of the point process and this is some region or set of points which we will denote by $\mathcal{S}$. This is analogous to what is sometimes called the *observation window* for $\mathbf{s}$ in the point process literature. The region $\mathcal{S}$ serves as a prior distribution for $\mathbf{s}_i$ (or, equivalently, the random effects distribution). In animal studies, as a description of where individuals that could be captured are located, it includes our study area, and should accommodate all individuals that could have been captured. In the practical application of SCR models, in most cases estimates of density will be relatively insensitive to choice of state-space which we discuss further in Chapter 5 and elsewhere.

1.5.5 Abundance and density

When the underlying point process is well defined, including a precise definition of the state-space, this in turn induces a precise definition of the parameter N, "population size," as the number of individual activity centers located within the prescribed state-space, and its direct linkage to density, D. That is, if $A(\mathcal{S})$ is the area of the state-space then

$$D = \frac{N}{A(\mathcal{S})}.$$

A deficiency with some classical methods of "adjustment" is that they attempted to prescribe something like a state-space—a "sampled area"—except absent any precise linkage of individuals with the state-space. SCR models formalize the linkage between individuals and space and, in doing so, provide an explicit definition of N associated with a well-defined spatial region, and hence density. That is, they provide a model in which N scales, as part of the model, with the size of the prescribed state-space. In a sense, the whole idea of SCR models is that by defining a point process and its state-space $\mathcal{S}$, this gives context and meaning to N which can be estimated directly for that specific state-space. Thus, it is fixing $\mathcal{S}$ that resolves the problem of "unknown area" that we have previously discussed.

1.6 Ecological focus of SCR models

Formulation of capture-recapture models conditional on the latent point process is the critical and unifying element of *all* SCR models. However, SCR models differ in how the underlying process model is formulated, and its complexity. Most of the development and application of SCR models has focused on their use to estimate density and touting the fact that they resolve certain specific technical problems related to the use of ordinary capture-recapture models. This is achieved with a simple process model being a basic point process of independently distributed points. At the same time, there are models of CR data that focus exclusively on *movement* modeling, or models with explicit dynamics (Ovaskainen, 2004; Ovaskainen et al., 2008).

Conceptually, these are akin to spatial versions of so-called Cormack-Jolly-Seber (CJS) models in the traditional capture-recapture literature, except they involve explicit mathematical models of movement based on diffusion or Brownian motion. Finally, there are now a very small number of papers that focus on *both* movement and density simultaneously (Royle and Young, 2008; Royle et al., 2011a; Royle and Chandler, 2012) or population dynamics and density (Gardner et al., 2012a).

A key thing is that these models, whether focused just on density, or just on movement, or both, are similar in terms of the underlying concepts, the latent structure, and the observation model. They differ primarily in terms of the ecological focus. Understanding movement is an important topic in ecology, but models that strictly focus on movement will be limited by two practical considerations: (1) most capture-recapture data, e.g., by camera trapping, produces only a few observations of each individual (between 1 and 5 would be typical). So there is not too much information about complex movement models. (2) Typically people have an interest in density of individuals and therefore we need models that can be extrapolated from the sample to the unobserved part of the population. That said, there are clearly some cases where more elaborate movement models should come into play. If one has some telemetry data in addition to SCR then there is additional information on fine-scale movements that should be useful.

1.7 Summary and outlook

Spatial capture-recapture models are an extension of traditional capture-recapture models to accommodate the spatial organization of both individuals in a population and the observation mechanism (e.g., locations of traps). They resolve problems which have been recognized historically and for which various ad hoc solutions have been suggested: heterogeneity in encounter probability due to the spatial organization of individuals relative to traps, the need to model trap-level effects on encounter, and that a well-defined sample area does not exist in most studies, and thus estimates of N using ordinary capture-recapture models cannot be related directly to density.

As we have shown already, SCR models are not simply an extension of a technique to resolve certain technical problems. Rather, they provide a coherent, flexible framework for making ecological processes explicit in models of individual encounter history data, and for studying animal populations processes such as individual movement, resource selection, space usage, population dynamics, and density. Historically, researchers studied these questions independently, using ostensibly unrelated study designs and statistical procedures. For example, resource selection function (RSF) models for resource selection, state-space models for movement, density using closed capture-recapture methods, and population dynamics with various "open" capture-recapture models. SCR can bring all of these problems together into a single unified framework for modeling and inference. Most importantly, spatial capture-recapture models promise the ability to integrate explicit ecological theories directly into the models so that we can directly test hypotheses about either space usage (e.g., Chapter 13), landscape connectivity (Chapter 12), movement, or spatial distribution (Chapter 11). We imagine that, in the near future, SCR models will include point process models that allow for interactions among individuals such as inhibition or clustering (Reich et al., in review). In the following chapters we develop a comprehensive synthesis and extension of spatial capture-recapture models as they presently exist, and we suggest areas of future development and needed research.

CHAPTER

Statistical Models and SCR

2

In the previous chapter we described the basics of capture-recapture methods and the advantages that spatial models have over traditional non-spatial models. We avoided statistical terminology like the plague so that we could focus on a few key concepts. Although it is critical to understand the non-technical motivation for this broad class of models, it is impossible to fully appreciate them, and apply them to real data, without a solid grasp of the fundamentals of statistical inference.

In this chapter, we present a brief overview of the basic statistical principles that are referenced throughout the remainder of this book. Emphasis is placed on the definition of a random variable, the common probability distributions used to model random variables, and how hierarchical models can be used to describe conditionally related random variables. For some readers, this material will be familiar, perhaps even elementary, and thus you may want to skip to the next chapter. However, our experience is that many basic statistics courses taken by ecologists do not emphasize the important subjects covered in this chapter. Instead, there seems to be much attention paid to minor details such as computing the number of degrees of freedom in various F-tests, which, although useful in some contexts, do not provide the basis for drawing conclusions from data and evaluating scientific hypotheses.

The material in the beginning of this chapter is explained in numerous other texts. Technical treatments that emphasize ecological problems are given by Williams et al. (2002), Royle and Dorazio (2008), and Link and Barker (2010), to name just a few. A very accessible introduction to some of the topics covered in this chapter is presented in Chapter 3 of MacKenzie et al. (2006). With all these resources, one might wonder why we bother rehashing these concepts here. Our motivation is twofold: first, we wish to develop this material using examples relevant to spatial capture-recapture, and second, we find that most introductory texts are not accompanied by code that can be helpful to the novice. We therefore attempt to present simple **R** code throughout this chapter so that those who struggle with equations and mathematical notation can learn by doing. As mentioned in the Preface, we rely on **R** because it provides tremendous flexibility for analyzing data and because it is free. We do not, however, try to explain how to use **R** because there are so many good references already, including Venables and Ripley (2002), Bolker (2008), and Venables et al. (2012).

After covering some basic concepts of hierarchical modeling, we end the chapter by describing spatial capture-recapture models using hierarchical modeling

Spatial Capture-Recapture. http://dx.doi.org/10.1016/B978-0-12-405939-9.00002-5

notation. This makes the concepts outlined in the previous chapter more precise, and it highlights the fact that SCR models include explicit models for the ecological processes of interest (e.g., spatial variation in density) and the observation process, which describes how individuals are encountered.

2.1 Random variables and probability distributions

2.1.1 Stochasticity in ecology

Few ecological processes can be described using purely deterministic models, and thus we need a formal method for drawing conclusions from data while acknowledging the stochastic nature of ecological systems. This is the role of statistical inference, which is founded on the laws of probability. For our purposes, it suffices to be familiar with a small number of concepts from probability theory—the most important of which is the concept of a random variable, say X. A random variable is a variable whose realized value is the outcome of some stochastic process. To be more precise, a random variable is characterized by a function that describes the probability of observing the value x. This probability function can be written $\Pr(X = x|\theta)$ where θ is a parameter, or set of parameters of the function. If X is discrete, e.g., binary or integer, then we call the probability function a probability mass function (pmf). If X is continuous, the function is called a probability density function (pdf).

To clarify the concept of a random variable, let X be the number of American shad (*Alosa sapidissima*) caught after $K = 20$ casts at the shad hole on Deerfield River in Massachusetts. Suppose that we had a good day and caught $x = 7$ fish. If there were no random variation at play, we would say that the probability of catching a fish, which we will call p, is $p = 7/20 = 0.35$, and we would always expect to catch 7 shad after 20 casts. In other words, our deterministic model is $x = 0.35 \times K$. In reality, however, we can be pretty sure that this deterministic model would not be very good. Even if we knew for certain that $p \equiv 0.35$, we would expect some variation in the number of fish caught on repeated fishing outings. To describe this variation, we need a model that acknowledges uncertainty (i.e., stochasticity), and specifically we need a model that describes the probability of catching x fish given K and p, $\Pr(X = x|K, p)$. Since X is discrete, not continuous, we need a pmf. Before contemplating which pmf is most appropriate in this case, we need to first mention a few issues related to notation.

Statisticians make things easier for themselves, and more complicated for everyone else, by using different notation for probability distributions. Sometimes you will see $\Pr(X = x|K, p)$ expressed as $f(X|K, p)$ or $f(X; K, p)$ or $p(X|K, p)$ or $\pi(X|K, p)$ or $\mathbb{P}(X|K, p)$ or $[X|K, p]$ or even just $[X]$! Just remember that these expressions all have the same meaning—they are all probability distributions that tell us the probability of observing any possible realization of the random variable X. In this book, we will almost always use bracket notation (the last two examples above) to represent arbitrary probability distributions. Hence, from here on out, when you see $[X|K, p]$, just remember that this is equivalent to the more traditional expression

$\Pr(X = x|K, p)$. In addition, from here on, to achieve a more concise presentation, we will no longer use uppercase letters to denote random variables and lowercase letters for realized values. Rather, we will define a random variable by some symbol (x, N, etc ...) and let the context determine whether we are talking about the random variable itself, or realized values of it. In some limited cases, we will want upper- and lowercase letters to represent different variables. For example, we will often let N denote population size and n denote the number of individuals actually detected.

When we wish to be specific about a probability distribution, we will do so in one of two ways, one mathematically precise and one symbolic. Before explaining these two options, let's choose a specific distribution as a model for the data in our example. In this case, the natural choice for $[x|K, p]$ is the binomial distribution, the mathematically precise representation of which is

$$[x|K, p] = \binom{K}{x} p^x (1 - p)^{K-x}. \tag{2.1.1}$$

The right-hand side of this equation is the binomial pmf (described in more detail in Section 2.2), and plugging in values for the parameters K, and p will return the probability of observing any realized value of the random variable x. This is precise, but it is also cumbersome to write repetitively, and it may make the eyes glaze over when seen too often. Thus, we will often simplify Eq. (2.1.1) using the symbolic notation:

$$x \sim \text{Binomial}(K, p). \tag{2.1.2}$$

The "∼" symbol is meant to represent a stochastic relationship and can be read "is distributed as." Another reason for using this notation is that it resembles the syntax of the **BUGS** language, which we will frequently use to conduct Bayesian inference.

Note that once we choose a probability distribution, we have chosen a model. In our example, we have specified our model as $x \sim \text{Binomial}(K, p)$, and because we are assuming that the parameters are known, we can make probability statements about future outcomes. Continuing with our fish example, we might want to know the probability of catching $x = 7$ again after $K = 20$ casts on a future fishing outing, assuming that we know $p = 0.35$. Evaluating the binomial pmf returns a probability of approximately 0.18, as show using this bit of **R** code:

```
> dbinom(7, 20, 0.35)
[1] 0.1844012
```

By definition, the pmf allows us to evaluate the probability of observing any x given $K = 20$ and $p = 0.35$, thus the distribution of the random variable can be visualized by evaluating it for all values of x that have non-negligible probabilities, as can be easily done in **R**:

```
plot(0:20, dbinom(0:20, 20, 0.35), type="h", ylab="Probability",
     xlab="Number of shad caught (X) after 20 casts")
```
[1]

the result of which is shown in Figure 2.1 with some extra details.

[1] We suppress the **R** command prompts (">") when showing code without output.

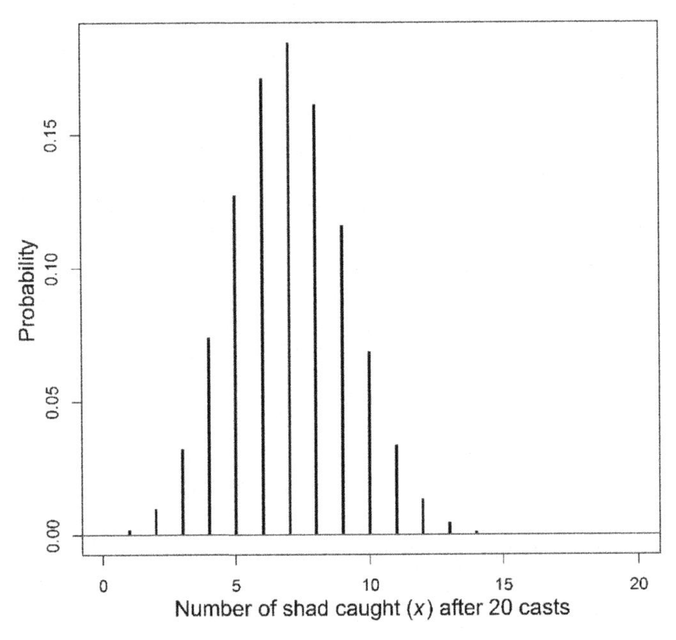

FIGURE 2.1

The binomial probability mass function with $N = 20$ and $p = 0.35$.

The purpose of this little example is to show that once we specify a model for the random variable(s) being studied, we can begin drawing conclusions, i.e., making inferences, about the processes of interest, even in the face of uncertainty. Probability distributions are essential to this process, and thus we need to understand them in more depth.

2.1.2 Properties of probability distributions

A pdf or a pmf is a function like any other function in the sense that it has one or more arguments whose values determine the result of the function. However, probability functions have a few properties that distinguish them from other functions. The first is that the function must be non-negative for all possible values of the random variable, i.e., $[x] \geq 0$. The second requirement is that the integral of a pdf must be unity, $\int_{-\infty}^{\infty}[x]dx = 1$, and similarly for a pmf, the summation over all possible values is unity, $\sum_x [x] = 1$. The following **R** code demonstrates this for the normal and binomial distributions:

```
> integrate(dnorm, -Inf, Inf, mean=0, sd=1)$value
[1] 1
> sum(dbinom(0:5, size=5, p=0.1))
[1] 1
```

This requirement is important to remember when one develops a non-standard probability distribution. For example, in Chapters 11 and 13, we work with a resource selection function whose probability density function is not one that is pre-defined in software packages such as **R** or **BUGS**.

Another feature of probability distributions is that they can be used to compute important summaries of random variables. The two most important summaries are the expected value, $\mathbb{E}(x)$, and the variance $\text{Var}(x)$. The expected value, or mean, can be thought of as the average of a very large sample from the specified distribution. For example, one way of approximating the expected values of a binomial distribution with $K = 20$ trials and $p = 0.35$ can be implemented in **R** using:

```
> mean(rbinom(10000, 20, 0.3))
[1] 6.9865
```

For most probability distributions used in this book, the expected values are known exactly, as shown in Table 2.1, and thus we don't need to resort to simulation methods. For instance, the expected value of the binomial distribution is exactly $\mathbb{E}(x) = Kp = 20 \times 0.35 = 7$. In this case, it happens to take an integer value, but this is not a necessary condition, even for discrete random variables.

A more formal definition of an expected value is the average of all possible values of the random variable, weighted by their probabilities. For continuous random variables, this weighted average is found by integration:

$$\mathbb{E}(x) = \int_{-\infty}^{\infty} x \times [x] \mathrm{d}x. \tag{2.1.3}$$

For example, if x is normally distributed with mean 3 and unit variance (variance equal to 1), we could find the expected value using the following code.

```
> integrate(function(x) x*dnorm (x, 3, 1), -Inf, Inf)
   3 with absolute error < 0.00033
```

Of course, the mean *is* the expected value of the normal distribution, so we didn't need to compute the integral but, the point is, that Eq. (2.1.3) is generic. For discrete random variables, the expected value is found by summation rather than integration:

$$\mathbb{E}(x) = \sum_{x} x \times [x], \tag{2.1.4}$$

where the summation is over all possible values of x. Earlier we approximated the expected value of the binomial distribution with $K = 20$ trials and $p = 0.35$ by taking a Monte Carlo average. Equation (2.1.4) let's us find the exact answer, using this bit of **R** code:

```
> sum(dbinom(0:100, 20, 0.35)*0:100)
[1] 7
```

Table 2.1 Common probability density functions (pdfs) and probability mass functions (pmfs) used throughout this book.

Distribution	Notation	pdf or pmf	Support	Mean $\mathbb{E}(x)$	Variance Var (x)
Discrete random variables					
Poisson	$x \sim \text{Pois}(\lambda)$	$\exp(-\lambda)\lambda^x/x!$	$x \in \{0,1,\ldots\}$	λ	λ
Bernoulli	$x \sim \text{Bern}(p)$	$p^x(1-p)^{1-x}$	$x \in \{0,1\}$	p	$p(1-p)$
Binomial	$x \sim \text{Bin}(N,p)$	$\binom{N}{x} p^x(1-p)^{N-x}$	$x \in \{0,1,\ldots,N\}$	Np	$Np(1-p)$
Multinomial	$\mathbf{x} \sim \text{Multinom}(N,\boldsymbol{\pi})$	$\binom{N}{x_1 \cdots x_k} \pi_1^{x_1} \cdots \pi_k^{x_k}$	$x_k \in \{0,1,\ldots,N\}$	$N\pi_k$	$N\pi_k(1-\pi_k)$
Continuous random variables					
Normal	$x \sim \text{N}(\mu,\sigma^2)$	$\frac{1}{\sigma\sqrt{2\pi}} \exp\left(-\frac{(x-\mu)^2}{2\sigma^2}\right)$	$x \in [-\infty,\infty]$	μ	σ^2
Uniform	$x \sim \text{Unif}(a,b)$	$1/(b-a)$	$x \in [a,b]$	$(a+b)/2$	$(b-a)^2/12$
Beta	$x \sim \text{Beta}(a,b)$	$\frac{\Gamma(a+b)}{\Gamma(a)+\Gamma(b)} x^{a-1}(1-x)^{b-1}$	$x \in [0,1]$	$a/(a+b)$	$\frac{ab}{(a+b)^2(a+b+1)}$
Gamma	$x \sim \text{Gamma}(a,b)$	$\frac{b^a}{\Gamma(a)} x^{a-1} \exp(-bx)$	$x \in [0,\infty]$	a/b	a/b^2
Multivariate Normal	$\mathbf{x} \sim \text{N}(\boldsymbol{\mu},\boldsymbol{\Sigma})$	$(2\pi)^{-\frac{k}{2}}\lvert\boldsymbol{\Sigma}\rvert^{-\frac{1}{2}} \exp(-\frac{1}{2}(\mathbf{x}-\boldsymbol{\mu})'$ $\times\boldsymbol{\Sigma}^{-1}(\mathbf{x}-\boldsymbol{\mu}))$	$x_k \in [-\infty,\infty]$	$\boldsymbol{\mu}$	$\boldsymbol{\Sigma}$

This is great. But of what use is it? One very important concept to understand is that when we fit models, we are often modeling changes in the expected value of some random variable. For example, in Poisson regression, we model the expected value of the random variable, which may be a function of environmental variables.

The ability to model the expected value of a random variable gets us very far, but we also need a model for the variance of the random variable. The variance describes the amount of variation around the expected value. Specifically, $\text{Var}(x) = \mathbb{E}((x - \mathbb{E}(x))^2)$. Clearly, if the variance is zero, the variable is not random as there is no uncertainty in its outcome. For some distributions, notably the normal distribution, the variance is a parameter to be estimated. Thus, in ordinary linear regression, we estimate both the expected value $\mu = \mathbb{E}(x)$, which may be a function of covariates, and the variance σ^2, or similarly the residual standard error σ. For other distributions, the variance is not an explicit parameter to be estimated, and instead, the mean to variance ratio is fixed. In the case of the Poisson distribution, the mean is equal to the variance, $\mathbb{E}(x) = \text{Var}(x) = \lambda$. A similar situation is true for the binomial distribution—the variance is determined by the two parameters K and p, $\text{Var}(x) = Kp(1-p)$. In our earlier example with $K = 20$ and $p = 0.35$, the variance is 4.55. Toying around with these ideas using random number generators may be helpful. Here is some code to illustrate some of these basic concepts:

```
> 20* 0.35*(1-0.35)                  # Exact variance, Var(x)
[1] 4.55
> x <- rbinom(100000, 20, 0.35)
> mean((x-mean (x))^2)               # Monte Carlo approximation
[1] 4.545525
```

2.2 Common probability distributions

We got a little ahead of ourselves in the previous sections by using the binomial and Poisson distributions without describing them in detail. A solid understanding of the binomial, Poisson, multinomial, uniform, and normal (or Gaussian) distributions is absolutely essential throughout the remainder of the book. We will occasionally make use of other distributions such as the beta, log-normal, gamma, Dirichlet, etc... that can be helpful when modeling capture-recapture data, but these distributions can be readily understood once you are comfortable with the more commonly used distributions described in this section.

2.2.1 The binomial distribution

The binomial distribution plays a critical role in ecology. It is used for purposes as diverse as modeling count data, survival probability, occurrence probability, and capture probability, just to name a few. To describe the properties of the binomial distribution, and related distributions, we will introduce a new example. Suppose we are conducting a bird survey at a site in which $N = 10$ chestnut-sided warblers

(*Setophaga pensylvanica*) occur, and each of these individuals has a detection probability of $p = 0.5$. The binomial distribution is the natural choice for describing the number of individuals that we would expect to detect (n) in this situation, and using our notation, we can write the model as: $n \sim \text{Binomial}(10, 0.5)$. When $p < 1$, we can expect that we will observe a different number of warblers on each of J replicate survey occasions. To see this, we simulate data under this simple model with $J = 3$:

```
> n <- rbinom(3, size=10, prob=0.5)    # Generate 3 binomial outcomes
> n                                    # Display the 3 values
[1]  6  4  8
```

The vector of counts will typically differ each time you issue this command; however, we know the probability of observing any value of n_j because it is defined by the binomial pmf. As we demonstrated earlier, in **R** this probability can be found using the `dbinom` function. For example, the probability of observing $n_j = 5$ is given by:

```
> dbinom(5, 10, 0.5)
```

This simply evaluates the function shown in Table 2.1. We could do the same more transparently, but less efficiently, using any of the following:

```
n <- 5; N<- 10; p <- 0.5
factorial(N)/(factorial(n)*factorial(N-n))*p^n*(1-p)^(N-n)
exp(lgamma(N+1) - (lgamma(n+1) + lgamma(N-n+1)))*p^n*(1-p)^(N-n)
choose(N, n)*p^n*(1-p)^(N-n)
```

Note that the last three lines of code differ only in how they compute the binomial coefficient $\binom{N}{n}$, which is the number of different ways we could observe $n = 5$ of the $N = 10$ chestnut-sided warblers at the site. The binomial coefficient, which is read "N choose n," is defined as

$$\binom{N}{n} = \frac{N!}{n!(N-n)!}. \tag{2.2.1}$$

Now that we know how to simulate binomial data and compute the probabilities of observing any particular outcome n, conditional on the parameters N and p, we can contemplate the relevance of the binomial distribution in spatial capture-recapture models. One important application of the binomial distribution is as a model of encounter frequencies. Indeed, one of the most important encounter models in SCR will be referred to as the "binomial encounter model," in which the number of times individual i is captured at "trap" j after K survey occasions is modeled as $y_{ij} \sim \text{Binomial}(K, p_{ij})$. Here, p_{ij} is the encounter probability determined, in part, by the distance between an animal's activity center and the trap location. This binomial encounter model is described in detail in Section 7.1. Another important application of the binomial distribution is as a prior for the population size parameter in Bayesian analyses, as is discussed in Chapter 4.

2.2.2 The Bernoulli distribution

Above, we showed three alternatives to `dbinom` for evaluating the binomial pmf. These three commands differed only in how they computed the binomial coefficient, which we needed because of the numerous ways in which we could observe $n = 5$ given $N = 10$. To conceptualize this, let y_i be a binary variable indicating whether individual i was detected or not. Hence, given that five individuals were detected, the vector of individual detections could be something like $\mathbf{y} = (0, 0, 1, 1, 1, 1, 1, 0, 0, 0)$, indicating that we detected individuals 3–7 but not 1–2 or 8–10. For $N = 10$ and $n = 5$, the binomial coefficient tells us that there are 252 possible vectors $\mathbf{y}$ with 5 ones. However, when $N \equiv 1$, this term drops from the pmf and the result is the pmf for the Bernoulli distribution. That is, the Bernoulli distribution is simply the binomial distribution when $N \equiv 1$. Alternatively, we could say that the binomial distribution is the outcome of N iid Bernoulli trials. We use the standard abbreviation "iid" to mean *independent, identically distributed.*

The utility of the Bernoulli distribution is evident when we imagine that not all of the chestnut-sided warblers have the same detection probability. Thus, if some individuals can be detected with probability 0.3 and others have a 0.7 detection probability, then the model $n \sim \text{Binomial}(N, p)$ is no longer an accurate description of system since p is no longer constant for all individuals.

To properly account for variation in p, we could redefine our model for the counts of chestnut-sided warblers as

$$\begin{aligned} y_{ik} &\sim \text{Bernoulli}(p_i), \\ n_k &= \sum_{i=1}^{N} y_{ik}. \end{aligned} \tag{2.2.2}$$

This states that individual i is detected with probability p_i, and the observed count is the sum of the N Bernoulli outcomes.

An important point is that the individual-specific data y_{ik} can only be observed if the individuals are uniquely distinguishable, such as when they are marked with color bands by biologists. In such cases, the Bernoulli distribution allows us to model variation in detection probability among individuals and thus would be preferable to the binomial distribution, which assumes that each of the N individuals has the same p. For this reason, the Bernoulli distribution, as simple as it is, is of paramount importance in capture-recapture models, including spatial capture-recapture models in which there is virtually always substantial and important variation in capture probability among individuals. Indeed, it could be said that the Bernoulli model is the canonical model in capture-recapture studies, and most of the different flavors of capture-recapture models differ primarily in how p_i is specified.

The Bernoulli pmf is given by $p^n(1 - p)^{1-n}$, which you could implement directly in **R**, or use the `dbinom` function with the `size` argument set to 1. For example, `dbinom(1, 1, 0.3)` returns the Bernoulli probability of observing $n = 1$ given $p = 0.3$.

2.2.3 The multinomial and categorical distributions

The binomial distribution is used when we are accumulating a binary response—that is, one in which there are two possible categories such as success/failure or captured/not-captured. The multinomial distribution is a multivariate extension of the binomial used when there are $G > 2$ categories. The multinomial distribution can be thought of as a model for placing N items in the G categories, which are also called bins or cells. Each bin has its own probability π_g and these probabilities must sum to one. In ecology, N is often population size or the number of individuals detected, but the definition of the G bins varies among applications. For example, in distance sampling, when the distance data are aggregated into intervals, the bins are the distance intervals, and the cell probabilities are functions of detection probability in each interval (Royle et al., 2004).

The multinomial distribution is widely used to model data from traditional, non-spatial capture-recapture studies. Earlier we let y_{ik} denote a binary random variable indicating if warbler i was detected on survey k. The vector of observations for an individual, $\mathbf{y}_i$, is often referred to as the individual's "encounter history." The number of possible encounter histories depends on K, the number of survey occasions. Specifically, there are 2^K possible encounter histories.[2] If we tabulate the number of individuals with each encounter history, the frequencies can be modeled using the multinomial distribution.

Going back to our chestnut-sided warbler example, suppose the 10 individuals are marked and we make $K = 2$ visits to the site such that there are $2^K = 4$ possible encounter histories: (11, 10, 01, 00), where, for example, "10" is the encounter history for an individual detected on the first visit but not the second. If $p = 1$, then the encounter history for each of the 10 individuals must be "11." That is, we would detect each individual on both occasions. In this case, the data would be: $\mathbf{h} = (10, 0, 0, 0)$, which indicates that all 10 warblers had the first encounter history. The corresponding cell probabilities would be $\boldsymbol{\pi} = (1, 0, 0, 0)$. What about the situation where $p < 1$, e.g., $p = 0.3$? In this case, the probability of observing the capture history "11" (detected on both occasions) is $p \times p = 0.3 \times 0.3 = 0.09$. The probability of observing "10" is $p \times (1 - p) = 0.21$. Following this logic, the vector of cell probabilities is $\boldsymbol{\pi} = (0.09, 0.21, 0.21, 0.49)$. We can simulate data under this model as follows:

```
> caphist.probs <- c("11"=0.09, "10"=0.21, "01"=0.21, "00"=0.49)
> drop(rmultinom(1, 10, caphist.probs))
11 10 01 00
 0  3  2  5
```

The result of our simulation is that zero individuals were observed with the capture history "11" and 5 individuals were observed with the capture history "00." The other

[2] When N is unknown, we can never observe the "all-0" encounter history, corresponding to an individual that is not detected, and thus the number of "observable" encounter histories is $2^K - 1$.

5 individuals were observed one out of the two occasions. This is not such a surprising outcome given $p = 0.3$.

As in non-spatial capture-recapture studies, the multinomial distribution turns out to be very important in spatial capture-recapture studies. However, in this case, N is not defined as population size. Rather, we use the multinomial distribution when an individual can only be captured in a single trap during an occasion. Thus $N = 1$ and the cell probabilities are the probabilities of being captured in each trap. A thorough discussion of this point can be found in Chapter 9. Another application of the multinomial distribution in SCR models is discussed in Chapter 11 where we discuss how to model the probability that an individual's activity center is located in one of the cells of a raster defining the spatial region of interest.

Just as the Bernoulli distribution is the elemental form of the binomial distribution (being the case $N = 1$), the categorical distribution is essentially equivalent to the multinomial distribution with size parameter $N \equiv 1$. The only difference is that, rather than returning a vector with a single element equal to 1, it returns the element *location* where the 1 occurs. For example, if $\mathbf{y} = (0, 0, 1, 0)$ is an outcome of a multinomial distribution with $N = 1$, then the categorical outcome would be 3 because the 1 is located in third position in the vector. Thus, in spatial capture-recapture models, we might use either the multinomial distribution with $N = 1$ or the categorical distribution. The various **BUGS** engines describe the categorical distribution by the declaration `dcat` and, in **R**, we can simulate categorical outcomes using the function which as so:

```
> which(rmultinom(1, 1, c(0.1, 0.7, 0.2)) == 1)
[1] 2
```

2.2.4 The Poisson distribution

The Poisson distribution is the canonical model for count data in ecology. More generally, the Poisson distribution is a model for random variables taking on non-negative, integer values. Although it is a simple model having just one parameter, $\lambda = \mathbb{E}(x) = \text{Var}(x)$, its applications are highly diverse, including as a model of spatial variation in abundance or as a model for the frequency of behaviors over time. Just as logistic regression is the standard generalized linear model (GLM) used to model binary data, Poisson regression is the default GLM for modeling count data and variation in λ.

The Poisson distribution is related to both the binomial and multinomial distributions, and the following three bits of trivia are occasionally worth knowing. First, it is the limit of the binomial distribution as $N \to \infty$ and $p \to 0$, which means that for high values of N and low values of p, Poisson$(N \times p)$ is approximately equal to Binomial(N, p). Second, if $\{n_1 \sim \text{Poisson}(\lambda_1), \ldots, n_K \sim \text{Poisson}(\lambda_K)\}$ then, conditional on the total, the vector of counts is multinomial, $\{n_1, \ldots, n_K\} \sim \text{Multinomial}\big(\sum_k n_k, \{\frac{\lambda_1}{\sum_k \lambda_k}, \ldots, \frac{\lambda_K}{\sum_k \lambda_k}\}\big)$. Third, the sum of two Poisson random variables $x_1 \sim \text{Poisson}(\lambda_1)$ and $x_2 \sim \text{Poisson}(\lambda_2)$ is also Poisson: $x_1 + x_2 \sim \text{Poisson}(\lambda_1 + \lambda_2)$.

The Poisson distribution has two important uses in spatial capture-recapture models: (1) as a prior distribution for the population size parameter N, and (2) as a model for the frequency of captures in a trap. In the first context, the Poisson prior for N results in a Poisson point process for the location of the N activity centers in the region of interest. This topic is discussed in Chapters 5 and 11. The second use of the Poisson distribution in spatial capture-recapture is to describe data from sampling methods in which an individual can be detected multiple times at a trap during a single occasion. For example, in camera trapping studies we might obtain multiple pictures of the same individual at a trap during a single sampling occasion. Thus, λ in this case would be defined as the expected number of detections or captures per occasion.

2.2.5 The uniform distribution

The lowly uniform distribution is a continuous distribution whose only two parameters are the lower and upper bounds that restrict the possible values of the random variable x. These bounds are almost always known, so there is typically nothing to estimate. Nonetheless, the uniform distribution is one of the most widely used distributions, especially among Bayesians who frequently use it to as a "non-informative" prior distribution for a parameter. For example, if we have a capture probability parameter p that we wish to estimate, but we have no prior knowledge of what value it may take in the range [0,1], we will often use the prior $p \sim \text{Uniform}(0, 1)$. This states that p is equally likely to take on any value between zero and one. Prior distributions are described in more detail in the next chapter.

Another common usage of the uniform distribution is as a prior for the coordinates of points in the real plane, i.e., in two-dimensional space. Such a use of the uniform distribution implies that a point process is "homogeneous," meaning that the location of one point does not affect the location of another point and that the expected density of points is constant throughout the region. Thus, to simulate a realization from a homogeneous Poisson point process in the unit square $[0, 1] \times [0, 1]$, we could use the following **R** code:

```
D <- 100   # points per unit area
A <- 1     # Area of unit square
N <- rpois(1, D*A)
plot(s <- cbind(runif(N), runif(N)))
```

where s is a matrix of coordinates with N rows and 2 columns. We will often represent the uniform point process using the following notation:

$$\mathbf{s} \sim \text{Uniform}(\mathcal{S}), \tag{2.2.3}$$

where $\mathcal{S}$ is some two-dimensional spatial region called the state-space of the random variable $\mathbf{s}$. It would be more correct to distinguish this two-dimensional uniform distribution from the univariate one. That is, it might be more clear to use notation such as $\mathbf{s} \sim \text{Uniform}_2(\mathcal{S})$ instead, but this is somewhat cumbersome, so we will opt for the former expression.

2.2.6 Other distributions

The other continuous distributions that are regularly encountered in SCR models are primarily used as priors in Bayesian analyses, and thus we will avoid a lengthy discussion of their properties. The normal distribution, also called the Gaussian distribution, is perhaps the most widely recognized and applied probability model in statistics, but it plays only a minor role in SCR models other than as a model for signal strength in acoustic SCR models (Efford et al., 2009b; Dawson and Efford, 2009, and see Section 9.4). Nonetheless, it is the canonical prior for any continuous random variable with infinite support, and thus it is often used as a prior when applying Bayesian methods. One common usage is as a prior for the β coefficients of a linear model defining some parameter as a function of covariates (usually on a transformed scale). An example, including a cautionary note, is provided in Section 3.5.1. Be aware that although the normal distribution is typically parameterized in terms of the variance parameter σ^2, in the **BUGS** language, the inverse of the variance, or precision, is used instead, $\tau = 1/\sigma^2$. In **R**, the `dnorm` function requires the standard deviation σ, rather than the variance σ^2.

The bivariate normal distribution is a generalization of the normal distribution and a special case of the multivariate normal distribution whose pdf is shown in Table 2.1. The bivariate normal distribution is used to model two (possibly) dependent continuous variables whose symmetric variance-covariance matrix is denoted $\mathbf{\Sigma}$. In SCR models, we most often use this model as a rudimentary description of movement outcomes about a home range center. If there is no correlation, then the model reduces to two independent normal draws along the coordinate axes. The following code generates bivariate normal outcomes with no correlation ($\rho = 0$), as well as outcomes in which the correlation is $\rho = 0.9$:

```
library(mvtnorm)
set.seed(3)
mu <- c(0,0)
Sigma <- matrix(c(1,.9,.9, 1), 2, 2)
X1 <- cbind(rnorm(50, mu[1], Sigma[1,1]),    # No correlation (rho=0)
            rnorm(50, mu[2], Sigma[2,2]))
X2 <- rmvnorm(50, mu, Sigma)                 # rho=0.9
```

Figure 2.2 shows the simulated points.

Several of the parameters in capture-recapture models do not have infinite support, but instead are positive valued, living between zero and infinity, or are probabilities restricted to the range [0, 1]. The beta distribution is the standard prior used for probabilities because it can be used to express either a lack of knowledge or very precise knowledge about a parameter. For example, a Beta(1, 1) distribution is equivalent to a Uniform(0, 1) distribution. However, unlike the uniform distribution, the beta distribution can be used as an informative prior; for example if published estimates of detection probability exist we can choose parameters of the beta distribution to reflect that. To gain some familiarity with the beta distribution, execute the following **R** commands:

```
curve(dbeta(x, 1, 1), col="black", ylim=c(0,5))
```

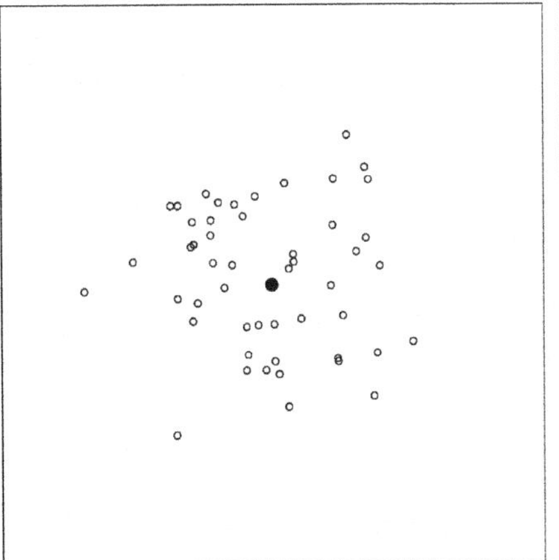
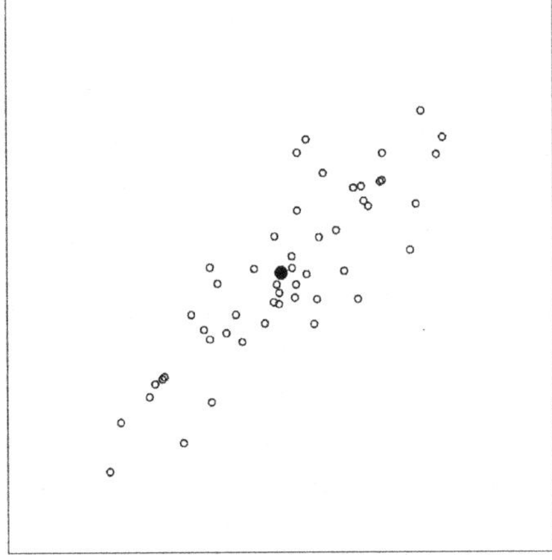

FIGURE 2.2

Two realized point patterns from the bivariate normal distribution.

```
curve(dbeta(x, 10, 10), col="blue", add=TRUE)
curve(dbeta(x, 10, 20), col="darkgreen", add=TRUE)
```

Other parameters in SCR models are continuous but positive-valued and can be modeled using the gamma distribution. As with the beta distribution, the gamma distribution is typically favored over the uniform distribution when one is interested in using an informative prior. It is also frequently used as a vague prior for the inverse of variance parameters, but it is wise to compare this prior to a uniform to assess its influence on the posterior.

2.3 Statistical inference and parameter estimation

If the parameters of a statistical model were known with absolute certainty, then it would be possible to use pdfs and pmfs to make direct probability statements about unknowns such as future outcomes. However, we almost never know the actual values of parameters, and instead we have to estimate them from observations (i.e., data). Our inferences must then acknowledge the uncertainty associated with our imperfect knowledge of the parameters. Doing so is most often accomplished using one of two approaches: classical (frequentist) inference or Bayesian inference. These two modes of inference regard the uncertainty about parameters in entirely different ways. In the next chapter, we will review some of the important concepts in Bayesian inference, so here, we will focus on the frequentist perspective.

Suppose we count oak trees at n sites, and the resulting data $\{y_1, \ldots, y_n\}$ can be assumed to be *iid* outcomes from some distribution, such as the Poisson with unknown parameter λ. We want to estimate this parameter. In classical inference, the only uncertainty about λ is that attributable to sampling. For instance, we can imagine repeatedly sampling the population (sites in this example) and obtaining sample-specific estimates of λ. If these estimates are produced using the method of maximum likelihood, and as n tends to infinity, the distribution of estimates, called the sampling distribution, will be normally distributed with $\mathbb{E}(\hat{\lambda}) = \lambda$. The standard deviation of the sampling distribution is called the standard error, which can also be estimated as part of the maximum likelihood procedure. Of course, we almost always have just a single sample of data, and hence a single $\hat{\lambda}$ and a single estimate of the standard error. However, under the assumption of a normally distributed sampling distribution, we can easily construct a confidence interval that will include the true value of λ with coverage probability $1-\alpha$, where α is a prescribed value like 0.05. An important point is that there is no uncertainty associated with the actual parameter—it is regarded as a fixed value, and hence probability is only used to characterize the estimator via its sampling distribution.

Maximum likelihood is heuristically a method of finding the most "likely" value of λ, given the observed data, and of characterizing the variance of the sampling distribution. Of course, it also applies to cases where the observations are multivariate, or the probability distribution is a function of multiple parameters. Endless numbers of textbooks and online resources are available for those interested in a detailed explanation of maximum likelihood. For our purposes, we wish to keep it simple and focus on *how* to do it. The first step is to define the likelihood function, which is the joint distribution of the data regarded as a function of the parameter(s). If the joint distribution of the observations is denoted by $[y_1, y_2, \ldots, y_n|\lambda]$, we usually denote the likelihood as $\mathcal{L}(\lambda|\mathbf{y}) = [\lambda|y_1, y_2, \ldots, y_n]$. If the observations are iid, the likelihood simplifies to

$$\mathcal{L}(\lambda|\mathbf{y}) = \prod_{i=1}^{n}[y_i|\lambda], \tag{2.3.1}$$

where $[y_i|\lambda]$ is a probability distribution, like those discussed in the previous sections. For example, if y_i is Poisson distributed, then $[y_i|\lambda] = \text{Poisson}(\lambda) = \frac{\lambda^{y_i}e^{-\lambda}}{y_i!}$. Although likelihoods are typically shown on the natural scale, we almost always maximize the logarithm of the likelihood to avoid computational problems that arise when multiplying very small probabilities. Thus, we rewrite Eq. (2.3.1) as

$$\ell(\lambda|\mathbf{y}) = \sum_{i=1}^{n}\log[y_i|\lambda]. \tag{2.3.2}$$

Here is some simple **R** code to simulate independent Poisson outcomes and estimate λ (as though we did not know it) using the method of maximum likelihood. Actually, we will minimize the negative log-likelihood because it is equivalent and is the default for **R**'s optimizers like `optim` and `nlm`.

```
> lambda <- 3                   # Actual parameter value
> y1 <- rpois(100, lambda)      # Realized values (data)
> negLogLike1 <- function (par) -sum(dpois(y1, par, log=TRUE))
> starting.value <- c('lambda'=1)
> optim(starting.value, negLogLike1)$par # MLE
  lambda
3.039844
```

Explicitly maximizing the likelihood, numerically, isn't actually necessary here because the MLE of λ is given by the mean of the observations. A more interesting example is when there are covariates of λ. For example, suppose λ is a function of elevation and vegetation height according to: $\log(\lambda_i) = \beta_0 + \beta_1 \text{ELEV}_i + \beta_2 \text{VEGHT}_i$. This is a standard Poisson regression problem, with likelihood:

$$\mathcal{L}(\boldsymbol{\beta}|\mathbf{y}) = \prod_i \text{Poisson}(y_i|\lambda_i). \tag{2.3.3}$$

This likelihood is almost identical to the previous one except that λ is now a function, and so we need to estimate the parameters of the function, i.e., the $\beta's$. Some code to fit this model to simulated data is shown here:

```
> nsites <- 100
> elevation <- rnorm(100)
> veght <- rnorm(100)
> beta0 <- 1
> beta1 <- -1
> beta2 <- 0
>lambda <- exp(beta0 + beta1*elevation + beta2*veght)
> y2 <- rpois(nsites, lambda)
> negLogLike2 <- function(pars) {
+     beta0 <- pars[1]
+     beta1 <- pars[2]
+     beta2 <- pars[3]
+     lambda <- exp(beta0 + beta1*elevation + beta2*veght)
+     -sum(dpois(y2, lambda, log=TRUE))
+ }
> starting.values <- c ('beta0'=0, 'beta1'=0, 'beta2'=0)
> optim(starting.values, negLogLike2)$par
      beta0        beta1        beta2
   0.98457756 -1.03025173 -0.01218292
```

We see that the maximum likelihood estimates (MLEs) are very close to the true parameter values.

In these examples, the parameters we estimated are called fixed effects by frequentists. Fixed effects are parameters that are not regarded as being random variables. A random effect, in contrast, is a parameter that can be regarded as the outcome of a random variable. For instance, we could entertain the idea that the intercept of our GLM differs among locations, and that its actual value is an outcome of a normal distribution with parameters μ and σ^2. In this case, β_{0i} would be a random effect, and our model could be written as:

$$\begin{aligned} y_i &\sim \text{Poisson}(\lambda_i), \\ \log(\lambda_i) &= \beta_{0i} + \beta_1 \text{ELEV}_i + \beta_2 \text{VEGHT}_i, \\ \beta_{0i} &\sim \text{Normal}(\mu, \sigma^2). \end{aligned}$$

This is an example of a mixed effects model or a hierarchical model. How do we estimate the parameters of a model that includes random effects? Earlier the likelihood function was written as the product of probabilities determined by a single pmf or pdf, $[y|\lambda]$, but now we have an additional random variable, and we are forced to think about conditional relationships, because y depends upon β_{0i}, and β_{0i}, depends upon other parameters, specifically μ and σ^2. This type of conditional dependence among parameters is the essence of hierarchical models, and statistical analysis of hierarchical models requires that we discuss joint distributions, marginal distributions, and conditional distributions. These concepts will be used extensively in Chapter 6 where we demonstrate how to estimate parameters of hierarchical models using maximum likelihood.

2.4 Joint, marginal, and conditional distributions

So far we have restricted our attention to situations in which we wish to make inference about a single random variable. However, in ecology, we often are interested in multiple random variables and how they are related. Let Y be a random variable that may or may not be independent of X (here again we will distinguish between random variables and realized values for conceptual clarity). Inference about these two random variables can be made using the joint, marginal, or conditional distributions—or, we may make use of all of them depending on the question being asked. In the case of discrete random variables, the joint distribution is the probability that X takes on the value x *and* that Y takes on the value y, which is written $[X = x, Y = y]$. To clarify this concept, let's go back to our original example where X was the number of fish caught after 20 casts, which we said was an *iid* binomial random variable. Now, let's suppose that X depends on the random variable Y, which is the number of other fishermen at the hole. Specifically, let's say that the probability of catching a fish p is related to Y according to $\text{logit}(p) = -0.6 - 2y$. Furthermore, let's make the intuitive assumption that the number of fishermen at the hole is a Poisson random variable with mean 0.6, i.e., $Y \sim \text{Poisson}(0.6)$. Our model is now fully specified, and so we can answer the question: "what is the probability of catching x fish *and* of there being y fishermen at the hole." This joint distribution is given by the product of the binomial pmf (with p determined by y) and the Poisson pmf with $\lambda = 0.6$. The following **R** code creates the joint distribution:

```
> X <- 0:20 # All possible values of X
> Y <- 0:10 # All possible values of Y
> lambda <- 0.6
> p <- plogis(-0.62 + -2*Y) # p as function of Y
> round(p,2)
[1]   0.35   0.07   0.01   0.00   0.00   0.00   0.00   0.00   0.00   0.00   0.00
> joint <- matrix(NA, length(X), length(Y))
> rownames(joint) <- paste("X=", X, sep="")
> colnames(joint) <- paste("Y=", Y, sep="")
>
> # Joint distribution [X,Y]
> for(i in 1:length(Y)) {
+     joint[,i] <- dbinom(X, 20, p[i]) * dpois(Y[i], lambda)
+ }
> round (joint,2)
        Y=0    Y=1    Y=2    Y=3   Y=4   Y=5   Y=6   Y=7   Y=8   Y=9  Y=10
X=0    0.00   0.08   0.08   0.02     0     0     0     0     0     0     0
X=1    0.00   0.12   0.02   0.00     0     0     0     0     0     0     0
X=2    0.01   0.08   0.00   0.00     0     0     0     0     0     0     0
X=3    0.02   0.04   0.00   0.00     0     0     0     0     0     0     0
X=4    0.04   0.01   0.00   0.00     0     0     0     0     0     0     0
X=5    0.07   0.00   0.00   0.00     0     0     0     0     0     0     0
X=6    0.09   0.00   0.00   0.00     0     0     0     0     0     0     0
X=7    0.10   0.00   0.00   0.00     0     0     0     0     0     0     0
X=8    0.09   0.00   0.00   0.00     0     0     0     0     0     0     0
X=9    0.06   0.00   0.00   0.00     0     0     0     0     0     0     0
X=10   0.04   0.00   0.00   0.00     0     0     0     0     0     0     0
X=11   0.02   0.00   0.00   0.00     0     0     0     0     0     0     0
X=12   0.01   0.00   0.00   0.00     0     0     0     0     0     0     0
X=13   0.00   0.00   0.00   0.00     0     0     0     0     0     0     0
X=14   0.00   0.00   0.00   0.00     0     0     0     0     0     0     0
X=15   0.00   0.00   0.00   0.00     0     0     0     0     0     0     0
X=16   0.00   0.00   0.00   0.00     0     0     0     0     0     0     0
X=17   0.00   0.00   0.00   0.00     0     0     0     0     0     0     0
X=18   0.00   0.00   0.00   0.00     0     0     0     0     0     0     0
X=19   0.00   0.00   0.00   0.00     0     0     0     0     0     0     0
X=20   0.00   0.00   0.00   0.00     0     0     0     0     0     0     0
```

This matrix tells us the probability of all possible combinations of x and y, and we see that the most likely value is $(X = 1, Y = 1)$, i.e., we will catch 1 fish and there will be 1 other fisherman. This matrix also demonstrates the law of total probability, which dictates that the sum of these probabilities must equal 1.

Perhaps most fishermen don't care about joint distributions, but a question that might be asked is "what is the probability of catching 1 fish today?" We know that this depends on the number of fishermen, but we don't know how many will show up today, so this is a different question than "what is most likely value of X and Y." This brings us to the marginal distribution, which is defined by

$$[X] = \sum_{Y}[X, Y] \qquad [Y] = \sum_{X}[X, Y],$$

for discrete random variables, and

$$[X] = \int_{-\infty}^{\infty} [X, Y]\mathrm{d}Y \qquad [Y] = \int_{-\infty}^{\infty} [X, Y]\mathrm{d}X,$$

for continuous random variables. The key idea here is that to get the marginal distribution of X, we have to contemplate all possible values of Y. Computing marginal distributions is a key step in maximizing likelihoods involving random effects, as will be demonstrated in Chapter 6. Here is some **R** code to compute the marginal distribution of X, i.e., the probability of catching $X = x$ fish:

```
> margX <- rowSums(joint)
> round(margX, 2)
 X=0  X=1  X=2  X=3  X=4  X=5  X=6  X=7  X=8  X=9  X=10 X=11 X=12 X=13 X=14
0.18 0.14 0.09 0.05 0.05 0.07 0.09 0.10 0.09 0.06 0.04 0.02 0.01 0.00 0.00
X=15 X=16 X=17 X=18 X=19 X=20
0.00 0.00 0.00 0.00 0.00 0.00
```

Bad news—the most likely value is $X = 0$. However, the chance of catching 1 fish is pretty similar.

The last type of question we can ask about these two random variables relates to their conditional distributions. The conditional probability distribution is the distribution of one variable, given a realized value of the other. In the case of two discrete random variables, the conditional distribution may be written as $[X = x|Y = y]$, i.e., the probability of X taking on the value x given the realized value of Y being y. For simplicity, we will write this as $[X|Y]$. Conditional distributions are defined as follows:

$$[X|Y] = \frac{[X, Y]}{[Y]} \qquad [Y|X] = \frac{[X, Y]}{[X]}.$$

That is, the conditional distribution of X given Y is the joint distribution divided by the marginal distribution of Y.

```
margY <- colSums (joint)
> XgivenY <- joint/matrix(margY, nrow(joint), ncol(joint), byrow=TRUE)
> round(XgivenY, 2)
       Y=0   Y=1   Y=2   Y=3  Y=4  Y=5  Y=6  Y=7  Y=8  Y=9  Y=10
X=0   0.00  0.25  0.82  0.97    1    1    1    1    1    1     1
X=1   0.00  0.36  0.16  0.03    0    0    0    0    0    0     0
X=2   0.01  0.25  0.02  0.00    0    0    0    0    0    0     0
X=3   0.03  0.11  0.00  0.00    0    0    0    0    0    0     0
X=4   0.07  0.03  0.00  0.00    0    0    0    0    0    0     0
X=5   0.13  0.01  0.00  0.00    0    0    0    0    0    0     0
X=6   0.17  0.00  0.00  0.00    0    0    0    0    0    0     0
X=7   0.18  0.00  0.00  0.00    0    0    0    0    0    0     0
X=8   0.16  0.00  0.00  0.00    0    0    0    0    0    0     0
X=9   0.12  0.00  0.00  0.00    0    0    0    0    0    0     0
X=10  0.07  0.00  0.00  0.00    0    0    0    0    0    0     0
X=11  0.03  0.00  0.00  0.00    0    0    0    0    0    0     0
X=12  0.01  0.00  0.00  0.00    0    0    0    0    0    0     0
X=13  0.00  0.00  0.00  0.00    0    0    0    0    0    0     0
X=14  0.00  0.00  0.00  0.00    0    0    0    0    0    0     0
X=15  0.00  0.00  0.00  0.00    0    0    0    0    0    0     0
X=16  0.00  0.00  0.00  0.00    0    0    0    0    0    0     0
X=17  0.00  0.00  0.00  0.00    0    0    0    0    0    0     0
X=18  0.00  0.00  0.00  0.00    0    0    0    0    0    0     0
X=19  0.00  0.00  0.00  0.00    0    0    0    0    0    0     0
X=20  0.00  0.00  0.00  0.00    0    0    0    0    0    0     0
```

Note that we have 11 probability distributions for X, one for each possible value of Y, and each pmf sums to unity as it should. Note also that if you show up at the hole and there are >2 fishermen, your chance of catching a fish is very low. Go home. These concepts are explained in more detail in other texts such as Casella and Berger (2002), Royle and Dorazio (2008), and Link and Barker (2010), but hopefully, the code shown here complements the equations and makes it easier for non-statisticians to understand these concepts.

The last point we wish to make in the section is that this simple example *is* a hierarchical model, and we can put the pieces together using the following notation:

$$Y \sim \text{Poisson}(0.6), \tag{2.4.1}$$

$$\text{logit}(p) = -0.6 + -2Y, \tag{2.4.2}$$

$$X|Y \sim \text{Binomial}(20, p). \tag{2.4.3}$$

From here on out, when you see such notation, you should immediately grasp the fact that Y is a random variable independent of X, but X depends upon Y through p. Now you have the tools to make probability statements about the random variables in this system. The one caveat faced in reality is that we typically do not know the values of the parameters, and instead we have to estimate them. Maximum likelihood methods for hierarchical models are covered in Chapter 6.

2.5 Hierarchical models and inference

The term hierarchical modeling (or hierarchical model) has become something of a buzzword over the last decade with hundreds of papers published in ecological journals using that term. So then, what exactly is a hierarchical model, anyhow? Obviously, this term stems from the root "hierarchy" which means:

Definition. *hierarchy* (noun)—a series of ordered groupings of people or things within a system.

In the case of a hierarchical model (hierarchical being the adjective form of hierarchy), the "things" are probability distributions, and they are ordered according to their conditional probability structure. Thus, a hierarchical model is *an ordered series of models, ordered by their conditional probability structure*.

A canonical hierarchical model in ecology is this elemental model of species occurrence or distribution (MacKenzie et al., 2002; Tyre et al., 2003; Kéry, 2011):

$$y_i|z_i \sim \text{Binomial}(K, z_i \times p),$$

$$z_i \sim \text{Bernoulli}(\psi),$$

where y_i = observation of presence/absence at a site i and z_i = occurrence status (z_i = 1 if a species occurs at site i and z_i = 0 if not). Note that if p = 1, then we would perfectly observe z and the model would no longer be hierarchical—it would be a simple logistic regression model. Note also that this hierarchical model has an important conceptual distinction between other types of classical multi-level models such as repeated measures on subjects, in that z_i is an actual state of nature. In that sense, z is a random variable that is the outcome of a "real" process. Royle and Dorazio (2008) used the term *explicit* hierarchical model to describe this type of model to distinguish from hierarchical models (*implicit* hierarchical models) where the latent variables don't correspond to an actual state of nature—but rather just soak up variation that is unmodeled by explicit processes. At best, latent variables in such models are surrogates for something of ecological relevance ("time effects," "space effects," etc.).

With these examples, we expand on our definition of a hierarchical model as we will use it in this book:

Definition. *Hierarchical Model*: A model with explicit component models that describe variation in the data due to (spatial/temporal) variation in *ecological process*, and due to *imperfect observation* of the process.

Most models considered in this book describe the encounter of individuals conditional on the "activity center" of the individual, which is a latent variable (i.e., unobserved random effect). The definition of an activity center will be context-dependent as discussed in Chapter 5, but often it can be thought of as an individual's home range center. The collection of these latent variables represents the outcome of an ecological process describing how individuals distribute themselves over the landscape. Moreover, how individuals are encountered in traps is, in some cases, the result of a model governing movement. As such, these models are examples of hierarchical models that contain formal model components representing both ecological process and also the observation of that process. That is, they are explicit hierarchical models (Royle and Dorazio, 2008) as opposed to implicit hierarchical models.

2.6 Characterization of SCR models

For the purposes of this book, an SCR model is any "individual encounter model" (not just "capture-recapture"!) where auxiliary spatial information is also obtained. To be more precise we could as well use the term "spatial capture and/or recapture" but that is slightly unwieldy and, besides, it also abbreviates to SCR. The class of SCR models includes traditional capture-recapture models with auxiliary spatial information and even some models that do not even require "recapture" (e.g., distance sampling). There is even a class of models (Chapter 18) which don't require capture or unique identification of individuals.

Conceptually, SCR models involve a collection of random variables, $\mathbf{s}$, $\mathbf{u}$, and y where $\mathbf{s}$ is the activity center, or home range center, $\mathbf{u}$ is the location of the individual at the time of sampling, which we may think of as a realization from some movement model, and y is the "response variable"—what the observer records. For example, $y = 1$ means "detected" and $y = 0$ means "not detected," but many other types of responses are possible (Chapter 9). A broad class of models for estimating density are unified by a hierarchical model involving explicit models for animal activity centers $\mathbf{s}$, movement outcomes $\mathbf{u}$, and encounter data y. In some cases, we don't observe y but rather summaries of y, say $n(y)$, yet it might be convenient in such cases to retain an explicit focus on y in terms of model construction. We thus introduce a sequence of models—a hierarchical model—to relate these random variables, which can be written as

$$[n(y)|y][y|\mathbf{u}][\mathbf{u}|\mathbf{s}][\mathbf{s}]. \tag{2.6.1}$$

Every model we talk about in this book has a subset of these components although we never fit the full model because we have not encountered a situation requiring that we do so. However, a detailed description of this model and its various components is the subject of this book, and we will not pretend to condense hundreds of pages of material into the next few paragraphs. But, we give a cursory overview here to whet the appetite and provide some indication of where we are going. Don't worry if some of this material doesn't sink in just yet—we will walk through it slowly in the subsequent chapters.

Let's begin with the model $[\mathbf{s}]$ that describes the distribution of the activity centers of each animal in the spatial region $\mathcal{S}$ (the state-space as we called it previously). As will be explained in Chapters 5 and 11, $[\mathbf{s}]$ is a spatial point process, which may be inhomogeneous if there exists spatial variation in density, or it may be homogeneous if density is constant throughout $\mathcal{S}$. In the latter case, we can write $[\mathbf{s}] = \text{Uniform}(\mathcal{S})$, which is to say that the N activity centers are uniformly distributed in the polygon $\mathcal{S}$. A point process is also a model for the number of individuals in the population N. So we could write $[\mathbf{s}|\mu]$ where μ is an intensity parameter defined as the number of points per unit area. In other words, μ is population density, and we often model population size as either $N \sim \text{Poisson}(\mu||\mathcal{S}||)$, where $||\mathcal{S}||$ is the area of the state-space; or, $N \sim \text{Binomial}(M, \psi)$, where $\psi = \mu||\mathcal{S}||)/M$ and M is some large integer used simply as a convenience measure when conducting Bayesian analysis. As it turns out, there is very little practical difference in the Poisson prior versus a binomial model for N (Chapter 11).

The model $[\mathbf{u}|\mathbf{s}]$ describes the locations of animals conditional on their activity center. In the original formulation of SCR models (Efford, 2004), this model component was intentionally ignored. Indeed when movement is not of direct interest, or when $\mathbf{s}$ is defined in a way not related to a home range center, it may be preferable to ignore this model component (Borchers, 2012). In other cases, we might use an explicit model, such as the bivariate normal model (Royle and Young, 2008).

The third component of the model, $[y|\mathbf{u}]$, describes how the observed data—the so-called capture-histories—arise conditional on the locations of animals. However, as mentioned previously, most SCR models do not contain a movement model, and thus, we typically entertain the model $[y|\mathbf{s}]$ instead of $[y|\mathbf{u}]$. This encounter model generally has at least two parameters, say p_0 and σ, describing the probability of capturing or detecting an individual given the distance between $\mathbf{s}$ and the trap. The most basic model is often called the half-normal model, although we typically refer to it as the Gaussian model since, in two-dimensional space, it is the kernel of a bivariate normal distribution. The model is $p_{ij} = p_0 \exp(-\|\mathbf{x}_j - \mathbf{s}_i\|^2/(2\sigma^2))$ where p_0 is the capture probability when the activity center occurs at the trap location $\mathbf{x}_j$, and σ is a spatial scale parameter determining how rapidly capture probability declines with distance. One common design leads to the model $[y_{ij}|\mathbf{s_i}] = \text{Bernoulli}(p_{ij})$. Chapters 5 and 9 describe many other possible encounter models.

When individuals are marked by biologists or have natural markings permitting individual recognition, y_{ij} is the observed data. However, if some or all of the individuals cannot be uniquely identified, then we cannot record this individual-specific encounter history data. Instead, the data might be simply the number of detections at a trap or perhaps binary detection/non-detection data at each trap on each survey occasion. We call this reduced information data $n(y)$, and Chapters 18 and 19 describe models for $[n(y)|y]$ that still allow for density estimation. The basic strategy is to view y as "missing data" and to use the spatial correlation in the counts, or other sources of information, to provide information about these latent encounter histories.

Equation (2.6.1) is a compact description of the the basic components of an SCR model, but it is also rather vague. The previous four paragraphs added enough extra detail so that we can now describe a specific SCR model. Perhaps the simplest SCR model is this:

$$\begin{aligned} N &\sim \text{Poisson}(\mu||\mathcal{S}||), \\ \mathbf{s}_i &\sim \text{Uniform}(\mathcal{S}), \\ y_{ijk}|\mathbf{s}_i &\sim \text{Bernoulli}(p(\|\mathbf{x}_j - \mathbf{s}_i\|)). \end{aligned} \tag{2.6.2}$$

These "assumptions" are statistical statements of three basic hypotheses that (1) population size N is Poisson distributed (2) activity centers are uniformly distributed in two-dimensional space, and (3) capture probability is a function of the distance between the activity and the trap. Each of these model components can be modified as needed to match specific hypotheses, study designs, and data structures. For example, spatial variation in abundance or density can be easily modeled as a function of habitat covariates (Chapter 11).

We realize that many of the model description in Eq. (2.6.2) may not be self-evident to some ecologists. However, it is absolutely essential that one can understand such a model description—not just for being able to read this book, but also for understanding any statistical model in ecology. One of the best ways of familiarizing oneself with this notation is to translate it into **R** code that simulates outcomes from the model. The following code is an example:

```
set.seed(36372)
Area <- 1                                    # area of state-space (unit square)
x <- cbind(rep(seq(.1,.9,.2), each=5),       # trap locations
           rep(seq(.1,.9,.2), times=5))
p0 <- 0.3                                    # baseline capture probability
sigma <- 0.05                                # Gaussian scale parameter
mu <- 50                                     # population density
N <- rpois(1, mu*Area)                       # population size
s <- cbind(runif(N, 0, 1),                   # activity centers in unit square
           runif(N, 0, 1))
K <- 5
y <- matrix (NA, N, nrow(x))                 # capture data
for(i in 1:N) {
  d.ij <- sqrt((x[,1] - s[i,1])^2 +          # distance between x and s[i]
               (x[,2] - s[i,2])^2)
p.ij <- p0*exp(-d.ij^2/ (2*sigma^2))         # capture probability
y[i,] <- rbinom(nrow(x), K, p.ij)            # capture history for animal i
}
```

Figure 2.3 shows the results of this simulation from a basic, yet very useful, SCR model.

Having briefly explained each of the model components in Eq. (2.6.1), and having shown how a subset of these components results in a basic SCR model, we can now discuss other relevant arrangements. Examples include: (1) classical distance sampling (Buckland et al., 2001; Borchers et al., 2002), (2) spatial capture-recapture models with fixed arrays of traps (Efford, 2004; Borchers and Efford, 2008; Royle et al., 2009a, b; Gardner et al., 2010a; Royle et al., 2011b), and (3) search-encounter models (Royle and Young, 2008; Royle et al., 2011a). We will now elaborate on some of these distinctions.

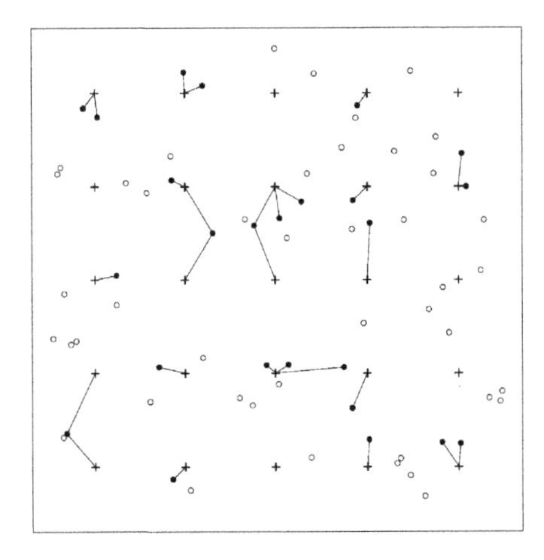

FIGURE 2.3

Population of $N = 69$ home range centers ($\mathbf{s}$, circles) and 25 trap locations ($\mathbf{x}$, crosses). Lines connect activity centers to the traps where the individuals were detected. As in many SCR models, movement outcomes ($\mathbf{u}$) are ignored.

1. **Distance sampling.** The last two, stages of the hierarchy are confounded (implicitly) and so analysis is based on the model $[y|\mathbf{u}][\mathbf{u}]$. The "process model" is that of "uniformity": $\mathbf{u} \sim \text{Uniform}(\mathcal{S})$.
2. **Spatial capture-recapture model with a fixed array of traps.** SCR models appear to have little in common with distance sampling because observations are made only at a pre-defined set of discrete locations—where traps are placed. However, the models are closely related in terms of our hierarchical representation above. In SCR models based on fixed arrays, we cannot estimate both $\Pr(y = 1|\mathbf{u})$ and $\Pr(\mathbf{u}|\mathbf{s})$—the probability that an individual "moves to $\mathbf{u}$" cannot be separated from the probability that it is detected given that it moves to $\mathbf{u}$, because of the fact that the observation locations are fixed by design. Formally, such SCR models confound $[y|\mathbf{u}]$ with $[\mathbf{u}|\mathbf{s}]$ so that the observation model arises as:

$$[y|\mathbf{s}] = \int_{\mathbf{u}} [y|\mathbf{u}][\mathbf{u}|\mathbf{s}]d\mathbf{u}.$$

   This confounding happens because SCR sampling is spatially biased—restricted to a fixed pre-determined set of locations. Conversely, distance sampling confounds $[\mathbf{u}|\mathbf{s}][\mathbf{s}]$ because, essentially, there is only a single realization of the encounter process. It is probably reasonable to assume that $\Pr(y = 1|\mathbf{u}) = 1$ or at least it is locally constant for most devices (e.g., cameras, etc.), and thus the detection model will have the interpretation in terms of movement (see Chapter 13).
3. **Search-encounter models.** What we call "search-encounter" models (Royle and Young, 2008; Royle et al., 2011a) are kind of a hybrid model combining features of SCR models and features of distance sampling. Like distance sampling they allow for encounters in continuous space which provide direct observations from $[\mathbf{u}|\mathbf{s}]$. Thus, the hierarchical model is fully identified. These models are described in Chapt. 15.

2.7 Summary and outlook

Spatial capture-recapture models are hierarchical models, and hierarchical models are models of multiple random variables that are conditionally related. It is therefore important that the basic rules of modeling random variables are understood, and we hope that this chapter has made some of the basic concepts accessible to ecologists with rudimentary background in statistics. If some of this material still seems difficult to grasp, we recommend working with the provided **R** code, which is perhaps the best way of making the equations more tangible.

In some respects, it is possible to understand the gist of SCR without knowing anything about marginal and conditional relationships. One can always fit models using canned software and interpret the output without understanding the guts of the model or the details of the estimation process. For some applied ecologists, this may be perfectly fine, and this book is meant to be useful for both statistical novices and ecologists with more advanced quantitative skills. In most chapters, we begin with a basic conceptual discussion, then we explain the technical details that require an understanding of the concepts in this chapter, and finally we end with one or more worked examples. For those not interested in the technical details, we recommend focusing on the chapter introductions and the examples. However, taking the time to understand the concepts presented in this chapter can only increase one's ability to tackle the unique and complex problems that often present themselves when modeling spatial and temporal aspects of population dynamics.

CHAPTER

GLMs and Bayesian Analysis

3

A major theme of this book is that spatial capture-recapture models are, for the most part, just generalized linear models (GLMs) wherein the covariate, distance between trap and home range center, is unobserved—and therefore regarded as a random effect. Outside of capture-recapture, such models are usually referred to as generalized linear mixed models (GLMMs) and, therefore, SCR models can be thought of as a specialized type of GLMM. Naturally then, we should consider analysis of these slightly simpler models in order to gain some experience and, hopefully, develop a better understanding of spatial capture-recapture models.

In this chapter, we consider classes of GL(M)Ms—Poisson and binomial (i.e., logistic regression) models—that will prove to be enormously useful in the analysis of capture-recapture models of all kinds. Many readers are likely familiar with these models already because they are among the most useful models in ecology and, as such, have received considerable attention in many introductory and advanced texts. We focus on them here in order to introduce readers to the analysis of such models in **R** and **WinBUGS** or **JAGS**, which we will translate directly to the analysis of SCR models in subsequent chapters.

Bayesian analysis is convenient for analyzing GL(M)Ms, because it allows us to work directly with the conditional model—i.e., the model that is conditional on the random effects, using computational methods known as Markov chain Monte Carlo (MCMC). We focus here on the use of **WinBUGS** because it is the most popular "**BUGS** engine." However, later in the book we transition to another popular **BUGS** engine known as **JAGS** (Plummer, 2003) which stands for *Just Another Gibbs Sampler*. For most of our purposes, the specification of models in either platform is the same, but **JAGS** is under active development at the present time while **WinBUGS** no longer is. While we use **BUGS** of one sort or another to do the Bayesian computations, we organize and summarize our data and execute **WinBUGS** or **JAGS** from within **R** using the packages `R2WinBUGS` (Sturtz et al., 2005), `R2jags` (Su and Yajima, 2011), or `rjags` (Plummer, 2011). Kéry (2010) provides an accessible introduction to the basics of Bayesian analysis and GL(M)Ms using **WinBUGS**. We don't want to be too redundant and so we avoid a detailed treatment of Bayesian methodology and software usage—instead just providing a cursory overview so that we can move on and attack the problems we're most interested in related to spatial capture-recapture. In addition, there are a number of texts that provide general introductions to Bayesian

Spatial Capture-Recapture. http://dx.doi.org/10.1016/B978-0-12-405939-9.00003-7

analysis, MCMC, and their applications in ecology including McCarthy (2007), Kéry (2010), Link and Barker (2010), and King et al. (2008).

While this chapter is about Bayesian analysis of GL(M)Ms, such models are routinely analyzed using likelihood methods, too. Later in this book (Chapter 6), we will use likelihood methods to analyze SCR models but, for now, we concentrate on providing a basic introduction to Bayesian analysis because that is the approach we will use in a majority of cases in later chapters.

3.1 GLMs and GLMMs

We have asserted already that SCR models work out most of the time to be variations of GL(M)Ms. You might therefore ask: What are these GLM and GLMM models, anyhow? These models are covered extensively in many very good applied statistics books and we refer the reader elsewhere for a detailed introduction. The classical references for GLMs are Nelder and Wedderburn (1972) and McCullagh and Nelder (1989). In addition, we think Kéry (2010), Kéry and Schaub (2012), and Zuur et al. (2009) are all accessible treatments. Here, we'll give the 1 min treatment of GL(M)Ms, not trying to be complete but rather only to preserve a coherent organization to the book.

The GLM is an extension of standard linear models allowing the response variable to have some distribution from the exponential family of distributions. This includes the normal distribution but also others such as the Poisson, binomial, gamma, exponential, and many more. In addition, GLMs allow the response variable to be related to the predictor variables (i.e., covariates) using a link function, which is usually nonlinear. The GLM consists of three components:

1. A probability distribution for the dependent (or response) variable y, from the exponential family of probability distributions.
2. A "linear predictor" $\eta = \beta_0 + \beta_1 x$, where x is a predictor variable (i.e., a covariate).
3. A link function g that relates the expected value of y, $\mathbb{E}(y)$, to the linear predictor, $\mathbb{E}(y) = \mu = g^{-1}(\eta)$. Therefore, $g(\mathbb{E}(y)) = \eta = \beta_0 + \beta_1 x$.

A key aspect of GLMs is that $g(\mathbb{E}(y))$ is assumed to be a linear function of the predictor variable (s), here x, with unknown parameters β_0 and β_1, to be estimated. In standard GLMs, the variance of y is a function V of the mean of y: $\text{Var}(y) = V(\mu)$. As an example, a Poisson GLM posits that $y \sim \text{Poisson}(\lambda)$ with $\mathbb{E}(y) = \lambda$ and usually the model for the mean, here at sampling unit i, is specified using the *log link function* by

$$\log(\lambda_i) = \beta_0 + \beta_1 x_i$$

The variance function is $\text{V}(y_i) = \lambda_i$. To see how a Poisson GLM works, use the **R** code below to simulate some data and then estimate the parameters:

```
> set.seed(13)
> n <- 100                  # set sample size
> beta0 <- -2               # set intercept term
> beta1 <- 1.5              # set coefficient
> x <- rnorm(n, 0,1)        # generate a predictor variable, x

> linpred <- beta0 + beta1*x    # calculate linear predictor of E(y)
> y <- rpois(n, exp(linpred))   # generate observations from model
```

The **R** function `glm` fits a GLM to the data we just generated and returns estimates of β_0 and β_1:

```
> glm(y~1 + x, family='poisson')      # the fit model
```

This produces the output:

```
Call: glm(formula = y~1 + x, family = "poisson")

Coefficients:
(Intercept)          x
   -2.007        1.446

[... some output deleted ...]
```

In this summary output, the maximum likelihood estimates (MLEs) of the regression parameters β_0 and β_1 are labeled "`Coefficients`." We see that these are not too different from the data-generating values (-2 and 1.5, respectively).

The binomial GLM posits that $y_i \sim \text{Binomial}(K, p)$ where K is the fixed sample size parameter and $\mathbb{E}(y_i) = K \times p_i$. Usually the model for the mean is specified using the *logit link function* according to

$$\text{logit}(p_i) = \beta_0 + \beta_1 x_i,$$

where $\text{logit}(p)\,\eta = \log(p/(1 - p))$. The inverse-logit function, consequently, is $\text{logit}^{-1}(\eta) = \exp(\eta)/(1 + \exp(\eta))$.

A GLMM is the extension of GLMs to accommodate "random effects." Often this involves adding a normal random effect to the linear predictor. One simple example is using a random intercept, $\boldsymbol{\alpha}$:

$$\log(\lambda_i) = \alpha_i + \beta_1 x_i,$$

where

$$\alpha_i \sim \text{Normal}(\mu, \sigma^2).$$

Many other probability distributions and formulations of the linear predictor might be considered. GLMMs are enormously useful in ecological modeling applications for modeling variation due to subjects, observers, spatial or temporal stratification, clustering, and dependence that arises from any kind of group structure and, of course, because SCR models prove to be a type of GLM with a random effect.

3.2 Bayesian analysis

Bayesian analysis is less familiar to many ecological researchers because they are often educated only in the classical statistical paradigm of frequentist inference. But advances in technology and increasing exposure to the benefits of Bayesian analysis are fast making Bayesians out of people or at least making Bayesian analysis an acceptable, general alternative to classical, frequentist inference.

Conceptually, the main thing about Bayesian inference is that it uses probability directly to characterize uncertainty about things we don't know. "Things," in this case, are parameters of models and, just as it is natural to characterize uncertain outcomes of stochastic processes using probability, it seems natural also to characterize information about unknown parameters using probability. At least this seems natural to us and, we think, most ecologists either explicitly adopt that view or tend to fall into that point of view naturally. Conversely, frequentists use probability in many different ways, but never to characterize uncertainty about parameters.[1] Instead, frequentists use probability to characterize the behavior of *procedures* such as estimators or confidence intervals (see below). It is surprising that people readily adopt a philosophy of statistical inference in which the things you don't know (i.e., parameters) should *not* be regarded as random variables, so that, as a consequence, one cannot use probability to characterize one's state of knowledge about them.

3.2.1 Bayes' rule

As its name suggests, Bayesian analysis makes use of Bayes' rule in order to make direct probability statements about model parameters. Given two random variables z and y, Bayes' rule relates the two conditional probability distributions $[z|y]$ and $[y|z]$ by the relationship:

$$[z|y] = [y|z][z]/[y]. \tag{3.2.1}$$

Bayes' rule itself is a mathematical fact and there is no debate in the statistical community as to its validity and relevance to many problems. Generally speaking, these distributions are characterized as follows: $[y|z]$ is the conditional probability distribution of y *given* z, $[z]$ is the marginal distribution of z and $[y]$ is the marginal distribution of y. In the context of Bayesian inference we usually associate specific meanings in which $[y|z]$ is thought of as "the likelihood," $[z]$ as the "prior," and so on. We leave this for later because here the focus is on this expression of Bayes' rule as a basic fact of probability.

As an example of a simple application of Bayes' rule, consider the problem of determining species presence at a sample location based on imperfect survey information. Let z be a binary random variable that denotes species presence ($z = 1$) or absence ($z = 0$), let $\Pr(z = 1) = \psi$, where ψ is usually called occurrence probability, "occupancy" (MacKenzie et al., 2002) or "prevalence." Let y be the *observed* presence

[1] To hear this will be shocking to some readers perhaps.

($y = 1$) or absence ($y = 0$) (or, strictly speaking, detection and non-detection), and let p be the probability that a species is detected in a single survey at a site given that it is present. Thus, $\Pr(y = 1|z = 1) = p$. The interpretation of this is that, if the species is present, we will only observe it with probability p. In addition, we assume here that $\Pr(y = 1|z = 0) = 0$. That is, the species cannot be detected if it is not present, which is a conventional view adopted in most biological sampling problems (but see Royle and Link, 2006). If we survey a site K times but never detect the species, then this clearly does not imply that the species is not present ($z = 0$) at this site, just that we failed to observe it. Thus, our degree of belief in $z = 0$ should be made with a probabilistic statement, namely the conditional probability $\Pr(z = 1|y_1 = 0, \ldots, y_K = 0)$. If the K surveys are independent so that we might regard y_k as *iid* Bernoulli trials, then the total number of detections, say y, is binomial with probability p, and we can use Bayes' rule to compute the probability that the species is present given that it is not detected in K samples, i.e., $\Pr(z = 1|y_1 = 0, \ldots, y_K = 0)$. In words, the expression we seek is:

$$\Pr(\text{present}|\text{not detected}) = \frac{\Pr(\text{not detected}|\text{present})\Pr(\text{present})}{\Pr(\text{not detected})}.$$

Mathematically, this is

$$\begin{aligned}\Pr(z = 1|y = 0) &= \frac{\Pr(y = 0|z = 1)\Pr(z = 1)}{\Pr(y = 0)} \\ &= \frac{(1 - p)^K \psi}{(1 - p)^K \psi + (1 - \psi)}.\end{aligned}$$

The denominator here, the probability of not detecting the species, is composed of two parts: (1) not observing the species given that it is present (this occurs with probability $(1 - p)^K \psi$) and (2) the species is not present (this occurs with probability $1 - \psi$). To apply this equation, suppose that $K = 2$ surveys are done at a wetland for a species of frog, and the species is not detected there. Suppose further that $\psi = 0.8$ and $p = 0.5$ are obtained from a prior study. Then the probability that the species is present at this site, even though it was not detected, is $(1-0.5)^2 \times 0.8/((1-0.5)^2 \times 0.8+(1-0.8)) = 0.5$. That is, there is a 50/50 chance that the site is occupied despite the fact that the species wasn't observed there.

In summary, Bayes' rule provides a simple linkage between the conditional probabilities $[y|z]$ and $[z|y]$, which is useful whenever we need to deduce one from the other.

3.2.2 Principles of Bayesian inference

Bayes' rule as a basic fact of probability is not disputed. What is controversial to some is the scope and manner in which Bayes' rule is applied by Bayesian analysts. Bayesian analysts assert that Bayes' rule is relevant, in general, to all statistical problems by regarding all unknown quantities of a model as realizations of random variables—this includes data, latent variables, and parameters. Classical (non-Bayesian) analysts sometimes object to regarding parameters as outcomes of random variables. Classically, parameters are thought of as "fixed but unknown" (using the terminology of classical statistics). Indeed, a common misunderstanding on the distinction between Bayesian and frequentist inference goes something like this: "in frequentist inference parameters are fixed but unknown but in a Bayesian analysis parameters are random." At best this is a sad caricature of the distinction and at worst it is downright wrong. In Bayesian analysis the parameters are also unknown and, in fact, there is a single data-generating value of each parameter, and so they are also fixed. The difference is that the fixed but unknown values are regarded as having been generated from some probability distribution. Specification of that probability distribution is necessary to carry out Bayesian analysis, but it is not required in classical frequentist inference.

To see the general relevance of Bayes' rule in the context of statistical inference, let y denote observations—i.e., data—and let $[y|\theta]$ be the observation model (often colloquially referred to as the "likelihood"). Suppose θ is a parameter of interest having (prior) probability distribution $[\theta]$ (also simply referred to as the prior). These are combined to obtain the posterior distribution using Bayes' rule, which is:

$$[\theta|y] = [y|\theta][\theta]/[y].$$

Asserting the general relevance of Bayes' rule to all statistical problems, we can conclude that the two main features of Bayesian inference are that: (1) parameters, θ, are regarded as realizations of a random variable and, as a result, (2) inference is based on the probability distribution of the parameters given the data, $[\theta|y]$, which is called the posterior distribution. This is the result of using Bayes' rule to combine the "likelihood" and the prior distribution. The key concept is regarding parameters as realizations of a random variable because, once you admit this conceptual view, it leads directly to the posterior distribution, a very natural quantity upon which to base inference about things we don't know, including parameters of statistical models. In particular, $[\theta|y]$ is a probability distribution for θ and therefore we can make direct probability statements to characterize uncertainty about θ.

The denominator of our invocation of Bayes' rule, $[y]$, is the marginal distribution of the data y. We note without further remark right now that, in many practical problems, this can be an enormous pain to compute. The main reason that the Bayesian paradigm has become so popular in the last 20 years or so is because methods have been developed for characterizing the posterior distribution that do not require that we possess a mathematical understanding of $[y]$. This means we never have to compute it or know what it looks like, or know anything specific about it.

While we can understand the conceptual basis of Bayesian inference merely by understanding Bayes' rule—that's really all there is to it—it is not so easy to understand the basis of classical frequentist inference. What is mostly coherent in frequentist inference is the manner in which procedures are evaluated—the performance of a given procedure is evaluated by "averaging over" hypothetical realizations of y, regarding the *estimator* as a random variable. For example, if $\hat{\theta}$ is an estimator of θ then the frequentist is interested in $\mathbb{E}_y(\hat{\theta}|y)$, which is used to characterize bias. If the expected value of $\hat{\theta}$, when averaged over realizations of y, is equal to θ, then $\hat{\theta}$ is unbiased.

The view of parameters as being random variables allows Bayesians to use probability to make direct probability statements about parameters. Frequentist inference procedures do not permit direct probability statements to be made about parameter values. Instead, the view of parameters as fixed constants and estimators as random variables leads to interpretations that are not so straightforward. For example, confidence intervals having the interpretation "95% probability that the interval contains the true value" and p-values being "the probability of observing an outcome of the test statistic as extreme or more than the one observed." These are far from intuitive interpretations to most people. Moreover, this is conceptually problematic to some because we will never get to observe the hypothetical realizations that characterize the performance of our procedure. Rather, we just have one realization.

While we do tend to favor Bayesian inference for the conceptual simplicity (parameters are random, posterior inference), we mostly advocate for a pragmatic nonpartisan approach to inference because, frankly, some of the frequentist methods are actually very convenient in certain situations, and will generally yield very similar inferences about parameters, as we will see in later chapters.

3.2.3 Prior distributions

The prior distribution $[\theta]$ is an important feature of Bayesian inference. As a conceptual matter, the prior distribution characterizes "prior beliefs" or "prior information" about a parameter. Indeed, an oft-touted benefit of Bayesian analysis is the ease with which prior information can be included in an analysis. However, more commonly, the prior is chosen to express a lack of prior information, even if previous studies have been done and even if the investigator does in fact know quite a bit about a parameter. This is because the manner in which prior information is embodied in a prior (and the amount of information) is usually very subjective and thus the result can wind up being very contentious; e.g., different investigators might report different results based on subjective assessments of prior information. Thus it is usually better to "let the data speak" and use priors that reflect absence of information beyond the data set being analyzed. An example for an uninformative prior is a Uniform(0, 1) for a probability, or a Uniform($-\infty$, ∞) (also called a "flat" or "improper" prior) for an unbounded continuous parameter. Alternatively, people use "diffuse priors"; these contain some information, but (ideally) not enough to exert meaningful influence on the posterior. An example for a diffuse prior could be a normal distribution with a large standard deviation.

But still the need occasionally arises to embody prior information or beliefs about a parameter formally into the estimation scheme. In SCR models we often have a parameter that is closely linked to "home range size" and thus auxiliary information on the home range size of a species can be used as prior information, which may improve parameter estimation (e.g., see Chandler and Royle (2013); Chapter 18).

At times the situation arises where a prior can inadvertently impose substantial effect on the posterior of a parameter, and that is not desirable. For example, we use data augmentation to deal with the fact that the population size N is an unknown parameter (Royle et al., 2007), which is equivalent to imposing a Binomial(M, ψ) prior on N for some integer M (see Section 4.2). One has to take care that M is sufficiently large so as to not affect the posterior distribution on N (see Figure 17.6, and Kéry and Schaub (2012, Chapter 5)). Another situation that we have to be careful of is that prior distributions are *not* invariant to transformation of the parameter, and therefore neither are posterior distributions (Link and Barker, 2010, Section 6.2.1). Thus, a prior that is ostensibly non-informative on one scale may be very informative on another scale. For example, if we have a flat prior on logit(p) for some probability parameter p, this is very different from having a Uniform(0, 1) prior on p. We show an example where this makes a difference in Chapter 5. Nonetheless, it is always possible to assess the influence of prior choice, and it is often the case (with sufficient data and a structurally identifiable model) that the influence of priors is negligible.

3.2.4 Posterior inference

In Bayesian inference, we are not focusing on estimating a single point or interval but rather on characterizing a whole distribution—the posterior distribution—from which one can report any summary of interest. A point estimate might be the posterior mean, median, mode, etc. In many applications in this book, we will compute 95% Bayesian confidence intervals using the 2.5% and 97.5% quantiles of the posterior distribution. For such intervals, it is correct to say $\Pr(L < \theta < U) = 0.95$. That is, "the probability that θ lies between L and U is 0.95."

As an example, suppose we conducted a Bayesian analysis to estimate detection probability (p) of some species at a study site, and we obtained a posterior distribution of Beta(20, 10) for the parameter p. The following **R** commands demonstrate how we make inferences based upon summaries of the posterior distribution:

```
> post.median <- qbeta(0.5, 20, 10)
[1] 0.6704151

> post.95ci <- qbeta(c(0.025, 0.975), 20, 10)
[1] 0.4916766 0.8206164
```

Thus, we can state that there is a 95% probability that p lies between 0.49 and 0.82. Figure 3.1 shows the posterior along with the summary statistics. It is not a subtle thing that such statements cannot be made using frequentist methods, although people tend to say it anyway and not really understand why it is wrong or even that it is wrong.

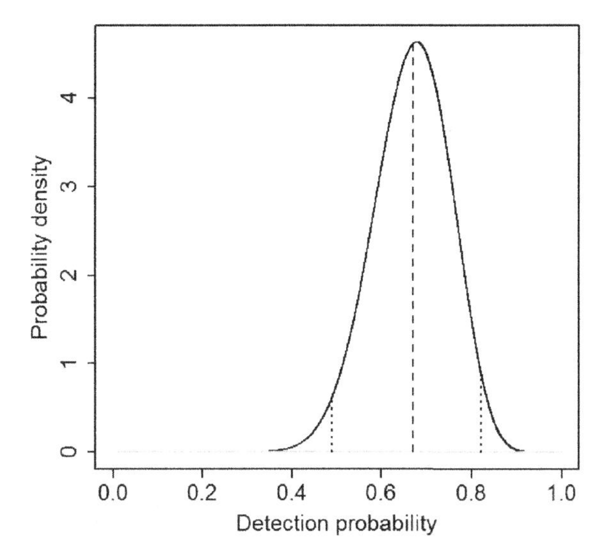

FIGURE 3.1

Probability density plot of a hypothetical posterior distribution of Beta(20, 10); dashed lines indicate mean and upper and lower 95% interval.

3.2.5 Small sample inference

The posterior distribution is an exhaustive summary of the state-of-knowledge about an unknown quantity. It is *the* posterior distribution—not an estimate of that thing. It is also not, usually, an approximation except to within Monte Carlo error (in cases where we use simulation to calculate it, see Section 3.5.2). One of the great virtues of Bayesian analysis which is not widely appreciated is that posterior inference is not "asymptotic," which is to say, valid only in a limiting sense as the sample size tends to infinity. Rather, posterior inference is valid for *any* sample size and, in particular, *the* sample size on hand. Conversely, almost all frequentist procedures are based on asymptotic approximations to the procedure that is being employed.

There seems to be a prevailing view in statistical ecology that classical likelihood-based procedures are virtuous because of the availability of simple formulas and procedures for carrying out inference, such as calculating standard errors, doing model selection by Akaike information criterion (AIC), and assessing goodness-of-fit. In large samples, this may be an important practical benefit, but the theoretical validity of these procedures cannot be asserted in most situations involving small samples. This is not a minor issue because it is typical in many wildlife sampling problems—especially in surveys of carnivores or rare/endangered species—to wind up with a small, sometimes extremely small, data set that is nevertheless extremely valuable (Foster and Harmsen, 2012). For examples: A recent paper (Hawkins and Racey, 2005) on the fossa (*Cryptoprocta ferox*) estimated an adult density of 0.18 adults per km^2

based on a sample size of 20 animals captured over 3 years. Sepúlveda et al. (2007) estimated density of the endangered southern river otter (*Lontra provocax*) based on 12 individuals captured over 3 years, Gardner et al. (2010a) estimated density from a study of the Pampas cat (*Leopardus colocolo*), a species for which very little is known, based on only 22 captured individuals over a 2-year study period, Trolle and Kéry (2005) reported only 9 individual ocelots captured and Jackson et al. (2006) captured 6 individual snow leopards (*Panthera uncia*) using camera trapping. Thus, almost all likelihood-based analysis of data on rare and/or secretive carnivores necessarily and flagrantly violate one of Le Cam's Basic Principles: "If you need to use asymptotic arguments, do not forget to let your number of observations tend to infinity" (Le Cam, 1990).

The biologist thus faces a dilemma with such data. On one hand, these data sets, and the resulting inference, are often criticized as being poor and unreliable. Or, even worse,[2] "the data set is so small, this is a poor analysis." On the other hand, such data may be all that is available for species that are extraordinarily important for conservation and management. The Bayesian framework for inference provides a valid, rigorous, and flexible framework that is theoretically justifiable in arbitrary sample sizes. This is not to say that one will obtain precise estimates of density or other parameters, just that your inference is coherent and justifiable from a conceptual and technical statistical point of view.

3.3 Characterizing posterior distributions by MCMC simulation

In practice, it is not really feasible to ever compute the marginal probability distribution $[y]$, the denominator resulting from application of Bayes' rule (Eq. (3.2.1)). For decades (even centuries!) this impeded the adoption of Bayesian methods by practitioners. Or, the few Bayesian analyses done were based on asymptotic normal approximations to the posterior distribution. While this was useful from a theoretical and technical standpoint and, practically, it allowed people to make the probability statements that they naturally would like to make, it was kind of a bad joke around the Bayesian water-cooler to, on one hand, criticize classical statistics for being, essentially, completely ad hoc in their approach to things but then, on the other hand, have to devise various approximations to what they were trying to characterize. The advent of Markov chain Monte Carlo (MCMC) methods has made it easier to calculate posterior distributions for just about any problem to sufficient levels of precision.

Broadly speaking, MCMC is a class of methods for drawing random samples (i.e., simulating from or just "sampling") from the target posterior distribution. Thus, even though we might not recognize the posterior as a named distribution or be able to analyze its features analytically, e.g., devise mathematical expressions for the mean and variance, we can use these MCMC methods to obtain a large sample from the

[2] Actual quote from a referee.

posterior and then use that sample to characterize features of the posterior. What we do with the sample depends on our intentions—typically we obtain the mean or median for use as a point estimate, and take a confidence interval based on Monte Carlo estimates of the quantiles.

3.3.1 What goes on under the MCMC hood

We will develop and apply MCMC methods in some detail for spatial capture-recapture models in Chapter 17. Here we provide a simple illustration of some basic ideas related to the practice of MCMC.

A type of MCMC method relevant to most problems is Gibbs sampling (Geman and Geman, 1984) which we address in more detail in Chapter 17. Gibbs sampling involves iterative simulation from the "full conditional" distributions (also called conditional posterior distributions). The full conditional distribution for an unknown quantity is the conditional distribution of that quantity given every other random variable in the model—the data and all other parameters (see Section 3.3.2 for rules of how to construct full conditionals). For example, for a normal regression model[3] with $y \sim \text{Normal}(\beta_0 + \beta_1(x - \bar{x}), \sigma^2)$ where, lets say, σ^2 is known, the full conditionals are, using "bracket notation,"

$$[\beta_0|y, \beta_1]$$

and

$$[\beta_1|y, \beta_0].$$

We might use our knowledge of probability to identify these mathematically. In particular, by Bayes' rule, $[\beta_0|y, \beta_1] = [y|\beta_0, \beta_1][\beta_0|\beta_1]/[y|\beta_1]$ and similarly for $[\beta_1|y, \beta_0]$. For example, if we have priors for $[\beta_0] = \text{Normal}(\mu_{\beta_0}, \sigma^2_{\beta_0})$ and $[\beta_1] = \text{Normal}(\mu_{\beta_1}, \sigma^2_{\beta_1})$ then some algebra reveals that

$$\left[\beta_0|y, \beta_1\right] = \text{Normal}\left(w\bar{y} + (1 - w)\mu_{\beta_0}, (\tau n + \tau_{\beta_0})^{-1}\right), \tag{3.3.1}$$

where n is the sample size, $\tau = 1/\sigma^2$ and $\tau_{\beta_0} = 1/\sigma^2_{\beta_0}$ (the inverse of the variance is sometimes called precision), and $w = \tau n/(\tau n + \tau_{\beta_0})$. We see in this case that the posterior mean is a *precision-weighted* sum of the sample mean $\bar{y}$ and the prior mean μ_{β_0}, and the posterior *precision* is the sum of the precision of the likelihood and that of the prior. These results are typical of many classes of problems. In particular, note that as the prior precision tends to 0, i.e., $\tau_{\beta_0} \rightarrow 0$, then the posterior of β_0 tends to $\text{Normal}(\bar{y}, \sigma^2/n)$. We recognize the variance of this distribution as that of the variance of the sampling distribution of $\bar{y}$ and its mean is in fact the MLE of β_0

[3] We center the independent variable here so that things look more familiar in the result.

for this model. The conditional posterior of β_1 has a very similar form:

$$[\beta_1|y,\beta_0] = \text{Normal}\left(\frac{\tau(\sum_i y_i(x_i-\bar{x})) + \tau_{\beta_1}\mu_{\beta_1}}{\tau\sum_i(x_i-\bar{x})^2 + \tau_{\beta_1}}, \left(\tau\sum_i(x_i-\bar{x})^2 + \tau_{\beta_1}\right)^{-2}\right), \tag{3.3.2}$$

which might look slightly unfamiliar, but note that if $\tau_{\beta_1} = 0$, then the mean of this distribution is the familiar $\hat{\beta}_1$, and the variance is, in fact, the sampling variance of $\hat{\beta}_1$. The MCMC algorithm for this model has us simulate in succession, repeatedly, from those two distributions. See Gelman et al. (2004) for more examples of Gibbs sampling for the normal model, and we also provide another example in Chapter 17. A conceptual representation of the MCMC algorithm for this simple model is:

```
Algorithm: Gibbs sampling for linear regression

  0. Initialize β0 and β1

  Repeat {

    1. Draw a new value of β0 from Eq. (3.3.1)

    2. Draw a new value of β1 from Eq. (3.3.2)

  }
```

As we just saw for this simple "normal-normal" model, it is sometimes possible to specify the full conditional distributions analytically. In general, when certain so-called conjugate prior distributions are used, which have an analytic form that, in a statistical sense, "matches" the likelihood, then the form of the full conditional distributions is also similar to that of the observation model. In this normal-normal case, the normal distribution for the mean parameters is the conjugate prior for the normal observation model, and thus the full conditional distributions are also normal. This is convenient because, in such cases, we can simulate directly from them using standard methods (or **R** functions). But, in practice, we don't really ever need to know such things because most of the time we can get by using a simple algorithm, called the Metropolis-Hastings (henceforth "MH") algorithm, to obtain samples from these full conditional distributions without having to recognize them as specific, named, distributions. This gives us enormous freedom in developing models and analyzing them without having to resolve them mathematically, because to implement the MH algorithm, we need only identify the full conditional distribution up to a constant of proportionality, that being the marginal distribution in the denominator (e.g., $[y|\beta_1]$ above).

We will talk about the Metropolis-Hastings algorithm shortly, and we will use it extensively in the analysis of SCR models (e.g., Chapter 17).

3.3.2 Rules for constructing full conditional distributions

The strategy for constructing full conditional distributions for devising MCMC algorithms can be reduced conceptually to a couple of basic steps summarized as follows:

(step 1) Identify all stochastic components of the model and collect their probability distributions.
(step 2) Express the full conditional in question as proportional to the product of all probability distributions identified in step 1.
(step 3) Remove the ones that don't have the focal parameter in them.
(step 4) Do some algebra on the result in order to identify the resulting probability distribution function (pdf) or mass function (pmf).

Of the four steps, the last of those is the main step that requires quite a bit of statistical experience and intuition because various algebraic tricks can be used to reshape the mess into something recognizable—i.e., a standard, named distribution. But step 4 is not necessary if we decide instead to use the Metropolis-Hastings algorithm as described below.

In the context of our simple linear regression model that we've been working with, to characterize $[\beta_0|y, \beta_1]$ we first apply step 1 and identify the model components as: $[y|\beta_0, \beta_1]$, with prior distributions $[\beta_0]$ and $[\beta_1]$. Step 2 has us write $[\beta_0|y, \beta_1] \propto [y|\beta_0, \beta_1][\beta_0][\beta_1]$. Step 3: We note that $[\beta_1]$ is not a function of β_0 and therefore we remove it to obtain $[\beta_0|y, \beta_1] \propto [y|\beta_0, \beta_1][\beta_0]$. Similarly, applying steps 2 and 3 for β_1 we obtain $[\beta_1|y, \beta_0] \propto [y|\beta_0, \beta_1][\beta_1]$. We apply step 4 and manipulate these algebraically to arrive at the result (which we provided in Eqs. (3.3.1) and (3.3.2)) or, alternatively, we can sample them indirectly using the Metropolis-Hastings algorithm, which we discuss next.

3.3.3 Metropolis-Hastings algorithm

The Metropolis-Hastings (MH) algorithm is a completely generic method for sampling from any distribution, say $[\theta]$. In our applications, $[\theta]$ will typically be the full conditional distribution of θ. While we sometimes use Gibbs sampling, we seldom use "pure" Gibbs sampling because full conditionals do not always take the form of known distributions we can sample from directly. In such cases, we use MH to sample from the full conditional distributions. When the MH algorithm is used to sample from full conditional distributions of a Gibbs sampler, the resulting hybrid algorithm is called *Metropolis-within-Gibbs*. In Section 3.6.3 we will construct such an algorithm for a simple class of models. We discuss both the Gibbs and the MH algorithm, as well as their hybrid in more depth in Chapter 17.

The MH algorithm generates candidate values for the parameter(s) we want to estimate from some proposal or candidate-generating distribution that may be conditional on the current value of the parameter, denoted by $h(\theta^*|\theta^{t-1})$. Here, θ^* is the *candidate* or proposed value and θ^{t-1} is the value of θ at the previous time step, i.e., at iteration $t-1$ of the MCMC algorithm. The proposed value is accepted with probability

$$r = \frac{[\theta^*]h(\theta^{t-1}|\theta^*)}{[\theta^{t-1}]h(\theta^*|\theta^{t-1})},$$

which is called the MH acceptance probability. This ratio can sometimes be >1 in which case we set it equal to 1. It is useful to note that $h()$ can be any probability distribution.

In the context of using the MH algorithm to do MCMC (in which case the target distribution is a full conditional or posterior distribution), an important fact is, no matter the choice of $h()$, we can compute the MH acceptance probability directly because the marginal distribution of y cancels from both the numerator and denominator of r. This is the magic of the MH algorithm.

3.4 Bayesian analysis using the BUGS language

We won't be too concerned with devising our own MCMC algorithms for every analysis, although we will do that a few times for fun. More often, we will rely on the freely available software packages **WinBUGS** or **JAGS** for doing this. We will always execute these **BUGS** engines from within **R** using the `R2WinBUGS` or, for **JAGS**, the `R2jags` or `rjags` packages. **WinBUGS** and **JAGS** are MCMC black boxes that take a pseudo-code description (i.e., written in the **BUGS** language) of all of the relevant stochastic and deterministic elements of a model and generate an MCMC algorithm for that model. But you never get to see the algorithm. Instead, **WinBUGS/JAGS** will run the algorithm and return the Markov chain output—the posterior samples of model parameters.

The great thing about using the **BUGS** language is that it forces you to become intimate with your statistical model—you have to write each element of the model down, admit (explicitly) all of the various assumptions, understand what the actual probability assumptions are and how data relate to latent variables and data and latent variables relate to parameters, and how parameters relate to one another.

While we normally use **WinBUGS**, we note that **OpenBUGS** is the current active development tree of the **BUGS** project. See Kéry (2010) and Kéry and Schaub (2012, especially Appendix 1) for more on practical analysis in **WinBUGS**. Those books should be consulted for a more comprehensive introduction to using **WinBUGS**. Recently we have migrated many of our analyses to **JAGS**, which we adopt later in the book. You can refer to Hobbs (2011) for an ecological introduction to **JAGS**. Next, we provide an example of a Bayesian analysis using **WinBUGS**.

3.4.1 Linear regression in WinBUGS

We provide a brief introductory example of a normal regression model using a small simulated data set. The following commands are executed from within your **R** workspace. First, simulate a covariate x and observations y having prescribed intercept, slope, and variance:

```
> x <- rnorm(10)
> mu <- -3.2 + 1.5*x
> y <- rnorm(10,mu,sd=4)
```

The **BUGS** model specification for a normal regression model is written within **R** as a character string input to the command `cat` and then dumped to a text file named `normal.txt`:

```
> cat ("
 model{
    for (i in 1:10){
       y[i] ~ dnorm(mu[i],tau)           # The likelihood
       mu[i] <- beta0 + beta1*x[i]       # The linear predictor
     }
    beta0 ~ dnorm(0,.01)                 # Prior distributions
    beta1 ~ dnorm(0,.01)
    sigma ~ dunif(0,100)
    tau <- 1/(sigma*sigma)               # Tau is the precision
}                                        #   and a derived parameter
",file="normal.txt")
```

Alternatively, you can write the model specifications directly within a text file and save it in your current working directory, but we do not usually take that approach in this book.

The **BUGS** dialects[4] parameterize the normal distribution in terms of the mean and inverse-variance, called the precision. Thus, `dnorm(0,.01)` implies a variance of 100. We typically use diffuse normal priors for mean parameters, β_0 and β_1 in this case, but sometimes we might use uniform priors with suitable bounds $-B$ and $+B$. Also, we typically use a Uniform(0, B) prior on standard deviation parameters (Gelman, 2006). Alternatively, we could use a gamma prior on the precision parameter τ. In a **BUGS** model file, every variable referenced in the model description has to be either data, which will be input (see below), a random variable, which must have a probability distribution associated with it using the tilde character "$\sim$" (a.k.a. "twiddle"), or it has to be a derived parameter connected to variables and data using an assignment arrow: "<-".

To fit the model, we need to describe various data objects to **WinBUGS**. In particular, we create an **R** list object called `data` which holds the data objects identified in the **BUGS** model file. In the example, the data consist of two objects, which exist as y and x in the **R** workspace and in the **WinBUGS** model definition. We also create an **R** function that produces a list of starting values, `inits`, that get sent to

[4]We use this to mean **WinBUGS**, **OpenBUGS**, and **JAGS**.

WinBUGS. In general, starting values are optional. We recommend to always provide reasonable starting values where possible, both for structural parameters and random effects.[5] Finally, we identify the names of the parameters (labeled correspondingly in the **WinBUGS** model specification) that we want **WinBUGS** to save the MCMC output for. In this example, we will "monitor" the parameters β_0, β_1, σ, and τ. **WinBUGS** is executed using the **R** command `bugs`. We set the option `debug=TRUE` if we want the **WinBUGS** GUI to stay open (useful for analyzing MCMC output and looking at the **WinBUGS** error log). Also, we set `working.dir=getwd()` so that **WinBUGS** output files and the log file are saved in the current **R** working directory (note that sometimes you will need to specify the place where you installed **WinBUGS** within the `bugs` call, using the `bugs.directory` argument). All of these activities together look like this:

```
>library(R2WinBUGS)    # "load" the R2WinBUGS package
>data <- list(y=y, x=x)
>inits <- function()
>list(beta1=rnorm(1),beta0=rnorm(1),sigma=runif(1,0,2))
>parameters <- c("beta0","beta1","sigma","tau")
>out <- bugs(data, inits, parameters, "normal.txt", n.thin=1, n.chains=2,
          n.burnin=2000, n.iter=6000, debug=TRUE,working.dir=getwd())
```

Note that the previously created objects defining data, initial values, and parameters to monitor are passed to the function `bugs`. In addition, various other things are declared: The number of parallel Markov chains (`n.chains`), the thinning rate (`n.thin`), the number of burn-in iterations (`n.burnin`), and the total number of iterations (`n.iter`). To develop a detailed understanding of the various parameters and settings used for MCMC, consult a basic reference such as Kéry (2010). We also come back to these issues in the following Section (3.5) and in Chapter 17. A common question is "how should my data be formatted?" That depends on how you describe the model in the **BUGS** language, and how your data are input into **R**. There is no unique way to describe any particular model and so you have some flexibility. We talk about data format further in the context of capture-recapture models and SCR models in Chapter 5 and elsewhere.

You should execute all of the commands given above and then close the **WinBUGS** GUI, and the data will be read back into **R** (or specify `debug=FALSE` in the `bugs` call). We don't want to give instructions on how to navigate and use the standalone **WinBUGS** GUI—but you can fire up **WinBUGS** and read the help files, or see Chapter 4 from Kéry (2010) for a brief introduction. The `print` command applied to the object `out` prints some basic summary output (this is slightly edited):

[5] While **WinBUGS** is reasonably robust to a wide range of more or less plausible starting values, **JAGS** is a lot more sensitive and especially with more complex models you might actually have to spend some time thinking about how to specify good starting values to get the model running (Appendix 1); we will come back to this issue when we use **JAGS**.

```
>print(out,digits=2)
Inference for Bugs model at "normal.txt", fit using WinBUGS,
 2 chains, each with 6000 iterations(first 2000 discarded)
 n.sims = 8000 iterations saved
          mean   sd   2.5%    25%    50%    75%  97.5% Rhat n.eff
beta0    -6.62 1.64  -9.77  -7.63  -6.64  -5.63  -3.29    1  4200
beta1     0.81 1.20  -1.63   0.09   0.80   1.54   3.24    1  5100
sigma     4.99 1.56   2.93   3.92   4.66   5.70   8.85    1  8000
tau       0.05 0.03   0.01   0.03   0.05   0.07   0.12    1  8000
deviance 58.72 3.21  55.06  56.35  57.85  60.26  67.15    1  6200

For each parameter, n.eff is a crude measure of effective sample size,
and Rhat is the potential scale reduction factor (at convergence, Rhat=1).

DIC info (using the rule, pD = Dbar-Dhat)
pD = 2.5 and DIC = 61.3
```

We see that parameter estimates are not particularly close to the true input values (-3.2 and 1.5 for β_0 and β_1, respectively, and 4 for σ) and have very wide credible intervals, which is most likely a result of the extremely low sample size ($n = 10$). In the **WinBUGS** output you see a column called "`Rhat`," as well as one called "`n.eff`." These are convergence diagnostics (the Brooks-Gelman-Rubin statistic and the effective sample size) and we will discuss those in Section 3.5.2. DIC is the deviance information criterion (Spiegelhalter et al. (2002), see Section 3.9), which some people use in a manner similar to AIC although it is recognized to have some problems in hierarchical models (Millar, 2009). We consider use of DIC in the context of SCR models in Chapter 8.

3.5 Practical Bayesian analysis and MCMC

The mere execution of a Bayesian analysis using the **BUGS** language, as demonstrated with the linear regression example, is fairly straightforward. There are, however, a number of really important practical issues to be considered in any Bayesian analysis and we cover some of these briefly here before we move onto implementing slightly more complex GL(M)Ms in a Bayesian framework.

3.5.1 Choice of prior distributions

Bayesian analysis requires that we choose prior distributions for all of the structural parameters of the model (we use the term "structural parameter" to mean all parameters that aren't customarily thought of as latent variables). We will strive to use priors that are meant to express little or no prior information—default or customary "non-informative" or diffuse priors. This will be Uniform(a, b) priors for parameters that have a natural bounded support and, for parameters that, in theory, can take any value on the real line we use either (1) diffuse normal priors, as we did in the linear regression example above; (2) improper uniform priors which have unbounded support, e.g., $[\theta] \propto 1$, or (3) sometimes even a bounded Uniform(a, b) prior, if that greatly improves the performance of **WinBUGS** or other software doing the MCMC for us. In **WinBUGS** a prior with low precision, τ, where $\tau = 1/\sigma^2$, such as Normal(0, .01) will typically be used. Of course $\tau = 0.01$ ($\sigma^2 = 100$) might be very informative for a regression parameter depending on its magnitude and scaling of x. Therefore, we recommend that predictor variables (covariates) *always* be standardized to have mean 0 and variance 1.

Lack of invariance of priors to transformation. Clearly there are a lot of choices for ostensibly non-informative priors, and the degree of non-informativeness depends on the parameterization. For example, a natural non-informative prior for the intercept of a logistic regression

$$\text{logit}(p_i) = \beta_0 + \beta_1 x_i$$

would be a very diffuse normal prior, $\beta_0 \sim$ Normal(0, Large) or even $\beta_0 \sim$ Uniform (−Large, Large). However, we might also use a prior on the parameter $p_0 = \text{logit}^{-1}(\beta_0)$, which is $\Pr(y = 1)$. Since p_0 is a probability, a natural choice is $p_0 \sim$ Uniform (0, 1). These priors are very different in their implications. For example, if we choose the normal prior for β_0 with variance Large $= 5^2$ and look at the implied prior for p_0 we have the result shown in Figure 3.2 which looks nothing like a Uniform(0, 1) prior. These two priors can affect results (see Section 4.4.2 for an illustration of this for a real data set), yet they are both sensible non-informative priors. Despite this, it is often the case that priors will have little or no impact on the results. Choice of priors and parameterization is very much problem-specific and often largely subjective. Moreover, it also affects the behavior of MCMC algorithms and therefore the analyst needs to pay some attention to this issue and possibly try different things out. Most standard Bayesian analysis books address issues related to specification and effect of prior distribution choice in some depth. Some good references include Kass and Wasserman (1996), Gelman (2006), and Link and Barker (2010).

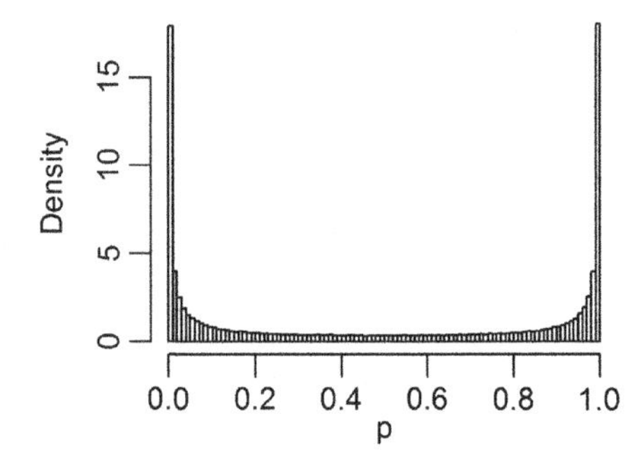

FIGURE 3.2

Implied prior for $p_0 = \exp(\beta_0)/(1 + \exp(\beta_0))$ if $\beta_0 \sim \text{Normal}(0,5^2)$.

3.5.2 Convergence and so forth

Once we have carried out an analysis by MCMC, there are many other practical issues that we have to confront. One characteristic of MCMC sampling is that Markov chains take some time to converge to their stationary distribution—in our case the posterior distribution for some parameter given data, $[\theta|y]$. Only when the Markov chain has reached its stationary distribution, the generated samples can be used to characterize the posterior distribution. Thus, one of the most important issues we need to address is "have the chains converged?" Since we do not know what the stationary posterior distribution of our Markov chain should look like (this is the whole point of doing an MCMC analysis), we effectively have no means to assess whether or not it has truly converged to this desired distribution. Most MCMC algorithms only guarantee that, eventually, the samples being generated will be from the target posterior distribution, but no one can tell us how long this will take. Also, you only know the part of your posterior distribution that the Markov chain has explored so far—for all you know the chain could be stuck in some part of the parameter space, while other regions remain completely undiscovered. Acknowledging that there is truly nothing we can do to ever prove convergence of our Markov chains, there are several things we can do to increase the degree of confidence we have about the convergence of our chains. Some problems are easily detected using simple plots, such as a time-series plot, where parameter values of each MCMC iteration are plotted against the number of iterations. Figure 3.3 shows the time-series plots for the three parameters—β_0, β_1, and σ—from our linear regression example, taken from the **WinBUGS** GUI before closing it to return to **R**.

Typically, a period of transience is observed in the early part of the MCMC algorithm, and this is usually discarded as the "burn-in" period. In our linear regression example, within the `bugs` call we set the burn-in period as 2000 iterations so these are automatically removed by **WinBUGS** and are not part of the output (but Figure 3.6 shows a time-series plot that starts at iteration 0 with a clearly visible burn-in period). The quick diagnostic to whether convergence has been achieved is that your Markov chains look "grassy"—this seems a reasonable statement for the plots in Figure 3.3. Another way to check convergence is to update the parameters some more and see if the posterior changes. If the chains have converged to the posterior, the posterior mean, confidence intervals, and other summaries should be relatively static as we continue to run the algorithm. Yet another option, and one generally implemented in **WinBUGS**, is to run several Markov chains and to start them off at different initial values that are overdispersed relative to the posterior distribution. Such initial values help to explore different areas of the parameter space simultaneously; if, after a while, all chains oscillate around the same average value, chances are good that they indeed converged to the posterior distribution. Gelman and Rubin came up with the so-called "R-hat" statistic ($\hat{R}$) or Brooks-Gelman-Rubin statistic that essentially compares within-chain and between-chain variance to check for convergence of multiple chains (Gelman et al., 1996). The $\hat{R}$ statistic should be close to 1 if the Markov chains have converged and sufficient posterior samples have been obtained. For the

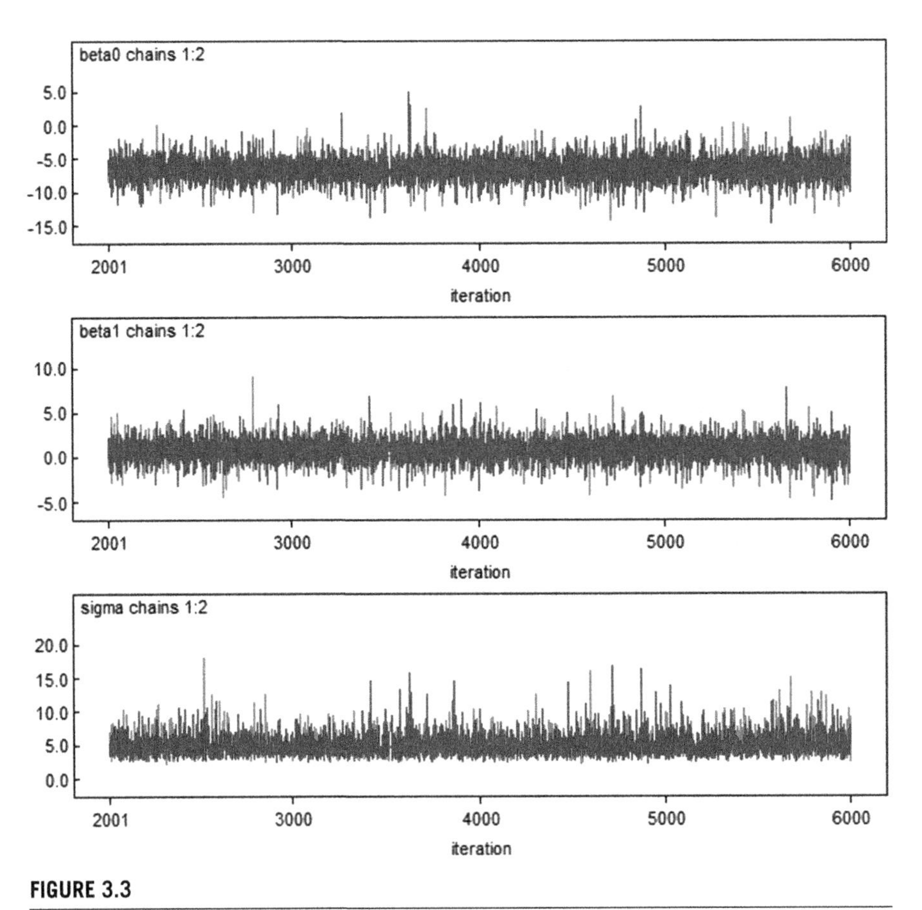

FIGURE 3.3

Time-series plots for parameters from a linear regression run in **WinBUGS** using two parallel Markov chains. In the plot, chains are effectively not distinguishable because they converge to the same stationary distribution.

linear regression example, we ran two parallel chains (also specified in the `bugs` call) and **WinBUGS** returns the $\hat{R}$ statistic for us as part of the summary model output. If you look back to Section 3.4.1 you see that $\hat{R} = 1$ for all parameters of the linear model. In practice, $\hat{R} \leq 1.2$ may be good enough for some problems. For some models you can't actually realize a low $\hat{R}$. E.g., if the posterior is a discrete mixture of distributions then you can be misled into thinking that your Markov chains have not converged when in fact the chains are just jumping back and forth in the posterior state-space. This happens in some of indicator variable model selection discussed in Chapter 8. Often, when there is little information about a parameter in the data, or when parameters are on the boundary of the parameter space, convergence will appear to be poor also. These kinds of situations are normally ok and you need to think really hard about the context of the model and the problem before you conclude that your MCMC algorithm is ill behaved.

Some models exhibit "poor mixing" of the Markov chains (or "slow convergence") in which case the samples might well be from the posterior (i.e., the Markov chains have converged to the proper stationary distribution) but simply mix or move around the posterior rather slowly. Poor mixing can happen for many reasons—when parameters are highly correlated (even confounded), or barely identified from the data, or the algorithms are terrible and probably other reasons as well. Slow mixing equates to high autocorrelation in the Markov chain—the successive draws are highly correlated, and thus we need to run the MCMC algorithm much longer to get an effective sample size that is sufficient for estimation, or to reduce the MC error (see below) to a tolerable level. A strategy often used to reduce autocorrelation is "thinning," where only every *m*th value of the Markov chain output is kept. However, thinning is necessarily inefficient from the standpoint of inference—you can always get more precise posterior estimates by using all of the MCMC output regardless of the level of autocorrelation (MacEachern and Berliner, 1994; Link and Eaton, 2011). Practical considerations might necessitate thinning, even though it is statistically inefficient. For example, in models with many parameters or other unknowns being tabulated, the output files might be enormous and unwieldy to work with. In such cases, thinning is perfectly reasonable. In many cases, how well the Markov chains mix is strongly influenced by parameterization, standardization of covariates, and the prior distributions being used. Some things work better than others, and the investigator should experiment with different settings and remain calm when things don't work out perfectly.

Is the posterior sample large enough? The subsequent samples generated from a Markov chain are not *independent* samples from the posterior distribution, due to the correlation among samples introduced by the Markov process,[6] and the sample size has to be adjusted to account for the autocorrelation in subsequent samples (see Chapter 8 in Robert and Casella (2010) for more details). This adjusted sample size is referred to as the effective sample size. Checking the degree of autocorrelation in your Markov chains and estimating the effective sample size your chain has generated should be part of evaluating your model output. **WinBUGS** will automatically return the effective sample size for all monitored parameters, as we saw in our linear regression example (the "n.eff" column of the summary output). If you find that your supposedly long Markov chain has only generated a very short effective sample, you should consider a longer run. What exactly constitutes a reasonable effective sample size is hard to say. A more palpable measure of whether you've run your chain for enough iterations is the time-series or Monte Carlo error—the "noise" introduced into your samples by the stochastic MCMC process. The MC error is printed by default in summaries produced in the **WinBUGS** GUI, which can be reproduced in **R** using `model$summary` (note that `model` refers to an output object created by the `bugs` call).

---

[6]In case you are not familiar with Markov chains, for T random samples $\theta^{(1)}, \ldots, \theta^{(T)}$ from a Markov chain, the distribution of $\theta^{(t)}$ depends only on the immediately preceding value, $\theta^{(t-1)}$.

```
> model$summary
              mean        sd      MCerror      2.5%     median     97.5%   start   sample
beta0     -6.64700   1.60300    0.0179400   -9.7140   -6.70800   -3.2730    2001     8000
beta1      0.82100   1.19000    0.0116800   -1.4900    0.82560    3.1800    2001     8000
deviance  58.66000   3.08800    0.0506800   55.0700   57.93000   66.8400    2001     8000
sigma      4.96800   1.52300    0.0248300    2.9350    4.68100    8.7410    2001     8000
tau        0.05074   0.02677    0.0003651    0.0131    0.04564    0.1162    2001     8000
```

When using **JAGS**, the `summary` command will automatically produce the MC error (which is called "Time-series SE" in **JAGS**). You want the MC error to be smallish relative to the magnitude of the parameter and what smallish means will depend on the purpose of the analysis. For a preliminary analysis you might settle for a few percent, whereas for a final analysis less than 1% is certainly called for. You can run your MCMC algorithm as long as it takes to achieve that. A consequence of the MC error is that even for the exact same model, results will usually be slightly different. Thus, as a good rule of thumb, you should avoid reporting MCMC results to more than 2 or 3 significant digits!

3.5.3 Bayesian confidence intervals

The 95% Bayesian confidence interval based on percentiles of the posterior is not a unique interval—there are many of them. The so-called "highest posterior density" (HPD) interval is defined as the narrowest interval that contains *at least* 95% of the posterior mass. As a result (of the *at least* clause), for discrete parameters, the 95% HPD is not often exactly 95% but usually slightly more conservative than nominal.

3.5.4 Estimating functions of parameters

A benefit of analysis by MCMC is that we can seamlessly estimate functions of parameters by simply tabulating the desired function of the simulated posterior draws. For example, if θ is the parameter of interest and let $\theta^{(i)}$ for $i = 1, 2, \ldots, M$ be the posterior samples of θ. Let $\eta = \exp(\theta)$, then a posterior sample of η can be obtained simply by computing $\exp(\theta^{(i)})$ for $i = 1, 2, \ldots, M$. Almost all SCR models in this book involve at least one derived parameter. For example, density D is a derived parameter, being a function of population size N and the area A of the underlying state-space of the point process (see Chapter 5).

Example: Finding the optimum value of a covariate. As another example of estimating functions of model parameters, suppose that the normal regression model from Section 3.4.1 had a quadratic response function of the form

$$\mathbb{E}(y_i) = \beta_0 + \beta_1 x_i + \beta_2 x_i^2.$$

Then the optimum value of x, i.e., that corresponding to the optimal expected response, can be found by setting the derivative of this function to 0 and solving for x. We find that

$$df/dx = \beta_1 + 2\beta_2 x = 0$$

yields that $x_{opt} = -\beta_1/(2\beta_2)$. We can just take our posterior draws for β_1 and β_2 and obtain a posterior sample of x_{opt} by this simple calculation applied to the posterior output. As an exercise, take the normal model above and simulate a quadratic response and then describe the posterior distribution of x_{opt}.

3.6 Poisson GLMs

The Poisson GLM (also known as "Poisson regression") is probably the most relevant and important class of models in all of ecology. The basic model assumes observations y_i; $i = 1, 2, ..., n$ follow a Poisson distribution with mean λ which we write

$$y_i \sim \text{Poisson}(\lambda).$$

Commonly y_i is a count of animals or plants at some point in space ("site") i, and λ might vary over sites as well. For example, i might index point count locations in a forest, survey route centers, or sample quadrats, or similar, and we are interested in how λ depends on site characteristics such as habitat. If covariates are available it is typical to model them as linear effects on the log mean. If x_i is some measured covariate associated with observation i, then

$$\log(\lambda_i) = \beta_0 + \beta_1 x_i.$$

While we only specify the mean of the Poisson model directly, the Poisson model (and all GLMs) has a "built-in" variance which is directly related to the mean. In this case, $\text{Var}(y) = \mathbb{E}(y_i) = \lambda_i$. Thus the model accommodates a linear increase in variance with the mean.

3.6.1 North American breeding bird survey data

As an example we consider a classical situation in ecology where counts of an organism are made at a collection of spatial locations. In this particular example, we have mourning dove (*Zenaida macroura*) counts made along North American Breeding Bird Survey (BBS) routes in Pennsylvania, USA. A route consists of 50 stops separated by 0.5 miles. For the purposes here we are defining y_i = route total count and the sample location will be marked by the center point of the BBS route. The survey is run annually and the data set we analyze is 1966–1998. BBS data can be obtained online at `http://www.pwrc.usgs.gov/bbs/`, but the particular chunk of data we will be using here is also included in the `scrbook` package (`data(bbsdata)`). We will make use of the whole data set shortly but for now we're going to focus on a specific year of counts (1990) for the sake of building a simple model. In 1990 there were 77 active routes; this data set contains rows which index the unique route, column 1 is the route ID, columns 2–3 are the route coordinates (longitude/latitude), column 4 is a habitat covariate "forest cover" (standardized, see below), and the remaining columns are the yearly counts. Years for which a survey was not conducted on a route are coded as "NA" in the data matrix. We imagine that this will be a typical format for many

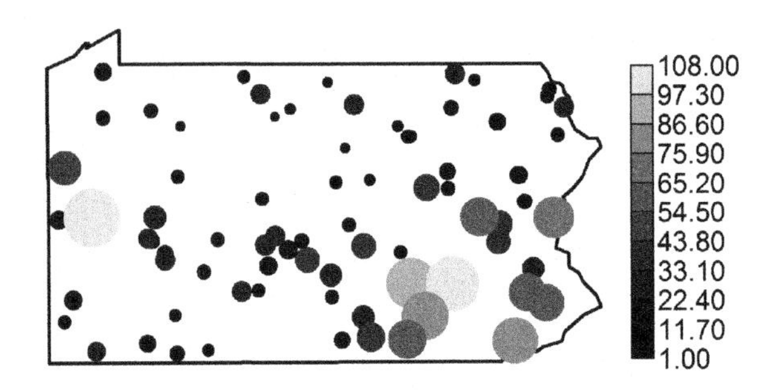

FIGURE 3.4

Mourning dove counts along North American Breeding Bird Survey routes in Pennsylvania (year = 1990). Plot symbol shading and circle size is proportional to raw count.

ecological studies, perhaps with more columns representing covariates. To read in the data and display the first few elements of the data frame containing the counts, do this:

```
> data(bbsdata)                        # loads data frame 'bbs'
> bbsdata$counts[1:2,1:6]

       X       lon      lat      habitat  X66  X67
1  72002   -80.445   41.501   -0.3871372   NA   24
2  72003   -80.347   41.214   -1.0171629   NA   NA
```

It is useful to display the spatial pattern in the observed counts. For that we use a spatial dot plot—where we plot the coordinates of the observations and mark the color of the plotting symbol based on the magnitude of the count. We have a special plotting function for that, which is called `spatial.plot`, and it is available with the supplemental **R** package `scrbook`. Actually, what we want to do here is plot the log-counts (+1 of course), which display a notable pattern that could be related to something (Figure 3.4). The **R** commands for obtaining this figure are:

```
> library(scrbook)
> data(bbsdata)
> library(maps)

> y <- bbsdata$counts[,"X90"] # Pick year 1990
> notna <- !is.na(y)
> y <- y[notna]
> locs <- bbsdata$counts[notna,c("lon","lat")]
> sz <- y/max(y)

> par (mar=c(3,3,3,6))
> plot(locs,pch=" ",axes=FALSE,xlim=range(locs[,1])+c(-.3,+.3),
    ylim= c(range(locs[,2]) + c(-.6,.6)), xlab=" ",ylab=" ")
> map('state', regions='pennsylvania', add=TRUE, lwd=2)
> spatial.plot(bbsdata$counts[notna,2:3], y, cx=1+sz*6, add=TRUE)
```

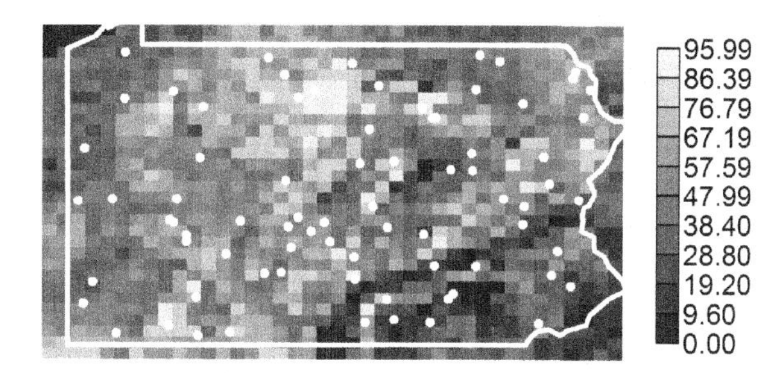

FIGURE 3.5

Forest cover (percent deciduous forest) in Pennsylvania. BBS route locations are shown by white dots.

We can ponder the potential effects that might lead to dove counts being high—corn fields, telephone wires, barn roofs along with misidentification of pigeons, these could all be correlated reasonably well with the observed count of mourning doves. Unfortunately we don't have any of that information. However, we do have a measure of forest cover (provided in the data frame `bbsdata$habitat`) which can be plotted using the `spatial.plot` function with the following **R** commands:

```
> habdata <- bbsdata$habitat
> map('state',regions="penn",lwd=2)
> I <- matrix(NA, nrow=30, ncol=40)
> I <- matrix(habdata[,"dfor"], ncol=40, byrow=FALSE)
> ux <- unique(habdata[,2])
> uy <- sort(unique(habdata[,3]))

> par(mar=c(3,3,3,6))
> plot(locs,pch=" ", axes=FALSE, xlim=range(locs[,1])+c(-.3,+.3),
     ylim=c(range(locs[,2]) + c(-.6,.6)), xlab=" ",ylab=" ")
> image(ux,uy,rot(I), add=TRUE, col=gray(seq(3,17,,10)/20))
> map('state', regions='pennsylvania', add=TRUE, lwd=3, col="white")
> image.scale(I, col=gray(seq(3,17,,10)/20))
> points(locs,pch=20, col="white")
```

The result appears in Figure 3.5. We see a prominent pattern that indicates high forest coverage in the central part of the state and low forest cover in the SE. Inspecting the previous figure of the raw counts suggests a relationship between counts and forest cover which is perhaps not surprising.

3.6.2 Poisson GLM in WinBUGS

Here we demonstrate how to fit a Poisson GLM in **WinBUGS** using the covariate x_i = forest cover along BBS route i. It is advisable that x_i be standardized in most cases as this will improve mixing of the Markov chains. We have pre-standardized the forest cover covariate for the BBS route locations, and so we don't have to worry about that here. To read the BBS data into **R** and get things set up for **WinBUGS** we issue the following commands:

```
> library(scrbook)
> data(bbsdata)

> y <- bbsdata$counts[,"X90"]              # Pick year 1990
> notna <- !is.na(y)
> y <- y[notna]

## Forest cover already standardized here:
> habitat <- bbsdata$counts[notna,"habitat"]
> M <- length(y)

> library(R2WinBUGS)                        # Load R2WinBUGS
> data <- list(y=y, M=M, habitat=habitat)   # Bundle data for WinBUGS
```

Now we write out the Poisson model specification in **WinBUGS** pseudo-code, provide initial values, identify parameters to be monitored, and then execute **WinBUGS**:

```
> cat("
model{
    for (i in 1:M){
      y[i] ~ dpois(lam[i])
      log(lam[i]) <- beta0+beta1*habitat[i]
     }
 beta0 ~ dunif(-5,5)
 beta1 ~ dunif(-5,5)
}
",file="PoissonGLM.txt")

> inits <- function() list (beta0=rnorm(1),beta1=rnorm(1))
> parameters <- c("beta0","beta1")
> out <- bugs(data, inits, parameters, "PoissonGLM.txt", n.thin=2,n.chains=2,
          n.burnin=2000,n.iter=6000,debug=TRUE,working.dir=getwd())
```

The **WinBUGS** output can be viewed in **R** using the `print` command:

```
print (out,digits=2)
Inference for Bugs model at "PoissonGLM.txt", fit using WinBUGS,
 2 chains, each with 6000 iterations(first 2000 discarded), n.thin = 2
 n.sims = 4000 iterations saved
            mean   sd    2.5%     25%     50%     75%    97.5%  Rhat  n.eff
beta0       3.15 0.02    3.10    3.13    3.15    3.17     3.20     1   4000
beta1      -0.50 0.02   -0.54   -0.51   -0.50   -0.48    -0.46     1   4000
deviance 1116.56 1.95 1115.00 1115.00 1116.00 1117.00  1122.00     1   4000
```

3.6.3 Constructing your own MCMC algorithm

At this point it might be helpful to suffer through an example of building a custom MCMC algorithm. Here, we develop an MCMC algorithm for the Poisson regression model, using a Metropolis-within-Gibbs sampling framework. Building MCMC algorithms is covered in more detail in Chapter 17 where you can also find step-by-step instructions for Metropolis-within-Gibbs samplers, should the following section move through all this material too quickly.

We will assume that the two parameters, β_0 and β_1, have diffuse normal priors, say $[\beta_0] = \text{Normal}(0, 100)$ and $[\beta_1] = \text{Normal}(0, 100)$ where each has *standard deviation* 100 (recall that **WinBUGS** parameterizes the normal in terms of $1/\sigma^2$). We need to assemble the relevant elements of the model, which are these two prior distributions and the likelihood $[\mathbf{y}|\beta_0, \beta_1] = \prod_i [y_i|\beta_0\beta_1]$, which is, mathematically, the product of the Poisson pmf evaluated at each y_i, given particular values of β_0 and β_1. Next, we need to identify the full conditionals $[\beta_0|\beta_1, \mathbf{y}]$ and $[\beta_1|\beta_0, \mathbf{y}]$. We use the all-purpose rule for constructing full conditionals (Section 3.3.2) to discover that:

$$[\beta_0|\beta_1, \mathbf{y}] \propto \left\{ \prod_i [y_i|\beta_0, \beta_1] \right\} [\beta_0].$$

Mathematically, the full conditional is of the form

$$[\beta_0|\beta_1, \mathbf{y}] \propto \left\{ \prod_i \exp(-\exp(\beta_0 + \beta_1 x_i)) \exp(\beta_0 + \beta_1 x_i)^{y_i} \right\} \exp\left(-\frac{\beta_0^2}{2 * 100}\right),$$

which you can program as an **R** function with arguments β_0, β_1, and $\mathbf{y}$ without difficulty. The full conditional for β_1 is:

$$[\beta_1|\beta_0, \mathbf{y}] \propto \left\{ \prod_i [y_i|\beta_0, \beta_1] \right\} [\beta_1],$$

which has a similar mathematical representation except the prior is expressed in terms of β_1 instead of β_0. Remember, we could replace the "$\propto$" with "=" if we put $[y|\beta_1]$ or $[y|\beta_0]$ in the denominator and retained all of the constants. But, in general, $[y|\beta_0]$ or $[y|\beta_1]$ will be quite a pain to compute and, more importantly, it is a constant as far as the operative parameters (β_0 or β_1, respectively) are concerned. Therefore, the MH acceptance probability will be the ratio of the full conditional evaluated at a candidate draw to that evaluated at the current value, and so the denominator required to change $\propto$ to $=$ winds up canceling from the MH acceptance probability.

Here we will use the so-called random walk candidate generator, which is a normal proposal distribution, so that, for example, $\beta_0^* \sim \text{Normal}(\beta_0^{t-1}, \delta^2)$ where δ is the standard deviation of the proposal distribution, which is just a tuning parameter that is set by the user and adjusted to achieve efficient mixing of chains (see Section 17.3.2). We remark also that calculations are often done on the log scale to preserve numerical integrity of things when quantities evaluate to small or large numbers, so keep in mind, for example, $a \times b = \exp(\log(a) + \log(b))$ for two positive numbers a and b. The "Metropolis within Gibbs" algorithm for a Poisson regression turns out to be remarkably simple and is given in Panel 3.1. It is also part of the `scrbook` package

and you can run 1,000 iterations of it by calling
`PoisGLMBBS (y=y, habitat=habitat, niter=1000)`
(note that `y` = point count data and `habitat` = forest cover have to be defined in your **R** workspace as shown in the previous analysis of these data).

```
> set.seed(2013)      # So we all get the same result

> out <- matrix(NA,nrow=1000,ncol=2)   # Matrix to store the output
> beta0 <- -1                          # Starting values
> beta1 <- -.8

# Begin the MCMC loop ; do 1000 iterations
> for(i in 1:1000){

# Update the beta0 parameter
lambda <- exp(beta0+beta1*habitat)
lik.curr <- sum(log(dpois(y,lambda)))
prior.curr <- log(dnorm(beta0,0,100))
beta0.cand <- rnorm(1,beta0,.05)        # generate candidate
lambda.cand <- exp(beta0.cand + beta1*habitat)
lik.cand <- sum(log(dpois(y,lambda.cand)))
prior.cand <- log(dnorm(beta0.cand,0,100))
mhratio <- exp(lik.cand +prior.cand - lik.curr-prior.curr)
if(runif(1)< mhratio)
     beta0 <- beta0.cand

# update the beta1 parameter
lik.curr <- sum(log(dpois(y,exp(beta0+beta1*habitat))))
prior.curr <- log(dnorm(beta1,0,100))
beta1.cand <-rnorm(1,beta1,.25)
lambda.cand <- exp(beta0+beta1.cand*habitat)
lik.cand <- sum(log(dpois(y,lambda.cand)))
prior.cand <- log(dnorm(beta1.cand,0,100))
mhratio <- exp(lik.cand + prior.cand - lik.curr - prior.curr)
if(runif(1)< mhratio)
     beta1 <- beta1.cand

out[i,] <- c(beta0,beta1)              # save the current values
}

> plot(out[,1],ylim=c(-1.5,3.3),type="l",lwd=2,ylab="parameter value",
     xlab="MCMC iteration")
> lines(out[,2],lwd=2,col="red")
```

PANEL 3.1

R code to run a Metropolis-within-Gibbs sampler on a Poisson regression model.

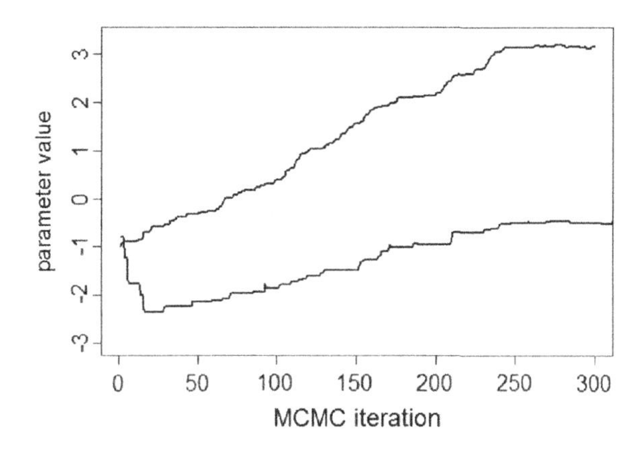

FIGURE 3.6

First 300 MCMC iterations for the Poisson GLM model parameters β_0 (top) and β_1 (bottom) using Metropolis-Hastings tuning parameters $\delta = 0.05$ for β_0 and 0.25 for β_1.

The first 300 iterations of the MCMC history of each parameter are shown in Figure 3.6. These chains are not very appealing but a couple of things are evident: We see that the burn-in takes about 250 iterations and that after that chains seem to mix reasonably well, although this is not so clear given the scale of the y-axis, which we have chosen to get both variables on the same graph. We generated 10,000 posterior samples, discarding the first 500 as burn-in, and the result is shown in Figure 3.7, this time on separate panels for each parameter. The "grassy" look of the MCMC history is diagnostic of Markov chains that are well mixing and we would generally be very satisfied with results that look like this.

Note that we used a specific set of starting values for these simulations. It should be clear that starting values closer to the mass of the posterior distribution might cause burn-in to occur faster. Note also that we have used a different prior than in our **WinBUGS** model specification given previously. We encourage you to evaluate whether this seems to affect the result.

3.7 Poisson GLM with random effects

In most of this book, we will be dealing with random effects in GLM-like models—similar to what are usually referred to as generalized linear mixed models (GLMMs). We provide a brief introduction of such a model by way of an example, extending our Poisson regression model to include a random effect.

The Log-Normal mixture: The classical situation involves a GLM with a normally distributed random effect that is additive on the linear predictor. For the Poisson case, we have:

$$\log(\lambda_i) = \beta_0 + \beta_1 x_i + \eta_i,$$

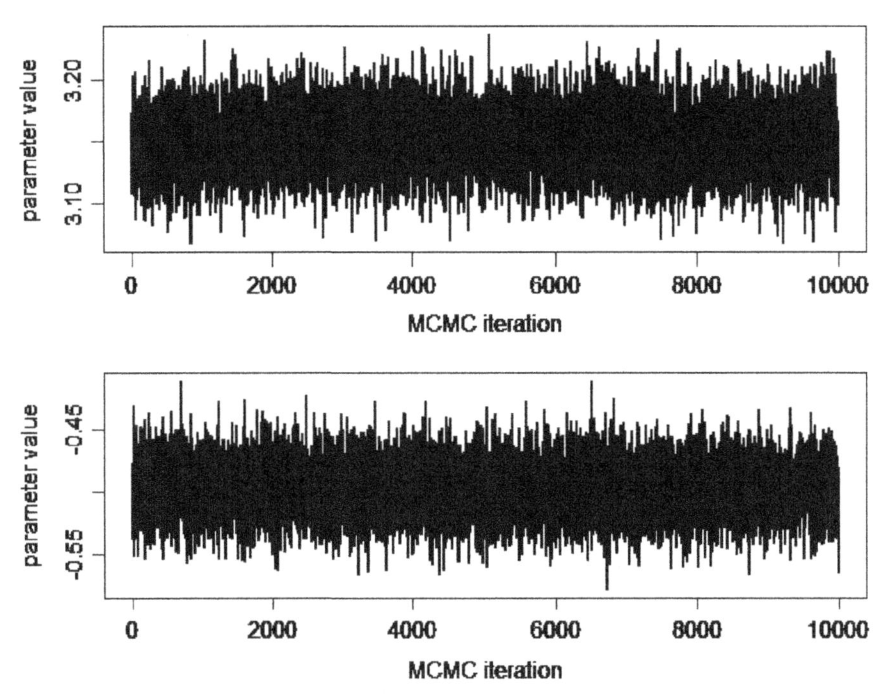

FIGURE 3.7

Nice, grassy plots of 10,000 MCMC iterations for the Poisson GLM model parameters β_0 (top) and β_1 (bottom) using Metropolis-Hastings tuning parameters $\delta = 0.05$ for β_0 and 0.25 for β_1.

where $\eta_i \sim \text{Normal}(0, \sigma^2)$. In this context, η represents an error term capturing variation in λ_i not accounted for by the covariates, or overdispersion. It is really amazingly simple to express this model in the **BUGS** language and have **WinBUGS** (or **JAGS**, etc.) draw samples from the posterior distribution. The code for analysis of the BBS dove counts is given as follows:

```
> library(scrbook)
### Grab the BBS Data as before
> data(bbsdata)
### Set random seed so that results are repeatable
> set.seed(2013)
### Dump the BUGS model into a file
> cat("
model{
  for (i in 1:M){     # Observation model, linear predictor, etc.
    y[i] ~ dpois(lam[i])
    log(lam[i]) <- beta0+beta1*habitat[i] + eta[i]
    frog[i] <- beta1*habitat[i] + eta[i]
    eta[i] ~ dnorm(0,tau)
  }
                        # Prior distributions:
```

```
beta0 ~ dunif(-5,5)
beta1 ~ dunif(-5,5)
sigma ~ dunif(0,10)
tau <- 1/(sigma*sigma)
}
",file="model.txt")

> data <- list("y","M","habitat")        # Define the data
> inits <- function()                     #    inits and parameters
  list (beta0=rnorm(1), beta1=rnorm(1), sigma=runif (1,0,4))
> parameters <- c("beta0","beta1","sigma","tau")

> library(R2WinBUGS)                      # Load and run R2WinBUGS
> out <- bugs(data, inits, parameters, "model.txt", n.thin=2,n.chains=2,
           n.burnin=1000, n.iter=5000, debug=TRUE)
```

This produces the posterior summary statistics given in Table 3.1. One thing we notice is that the posterior standard deviations of the regression parameters are much higher, a result of the extra-Poisson variation allowed for by this model. We would also notice much less precise predictions of hypothetical new observations.

3.8 Binomial GLMs

Another extremely important class of models in ecology are binomial models. We use binomial models for count data whenever the observations are counts or frequencies and it is natural to condition on a "sample size," say K, the maximum frequency possible in a sample. The random variable, $y \leq K$, is then the frequency of occurrences out of K "trials." The parameter of the binomial models is p, often called "success probability", which is related to the expected value of y by $\mathbb{E}(y) = pK$. Usually we are interested in modeling covariates that affect the parameter p, and such models are called binomial GLMs, binomial regression models or logistic regression, although logistic regression really only applies when the logistic link is used to model the relationship between p and covariates (see below).

Table 3.1 Posterior summaries for Poisson GLMM containing a normal random effect and a habitat effect for mourning dove counts on BBS routes in PA, 1990. Model was fit using **WinBUGS**, 2 chains, each with 5,000 iterations (first 1,000 discarded), n.thin = 2, n.sims = 4,000 iterations saved.

Parameter	Mean	SD	2.5%	25%	50%	75%	97.5%	Rhat	n.eff
β_0	2.98	0.08	2.82	2.93	2.98	3.03	3.12	1.00	1400
β_1	−0.53	0.07	−0.68	−0.58	−0.53	−0.49	−0.38	1.01	350
σ	0.60	0.06	0.49	0.56	0.59	0.64	0.73	1.00	2000
τ	2.88	0.57	1.88	2.47	2.86	3.24	4.12	1.00	2000
Deviance	445.94	12.18	424.00	437.40	445.20	453.90	471.50	1.00	4000

One of the most typical binomial GLMs occurs when the sample size equals 1 and the outcome, y, is "presence" ($y = 1$) or "absence" ($y = 0$) of a species. In this case, y has a Bernoulli distribution. This is a classical species distribution modeling situation. A special situation occurs when presence/absence is observed with error (MacKenzie et al., 2002; Tyre et al., 2003). In that case, $K > 1$ samples are usually needed for effective estimation of model parameters.

In standard binomial regression problems the sample size is fixed by design but interesting models also arise when the sample size is itself a random variable. These are the N-mixture models (Royle, 2004b; Kéry et al., 2005; Royle and Dorazio, 2008; Kéry, 2010) and related models (in this case, N being the sample size, which we labeled K above).[7] Another situation in which the binomial sample size is fixed is closed population capture-recapture models in which a population of individuals is sampled K times. The number of times each individual is encountered is a binomial outcome with parameter (encounter probability) p, based on a sample of size K. In addition, the total number of unique individuals observed, n, is also a binomial random variable based on population size N. We consider such models in Chapter 4.

3.8.1 Binomial regression

In binomial models, covariates are modeled on a suitable transformation (the link function) of the binomial success probability, p. Let x_i denote some measured covariate for sample unit i, and let p_i be the success probability for unit or subject i. The standard choice is the logit link function (3.1) but there are many other possible link functions. We sometimes use the complementary log-log (= "cloglog") link function in ecological applications because it is natural in some cases when the response should scale in relation to area or effort (Royle and Dorazio, 2008, p. 150). As an example, the "probability of observing a count greater than 0" under a Poisson model is $\Pr(y > 0) = 1 - \exp(-\lambda)$. In that case, for the ith observation,

$$\text{cloglog}(p_i) = \log(-\log(1 - p_i)) = \log(\lambda_i)$$

so that if you have covariates in your linear predictor for $\mathbb{E}(y)$ under a Poisson model then they are linear on the complementary log-log scale of p. In models of species occurrence it seems natural to view occupancy as being derived from local abundance N (Royle and Nichols, 2003; Royle and Dorazio, 2006; Dorazio, 2007). Therefore, models of local abundance in which $N_i \sim \text{Poisson}(A_i\lambda_i)$ for a habitat patch of area A_i implies a model for occupancy ψ_i of the form

$$\text{cloglog}(\psi_i) = \log(A_i) + \log(\lambda_i).$$

We will use the cloglog link in some analyses of SCR models in Chapter 5 and elsewhere.

[7] Some of the jargon is actually a little bit confusing here because the binomial index is customarily referred to as "sample size" but in the context of N-mixture models N is actually the "population size."

3.8.2 North American waterfowl banding data

The standard binomial modeling problem in ecology is that of modeling species distributions, where $K = 1$ and the outcome is occurrence ($y = 1$) or not ($y = 0$) of some species. Such examples abound in books (e.g., Royle and Dorazio (2008, Chapter 3), Kéry (2010, Chapter 21), Kéry and Schaub (2012, Chapter 13)) and in scientific papers. Therefore, instead, we will consider an example involving band returns of waterfowl in the upper great plains, which were analyzed by Royle and Dubovsky (2001).

For these data, y_{it} is the number of mallard (*Anas platyrhynchos*) bands recovered out of B_{it} birds banded at some location $\mathbf{s}_i$ in year t. In this case B_{it} is fixed. Thinking about recovery rate as being proportional to harvest rate, we use these data to explore geographic gradients in recovery rate resulting from variability in harvest pressure experienced by different populations. As such, we fit a basic binomial GLM with a linear response to geographic coordinates (including an interaction term). There are few structural differences between this model and the Poisson GLM fitted previously. The main things are due to the data structure (we have a matrix here instead of a vector) and otherwise we change the distributional assumption to binomial (specified with `dbin`) and then use the `logit` function to relate the parameter p_{it} to the covariates.

Dummy variables in BUGS: In the mallard example, we model the band recovery probability p_{it} not only as a linear function (on the logit scale) of geographic location, but also allow for variation in p_{it} with year, $t; t = 1, 2, \ldots, T$. In this particular example there are $T = 5$ years of data and we could describe the full mallard model with a formula in terms of "dummy variables." Dummy variables are binary variables, one variable for each level of the categorical variable they describe, such that variable for level t takes on the value 1 if the observation belongs to level t and 0 otherwise. So, the mallard model in terms of dummy variables for "year" looks like this:

$$\begin{aligned} y_{it} &\sim \text{Binomial}(B_{it}, p_{it}), \\ \text{logit}(p_{it}) &= \beta_0 + \beta_1 x_{2,it} + \beta_2 x_{3,it} + \beta_3 x_{4,it} + \beta_4 x_{5,it} + \beta_5 \text{LAT}_i + \beta_6 \text{LON}_i \\ &\quad + \beta_7 \text{LAT}_i \text{LON}_i. \end{aligned}$$

Here, x_2 to x_5 are the dummy variable vectors of length T that take on the value of 1 when t corresponds to the respective year and 0 otherwise; β_0 is the common intercept term and corresponds to $t = 1$; $\beta_1 - \beta_4$ describe the difference in p_{it} for each t relative to $t = 1$.

There is a more concise way of implementing such a model with a categorical covariate in **BUGS**, namely, by using indexing instead of dummy variables.[8] Essentially, instead of estimating the difference in p relative to category 1, we estimate a separate intercept term for each category, so that we have five different β_0 parameters indexed by t. This reduces the linear predictor to:

$$\text{logit}(p_{it}) = \beta_{0,t} + \beta_5 \text{LAT}_i + \beta_6 \text{LON}_i + \beta_7 \text{LAT}_i \text{LON}_i.$$

[8]Actually, in some cases a model may mix or converge better depending on whether you choose a dummy variable or an indexing description of it, although they are structurally equivalent (Kéry, 2010).

The model can be implemented in the **BUGS** language for the mallard banding data using the following **R** script, provided in the `scrbook` package (see `help (mallard)`):

```
> library(scrbook)
> data(mallard)    # Load mallard data

> cat ("
model{
  for(t in 1:5){
   for(i in 1:nobs){
    y[i,t] ~ dbin(p[i,t], B[i,t])
    pl[i,t] <- beta0[t]+beta1*X[i,1]+beta2*X[i,2]+beta3*X[i,1]*X[i,2]
    p[i,t] <- exp(pl[i,t])/(1+exp(pl[i,t]))
   }
}
beta1 ~ dnorm(0,.001)
beta2 ~ dnorm(0,.001)
beta3 ~ dnorm(0,.001)
for(t in 1:5){
  beta0[t] ~ dnorm(0,.001)
      }
}
  ",file="BinomialGLM.txt")

> library(R2WinBUGS)
> data <- list(B=mallard$bandings, y=mallard$recoveries,
              X=mallard$locs, nobs=nrow(mallard$locs))
> inits <- function(){list(beta0=rnorm(5),beta1=0,beta2=0,beta3=0)}
> parms <- c('beta0','beta1','beta2','beta3')
> out <- bugs(data, inits, parms,"BinomialGLM.txt",
              n.chains=3, n.iter=2000, n.burnin=1000, n.thin=2, debug=TRUE)
```

Look at the posterior summaries of model parameters in Table 3.2. The basic result suggests a negative east-west gradient and a positive south to north gradient of band recovery probabilities, but no interaction. A map of the response surface is shown in Figure 3.8.

3.9 Bayesian model checking and selection

In general terms, model checking—or assessing the adequacy of the model—and model selection are quite thorny issues and, despite contrary and, sometimes, strongly held belief among practitioners, there are no really definitive, general solutions to either problem. We're against dogma on these issues and think people need to be open-minded about such things and recognize that models can be useful whether or not they pass certain statistical tests. Some models are intrinsically better than others because they make more biological sense or foster understanding or achieve some objective that some bootstrap or other goodness-of-fit test can't decide for you. That said, it gives you some confidence if your model seems adequate in a purely

Table 3.2 Posterior summaries for the binomial GLM of mallard band recovery rate. Model contains year-specific intercepts ($\beta_{0,t}$) and a linear response surface with interaction. Model was fit using **WinBUGS**, and posterior summaries are based on 3 chains, each with 2,000 iterations (first 1,000 discarded), n.thin = 2, and n.sims = 1,500 iterations saved.

Parameter	Mean	SD	2.5%	50%	97.5%	Rhat	n.eff
$\beta_0[1]$	−2.346	0.036	−2.417	−2.346	−2.277	1.001	1500
$\beta_0[2]$	−2.356	0.032	−2.420	−2.356	−2.292	1.001	1500
$\beta_0[3]$	−2.220	0.035	−2.291	−2.219	−2.153	1.001	1500
$\beta_0[4]$	−2.144	0.039	−2.225	−2.143	−2.068	1.000	1500
$\beta_0[5]$	−1.925	0.034	−1.990	−1.924	−1.856	1.004	570
β_1	−0.023	0.003	−0.028	−0.023	−0.018	1.001	1500
β_2	0.020	0.006	0.009	0.020	0.031	1.001	1500
β_3	0.000	0.001	−0.002	0.000	0.002	1.001	1500
Deviance	1716.001	4.091	1710.000	1715.000	1726.000	1.001	1500

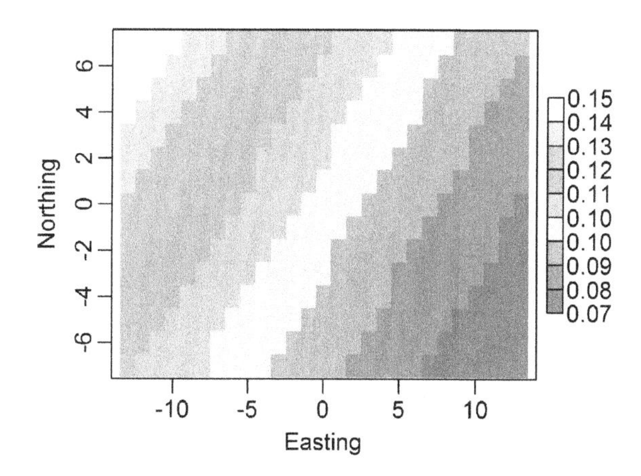

FIGURE 3.8

Predicted recovery rates of mallard bands in the upper great plains of North America. Note the negative gradient from the NW to the SE.

statistical sense. We provide a very brief overview of concepts here, but see also Kéry (2010) and Link and Barker (2010) for specific context related to Bayesian model checking and selection.

3.9.1 Goodness-of-fit

Goodness-of-fit testing is an important element of any analysis because our model represents a general set of hypotheses about the ecological and observation processes that generated our data. Thus, if our model "fits" in some statistical or scientific sense, then we believe it to be consistent with the hypotheses that went into the model. More formally, we would conclude that the data are *not inconsistent* with the hypotheses, or that the model appears adequate. If we have enough data, then of course we will reject any set of statistical hypotheses. Conversely, we can always come up with a model that fits by making the model extremely complex. Despite this paradox, it seems to us that simple models that you can understand should usually be preferred even if they don't fit, for example if they embody essential mechanisms central to our understanding of things, or if we think that some contributing factors to lack-of-fit are minor or irrelevant to the scientific context and intended use of the model. Yet the tension is there to obtain fitting models, and this comes naturally at the expense of models that can be easily interpreted and studied and effectively used. Unfortunately, conducting a goodness-of-fit test is not always so easy to do. And, moreover, it is never really easy (or especially convenient) to decide if your goodness-of-fit test is worth anything. It might have 0 power! Despite this, we recommend attempting to assess model fit in real applications, as a general rule, and we provide some basic guidance here and some more specific to SCR models in Chapter 8.

To evaluate goodness-of-fit in Bayesian analyses, we will most often use the Bayesian p-value (Gelman et al., 1996). The basic idea is to define a fit statistic or "discrepancy measure", D, and compare the posterior distribution of that statistic to the posterior predictive distribution of that statistic for hypothetical perfect data sets for which the model is known to be correct. For example, with count frequency data, a standard measure of fit is the sum of squares of the "Pearson residuals,"

$$D(y_i, \theta) = \frac{(y_i - \mathbb{E}(y_i))}{\sqrt{\text{Var}(y_i)}}.$$

The fit statistic based on the squared residuals computed from the observations is

$$T(\mathbf{y}, \theta) = \sum_i D(y_i, \theta)^2,$$

which can be computed at each iteration of an MCMC algorithm given the current values of parameters that determine the response distribution. At the same time (i.e., at each MCMC iteration), the equivalent statistic is computed for a "new" data set, say $\mathbf{y}^{new}$, simulated using the current parameter values. From the new data set, we compute the same fit statistic:

$$T(\mathbf{y}^{new}, \theta) = \sum_i D(y_i^{new}, \theta)^2$$

and the Bayesian p-value is simply the posterior probability $\Pr(T(\mathbf{y}^{new}) > T(\mathbf{y}))$, which should be close to 0.50 for a good model—one that "fits" in the sense that the observed data set is consistent with realizations simulated under the model being fitted to the observed data. In practice we judge "close to 0.50" as being "not too close to 0 or 1" and, as always, closeness is somewhat subjective. We're happy with anything $>.1$ and $<.9$ but might settle for $>.05$ and <0.95. Another useful fit statistic is the Freeman-Tukey statistic, in which

$$D(\mathbf{y}, \theta) = \sum_i (\sqrt{y_i} - \sqrt{\mathbb{E}(y_i)})^2$$

(Brooks et al., 2000), where y_i is the observed value at i and $\mathbb{E}(y_i)$ is its expected value. In summary, you can see that the Bayesian p-value is easy to compute, and it is widely used as a result.

3.9.2 Model selection

In ecology, scientific hypotheses are often manifest as different models or parameters of a model, and so evaluating the importance of different models is fundamental to many ecological studies. For Bayesian model selection we typically use three different methods: First is, let's say, common sense. If a variable should plausibly be relevant to explaining the data-generating processes, and it has posterior mass concentrated away from 0, then it seems like it should be regarded as important—that is, it is "significant." This approach seems to have fallen out of favor in ecology over the last 10 or 15 years but in many situations it is a reasonable thing to do.

For regression problems we sometimes use the indicator variable method of Kuo and Mallick (1998), in which we introduce a set of binary variables w_k for variable k, and express the model as, e.g., for a single covariate model:

$$\mathbb{E}(y_i) = \beta_0 + w_1\beta_1 x_i,$$

where w_1 is given a Bernoulli prior distribution with some prescribed probability. E.g., $w_1 \sim \text{Bernoulli}(0.50)$ to provide a prior probability of 0.50 that variable x should be an element of the linear predictor. The posterior probability of the event $w_1 = 1$ is a gage of the importance of the variable x, i.e., high values of $\Pr(w_1 = 1)$ indicate stronger evidence to support that "x is in the model" whereas values of $\Pr(w_1 = 1)$ close to 0 suggest that x is less important. Expansion of the model to include the binary variable w_1 defines a set of two distinct models for which we can directly compute the posterior probabilities, merely by tallying up the posterior frequency of w_1. See Royle and Dorazio (2008, Chapter 3), for an example in the context of logistic regression.

This approach even seems to work sometimes with fairly complex hierarchical models of a certain form. For example, Royle (2008) applied it to a random effects model to evaluate the importance of the random effect component of the model. The main problem, which is really a general problem in Bayesian model selection, is that its effectiveness and results will typically be highly sensitive to the prior distribution on the structural parameters (e.g., see Royle and Dorazio (2008, Table 3.6)). The reason for this is obvious: If $w_1 = 0$ for the current iteration of the MCMC algorithm, so that β is sampled from the prior distribution, and the prior distribution is very diffuse, then extreme values of β are likely. Consequently, when the current value of

β is far away from the mass of the posterior when $w_1 = 1$, then the Markov chain may only jump from $w_1 = 0$ to $w_1 = 1$ infrequently. One seemingly reasonable solution to this problem is to fit the full model to obtain posterior distributions for all parameters, and then use those as prior distributions in a "model selection" run of the MCMC algorithm (Aitkin, 1991). This seems preferable to more-or-less arbitrary restriction of the prior support to improve the performance of the MCMC algorithm.[9]

A third method that we advocate is subject-matter context. It seems that there are some situations—some models—where one should not have to do model selection because a specific model may be necessitated by the biological context of the problem, thus rendering a formal hypothesis test pointless (Johnson, 1999). Certain aspects of SCR models are such an example. In SCR models, we will see that "spatial location" of individuals is an element of the model. The simpler, reduced, model is an ordinary capture-recapture model which is not spatially explicit (i.e., Chapter 4), but it seems silly and pointless to think about actually using the reduced model even if we could concoct some statistical test to refute the more complex model. The simpler model is manifestly wrong but, more importantly, not even a plausible data-generating model! Other examples are when effort, area, or sample rate is used as a covariate. One might prefer to have such things in models regardless of whether or not they pass some statistical litmus test.

Many problems can be approached using one of these methods. In later chapters (especially Chapter 8) we will address model selection in specific contexts and we hope those will prove useful for a majority of the situations you might encounter.

3.10 Summary and outlook

GLMs and GLMMs are the most useful statistical methods in all of ecology. The principles and procedures underlying these methods are relevant to nearly all modeling and analysis problems in every branch of ecology. Therefore, understanding how to analyze these models is an essential skill for the quantitative ecologist to possess. If you understand and can conduct classical likelihood and Bayesian analysis of Poisson and binomial GL(M)Ms, then you will be successful analyzing and understanding more complex classes of models that arise. We will see shortly that spatial capture-recapture models are a type of GL(M)M and thus having a basic understanding of the conceptual origins and formulation of GL(M)Ms and their analysis is extremely useful.

We note that GL(M)Ms are routinely analyzed by likelihood methods but we have focused on Bayesian analysis here in order to develop the tools that are less familiar to most ecologists, and that we will apply in much of the remainder of the book. In particular, Bayesian analysis of models with random effects is relatively straightforward because the models are easy to analyze conditional on the random effect, using MCMC. Thus, we will often analyze SCR models in later chapters by MCMC,

[9]See O'Hara and Sillanpää (2009) and Tenan et al. (2013).

explicitly adopting a Bayesian inference framework. In that regard, the various **BUGS** engines (**WinBUGS**, **OpenBUGS**, **JAGS**; see also Appendix 1) are enormously useful because they provide an accessible platform for carrying out analyses by MCMC by just describing the model, and not having to worry about how to actually build MCMC algorithms. That said, the **BUGS** language is more important than just to the extent that it enables one to do MCMC—it is useful as a modeling tool because it fosters understanding, in the sense that it forces you to become intimate with your model. You have to think about and write down all of the probability assumptions, and the relationships between observations and latent variables and parameters in a way that is ecologically sensible and statistically coherent. Because of this, it focuses your thinking on *model construction*, as Kéry says in his **WinBUGS** book (Kéry, 2010), "**WinBUGS** frees the modeler in you."

While we have emphasized Bayesian analysis in this chapter, and make primary use of it through the book, we will provide an introduction to likelihood analysis in Chapter 6 and use those methods also from time to time. Before getting to that, however, it will be useful to talk about more basic, conventional closed population capture-recapture models and such models are the topic of the next chapter.

CHAPTER

Closed Population Models

4

In this chapter we introduce ordinary *non-spatial* capture-recapture (CR) models for estimating population size in closed populations. A closed population is one whose size, N, does not change during the study. Two forms of closure are often discussed: demographic closure, meaning that no births or deaths occur, and geographic closure, which states that no individuals move onto or off of the sampled area during the study. Although few populations are actually closed except during very short time intervals, closed population CR models serve as the basis for the development of the rest of the models presented in this book, including the models for open populations discussed in Chapter 16.

We begin with the most basic capture-recapture model, colloquially referred to as "model M_0" (Otis et al., 1978), in which encounter probability is strictly constant in all respects (across individuals, and occasions). This allows us to highlight the basic structure of closed population models as binomial GLMs. We then consider some important extensions of ordinary closed population models that accommodate various types of "individual effects"—either in the form of explicit, observed covariates (sex, age, body mass) or unstructured "heterogeneity" in the form of an individual random effect, which represent unobserved or unmeasured covariates. A special type of individual covariate model is distance sampling, which could be thought of as the most primitive spatial capture-recapture model. All of these different types of closed population models are closely related to binomial (or logistic) regression-type models. In fact, when N is known, they are precisely logistic regression models.

We emphasize Bayesian analysis of capture-recapture models and we accomplish this using a method related to classical "data augmentation" from the statistics literature (e.g., Tanner and Wong, 1987). This is a general concept in statistics but, in the context of capture-recapture models where N is unknown, it has a consistent implementation across classes of capture-recapture models and one that is really convenient from the standpoint of doing MCMC (Royle et al., 2007; Royle and Dorazio, 2012). We use data augmentation throughout this book and thus emphasize its conceptual and technical origins and demonstrate applications to closed population models. We refer the reader to Kéry and Schaub (2012, chapter 6) for a complementary development of Bayesian analysis of ordinary, i.e., non-spatial closed population models.

Spatial Capture-Recapture. http://dx.doi.org/10.1016/B978-0-12-405939-9.00004-9

4.1 The simplest closed population model: model M_0

To start looking at the simplest capture-recapture model, let's suppose there exists a population of N individuals which we subject to repeated sampling, say over K "occasions," such as trap nights, where individuals are captured, marked, released, and subsequently recaptured. We suppose that individual encounter histories are obtained, and these are of the form of a sequence of 0's and 1's indicating capture ($y = 1$) or not ($y = 0$) during any sampling occasion. As an example, suppose $K = 5$ sampling occasions, then an individual captured during occasions 2 and 3 but not otherwise would have an encounter history $\mathbf{y} = (0, 1, 1, 0, 0)$. Thus, the observation $\mathbf{y}_i$ for each individual ($i = 1, 2, \ldots, N$) is a vector having elements denoted by y_{ik} for $k = 1, 2, \ldots, K$. Usually, this is organized as a row of a matrix with elements y_{ik}, see Table 4.1. Except where noted explicitly, we suppose that observations are independent within individuals and among individuals. Formally, this allows us to say that y_{ik} are independent and identically distributed ("iid") Bernoulli random variables and we may write $y_{ik} \sim \text{Bernoulli}(p)$. Consequently, for this very simple model in which p is constant (i.e., there are no individual or temporal covariates that affect p) the original binary detection variables can be aggregated into the total number of encounters for each individual, $y_i = \sum_k y_{ik}$, and the observation model changes from a Bernoulli distribution to a binomial distribution based on a sample of size K. That is

$$y_i = \sum_k y_{ik} \sim \text{Binomial}(K, p)$$

for every individual in the population $i = 1, 2, \ldots, N$, where N is the number of individuals in the population (i.e., population size).

Table 4.1 A toy capture-recapture data set with $n = 6$ observed individuals and $K = 5$ sample occasions. Under a model with constant encounter probability, the binary detection history data can be summarized by the detection frequency (the total number of detections, y_i), which is shown in the rightmost column.

	Sample Occasion					
Indiv i	**1**	**2**	**3**	**4**	**5**	**y_i**
1	1	0	0	1	0	2
2	0	1	0	0	1	2
3	1	0	0	1	0	2
4	1	0	1	0	1	3
5	0	1	0	0	0	1
$n = 6$	1	0	0	0	0	1

We emphasize the central importance of the basic Bernoulli encounter model—an individual is either encountered in a sample, or not—which forms the cornerstone of almost all of classical capture-recapture models, including many spatial capture-recapture models discussed in this book. Evidently, the basic capture-recapture model is a simplistic version of a logistic-regression model with only an intercept term ($\text{logit}(p) = \text{constant}$). To say that all capture-recapture models are just logistic regressions is a slight oversimplification. In fact, we are proceeding here as if we knew N. In practice we don't, of course, and estimating N is actually the central objective. But, by proceeding as if N were known, we can specify a simple model using standard methods that you are already familiar with (i.e., GLMs - see Chapter 3) and then deal with the fact that N is unknown latter.

Assuming individuals in the population are encountered independently, the joint probability distribution of the observations is the product of N binomials

$$\Pr(y_1, \ldots, y_N | p) = \prod_{i=1}^{N} \text{Binomial}(y_i | K, p). \tag{4.1.1}$$

We emphasize that this expression is conditional on N, in which case we get to observe the $y_i = 0$ observations and the resulting data are just *iid* binomial counts. Because this is a binomial regression model of the variety described in Chapter 3, fitting this model using a **BUGS** engine poses no difficulty.

Equation (4.1.1) can be simplified even further if we reformat the observations as encounter frequencies. Specifically, let n_k denote the number of individuals captured exactly k times after K survey occasions, $n_k = \sum_{i=1}^{N} I(y_i = k)$ where $I()$ is the indicator function evaluating to 1 if its argument is true and 0 otherwise. For sake of illustration, we converted the data from Table 4.1 to this format (Table 4.2). What is important to note is that if we know N, then we know n_0, i.e., the number of individuals not captured. In this case, an alternative and equivalent expression to Eq. (4.1.1) is

$$\Pr(y_1, \ldots, y_N | p) = \prod_{k=0}^{K} \pi_k^{n_k} \tag{4.1.2}$$

where $\pi_k = \Pr(y = k)$ under the binomial model with parameter p and sample size K. The essential problem in capture-recapture, however, is that N is *not* known because the number of uncaptured individuals (n_0) is unknown. Consequently, the observed

Table 4.2 Data from Table 4.1 formatted as capture frequencies. Since N is unknown, the number of individuals not captured (n_0) is also unknown.

	k					
	0	**1**	**2**	**3**	**4**	**5**
Number of individuals captured k times (n_k)	$N-6$	2	3	1	0	0

capture frequencies n_k are no longer independent because n_0 is a function of the other frequencies, $n_0 = N - \sum_{k=1}^{K} n_k$. Hence, their joint distribution is multinomial (e.g., see Illian et al. (2008, p. 61)):

$$(n_0, n_1, \ldots, n_K) \sim \text{Multinomial}(N, \pi_0, \pi_1, \ldots, \pi_K) \quad (4.1.3)$$

We gave a general overview of the multinomial distribution in Section 2.2. The multinomial distribution is the standard model for discrete responses that can fall into a fixed number ($K+1$ in this case) of possible categories. In the context of capture-recapture, the multinomial posits a population of N individuals with $K+1$ possible outcomes defined by the possible encounter frequencies: encountered $y = 1, 2, \ldots, K$ times, or not encountered at all. These possible outcomes occur with probabilities π_k, which we refer to as "cell probabilities," or in the specific context of capture-recapture, encounter history probabilities.

To fit the model in which N is *unknown*, we can regard n_0 as a parameter and maximize the multinomial likelihood directly (see Section 2.3 on parameter estimation). Direct likelihood analysis of the multinomial model is straightforward, but that is not always sufficiently useful in practice because we seldom are concerned with models for the aggregated encounter history frequencies, which entail that capture probabilities are the same for all individuals. In many instances, including for spatial capture-recapture (SCR) models, we require a formulation of the model that can accommodate individual-level covariates to account for differences in detection among individuals, which we address subsequently in this chapter, and in Chapter 7.

4.1.1 The core capture-recapture assumptions

This basic capture-recapture model—model M_0—comes with a host of specific biological and statistical assumptions. In addition to the basic assumption of population closure, Otis et al. (1978) list the following:

1. animals do not lose their marks during the experiment,
2. all marks are correctly noted and recorded at each trapping occasion, and
3. each animal has a constant and equal probability of capture on each trapping occasion.

The remainder of their classic work is dedicated to relaxing assumption 3. While assumptions 1 and 2 are undoubtedly necessary for inference from basic CR methods to be valid, and while they are also assumed by most of the models we present in the following chapters, we refrain from repeatedly making such statements. Our opinion is that all model assumptions are apparent when a model is clearly specified, and it is both redundant and impossible to list all the things not allowed by the model. For example, closed population models also assume that other sources of error do not occur, but it is not necessary to enumerate each possibility. Rather, it is necessary to make clear statements such as

$$y_i \overset{iid}{\sim} \text{Bernoulli}(p) \quad \text{for } i = 1, \ldots, N.$$

This simple model description carries a tremendous amount of information, and it leaves very little left to say with respect to assumptions. Although we will not always show the *iid* symbol, it will be assumed unless otherwise noted, and this assumption is critical for valid inference. It implies that the encounter of one individual does not affect the encounter of another individual, and encounter does not affect future encounter. Under this assumption, it is easy to write down the likelihood of the parameters and obtain parameter estimates; however, whether or not it is true depends upon biological and sampling issues. If this assumption is deemed false, the model can be discarded in favor of a more realistic alternative. However, once we have settled on our model, statistical inference proceeds by assuming the model is truth—not an approximation to truth—but actual truth.

In spite of the fact that we assume that all models are truth, but we acknowledge that all models are wrong due to their assumptions, assumptions should not be viewed as a necessary evil. In fact, one way to view assumptions is as embodiments of our ecological hypotheses. If we make these assumptions too complex or too specific, then we will never be able to study general phenomena that hold true across space and time. Furthermore, in practice, we will rarely have enough data to estimate the parameters of highly complex models.

4.1.2 Conditional likelihood

We saw that the closed population model is a simple logistic regression model if N is known and, when N is unknown, the model is multinomial with index or sample size parameter N that can be estimated. This multinomial model, being conditional on N, is sometimes referred to as the "joint likelihood" the "full likelihood," or the "unconditional likelihood" (Sanathanan, 1972; Borchers et al., 2002). This formulation differs from the so-called "conditional likelihood" approach in which the likelihood of the observed encounter histories is devised conditional on the event that an individual is captured at least once. To construct this likelihood, we have to recognize that individuals appear or not in the sample based on the value of the random variable y_i, that is, if and only if $y > 0$. The observation model is therefore based on $\Pr(y|y > 0)$. For the simple case of model M_0, the resulting conditional distribution is a "zero truncated" binomial distribution which accounts for the fact that we cannot observe the value $y = 0$ in the data set. Both the conditional and unconditional models are legitimate modes of analysis in all capture-recapture types of studies. They provide equally valid descriptions of the data and, for many practical purposes provide equivalent inferences, at least in large sample sizes (Sanathanan, 1972).

In this book we emphasize Bayesian analysis of capture-recapture models using data augmentation (described in Section 4.2 below), which produces yet a third distinct formulation of capture-recapture models based on the zero-*inflated* binomial distribution that we describe in the next section. Thus, there are three distinct formulations of the model—or modes of analysis—for analyzing all capture-recapture models based on the (1) binomial model for the joint or unconditional specification; (2) zero-truncated binomial that arises "conditional on n"; and (3) the zero-inflated

Table 4.3 Modes of analysis of capture-recapture models. Closed population models can be analyzed using the joint or "full likelihood" which contains N as an explicit parameter, the conditional likelihood, which does not involve N, or by data augmentation which replaces N with ψ. Each approach yields a distinct likelihood.

Mode of Analysis	Parameters in Model	Statistical Model
Joint likelihood	p, N	Multinomial with index N
Conditional likelihood	p	Zero-truncated binomial
Data augmentation	p, ψ	Zero-inflated binomial

binomial that arises under data augmentation. Each formulation has distinct model parameters (shown in Table 4.3 for model M_0).

4.2 Data augmentation

We consider a method of analyzing closed population models using parameter-expanded data augmentation (PX-DA), which we abbreviate to "data augmentation" or DA, which is useful for Bayesian analysis and, in particular, analysis of models using the various **BUGS** engines and other Bayesian model fitting software. Data augmentation is a general statistical concept that is widely used in statistics in many different settings. The classical reference is Tanner and Wong (1987), but see also Liu and Wu (1999). Data augmentation can be adapted to provide a very generic framework for Bayesian analysis of capture-recapture models with unknown N. This idea was introduced for closed populations by Royle et al. (2007), and has subsequently been applied in a number of different contexts including individual covariate models, open population models (Royle and Dorazio, 2008, 2012; Gardner et al., 2010a), spatial capture-recapture models (Royle and Young, 2008; Royle et al., 2009a; Gardner et al., 2009), and many others. Kéry and Schaub (2012, Chapters 6 and 10) provide a good introduction to data augmentation in the context of closed and open population models.

Conceptually, the technique of data augmentation represents a reparameterization of the "complete data" model—i.e., that conditional on N. The reparameterization is achieved by embedding this data set into a larger data set having $M > N$ "rows" (individuals) and re-expressing the model conditional on M instead of N. The great thing about data augmentation is that we do not need to know N for this reparameterization. Although this has a whiff of arbitrariness or even outright ad hockery, as a practical matter it is always possible to choose M easily for a given problem and context, and results will be insensitive to choice of M.[1] Then, under data augmentation, analysis is focused on the "augmented data set." That is, we analyze the bigger data set—the one having M rows—with an appropriate model that accounts for the augmentation. This is achieved by a Bernoulli sampling process that determines whether

[1] Unless the data set is sufficiently small that parameters are weakly identified.

an individual in M is also a member of N. Inference is focused on estimating the proportion $\psi = E[N]/M$, where ψ is the "data augmentation parameter."

4.2.1 DA links occupancy models and closed population models

There is a close correspondence between so-called "occupancy" models and closed population models (see Royle and Dorazio, 2008, Section 5.6). In occupancy models (MacKenzie et al., 2002; Tyre et al., 2003) the sampling situation is that M sites, or patches, are sampled multiple times to assess whether a species occurs at the sites. This yields encounter data such as that illustrated in the left panel of Table 4.4. The important problem is that a species may occur at a site, but go undetected, yielding an all-zero encounter history for the site, which in the case of occupancy studies, are *observed*. However, some of the all-zero sites will typically correspond to sites where the species in fact *does* occur. Thus, while the zeros are observed, there are too many of them and, in a sense, the inference problem is to partition the zeros into "structural" (fixed) and "sampling" (or stochastic) zeros, where the former are associated with unoccupied

Table 4.4 Hypothetical occupancy data set (left), capture-recapture data in standard form (center), and capture-recapture data augmented with all-zero capture histories (right).

Occupancy Data				Capture-recapture				Augmented C-R			
Site	k = 1	k = 2	k = 3	ind	k = 1	k = 2	k = 3	ind	k = 1	k = 2	k = 3
1	0	1	0	1	0	1	0	1	0	1	0
2	1	0	1	2	1	0	1	2	1	0	1
3	0	1	0	3	0	1	0	3	0	1	0
4	1	0	1	4	1	0	1	4	1	0	1
5	0	1	1	5	0	1	1	5	0	1	1
.	0	1	1	.	0	1	1	.	0	1	1
.	1	1	1	.	1	1	1	.	1	1	1
.	1	1	1	.	1	1	1	.	1	1	1
	1	1	1	.	1	1	1	.	1	1	1
n	1	1	1	n	1	1	1	n	1	1	1
.	0	0	0					.	0	0	0
.	0	0	0					.	0	0	0
	0	0	0						0	0	0
	0	0	0						0	0	0
	0	0	0						0	0	0
	0	0	0					N	0	0	0
.	0	0	0					.	0	0	0
.	0	0	0						0	0	0
M	0	0	0					.	0	0	0
								.	.	.	.
								.	.	.	.
								.	.	.	.
								M	0	0	0

sites and the latter with occupied sites where the species went undetected. More formally, inference is focused on the parameter ψ, the probability that a site is occupied.

In contrast to occupancy studies, in classical closed population studies, we observe a data set as in the middle panel of Table 4.4 where *no* zeros are observed. The inference problem is, essentially, to estimate how many sampling zeros there are—or should be—in a "complete" data set. This objective (how many sampling zeros?) is precisely the same for both occupancy and CR methods if an upper limit M is specified for the closed population model. The only distinction being that, in occupancy models, M is set by design (i.e., the number of sites in the sample), whereas a natural choice of M for capture-recapture models may not be obvious. However, the choice of M implies a uniform prior for N on the integers $[0, M]$ (Royle et al., 2007). Thus, one can analyze capture-recapture models by adding $M - n$ all-zero encounter histories to the data set and regarding the augmented data set, essentially, as a site-occupancy data set, where the occupancy or data augmentation parameter (ψ) takes the place of the abundance parameter (N).

Thus, the heuristic motivation of data augmentation is to fix the size of the data set by adding *too many* all-zero encounter histories to create the data set shown in the right panel of Table 4.4, and then analyze the augmented data set using an occupancy-type model that includes both "unoccupied sites" (in capture-recapture, augmented individuals that are not members of the real population that was sampled) as well as "occupied sites" at which detections did not occur (in capture-recapture, individuals that are members of the population but were undetected by sampling). We call these $M - n$ all-zero histories "potential individuals" because they exist to be recruited (in a non-biological sense) into the population, for example during an analysis by MCMC.

To analyze the augmented data set, we recognize that it is a zero-inflated version of the known-N data set. That is, some of the augmented all-zero rows are sampling zeros (corresponding to actual individuals that were missed) and some are structural zeros, which do not correspond to individuals in the population. For a basic closed population model, the resulting likelihood under data augmentation—that is, for the data set of size M—is a simple zero-inflated binomial likelihood. The zero-inflated binomial model can be described "hierarchically," by introducing a set of binary latent variables, $z_1, z_2, \ldots, z_M$, to indicate whether each individual i is ($z_i = 1$) or is not ($z_i = 0$) a member of the population of N individuals exposed to sampling. We assume that $z_i \sim \text{Bernoulli}(\psi)$ where ψ is the probability that an individual in the data set of size M is a member of the sampled population—in the sense that $1 - \psi$ is the probability of a structural zero in the augmented data set. The zero-inflated binomial model that arises under data augmentation can be formally expressed by the following set of assumptions (we include typical priors for a Bayesian analysis):

$$
\begin{aligned}
y_i | z_i = 1 &\sim \text{Binomial}(K, p) \\
y_i | z_i = 0 &\sim I(y = 0) \\
z_i &\overset{iid}{\sim} \text{Bernoulli}(\psi)
\end{aligned}
$$

$$\psi \sim \text{Uniform}(0, 1)$$
$$p \sim \text{Uniform}(0, 1)$$

for $i = 1, \ldots, M$, where $I(y = 0)$ is a point mass at $y = 0$. It is sometimes convenient to express the conditional-on-z observation model concisely in just one step:

$$y_i | z_i \sim \text{Binomial}(K, z_i p)$$

and we understand this to mean, if $z_i = 0$, then y_i is necessarily 0 because its success probability is $z_i p = 0$.

Note that, under data augmentation, N is no longer an explicit parameter of this model. In its place, we estimate ψ and functions of the latent variables z. In particular, under the assumptions of the zero-inflated model, $z_i \sim \text{Bernoulli}(\psi)$; therefore, N is a function of these latent variables:

$$N = \sum_{i=1}^{M} z_i.$$

Further, we note that the latent z_i parameters *can be* removed from the model by integration, in which case the joint probability distribution of the data is

$$\Pr(y_1, \ldots, y_M | p, \psi) = \prod_{i=1}^{M} \left(\psi \times \text{Binomial}(y_i | K, p) + I(y_i = 0)(1 - \psi)\right) \tag{4.2.1}$$

Interpreted as a likelihood, we can directly maximize this expression to obtain the MLEs of the structural parameters ψ and p or those of other more complex models (e.g., see Royle, 2006). We could estimate these parameters and then use them to obtain an estimator of N using the so-called "Best unbiased predictor" (see Royle and Dorazio, 2012). Normally, however, we will analyze the model in its "conditional-on-z" form using methods of MCMC either in the **BUGS** engines or using our own MCMC algorithms (Chapter 17).

4.2.2 Model M_0 in BUGS

It is helpful to understand data augmentation by seeing what its effect is on implementing model M_0. For this model, in which we can aggregate the encounter data to individual-specific encounter frequencies, the augmented data are given by the vector of frequencies $(y_1, \ldots, y_n, 0, 0, \ldots, 0)$ where the augmented values of $y = 0$ represent the encounter frequency for potential individuals $y_{n+1}, \ldots, y_M$. The zero-inflated model of the augmented data combines the model of the latent variables, $z_i \sim \text{Bernoulli}(\psi)$. The **BUGS** model description of the closed population model M_0 is shown in Panel 4.1. The last line of the model specification provides the expression for computing N from the data augmentation variables z_i. Note that, to improve readability of code snippets (especially of large ones), we will sometimes deviate from

```
model{
p  ~ dunif(0,1)
psi ~ dunif(0,1)

# nind = number of individuals captured at least once
#   nz = number of uncaptured individuals added for DA
for(i in 1:(nind+nz)){
    z[i] ~ dbern(psi)
   mu[i] <- z[i]*p
    y[i] ~ dbin(mu[i],K)
 }

N<-sum(z[1:(nind+nz)])
}
```

PANEL 4.1

BUGS description of model M_0 under data augmentation. Here `y`, `K`, `nind`, and `nz` are provided as data. The population size, N, is computed as a function of the data augmentation variables z.

our standard notation a bit. In this case we use `nind` for n (the number of encountered individuals), and `M = nind + nz` is the total size of the augmented data set. In other cases we might also use `nocc` in place of K and `ntraps` in place of J. We find that word definitions make code easier to understand, especially without having to read surrounding text.

Specification of a more general model in terms of the individual encounter observations y_{ik} is not much more difficult than for the individual encounter frequencies. We define the observation model by a double loop and change the indexing of quantities accordingly, i.e.,

```
for(i in 1:(nind+nz)){
    z[i] ~ dbern(psi)
  for(k in 1:K){
      mu[i,k] <- z[i]*p
      y[i,k] ~ dbin(mu[i,k],1)
   }
}
```

In this manner, it is straightforward to incorporate covariates on p for both individuals and sampling occasions (see discussion of this below and Chapter 7) as well as to devise other extensions of the model, including models for open populations (Chapter 16).

4.2.3 Remarks on data augmentation

Data augmentation may seem like a strange and mysterious black-box, and likely it is unfamiliar to most people, even to many of those with substantial experience with capture-recapture models. However, it really is just a formal reparameterization of capture-recapture models in which N is marginalized out of the ordinary (conditional-on-N) model (by summation over a binomial prior). As a result, we could refer to the resulting model as the "binomial-integrated likelihood" to reflect that an estimator could be obtained from the ordinary likelihood, integrated over a binomial prior. Other such "integrated likelihood" models are sensible. For example, we could place a Poisson prior on N with mean Λ and marginalize N over the Poisson prior. This produces a likelihood in which Λ replaces N, instead of ψ replacing N. We note that this type of marginalization (over a Poisson prior) is done by the **R** package `secr` for analysis of spatial capture-recapture models (see Section 6.5.3).

We emphasize the motivation for data augmentation being that it produces a data set of fixed size, so that the parameter dimension in any capture-recapture model is also fixed. As a result, MCMC is a relatively simple proposition using standard Gibbs sampling. And, in particular, capture-recapture models become trivial to implement in **BUGS**. Consider the simplest context—analyzing model M_0. In this case, DA converts model M_0 to a basic occupancy model, and the parameters p and ψ have known full conditional distributions (in fact, beta distributions) that can be sampled from directly. Furthermore, the data augmentation variables, i.e., the collection of z's, can be sampled from Bernoulli full conditionals. MCMC is not much more difficult for complicated models—sometimes the hyperparameters need to be sampled using a Metropolis-Hastings step (e.g., Chapter 17), but nothing more sophisticated than that is required.

Potential sensitivity of parameter estimates to M might be cause for some concern. The guiding principle is that it should be chosen large enough so that the posterior for N is not truncated, but it should not be too large due to the increased computational burden. It seems likely that the properties of the Markov chains should be affected by M and so some optimal choice of M might exist (Gopalaswamy, 2012). Formal analysis of this is needed.

There are other approaches to analyzing models with unknown N, using reversible jump MCMC (RJMCMC) or other so-called "trans-dimensional" (TD) algorithms (King and Brooks, 2001; Durban and Elston, 2005; King et al., 2008; Schofield and Barker, 2008; Wright et al., 2009). What distinguishes DA from RJMCMC and related TD methods is that DA is used to create a distinctly new model that is unconditional on N and we (usually) analyze the unconditional model. The various TD/RJMCMC approaches seek to analyze the conditional-on-N model in which the dimension of the parameter space is a function of N, and will therefore typically vary at each iteration of the MCMC algorithm. TD/RJMCMC approaches might appear to have the advantage that one can model N explicitly or consider alternative priors for N. However, despite that N is removed as an explicit parameter in DA, it is possible to develop hierarchical models that involve structure on N (Converse and Royle, 2012;

Royle et al., 2012b; Royle and Converse, in review), which we consider in Chapter 14. Furthermore, data augmentation is often easier to implement than RJMCMC, and the details of the DA implementation are the same for all capture-recapture problems.

4.2.4 Example: Black bear study on Fort Drum

To illustrate the analysis of model M_0 using data augmentation, we use a data set collected at Fort Drum Military Installation in upstate New York. The data were collected by P.D. Curtis and M.T Wegan of Cornell University and their colleagues at the Fort Drum Military Installation. These data have been analyzed in various forms by Wegan (2008), Gardner et al. (2009) and Gardner et al. (2010b). The specific data used here are encounter histories on 47 individuals obtained from an array of 38 baited "hair snares" (Figure 4.1) during June and July 2006. Barbed wire traps were baited and checked for hair samples each week for 8 weeks, thus we distinguished $K = 8$ weekly sample occasions. The data are provided in the **R** package `scrbook`, can be loaded by typing `data(beardata)` at the **R** prompt, and the analysis can be set up and run as follows (also see `?beardata` for the commands to do the analysis). Here, the data were augmented with 128 all-zero encounter histories, resulting in a total data set size of $M = 175$.

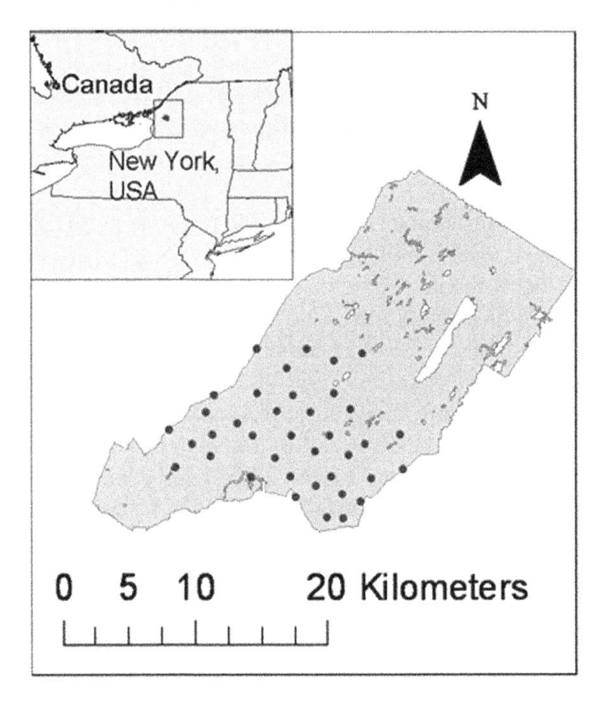

FIGURE 4.1

Fort Drum Black bear study area and the 38 baited hair snare locations operated for 8 weeks during June and July, 2006.

```
> library(scrbook)
> data(beardata)          # load the bear data and extract components
> trapmat <- beardata$trapmat
> nind <- dim(beardata$bearArray)[1]
> K <- dim(beardata$bearArray)[3]
> ntraps <- dim(beardata$bearArray)[2]
> M <- 175
> nz <- M-nind
> Yaug <- array(0, dim = c(M,ntraps,K))

> Yaug[1:nind,,] <- beardata$bearArray
> y <- apply(Yaug,c(1,3),sum)  # summarize by ind x rep
> y[y > 1] <- 1                # toss out multiple encounters per occasion
                               # b/c traditional CR models ignore space
```

The raw data object, beardata$bearArray, is a three-dimensional array nind × ntraps × K of individual encounter events (i.e., $y_{ijk} = 1$ if individual i was encountered in trap j during occasion k, and 0 otherwise). For fitting model M_0 (or M_h, see Section 4.4), it is sufficient to reduce the data to individual encounter frequencies which we have re-labeled "y" above. The **BUGS** model file along with commands to fit the model are as follows:

```
> set.seed(2013)                  # to obtain the same results each time
> library(R2WinBUGS)              # load R2WinBUGS, set-up:
> data0 <- list(y = y, M = M, K = K)     # data ....
> params0 <- c('psi','p','N')            # parameters ....
> zst <- c(rep(1,nind),rbinom(M-nind, 1, .5))     # inits .....
> inits <- function(){   list(z = zst, psi = runif(1), p = runif(1)) }

> cat("
model{

psi ~ dunif(0, 1)
p ~ dunif(0,1)

for (i in 1:M){
   z[i] ~ dbern(psi)
   for(k in 1:K){
     tmp[i,k] <- p*z[i]
     y[i,k] ~ dbin(tmp[i,k],1)
       }
    }
 N<-sum(z[1:M])
}
",file="modelM0.txt")

## Run the model:
> fit0 <- bugs(data0, inits, params0, model.file="modelM0.txt",n.chains = 3,
       n.iter = 2000, n.burnin = 1000, n.thin = 1,debug = TRUE,working.directory = getwd())
```

This produces the following posterior summary statistics:

```
> print(fit0,digits = 2)
Inference for Bugs model at "modelM0.txt", fit using WinBUGS,
 3 chains, each with 2000 iterations(first 1000 discarded)
 n.sims  = 3000 iterations saved
```

```
           mean     sd    2.5%     25%     50%     75%  97.75%  Rhat  n.eff
psi        0.29   0.04    0.22    0.26    0.29    0.31    0.36     1   3000
p          0.30   0.03    0.25    0.28    0.30    0.32    0.35     1   3000
N         49.94   1.99   47.00   48.00   50.00   51.00   54.00     1   3000
deviance 489.05  11.28  471.00  480.45  488.80  495.40  513.70     1   3000

[... some output deleted ...]
```

WinBUGS did well in choosing an MCMC algorithm for this model—we have $\hat{R} = 1$ for all parameters and an effective sample size of 3,000, equal to the total number of posterior samples. We see that the posterior mean of N under this model is 49.94 and a 95% posterior interval is (48, 54). We revisit these data later in the context of more complex models.

In order to obtain an estimate of density, D, we need an area to associate with the estimate of N, and in Chapter 1 we already went through a number of commonly used procedures to conjure up such an area, including buffering the trap array by the home range radius, often estimated by the mean maximum distance moved (MMDM) (Parmenter et al., 2003), 1/2 MMDM (Dice, 1938) or directly from telemetry data (Wallace et al., 2003). Typically, the trap array is defined by the convex hull around the trap locations, and this is what we applied a buffer to. We computed the buffer by using a telemetry-based estimate of the mean female home range radius (2.19 km) (Bales et al., 2005) instead of using an estimate based on our relatively sparse recapture data. For the Fort Drum study, the convex hull has an area of 157.135 km^2, and the buffered convex hull has an area of 277.011 km^2. To create this we used functions contained in the **R** package `rgeos` (Bivand and Rundel, 2011) and created a utility function `bcharea`, which is in our **R** package `scrbook`. The commands are as follows:

```
> library(rgeos)

> bcharea <- function(buff,traplocs){
    p1 <- Polygon(rbind(traplocs,traplocs[1,]))
    p2 <- Polygons(list(p1 = p1),ID = 1)
    p3 <- SpatialPolygons(list(p2 = p2))
    p1ch <- gConvexHull(p3)
    bp1 <- (gBuffer(p1ch, width = buff))
    plot(bp1, col='gray')
    plot(p1ch, border='black', lwd = 2, add = TRUE)
    gArea(bp1)
  }

> bcharea(2.19,traplocs = trapmat)
```

The resulting buffered convex hull is shown in Figure 4.2.

To conjure up a density estimate under model M_0, we compute the appropriate posterior summary of the ratio of N and the prescribed area (277.011 km^2):

```
> summary(fit0$sims.list$N/277.011)

   Min.  1st Qu.  Median    Mean  3rd Qu.    Max.
 0.1697   0.1733  0.1805  0.1803   0.1841  0.2130

> quantile (fit0$sims.list$N/277.011,c(0.025,0.975))

     2.5%      97.5%
0.1696684  0.1949381
```

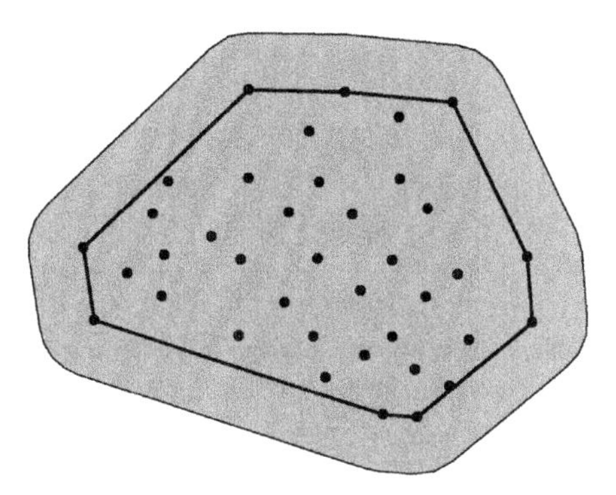

FIGURE 4.2

Convex hull of the bear hair snare array at Fort Drum, NY, buffered by mean female home range radius (2.19 km).

which yields a density estimate of about 0.18 ind/km^2, and a 95% Bayesian confidence interval of (0.170, 0.195). Our estimate of density should be reliable if we have faith in our stated value of the "sampled area." Clearly, though, this is largely subjective, and not something we can formally evaluate (or estimate) from the data based on model M_0.

4.3 Temporally varying and behavioral effects

The purpose of this chapter is mainly to emphasize the central importance of the binomial model in capture-recapture and so we have considered models for individual encounter frequencies—the number of times individuals are encountered out of K occasions. Sometimes we can't aggregate the encounter data for each individual, such as when encounter probability varies over occasions. Time-varying responses that are relevant in many capture-recapture studies are "effort" such as amount of search time, number of observers, or trap nights, or encounter probability varying over time, as a function of date or season (Kéry et al., 2011) due to species behavior. A common situation in many animal studies is that in which there exists a "behavioral response" to trapping (even if the animal is not physically trapped).

Behavioral response is an important concept in animal studies because individuals might learn to come to baited traps or avoid traps due to trauma related to being encountered. There are a number of ways to parameterize a behavioral response to encounter. The distinction between persistent and ephemeral was made by Yang and Chao (2005) who considered a general behavioral response model of the form:

$$\text{logit}(p_{ik}) = \alpha_0 + \alpha_1 y_{i,k-1} + \alpha_2 x_{ik}$$

where x_{ik} is a covariate indicator variable of previous capture (i.e., $x_{ik} = 1$ if captured in any previous period). Therefore, encounter probability changes depending on whether an individual was captured in the immediate previous period (a Markovian or ephemeral behavioral response; (Yang and Chao, 2005)), described by the term $\alpha_1 y_{i,k-1}$ or in *any* previous period (persistent behavioral response), described by the term $\alpha_2 x_{ik}$. Because spatial capture-recapture models allow us to include trap-specific covariates, we can describe a third type of behavioral response—a local behavioral response that is trap-specific (Royle et al., 2011b). In this local behavioral response, the encounter probability is modified for an individual trap depending on previous capture in that trap. Models with temporal effects are easy to describe and analyze in the **BUGS** language and we provide a number of examples in Chapter 7 and elsewhere.

4.4 Models with individual heterogeneity

Models in which encounter probability varies by individual have a long history in capture-recapture and, indeed, this so-called "model M_h" ("h" for heterogeneity) is one of the elemental capture-recapture models (Otis et al., 1978). Conceptually, we imagine that the individual-specific encounter probability parameters, p_i, are random variables distributed according to some probability distribution, $[p|\theta]$. We denote this basic model assumption as $p_i \sim [p|\theta]$. This type of model is similar in concept to extending a GLM to a GLMM but in the capture-recapture context N is unknown. The basic class of models is often referred to as model M_h, but really this is a broad class of models, each being distinguished by the specific distribution assumed for p_i. There are many different varieties of model M_h including parametric and various non-parametric approaches (Burnham and Overton, 1978; Norris and Pollock, 1996; Pledger, 2004). One important practical matter is that estimates of N can be extremely sensitive to the choice of heterogeneity model (Fienberg et al., 1999; Dorazio and Royle, 2003; Link, 2003). Indeed, Link (2003) showed that in some cases it's possible to find models that yield precisely the same expected data, yet produce wildly different estimates of N. In that sense, N for most practical purposes is not identifiable across classes of different heterogeneity models, and this should be understood before fitting any such model. One solution to this problem is to seek to model explicit factors that contribute to heterogeneity, e.g., using individual covariate models (Section 4.5 below). Indeed, spatial capture-recapture models do just that, by modeling heterogeneity due to the spatial organization of individuals in relation to traps or other encounter mechanism. For additional background and applications of model M_h see Royle and Dorazio (2008, Chapter 6) and Kéry and Schaub (2012, Chapter 6).

We will work with a specific type of model M_h here, which is a natural extension of the basic binomial observation model of model M_0 so that

$$\text{logit}(p_i) = \mu + \eta_i$$

where μ is a fixed parameter (the mean) to be estimated, and η_i is an individual random effect assumed to be normally distributed:

$$\eta_i \sim \text{Normal}(0, \sigma_p^2)$$

We could as well combine these two steps and write $\text{logit}(p_i) \sim \text{Normal}(\mu, \sigma_p^2)$. This "logit-normal mixture" was analyzed by Coull and Agresti (1999) and others. It is also natural to consider a beta distribution for p_i (Dorazio and Royle, 2003), and so-called "finite-mixture" models are also popular (Norris and Pollock, 1996; Pledger, 2004). In the latter, individuals are assumed to belong to a finite number of latent classes, each of which has its own capture probability.

Model M_h has important historical relevance to spatial capture-recapture situations (Karanth, 1995) because investigators recognized that the juxtaposition of individuals with the array of trap locations should yield heterogeneity in encounter probability, and thus it became common to use some version of model M_h to estimate N using data from spatial arrays of traps. While this doesn't resolve the problem of not knowing the effective sample area, it does yield an estimator that accommodates the heterogeneity in p induced by the spatial aspect of capture-recapture studies. To see how this juxtaposition induces heterogeneity, we have to understand the relevance of movement in capture-recapture models. Imagine a quadrat that can be uniformly searched by a crew of biologists for some species of reptile (e.g., Royle and Young, 2008). Further, suppose that the species exhibits some sense of spatial fidelity in the form of a home range or territory, and individuals move about their home range in some kind of random fashion. Figure 4.3 shows a sample quadrat searched repeatedly over a period of time (home range centroids are given by the solid dots). Heuristically, we imagine that each individual in the vicinity of the study area is liable to experience variable exposure to encounter due to the overlap of its home range with the sampled area—essentially the long-run proportion of times the individual is within the sample plot boundaries, say ϕ. We might model the exposure or *availability* of an individual

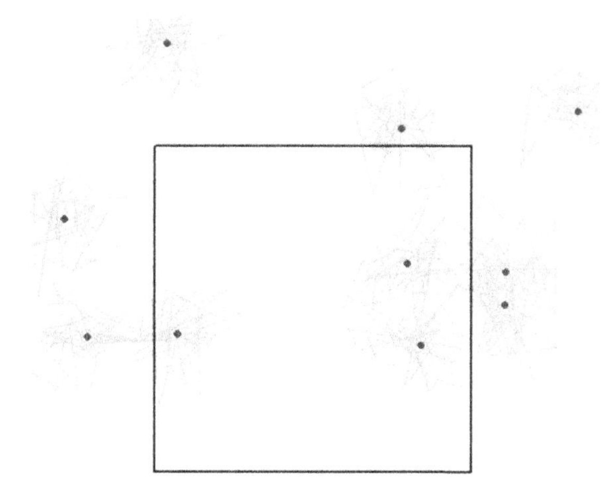

FIGURE 4.3

A quadrat searched for lizards over some period of time. The locations (simulated data) for each of 10 lizards are connected by lines—the dots are the activity centers.

to capture by supposing that $a_i = 1$ if individual i is available to be captured (i.e., within the survey plot) during any sample, and 0 otherwise. Then, $\Pr(a_i = 1) = \phi$. In the context of spatial studies, it is natural that ϕ should depend on *where* an individual lives, i.e., it should be individual-specific, ϕ_i (Chandler et al., 2011). This system describes, precisely, that of "random temporary emigration" (Kendall et al., 1997), where ϕ_i is the individual-specific probability of being available for capture.

Conceptually, SCR models aim to deal with this problem of variable exposure to sampling due to movement in the proximity of the trapping array explicitly and formally with auxiliary spatial information. If individuals are detected with probability p_0, *conditional* on $a_i = 1$, then the marginal probability of detecting individual i is

$$p_i = p_0\phi_i$$

so we see clearly that individual heterogeneity in encounter probability is induced as a result of the juxtaposition of individuals (i.e., their home ranges) with the sample apparatus and the movement of individuals about their home range.

4.4.1 Analysis of model M_h

If N is known, it is worth taking note of the essential simplicity of model M_h as a binomial GLMM. This type of model is widely applied throughout statistics using standard methods of inference based either on integrated likelihood (Laird and Ware, 1982; Berger et al., 1999), which we discuss in Chapter 6, or standard Bayesian methods. However, because N is not known, inference is somewhat more challenging. We address that here using Bayesian analysis based on data augmentation. Although we use data augmentation in the context of Bayesian methods here, we note that heterogeneity models formulated under DA are easily analyzed by conventional likelihood methods as zero-inflated binomial mixtures (Royle, 2006). More traditional analysis of model M_h based on integrated likelihood, without using data augmentation, has been considered by Coull and Agresti (1999), Dorazio and Royle (2003), and others.

As with model M_0, we have the Bernoulli model for the zero-inflation variables: $z_i \sim \text{Bernoulli}(\psi)$ and the model of the observations expressed conditional on these latent variables z_i. For $z_i = 1$, we have a binomial model with individual-specific p_i:

$$y_i|z_i = 1 \sim \text{Binomial}(K, p_i)$$

and otherwise $y_i|z_i = 0 \sim I(y = 0)$, i.e., a point mass at $y = 0$. Further, we prescribe a distribution for p_i. Here we assume

$$\text{logit}(p_i) \sim \text{Normal}(\mu, \sigma^2)$$

For prior distributions we assume $p_0 = \text{logit}^{-1}(\mu) \sim \text{Uniform}(0, 1)$ and, for the standard deviation $\sigma \sim \text{Uniform}(0, B)$ for some large B. Another common default prior is to assume $\tau = 1/\sigma^2 \sim \text{Gamma}(.1, .1)$.

4.4.2 Analysis of the Fort Drum data with model M_h

Here we provide an analysis of the Fort Drum bear survey data using the logit-normal heterogeneity model, and we used data augmentation to produce a data set of $M = 700$ individuals. We have so far mostly used **WinBUGS** but we are now transitioning to **JAGS** run from within **R** using the packages `R2jags` (Su and Yajima, 2011) or `rjags` (Plummer, 2009). The function `jags` from the `R2jags` package runs essentially like the `bugs` function which we demonstrate here for setting up and running model M_h for the Fort Drum bear data:

```
[... get data as before ....]

> set.seed(2013)

> cat("
model{
 p0 ~ dunif(0,1)              # prior distributions
 mup <- log(p0/(1-p0))
 sigmap ~ dunif(0,10)
 taup <- 1/(sigmap*sigmap)
 psi ~ dunif(0,1)

 for(i in 1:(nind + nz)){
   z[i] ~ dbern(psi)           # zero inflation variables
   lp[i] ~ dnorm(mup,taup) # individual effect
   logit(p[i]) <- lp[i]
   mu[i] <- z[i]*p[i]
   y[i] ~ dbin(mu[i],K)        # observation model
 }

 N <- sum(z[1:(nind + nz)])
}
",file="ModelMh.txt")

> data1 <- list(y = y, nz = nz, nind = nind, K = K)
> params1 <- c('p0','sigmap','psi','N')
> inits <- function(){ list(z = as.numeric(y>=1), psi=.6, p0 = runif(1),
          sigmap = runif(1,.7,1.2),lp = rnorm(M,-2)) }
> library(R2jags)
> wbout <- jags(data1, inits, params1, model.file = "modelMh.txt", n.chains = 3,
            n.iter = 1010000, n.burnin = 10000, working.directory = getwd())
```

We provide an **R** function `modelMhBUGS` in the package `scrbook` that will fit the model using either **JAGS** or **WinBUGS** as specified by the user. In addition, for fun, we construct our own MCMC algorithm using a Metropolis-within-Gibbs algorithm for model M_h in Chapter 17, where we also develop MCMC algorithms for spatial capture-recapture models. Using `modelMhBUGS`, we ran 3 chains of 1 *million* iterations (mixing is poor for this model and this data set), which produced the posterior distribution for N shown in Figure 4.4. Posterior summaries of parameters are given in Table 4.5.

We used $M = 700$ for this analysis and we note that while the posterior mass of N is concentrated away from this upper bound (Figure 4.4), the posterior has an extremely long right tail, with some MCMC draws at the upper boundary $N = 700$,

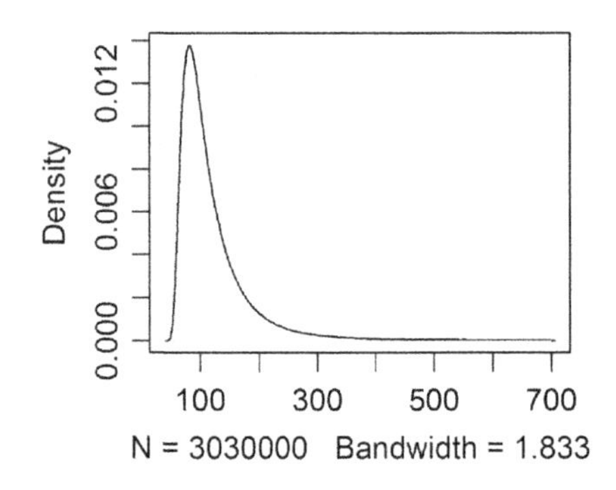

FIGURE 4.4

Posterior of N for Fort Drum black bear study data under the logit-normal version of model M_h.

Table 4.5 Posterior summaries from model M_h fitted to the Fort Drum black bear data. $p_0 = \text{expit}(\alpha_0)$. Results were obtained using **JAGS** running 3 chains, each with 1,010,000 iterations, discarding the first 10,000 for a total of three *million* posterior samples.

Parameter	Mean	SD	2.4%	50%	97.5%	Rhat	n.eff
p_0	0.072	0.056	0.002	0.060	0.203	1.008	540
σ_p	2.096	0.557	1.215	2.025	3.373	1.003	820
ψ	0.176	0.101	0.084	0.147	0.458	1.006	650
N	122.695	69.897	62.000	102.000	319.000	1.006	630

suggesting that an even higher value of M may be called for. To characterize the posterior distribution of density, we produce the relevant summaries of the posterior distribution of $D = N/277.11$ (recall the buffered area of the convex hull is 277.11 km^2):

```
> summary(wbout$sims.list$N/277.11)

    Min.  1st Qu.  Median    Mean  3rd Qu.    Max.
  0.1696   0.2959  0.3681  0.4428   0.4944  2.5260

> quantile(wbout$sims.list$N/277.11,c(0.025,0.50,0.975))

      2.5%        50%      97.5%
 0.2237379  0.3680849  1.1511674
```

Therefore, the point estimate, characterized by the posterior median, is around 0.37 bears/km^2 and a 95% Bayesian credible interval is (0.224, 1.151).

4.4.3 Comparison with MLE

The posterior of N is highly skewed; therefore, we see that the posterior mean ($N = 122.7$) is considerably higher than the posterior median ($N = 102$). Further, it may be surprising that these posterior summaries do not compare well with the MLE. To obtain the MLE of $\log(n_0)$, the logarithm of the number of uncaptured individuals, we used the **R** code contained in Panel 6.1 from Royle and Dorazio (2008). We found $\widehat{\log(n0)} = 3.86$ and therefore $\hat{N} = \exp(3.86) + 47 = 94.47$, which is larger than the mode shown in Figure 4.4. To see this, we compute the posterior mode, by finding the posterior value of N with the highest mass. Because N is discrete, we can use the `table()` function in **R** and find the most frequent value.[2] If we want to smooth out some of the Monte Carlo error a bit, we can use a smoother of some sort applied to the tabled posterior frequencies of N. Here we use a smoothing spline (**R** function `smooth.spline`) with the degree of smoothing chosen by cross-validation (the `cv = TRUE` argument):

```
> N <- table(wbout$BUGSoutput$sims.list$N)
> xg <- as.numeric(names(N))

> sp <- smooth.spline(xg,N,cv = TRUE)

> sp

Call:
smooth.spline (x  = xg, y  = N, cv  = TRUE)

Smoothing Parameter spar = 0.09339815 lambda = 8.201724e-09 (17 iterations)
Equivalent Degrees of Freedom (Df): 121.1825
Penalized Criterion: 2544481
PRESS: 5903.4
```

We obtain the mode of the smoothed frequencies as follows:

```
sp$x[sp$y==max(sp$y)]
[1] 82
```

We don't dwell too much on the difference between the MLE and features of the posterior, but we do note here that the posterior distribution for the parameters of this model, for the Fort Drum data set, is very sensitive to the prior distributions. In the present case, the use of a Uniform(0, 1) prior for $p_0 = \text{logit}^{-1}(\mu)$ is somewhat informative—in particular, it is not at all "flat" on the scale of μ, and this affects the posterior. We generally recommend use of a Uniform(0, 1) prior for $\text{logit}^{-1}(\mu)$ in such models. That said, we were surprised at this result, and we experimented with other prior configurations including putting a flat prior on μ directly. This kind of small sample instability has been widely noted in model M_h (Fienberg et al., 1999; Dorazio and Royle, 2003), as has extreme sensitivity to the specific form of model M_h (Link, 2003). In summary, while the mode is well defined, the data set is relatively sparse and hence inferences are poor and sensitive to model choice.

[2]For a continuous random variable we can use the function `density()` to smooth the posterior samples and obtain the mode.

4.5 Individual covariate models: toward spatial capture-recapture

A standard situation in capture-recapture is when a covariate that is thought to influence encounter probability is measured for each individual. These are often called "individual covariate models" but, in keeping with the classical nomenclature on closed population models, Kéry and Schaub (2012) referred to this class of models as "model M_x" (the x here being an explicit covariate). As with other closed population models, we begin with the basic binomial observation model:

$$y_i \sim \text{Binomial}(K, p_i).$$

To model the covariate, we use a logit model for encounter probability of the form:

$$\text{logit}(p_i) = \alpha_0 + \alpha_1 x_i, \tag{4.5.1}$$

where x_i is the covariate value for individual i and the parameters $\boldsymbol{\alpha} = (\alpha_0, \alpha_1)$ are the regression coefficients. Classical examples of covariates influencing detection probability are type of animal (juvenile/adult or male/female), a continuous covariate such as body mass, or a discrete covariate such as group or cluster size. For example, in models of aerial survey data, it is natural to model the detection probability of a group as a function of the observation-level individual covariate, "group size" (Royle, 2008; Langtimm et al., 2011).

Model M_x is similar in structure to model M_h, except that the individual effects are *observed* for the n individuals that appear in the sample. These models are important here because spatial capture-recapture models can be described precisely as a form of model M_x, where the covariate describes *where* the individual is located in relation to the trapping array. Specifically, SCR models *are* individual covariate models, but where the individual covariate is only observed imperfectly (or partially observed) for each captured individual. Unlike model M_h, in SCR models (and model M_x) we do have some direct information about the latent variable, which comes from the spatial locations/distribution of individual recaptures.

Traditionally, estimation of N in model M_x is achieved using methods based on ideas of unequal probability sampling (i.e., Horvitz-Thompson estimation[3]; Huggins (1989); Alho (1990); Borchers et al. (2002)). An estimator of N is

$$\hat{N} = \sum_{i=1}^{n} \frac{1}{\tilde{p}_i},$$

where $\tilde{p}_i$ is the probability that individual i appeared in the sample. This quantity is $\tilde{p}_i = \Pr(y_i > 0)$ and, in closed population capture-recapture models, it can be computed as:

$$\Pr(y_i > 0) = 1 - (1 - p_i)^K,$$

[3]For a quick summary of the idea see: http://en.wikipedia.org/wiki/Horvitz-Thompson_estimator.

where p_i is a function of parameters α_0 and α_1 according to Eq. (4.5.1). In practice, parameters are estimated from the conditional likelihood of the observed encounter histories which is, for observation y_i,

$$\mathcal{L}_c(\boldsymbol{\alpha}|y_i) = \frac{\text{Binomial}(y_i|\boldsymbol{\alpha})}{\tilde{p}_i}. \tag{4.5.2}$$

This derives from a straightforward application of the law of total probability. Conceptually, we partition $\Pr(y)$ according to $\Pr(y) = \Pr(y|y > 0)\Pr(y > 0) + \Pr(y|y = 0)\Pr(y = 0)$. For any positive value of y the second term is necessarily 0, and so we rearrange to obtain $\Pr(y|y > 0) = \Pr(y)/\Pr(y > 0)$ which, in the specific case where $\Pr(y)$ is the binomial probability mass function (pmf), produces Eq. (4.5.2).

Here, we take a formal model-based approach to Bayesian analysis of such models based on the joint likelihood using data augmentation. Classical likelihood analysis of the so-called "full likelihood" is covered by Borchers et al. (2002). For Bayesian analysis of model M_x, because the individual covariate is unobserved for the $n_0 = N - n$ uncaptured individuals, we require a model to describe variation in x among individuals, essentially allowing the sample to be extrapolated to the population. For example, if we have a continuous trait measured on each individual, then we might assume that x has a normal distribution:

$$x_i \sim \text{Normal}(\mu, \sigma^2)$$

Data augmentation can be applied directly to this class of models. In particular, reformulation of the model under DA yields a basic zero-inflated binomial model of the following form, for each $i = 1, 2, \ldots, M$:

$$\begin{aligned} z_i &\sim \text{Bernoulli}(\psi) \\ y_i|z_i = 1 &\sim \text{Binomial}(K, p_i(x_i)) \\ y_i|z_i = 0 &\sim I(y = 0) \\ x_i &\sim \text{Normal}(\mu, \sigma^2) \end{aligned}$$

As with the previous models, implementation is trivial in the **BUGS** language. The **BUGS** specification is very similar to that for model M_h, but we require the distribution of the covariate to be specified, along with priors for the parameters of that distribution.

4.5.1 Example: location of capture as a covariate

Here we consider a special type of model M_x that is particularly relevant to spatial capture-recapture. Intuitively, some measure of distance from home range center to traps for an individual should be a reasonable covariate to explain heterogeneity in encounter probability, i.e., individuals with more exposure to traps should have higher encounter probabilities and vice versa. So we can imagine *estimating* such a quantity, say average distance from home range center to "the trap array", and then using it as an individual covariate in capture-recapture models. A version of this idea was put forth by Boulanger and McLellan (2001) (see also Ivan, 2012), but using the Huggins-Alho estimator and with covariate "distance from home range center to edge of the trapping array", where the home range center is estimated by the average capture location. This is intuitively appealing because we can imagine, in some kind of an ideal situation where we have a dense grid of traps over some geographic region, that the average location of capture would be a decent estimate (heuristically) of an individual's home range center. We provide an example of this type of approach using a fully model-based analysis of the version of model M_x described above, analyzed by data augmentation. We take a slightly different approach than that adopted by Boulanger and McLellan (2001). By analyzing the full likelihood and placing a prior distribution on the individual covariate, we will resolve the problem of having an ill-defined sample area. After you read later chapters of this book, it will be apparent that SCR models represent a formalization of this heuristic procedure.

For our purposes here, we define the scalar individual covariate x_i to be the distance from the average encounter location of individual i, say $\mathbf{s}_i$, to the centroid of the trap array, $\mathbf{x}_0$: $x_i = ||\mathbf{s}_i - \mathbf{x}_0||$. Note that $||\mathbf{u}||$ is standard notation for Euclidean norm or magnitude of the vector $\mathbf{u}$, and we use it throughout the book. In practice, people have used distance from edge of the trap array but that is less easy to quantify, as "edge" itself is not precisely defined. Conceptually, individuals in the middle of the array should have a higher probability of encounter and, as x_i increases, p_i should therefore decrease. Note that we have defined $\mathbf{s}_i$ in terms of a sample quantity—the observed mean encounter location—which, while ad hoc, is consistent with the use of individual covariate models in the literature. For an expansive, dense trapping grid we might expect the sample mean encounter location to be a good estimate of home range center, but clearly this is biased for individuals that live around the edge (or off) the trapping array.

A key point is that $\mathbf{s}_i$ is missing for each individual that is not encountered and consequently x_i is also missing. Therefore, it is a latent variable, and we need to specify a probability distribution for it. As a measurement of distance we know it must be positive-valued, and it seems sensible that an individual located extremely far from the array of traps would not be captured. Therefore, let's assume that x_i is uniformly distributed from 0 to some large number, say B, beyond which it would be difficult to imagine an individual being captured by the trap array:

$$x_i \sim \text{Uniform}(0, B)$$

where B is a specified constant, which we may choose to be arbitrarily large. For example, B should be at least a home range diameter past the furthest trap from the centroid of the array.

4.5.2 Example: Fort Drum black bear study

We have to do a little bit of data processing to fit this individual covariate model to the Fort Drum data. We need to compute the individual covariate x_i (distance from the centroid of the trapping array) using the **R** function `spiderplot` provided in

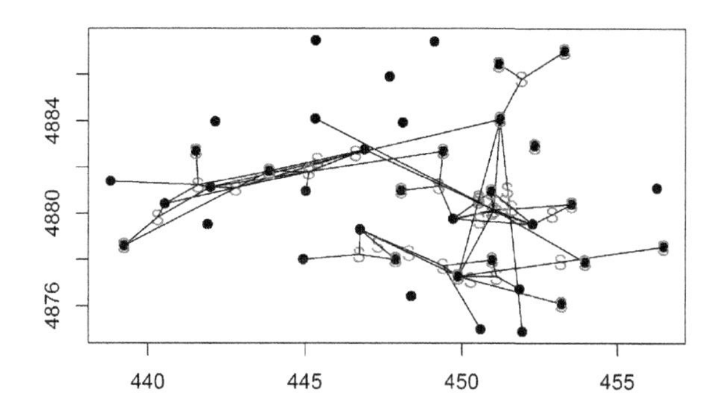

FIGURE 4.5

Spider plot of the Fort Drum black bear study data. The black dots represent the 47 trap locations with the "S" symbols being the average capture location of each bear, i.e., its estimated home range center. All traps in which a bear was captured are connected to its estimated home range center with a line.

scrbook. This function also produces the keen plot shown in Figure 4.5, which we call a "spider plot." The **R** commands for obtaining the individual covariate "distance from trap centroid" (the variable xcent returned by spiderplot) and making the spider plot are as follows:

```
> library(scrbook)
> data(beardata)
> toad<-spiderplot(beardata$bearArray,beardata$trapmat)
> xcent<-toad$xcent
```

For the analysis of these data using the individual covariate "distance from centroid" we used $x_i \sim \text{Uniform}(0, B)$ with $B = 11.5$ km, which is about the distance from the array center to the furthest trap. Once we choose a value for B, the direct implication is that the population size parameter, N, applies to the area within 11.5 kms of the trap centroid. Therefore, the model associates a precise area within which the population of N individuals resides. We will see shortly that N does, in fact, scale with our choice of B to reflect the changing area over which the N individuals of the model reside. The **BUGS** model specification and **R** commands to package the data and fit the model are as follows:

```
cat("
model{
  p0 ~ dunif(0,1)                 # prior distributions
  alpha0 <- log(p0/(1-p0))
  psi ~ dunif(0,1)
  alpha1 ~ dnorm(0,.01)

for(i in 1:(nind + nz)){
  xcent[i] ~ dunif(0,B)
  z[i] ~ dbern(psi)             # DA variables
```

Table 4.6 Posterior summaries from the individual covariate model (model M_x) with covariate "distance from the centroid of the trap array," fitted to the Fort Drum black bear data. Parameter $p_0 = \text{expit}(\alpha_0)$. Results were obtained using **WinBUGS** running 3 chains, each with 11,000 iterations, discarding the first 1,000 for a total of 30,000 posterior samples.

Parameter	Mean	SD	2.4%	50%	97.5%	Rhat	n.eff
p_0	0.54	0.07	0.40	0.54	0.67	1	1100
ψ	0.34	0.05	0.25	0.34	0.44	1	3500
N	58.92	5.49	50.00	58.00	71.00	1	1900
α_0	−0.25	0.06	−0.36	−0.25	−0.12	1	780

```
   lp[i] <- alpha0  + alpha1*xcent[i]   # individual effect
   logit(p[i]) <- lp[i]
   mu[i] <- z[i]*p[i]
   y[i] ~ dbin(mu[i],K)             # observation model
 }

N <- sum(z[1:(nind + nz)])
}
",file="modelMcov.txt")

> data2 <- list(y = y,nz = nz, nind = nind, K = K, xcent = xcent,B = 11.5)
> params2 <- c('p0','psi','N','beta')
> inits <- function() {list(z = zst, psi = psi, p0 = runif(1), alpha1 = rnorm(1) ) }
> fit2 <- bugs(data2, inits, params2, model.file="modelMcov.txt",
            n.chains = 3, n.iter = 11000, n.burnin = 1000, n.thin = 1)
```

This produces the posterior summary statistics in Table 4.6. We note that the estimated N is much lower than obtained by model M_h but there is a good explanation for this which we discuss in the next section. That issue notwithstanding, it is worth pondering how this model could be an improvement (conceptually or technically) over some other model/estimator including M_0 and M_h considered previously. Well, for one, we have accounted formally for heterogeneity due to spatial location of individuals relative to exposure to the trap array, characterized by the centroid of the array. Moreover, we have done so using a model that is based on an explicit mechanism, as opposed to a phenomenological one such as model M_h. In addition, and importantly, using our new model, *the estimated N applies to an explicit area which is defined by our prescribed value of B*. That is, this area is a fixed component of the model and the parameter N therefore has explicit spatial context, as the number of individuals with home range centers located less than B from the centroid of the trap array. As such, the implied "effective area" of the trap array for a given B is a precisely defined quantity—the area of a circle with radius B.

4.5.3 Extension of the model

The model developed in the previous section is not a very good model for one important reason: imposing a uniform prior distribution on x implies that density is *not constant* over space. In particular, this model implies that density *decreases* as we move away from the centroid of the trap array. That is, $x_i \sim \text{Uniform}(0, B)$ implies constant N in each distance band from the centroid but obviously the *area* of each distance band is increasing. This is one reason we have a lower estimate of density than that obtained previously from model M_h (Section 4.4.2) and also why, if we were to increase B, we would see density continue to decrease.

Fortunately, we are not restricted to use this specific distribution for the individual covariate. Clearly, it is a bad choice and, therefore, we should think about whether we can choose a better distribution for x—one that doesn't imply a decreasing density as distance from the centroid increases. Conceptually, what we want to do is impose a prior on distance from the centroid, x, such that abundance should be proportional to the amount of area in each successive distance band as you move farther away from the centroid, so that density is *constant*. In fact, theory exists that tells us we should choose $[x] = 2x/B^2$. This can be derived by noting that $F(x) = \Pr(X < x) = (\pi x^2)/(\pi B^2)$. Then, $f(x) = dF/dx = 2x/(B^2)$. This is a sort of triangular distribution in density induced because the incremental area in each additional distance band increases linearly with radius (i.e., distance from centroid). This can be verified empirically as follows:

```
> u <- runif (10000,-1,1)
> v <- runif (10000,-1,1)
> d <- sqrt (u*u + v*v)
> hist (d[d < 1])
> hist (d[d < 1],100)
> hist (d[d < 1],100,probability = TRUE)
> abline (0,2)
```

It would be useful if we could describe this distribution directly in **BUGS** but there is not a built-in way to do so. However, we can implement a discrete version of the pdf.[4] To do this, we break the interval $[0, B]$ into L distance classes of width δ, with probabilities proportional to $2x$. In particular, if we denote the cut-points by $g_1 = 0, g_2, \ldots, g_{L+1} = B$ and the interval midpoints are $m_i = g_{i+1} - \delta/2$. Then the interval probabilities are, approximately,[5] $p_i = \delta(2m_i/B^2)$, which we can compute once and then pass to **BUGS** as data. The **R** commands for doing all of this (noting that we have already loaded and processed the Fort Drum bear data) are given in the following **R/BUGS** script:

```
> delta <-.2
> xbin <- xcent%/%delta  + 1          # Put x in bins
> midpts <- seq (delta,Dmax,delta)
> xprobs <- delta* (2*midpts/ (B*B))
> xprobs <- xprobs/sum (xprobs)

> cat ("
```

[4]We might also be able to use what is referred to in **WinBUGS** jargon as the "zeros trick" (see *Advanced BUGS tricks* in the manual) although we haven't pursued this approach.

[5]This is just length × width, the area of small rectangles approximating the integral.

```
model{
p0 ~ dunif (0,1)                            # Prior distributions
alpha0 <- log (p0/(1-p0))
psi ~ dunif (0,1)
alpha1 ~ dnorm (0,.01)

for (i in 1: (nind + nz)){
  xbin[i] ~ dcat (xprobs[])
  z[i] ~ dbern (psi)                        # DA variables
  lp[i] <- alpha0  + alpha1*xbin[i]*delta
  logit (p[i]) <- lp[i]
  mu[i] <- z[i]*p[i]
  y[i] ~ dbin (mu[i],K)                     # Observation model
 }

N <- sum (z[1: (nind + nz)])                # N is derived
}
",file="modelMcov.txt")
```

In the model description, the variable x (observed distance from centroid of the trap array) has been rounded or binned (placed into a distance bin) so that the discrete version of the pdf of x can be used, as described previously. The new variable labeled `xbin` is then the *integer category* in units of δ from 0. Thus, to convert back to distance in the expression for `lp[i]`, `xbin[i]` has to be multiplied by δ. To fit the model, keeping in mind that the data objects required below have been defined in previous analyses of this chapter, we do this:

```
> data2 <- list (y = y, nz = nz, nind = nind, K = K, xbin = xbin, xprobs = xprobs,
                 delta = delta)
> params2 <- c('p0','psi','N','alpha1')
> inits <- function() { list (z = z, psi = psi, p0 = runif (1),alpha1 = rnorm(1) ) }
> fit <- bugs (data2, inits, params2, model.file="modelMcov.txt",
          working.directory = getwd (), debug = FALSE, n.chains = 3,
          n.iter = 11000, n.burnin = 1000, n.thin = 2)
```

By specification of B, this model induces a clear definition of area in which the population of N individuals resides. The parameter N of the model is the population size that applies to the particular value of B and, as such, we will see that N scales with our choice of B. This might be disconcerting to some—we can get whatever value of N we want by changing B! However, it is intuitively reasonable that, as we increase the area under consideration, there should be more individuals in it. Fortunately, we find empirically, that while N is highly sensitive to the prescribed value of B, density appears invariant to B as long as B is sufficiently large. We fit the model for a set of values of B from $B = 12$ (restricting values of x to be in close proximity to the trap array) on up to 20. The results are given in Table 4.7.

We see that the posterior mean and SD of density (individuals/km^2) appear insensitive to choice of B once we reach about $B = 17$ or so. The estimated density of $0.25/\text{km}^2$ is actually quite a bit lower than we reported using model M_h for which no relevant "area" quantity is explicit in the model (and so we had to make one up). Using MLEs of N in conjunction with buffer strips (see Table 1.1) our estimates were

Table 4.7 Posterior summaries of Fort Drum bear hair snare data using the individual covariate model, for different values of B, the upper limit of the uniform distribution of 'distance from centroid of the trap array'.

B	Density (post. mean)	Posterior SD
12	0.230	0.038
15	0.244	0.041
17	0.249	0.044
18	0.249	0.043
19	0.250	0.043
20	0.250	0.044

in the range of 0.32–0.43. On the other hand, our estimate of $\hat{D} = 0.25$ here (based on the posterior mean) is higher than that reported from model M_0 using the buffered area ($\hat{D} = 0.18$). There is no basis really for comparing or contrasting these various estimates. In particular, application of models M_0 and M_h are distinctly *not* spatially explicit models—the area within which the population resides is not defined under either model. There is therefore no reason at all to think that the estimates produced under either closed population model, based on a buffered "trap area," are justifiable by any theory. In fact, we would get exactly the same estimate of N no matter what we declare the area to be. On the other hand, the individual covariate model uses an explicit model for "distance from centroid" which is a reasonable and standard null model—it posits, in the absence of direct information, that individual home range centers are randomly distributed in space and that probability of detection depends on the distance between home range center and the centroid of the trap array. Under this definition of the system, we see that density is invariant to the choice of area, which seems like a desirable feature.

4.5.4 Invariance of density to B

Under model M_x, and under models that we consider in later chapters, a general property of the estimators is that while N increases with the prescribed area of the model (defined by B in this model), we expect that density estimators should be invariant to this area. In the model used above, we note that $\text{Area}(B) = \pi B^2$ and $\mathbb{E}(\text{N}(B)) = \lambda\text{Area}(B)$ and thus $\mathbb{E}(\text{Density}(B)) = \lambda$, i.e., constant. This should be interpreted as the *prior* density. Absent data, realizations under the model will have density, λ, regardless of what B is prescribed to be. As we verified empirically above, posterior summaries of density are also invariant to B as long as the prescribed area is sufficiently large.

4.5.5 Toward fully spatial capture-recapture models

While the use of an individual covariate model resolves two important problems inherent in almost all capture-recapture studies (induced heterogeneity and absence of a precise relationship between N and area), it is not ideal for all purposes because it does not make full use of the spatial information in the data set, i.e., the trap locations and the locations of each individual encounter, so that we cannot use this model to model trap-specific effects (e.g., trap effort or type). Moreover, we applied this model to "data" being the average observed encounter location, and equated that summary to the home range center $\mathbf{s}_i$. Intuitively, taking the average encounter location as an estimate of home range center makes sense, but more so when the trapping grid is dense and expansive relative to typical home range sizes, which might not be reasonable in practice. Additionally, this approach also ignored the variable precision with which each $\mathbf{s}_i$ is estimated. Finally, it ignores that estimates of $\mathbf{s}_i$ around the "edge" (however we define that) are biased because the observations are truncated—we can only observe locations interior to the array.

However, there is hope to extend this model in order to resolve these remaining deficiencies. In the next chapter we provide a further extension of this individual covariate model that definitively resolves the *ad hoc* nature of the approach we took here. In that chapter we build a model in which $\mathbf{s}_i$ are regarded as latent variables and the observation locations (i.e., trap-specific encounters) are linked to those latent variables with an explicit model. We note that the model fitted previously could be adapted easily to deal with $\mathbf{s}_i$ as a latent variable, simply by adding a prior distribution for $\mathbf{s}_i$. This is actually easier, and less *ad hoc* in a number of respects, and you should try it out.

4.6 Distance sampling: a primitive SCR model

Distance sampling is a class of methods for estimating animal density from measurements of distance from an observer to individual animals (or groups). The basic assumption is that detection probability is a function of distance. Distance sampling is one of the most popular methods for estimating animal abundance (Burnham et al., 1980; Buckland et al., 2001; Buckland, 2004) because, unlike ordinary closed population models, distance sampling provides explicit estimates of *density*. In terms of methodological context, the distance sampling model is a special case of a closed population model with an individual covariate. The covariate in this case, x, is the distance between an individual's location say $\mathbf{u}$ and the observation location or transect. In fact, distance sampling is precisely an individual-covariate model, except that observations are made at only $K = 1$ sampling occasion. Distance sampling eliminates the need to explicitly identify individuals (except they need to be *distinguished* from other individuals) repeatedly and so distance sampling can be applied to unmarked populations. This first and most basic spatial capture-recapture model has been used routinely for decades and, formally, it is a spatially explicit model in the sense that it describes, explicitly, the spatial organization of individual locations (although this is not always stated explicitly) and, as a result, somewhat general models of how individuals are distributed in space can be specified (Hedley et al., 1999; Royle et al., 2004; Johnson, 2010; Niemi and Fernández, 2010; Sillett et al., 2012).

As with other models we've encountered in this chapter, the distance sampling model, under data augmentation, includes a set of M zero-inflation variables z_i and a binomial observation model expressed conditional on z (binomial for $z_i = 1$, and fixed zeros for $z_i = 0$). In distance sampling we pay for having only a single sample occasion (i.e., $K = 1$) by requiring constraints on the model of detection probability, normally imposed as the assumption that detection probability is 1.0 when distance equals 0. A standard model for detection probability is the "half-normal" model:

$$p_i = \exp(-\alpha_1 x_i^2)$$

for $\alpha_1 > 0$, where x_i denotes the distance at which the ith individual is detected relative to some reference location. This encounter probability model is more often written with $\alpha_1 = 1/2\sigma^2$. If $K > 1$ then an intercept in this model, say α_0, is identifiable and such models are usually called "capture-recapture distance sampling" (Alpízar-Jara and Pollock, 1996; Borchers et al., 1998).

As with previous examples, we require a distribution for the individual covariate x_i. The customary choice for transect sampling is

$$x_i \sim \text{Uniform}(0, B)$$

wherein $B > 0$ is a known constant, being the upper limit of data recording by the observer (i.e., the transect half-width). Specification of this distance sampling model in the **BUGS** language is shown in Panel 4.2, taken from Royle and Dorazio (2008).

As with the individual covariate model in the previous section, the distance sampling model can be equivalently specified by putting a prior distribution on individual *location* instead of distance between individual and observation point (or transect). Thus we can write the general distance sampling model as

$$p_i = h(||\mathbf{u}_i - \mathbf{x}_0||, \alpha_1)$$

along with

$$\mathbf{u}_i \sim \text{Uniform}(\mathcal{S})$$

```
alpha1 ~ dunif(0,10)             # Prior distributions
psi ~ dunif(0,1)

for(i in 1:(nind+nz)){
   z[i] ~ dbern(psi)             # DA variables
   x[i] ~ dunif(0,B)             # B=strip width
   p[i] <- exp(logp[i])          # Detection function
   logp[i] <-    - alpha1*(x[i]*x[i])
   mu[i] <- z[i]*p[i]
   y[i] ~ dbern(mu[i])           # Observation model
 }

N <- sum(z[1:(nind+nz)])         # N is a derived parameter
D <- N/striparea                 # D = N/total area of transects
```

PANEL 4.2

Distance sampling model in **BUGS** for a line transect situation, using a half-normal detection function.

where $\mathbf{x}_0$ is a fixed point (or line) and $\mathbf{u}_i$ is the individual's location, which is observed for the sample of n individuals. In practice, it is easier to record distance instead of location, so the model is seldom described in terms of location. Basic math can be used to argue that if individuals have a uniform distribution in space, then the distribution of Euclidean distance is also uniform. In particular, if a transect of length L is used and x is distance to the transect, then $F(x) = \Pr(X \leq x) = (L \times x)/(L \times B) = x/B$ and $f(x) = dF/dx = (1/B)$. For measurements of radial distance, we provided the analogous argument in the previous section.

The preceding paragraph makes it clear that distance sampling is a special case of spatial capture-recapture models, such as those derived from model M_x of the previous section, where the encounter probability is related directly to *distance*, which is a reduced information summary of *location*, $\mathbf{u}$. Other intermediate forms of SCR/DS models can be described (Royle et al., 2011a). In the context of our general characterization of SCR models (Chapter 2.6), we suggested that every SCR model can be described, conceptually, by a hierarchical model of the form:

$$[y|\mathbf{u}][\mathbf{u}|\mathbf{s}][\mathbf{s}].$$

Distance sampling ignores the part of the model pertaining to $\mathbf{s}$, and deals only with the model components for the observed data $\mathbf{u}$.[6] Thus, we are left with a hierarchical model of the form

$$[y|\mathbf{u}][\mathbf{u}].$$

In contrast, as we will see in the next chapters, many SCR models (Chapter 5) ignore $\mathbf{u}$ and condition on $\mathbf{s}$, which is not observed:

$$[y|\mathbf{s}][\mathbf{s}]$$

Since $[\mathbf{u}]$ and $[\mathbf{s}]$ are both assumed to be uniformly distributed, these are equivalent models! The main differences have to do with interpretation of model components and whether or not the variables are observable (in distance sampling they are).

So why bother with SCR models when distance sampling yields density estimates and accounts for spatial heterogeneity in detection? For one, imagine trying to collect distance sampling data on species such as jaguars or tigers! Clearly, distance sampling requires that one can collect large quantities of distance data, which is not always possible. For tigers, it is much easier, efficient, and safer to employ camera traps and then apply SCR models. Furthermore, as we will see in Chapter 15, SCR models can make use of distance data, allowing us to study distribution, movement, and density. Thus, SCR models are more general and versatile than distance sampling models, and can accommodate data from virtually all animal survey designs.

4.6.1 Example: Sonoran desert tortoise study

We illustrate the application of distance sampling models using data on the Sonoran desert tortoise (*Gopherus agassizii*), shown in Figure 4.6, collected along transects in

[6]Equivalently, we could also say that $[\mathbf{u}]$ in the distance sampling model is $[\mathbf{u}] = \int [\mathbf{u}|\mathbf{s}][\mathbf{s}]ds$.

FIGURE 4.6

Desert tortoise in its native habitat (*Photo credit: Erin Zylstra, Univ. of Arizona*).

southern Arizona (see Zylstra et al. (2010) for details). The data are from 120 square transects having four 250-m sides, although we ignore this detail in our analysis here and regard them as 1 km transects, and we pooled the detection data from all 120 transects. The histogram of encounter distances from the 65 encountered individuals is shown in Figure 4.7.

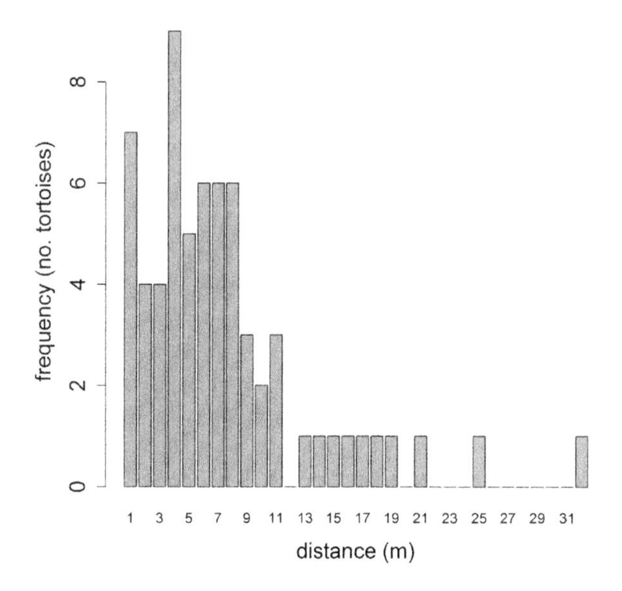

FIGURE 4.7

Distance histogram of $n = 65$ Sonoran desert tortoise detections from a total of 120 km of survey transect.

Table 4.8 Posterior summaries from the Sonoran desert tortoise distance sampling data. Results were obtained using **WinBUGS** running 3 chains, each with 3,000 iterations and the first 1,000 discarded, thinning by 2.

Parameter	Mean	SD	2.4%	50%	97.5%	Rhat	n.eff
α_1	0.01	0.00	0.00	0.01	0.01	1.02	130
σ	9.12	0.77	7.77	9.07	10.77	1.02	130
N	516.67	54.71	415.00	516.00	632.00	1.02	100
D	0.54	0.06	0.43	0.54	0.66	1.02	100
ψ	0.61	0.07	0.49	0.61	0.75	1.02	96

Commands for reading in and organizing the data for analysis using **WinBUGS** are given in the help file `?tortoise` provided with the `scrbook` package. To compute density, the total sampled area of the transects `striparea` is input as data, and computed as: 120 (transects) multiplied by the length (1,000 m) and half-width ($B = 40$ m), then multiplied by 2, and divided by 10,000 to convert to units of individuals/ha. We also provide commands for analyzing the data with `unmarked` (Fiske and Chandler, 2011) using hierarchical distance sampling models (Royle et al., 2004).

Posterior summaries for the tortoise data are given in Table 4.8. Estimated density (posterior mean) is 0.54 individuals/ha and the estimated scale parameter of the distance function (posterior mean) is $\sigma = 9.12$ m. The posterior mass of the data augmentation parameter ψ is located away from the upper bound $\psi = 1$ and so the degree of data augmentation appears sufficient.

4.7 Summary and outlook

Traditional closed population capture-recapture models are closely related to binomial generalized linear models. Indeed, the only real distinction is that in capture-recapture models, the population size parameter N is unknown. This requires special consideration in the analysis of capture-recapture models. The classical approach to inference recognizes that the observations don't have a standard binomial distribution but, rather, a truncated binomial (from which the so-called *conditional likelihood* derives) since we only have encounter frequency data on observed individuals. If instead we analyze capture-recapture models using data augmentation, which arises under a discrete Uniform$(0, M)$ prior for N, the observations can be modeled using a zero-inflated binomial distribution. The analysis of such zero-inflated models is practically convenient, especially using the **BUGS** variants.

Spatial capture-recapture models considered in the rest of the chapters of this book are closely related to individual covariate models (model M_x). Spatial capture-recapture models arise naturally by defining individual covariates based on observed locations of individuals—we can think of using some function of mean encounter location as an individual covariate. We did this in a novel way, by using distance to the centroid of the trapping array as a covariate. We analyzed the *full likelihood* using data augmentation, and placed a prior distribution on the individual covariate which was derived from an assumption that individual locations are, a priori, uniformly distributed in space. This assumption provides for invariance of the density estimator to the choice of area (induced by maximum distance from the centroid of the trap array). The model addressed some important problems in the use of closed population models: it allows for heterogeneity in encounter probability due to the spatial juxtaposition of individuals with the array of traps, and it also provides a direct estimate of density because area is a feature of the model (via the prior on the individual covariate). The model is still not completely general, however, because it does not make full use of the spatial encounter histories, which provide direct information about the locations and density of individuals.

A specific individual covariate model that is in widespread use is classical distance sampling. The model underlying distance sampling is precisely a special kind of SCR model—but one without replicate samples. Understanding distance sampling and individual covariate models more broadly provides a solid basis for understanding and analyzing spatial capture-recapture models. In fact if, instead of placing an explicit model on *distance* in the classical distance sampling model, we were to place the prior distribution on *location*, **s**, of each individual, then the form of the distance sampling model more closely resembles the SCR model we introduce in the next chapter.

PART

Basic SCR Models II

CHAPTER

5 Fully Spatial Capture-Recapture Models

In the previous chapter, we discussed models that could be viewed as primitive spatial capture-recapture models. We looked at a basic distance sampling model, and we also considered classical individual covariate models, in which we defined a covariate to be the distance from the (estimated) home range center to the center of the trap array. The individual covariate model that we conjured up was "spatial" in the sense that it included some characterization of where individuals live but, on the other hand, only a primitive or no characterization of trap location. That said, there is only a small step from that model to spatial capture-recapture models, which we consider in this chapter, and which fully recognize the spatial attribution of both individual animals *and* the locations of encounter devices.

Capture-recapture models must accommodate the spatial organization of individuals and the encounter devices, because the encounter process occurs at the level of individual traps. Failure to consider the trap-specific data is one of the key deficiencies with classical ad hoc approaches that aggregate encounter information to the resolution of the entire trap array. We have previously addressed some problems that this causes, including induced heterogeneity in encounter probability, imprecise notation of "sample area", and not being able to accommodate trap-specific effects or trap-specific missing values. In this chapter, we resolve these issues by developing our first fully spatial capture-recapture model. This model is not too different from that considered in Section 4.5 but, instead of defining the individual covariate to be distance to the centroid of the array, we define J individual covariates—the distance to *each* trap. And, instead of using estimates of individual locations $\mathbf{s}$, we consider a fully hierarchical model in which we regard $\mathbf{s}$ as a latent variable and impose a prior distribution on it.

In this chapter, we investigate the basic spatial capture-recapture model, which we refer to as "model SCR0", and address some important considerations related to its analysis in **BUGS**. We demonstrate how to summarize posterior output for the purposes of producing density maps or spatial predictions of density. The key aspect of the SCR models considered in this chapter is the formulation of a model for encounter probability that is a function of distance between individual home range center and trap locations. We also discuss how encounter probability models are related to explicit models of space usage or "home range area". Understanding this allows us to compute, for example, the area used by an individual during some prescribed time. While it is intuitive that SCR models should be related to some model of space usage, this has not been discussed much in the literature.

Spatial Capture-Recapture. http://dx.doi.org/10.1016/B978-0-12-405939-9.00005-0

5.1 Sampling design and data structure

In our development here, we will assume a standard sampling design in which an array of J traps is operated for K sample occasions (say, nights), producing encounters of n individuals. Because sampling occurs by traps and over time, the most general data structure yields temporally *and* spatially indexed encounter histories for *each individual*. Thus, a typical data set will include an encounter history *matrix* for each individual, indicating which trap the individual was captured in, during each sample occasion. For example, suppose we sample at 4 traps over 3 nights. A plausible data set for a single individual captured one time in trap 1 on the first night and one time in trap 3 on the third night is:

```
        night1  night2  night3
trap1     1       0       0
trap2     0       0       0
trap3     0       0       1
trap4     0       0       0
```

This data structure would be obtained for *each* of the $i = 1, 2, \ldots, n$ captured individuals.

We develop models in this chapter for passive detection devices such as "hair snares" or other DNA sampling methods (e.g., Kéry et al., 2010; Gardner et al., 2010b) and related types of sampling devices in which (i) devices ("traps") may capture any number of individuals (i.e., they don't fill up); (ii) an individual may be captured in more than one trap during each occasion, but (iii) individuals can be encountered at most once by each trap during any occasion. Hair snares for sampling DNA from bears and other species function according to these rules. An individual bear wandering about its territory might come into contact with >1 devices; a device may encounter multiple bears; however, in practice, it will often not be possible to attribute multiple visits of the same individual to the same snare during a single occasion (e.g., night) to distinct encounter events. Thus, an individual may be captured at most 1 time in each trap during any occasion. While this model, which we refer to as SCR0, is most directly relevant to hair snares and other DNA sampling methods for which multiple detections of an individual are not distinguishable, we will also make use of the model for data that arise from camera trapping studies. In practice, with camera trapping, individuals might be photographed several times in a night, but it is common to distill such data into a single binary encounter event for reasons discussed later in Chapter 9.

The statistical assumptions we make to build a model for these data are that individual encounters within and among traps are independent, and this allows us to regard individual- and trap-specific encounters as *independent* Bernoulli trials (see next section).

5.2 The binomial observation model

We begin by considering the simple model in which there are no time-varying or trap-specific covariates that influence encounter, there are no explicit individual-specific

Table 5.1 Hypothetical spatial capture-recapture data set showing 6 individuals captured in 4 traps. Each entry is the number of captures out of $K = 3$ nights of sampling.

Individual	Trap 1	Trap 2	Trap 3	Trap 4
1	1	0	0	0
2	0	2	0	0
3	0	0	0	1
4	0	1	0	0
5	0	0	1	1
6	1	0	1	0

covariates, and there are no covariates that influence density. In this case, we can aggregate the binary encounters over the K sample occasions and record the total number of encounters out of K. We will denote these individual- and trap-specific encounter frequencies by y_{ij} for $i = 1, 2, \ldots, n$ captured individuals and $j = 1, 2, \ldots, J$ traps. For example, suppose we observe 6 individuals in sampling at 4 traps over 3 nights of sampling, then a plausible data set is the 6×4 matrix of encounters (out of 3 sampling occasions) shown in Table 5.1. We assume that y_{ij} are mutually independent outcomes of a binomial random variable, which we express as:

$$y_{ij} \sim \text{Binomial}(K, p_{ij}). \tag{5.2.1}$$

This is the basic model underlying standard closed population models (Chapter 4) except that, in the present case, the encounter frequencies are individual- *and* trap-specific, and encounter probability p_{ij} depends on both individual *and* trap.

As we did in Section 4.5, we will make explicit the notion that p_{ij} is defined conditional on *where* individual i lives. Naturally, we think about defining an individual home range and then relating p_{ij} explicitly to a summary of its location relative to each trap. In what follows, we define $\mathbf{s}_i$, a two-dimensional spatial coordinate, to be the home range or activity center of individual i (Efford, 2004; Borchers and Efford, 2008; Royle and Young, 2008). Then, the SCR model postulates that encounter probability, p_{ij}, is a decreasing function of distance between $\mathbf{s}_i$ and the location of trap j, $\mathbf{x}_j$ (also a two-dimensional spatial coordinate). A standard approach for modeling binomial counts is the logistic regression, where we express the dependence of p_{ij} on distance according to:

$$\text{logit}(p_{ij}) = \alpha_0 + \alpha_1 \|\mathbf{x}_j - \mathbf{s}_i\|, \tag{5.2.2}$$

where $\|\mathbf{x}_j - \mathbf{s}_i\|$ is the distance between $\mathbf{s}_i$ and $\mathbf{x}_j$. We sometimes write $\|\mathbf{x}_j - \mathbf{s}_i\| = \text{dist}(\mathbf{x}_j, \mathbf{s}_i) = d_{ij}$. Alternatively, a popular model is

$$p_{ij} = p_0 \exp\left(-\frac{1}{2\sigma^2}\|\mathbf{x}_j - \mathbf{s}_i\|^2\right), \tag{5.2.3}$$

which is similar to the "half-normal" model in distance sampling, except with an intercept $p_0 \leq 1$, which can be estimated in SCR studies. Because it is the kernel of a bivariate normal, or Gaussian, probability density function for the random variable "individual location", we will refer to it as the "(bivariate) normal" or "Gaussian" model although the distance sampling term "half-normal" is widely used. In the context of two-dimensional space, the model is clearly interpretable as a primitive model of movement outcomes or space usage (we discuss this in Section 5.4).

There are a large number of standard detection models commonly used (see Chapter 7). All other standard models that relate encounter probability to **s** will also have a parameter that multiplies distance in some non-linear function. To be consistent with parameter naming across models, we will sometimes parameterize any encounter probability model so that the coefficient on distance (or distance squared) is α_1. So, for the Gaussian model, $\alpha_1 = 1/(2\sigma^2)$. A characteristic of the common parametric forms is that they are monotone decreasing with distance, but vary in their behavior as they approach distance = 0. We show the standard Gaussian, Gaussian hazard, negative exponential, and logistic models in Figure 5.1. The negative exponential model has $p_{ij} = p_0 \exp(-\alpha_1 \|\mathbf{x}_j - \mathbf{s}_i\|)$ and the Gaussian hazard model has $p_{ij} = 1 - \exp(-\lambda_0 g(\mathbf{x}_j, \mathbf{s}_i))$ where $g(\mathbf{x}_j, \mathbf{s}_i)$ is the Gaussian kernel and $\lambda_0 > 0$ is the baseline encounter rate. Whatever model we choose for encounter probability, we should always keep in mind that the activity center for individual i, $\mathbf{s}_i$, is an unobserved random variable. To be precise about this in the model, we should express the observation model as

$$y_{ij}|\mathbf{s}_i \sim \text{Binomial}(K, p(\mathbf{s}_i; \boldsymbol{\alpha}))$$

where $\boldsymbol{\alpha}$ represents all parameters of the model. Sometimes, for notational simplicity, we abbreviate this by omitting some or all of the arguments to p.

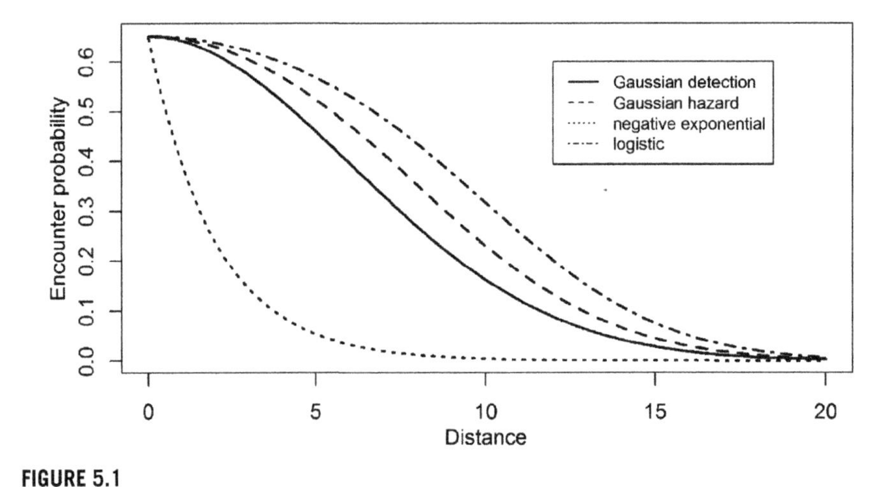

FIGURE 5.1

Some common encounter probability models showing the characteristic monotone decrease of encounter probability with distance between activity center and trap location.

5.2.1 Definition of home range center

We define an individual's home range as *the area used by an organism during some time period* which has a clear meaning for most species regardless of their biology. We therefore define the home range center (or activity center) to be the centroid of the space that individual occupied (or used) during the period in which traps were active. Thinking about it in that way, the activity center could even be observable (almost) as the average of a very large number of radio fixes over the course of a survey period or a season. Thus, this practical version of a home range center in terms of space usage is a well-defined construct regardless of whether one thinks that home range itself is a meaningful concept. We use the terms *home range center* and *activity center* interchangeably, and we recognize that it is a transient quantity that applies only to a well-defined period of study.

5.2.2 Distance as a latent variable

If we knew precisely every $\mathbf{s}_i$ in the population (and population size N), then the model specified by Eqs. (5.2.1) and (5.2.2) would be just an ordinary logistic regression type of a model (with covariate d_{ij}) which we learned how to fit using **WinBUGS** previously (Chapter 3). However, the activity centers are unobservable even in the best possible circumstances. In that case, d_{ij} is an unobserved variable, analogous to the situation in classical random effects models. Therefore we need to extend the model to accommodate these random variables with an additional model component—the random effects distribution. The customary assumption is the so-called "uniformity assumption," which states that the $\mathbf{s}_i$ are uniformly distributed over space (the obvious next question: "which space?" is addressed below). This uniformity assumption amounts to a uniform prior distribution on $\mathbf{s}_i$, i.e., the pdf of $\mathbf{s}_i$ is constant, which we may express

$$\Pr(\mathbf{s}_i) \propto \text{constant}. \tag{5.2.4}$$

As it turns out, this assumption is usually not precise enough to fit SCR models in practice for reasons we discuss shortly. We will give another way to represent this prior distribution that is more concrete, but depends on specifying the "state-space" of the random variable $\mathbf{s}_i$. The term state-space is a technical way of saying "the space of all possible outcomes" of the random variable.

5.3 The binomial point process model

In the SCR model, the individual activity centers are unobserved and thus we treat them as random effects. Specifically, the collection of individual activity centers $\mathbf{s}_1, \ldots, \mathbf{s}_N$ represents a realization of a *binomial point process* (Illian et al., 2008, p. 61). The binomial point process (BPP) is analogous to a Poisson point process in the sense that it represents a "random scatter" of points in space—except that the total number of points is *fixed*, whereas, in a Poisson point process, it is random (having a Poisson distribution). As an example, we show in Figure 5.2 locations of 20 individual

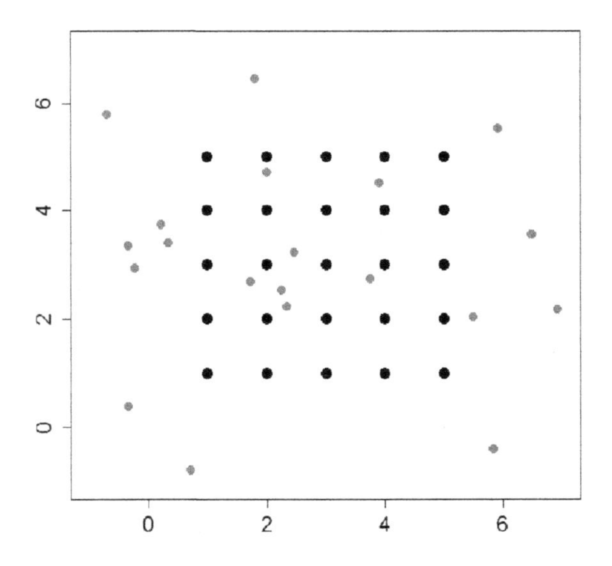

FIGURE 5.2

Realization of a binomial point process with $N = 20$ (small dots). The large dots represent trap locations.

activity centers (black dots) in relation to a grid of 25 traps. For a Poisson point process the number of such points in the prescribed state-space is random whereas we will often be interested in simulating a fixed numbers of points, e.g., for evaluating the performance of procedures, e.g., how well does our estimator perform when $N = 50$?

It is natural to consider a binomial point process in the context of capture-recapture models because it preserves N in the model and thus preserves the direct linkage with closed population models. In fact, under the binomial point process model, model M_0 and other closed models are simple limiting cases of SCR models, i.e., they arise as the coefficient on distance (α_1 above) tends to 0.

One consequence of having a fixed N in the BPP model is that the model is not strictly a model of "complete spatial randomness." This is because, if one forms counts $n(A_1), \ldots, n(A_k)$ in any set of disjoint regions of the state-space, say $A_1, \ldots, A_k$, then these counts are *not* independent. In fact, they have a multinomial distribution (see Illian et al., 2008, p. 61). Thus, the BPP model introduces a slight bit of dependence in the distribution of points. However, in most situations this will have no practical effect on any inference or analysis and, as a practical matter, we will usually regard the BPP model as one of spatial independence among individual activity centers, because each activity center is distributed independently of each other activity center. Despite this independence, we see in Figure 5.2 that *realizations* of randomly distributed points will typically exhibit distinct non-uniformity. Thus, independent, uniformly distributed points will almost never appear regularly, uniformly, or systematically distributed. For this reason, the basic binomial (or Poisson) point process models are enormously useful in practical settings, since they allow for a range of distribution patterns without violating the assumption of spatial randomness. This is because, for SCR models, we actually have a little bit of data on a sample of individuals and thus the resulting posterior point pattern can deviate strongly from uniformity, a point we come back to repeatedly in this book. The uniformity hypothesis is only a *prior* distribution which is directly affected by the quantity and quality of the observed data and may appear distinctly non-uniform. In addition, we can build more flexible models for the point process, a topic we take up in Chapter 11.

5.3.1 The state-space of the point process

Shortly we will focus on Bayesian analysis of model SCR0 with N known so that we can gain some basic experience with important elements of the model and its analysis. To do this, we note that the individual activity centers $\mathbf{s}_i, \ldots, \mathbf{s}_N$ are unknown quantities, and we will need to be able to simulate each $\mathbf{s}_i$ in the population from the posterior distribution. In order to simulate the $\mathbf{s}_i$, it is necessary to describe precisely the region over which they are distributed. This is the quantity we referred to above as the state-space, which is sometimes called the *observation window* in the point process literature. We denote the state-space henceforth (throughout this book) by $\mathcal{S}$ representing a region or a set of points comprising the potential values (the support) of the random variable $\mathbf{s}$. Thus, an equivalent explicit statement of the "uniformity assumption" is

$$\mathbf{s}_i \sim \text{Uniform}(\mathcal{S}),$$

where $\mathcal{S}$ is a precisely defined region, e.g., in Figure 5.2, $\mathcal{S}$ is the square defined by $[-1, 7] \times [-1, 7]$. Thus, each of the $N = 20$ points was generated by randomly selecting each coordinate on the line $[-1, 7]$. When points are distributed uniformly over some region, the point process is usually called a *homogeneous point process*.

5.3.1.1 Prescribing the state-space

Evidently, to define the model, we need to define the state-space, $\mathcal{S}$. How can we possibly do this objectively? Prescribing any particular $\mathcal{S}$ seems like the equivalent of specifying a "buffer" which we have criticized as being ad hoc. How is it, then, that the choice of a state-space is *not* ad hoc? As we observed in Section 4.5, it is true that N increases with $\mathcal{S}$, but only at the same rate as the area of $\mathcal{S}$ increases under the prior assumption of constant density. As a result, we say that density is invariant to $\mathcal{S}$ *as long as $\mathcal{S}$ is sufficiently large* to include all animals with non-negligible probability of encounter (see next sub-section). Thus, while choice of $\mathcal{S}$ is (or can be) essentially arbitrary, once $\mathcal{S}$ is chosen, it defines the population being exposed to sampling, which scales appropriately with the size of the state-space.

For our simulated system developed previously in this chapter, we defined the state-space to be a square within which our trap array was centered. For many practical situations this might be an acceptable approach to defining the state-space, i.e., just a rectangle around the trap array. Although defining the state-space to be a regular polygon has computational advantages (e.g., we can implement this more efficiently in **BUGS** and cannot for irregular polygons), a regular polygon induces an apparent problem of admitting into the state-space regions that are distinctly non-habitat (e.g., oceans, large lakes, ice fields, etc.). It is difficult to describe complex regions in mathematical terms that can be used in **BUGS**. As an alternative, we can provide a representation of the state-space as a discrete set of points, which the **R** package `secr` (Efford, 2011a) permits (`secr` uses the term "mask" for what we call the state-space). Defining the state-space by a discrete set of points is handy because it allows specific points to be deleted or not, depending on whether they represent available or suitable habitat (see Section 5.10). We can also define the state-space as an arbitrary collection of polygons stored as a GIS shapefile which can be analyzed easily by MCMC in **R** (see Section 17.7), but not so easily in the **BUGS** engines. In Section 5.10, we provide an analysis of wolverine camera trapping data, in which we define the state-space to be a regular continuous polygon (a rectangle).

5.3.1.2 *Invariance to the state-space*

We will assert for all models we consider in this book that density is invariant to the size and extent of $\mathcal{S}$, if $\mathcal{S}$ is sufficiently large, and as long as our model relating p_{ij} to $\mathbf{s}_i$ is a decreasing function of distance. We can demonstrate this easily by drawing an analogy with a 1-d case involving distance sampling on a transect. Let y_j be the number of individuals captured in some interval $[d_{j-1}, d_j)$, and define the maximum observation distance $d_J = B$ for some large value of B. The observations from a survey are $y_1, \ldots, y_J$, and the likelihood is a multinomial likelihood, so the log-likelihood is of the form

$$\log L(y_1, \ldots, y_J) = \sum_{j=1}^{J} y_j \log(\pi_j),$$

where π_j is the probability of detecting an individual in distance class j, which depends on parameters of the detection function (the manner of which is not relevant to the present discussion). Choosing B sufficiently large guarantees that $\mathbb{E}(y_J) = 0$ and therefore the observed frequency in the "last cell" contributes nothing to the likelihood, in regular situations in which the detection function decays monotonically with distance and prior density is constant. We can think of B as being related to the state-space in an SCR model, as the width of a rectangular state-space with area $B \times L$, L being the length of the transect. Thus, if we choose B large enough, then we ensure that the expected trap frequencies beyond B will be 0, and contribute nothing to the likelihood.

Sometimes our estimate of density can be affected by choosing an $\mathcal{S}$ that is small relative to animal movement or home range size. However, this might be reasonable if $\mathcal{S}$ is naturally well defined. As we discussed in Chapter 1, *$\mathcal{S}$ is part of the model*, and thus it is sensible that estimates of density might be sensitive to its definition in problems where it is natural to restrict $\mathcal{S}$. One could imagine, however, in specific cases (e.g., a small population with well-defined habitat preferences), that a problem could arise because changing the state-space based on differing opinions and GIS layers might have substantial affects on the density estimate. But this is a real biological problem, and a natural consequence of the spatial formalization of capture-recapture models—a feature, not a bug or some statistical artifact—and it should be resolved with better information, research, and thinking. For situations where there is not a natural choice of $\mathcal{S}$, we should default to choosing $\mathcal{S}$ to be very large relative to typical home range size of the species being studied in order to achieve invariance or, otherwise, evaluate sensitivity of density estimates by trying a couple of different choices of $\mathcal{S}$. This is a standard "sensitivity to prior" argument that Bayesians always have to be conscious of. We demonstrate this in our analysis of Section 5.9 below. As an additional practical consideration, we note that the area of the state-space $\mathcal{S}$ affects data augmentation. If you increase the size of $\mathcal{S}$, then there are more individuals to account for and therefore the size of the augmented data set M must increase, which has computational implications.

5.3.2 Connection to model M_h and distance sampling

SCR models are closely related to "model M_h" and distance sampling. In SCR models, heterogeneity in encounter probability is induced by both the effect of distance in the model for detection probability and from specification of the state-space. To understand this, suppose activity centers have the uniform distribution:

$$\mathbf{s} \sim \text{Uniform}(\mathcal{S})$$

and encounter probability is a function of $\mathbf{s}$, denoted by $p(\mathbf{s}) = p(y = 1|\mathbf{s})$. For example, under Eq. (5.2.2) we have that

$$p(\mathbf{s}) = \text{logit}^{-1}(\alpha_0 - \alpha_1 \|\mathbf{x}_j - \mathbf{s}_i\|)$$

and we can work out, either analytically or empirically, what is the implied distribution of p for a population of individuals. Figure 5.3 shows a histogram of p for a

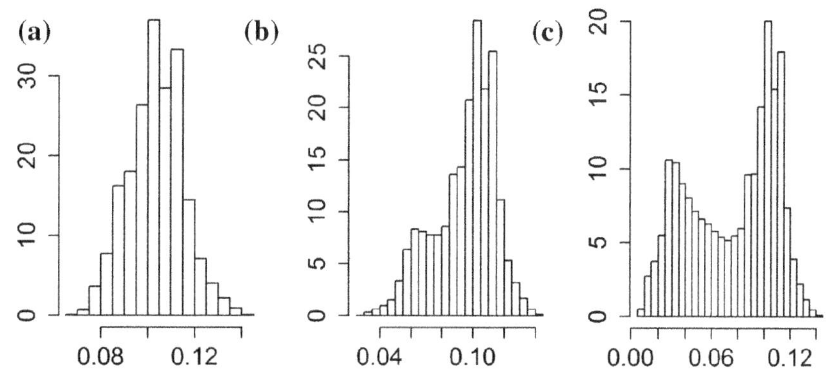

FIGURE 5.3

Implied distribution of p_i for a population of individuals as a function of the size of the state-space buffer around the trap array. The state-space buffer is 0.2, 0.5, and 1.0 for panels (a), (b), (c), respectively. In each case, the trap array is fixed and centered within a square state-space.

hypothetical population of 100,000 individuals on a state-space enclosing our 5×5 trap array above, under the logistic model for distance given by Eq. (5.2.2), with buffers of 0.2, 0.5, and 1.0. We see the mass shifts to the left as the buffer increases, implying more individuals with lower encounter probabilities, as their home range centers increase in distance from the trap array.

Another way to understand this is by representing $\mathcal{S}$ as a set of discrete points on a grid. In the coarsest possible case where $\mathcal{S}$ is a single arbitrary point, then every individual has exactly the same p. As we increase the number of points in $\mathcal{S}$, more distinct values of p are possible. Indeed, when $\mathcal{S}$ is characterized by discrete points, then SCR models are precisely a type of finite-mixture model (Norris and Pollock, 1996; Pledger, 2004), except that, in the case of SCR models, we have some information about which group an individual belongs to (i.e., where their activity centers are), as a result of which traps it is captured in.

It is also worth reemphasizing that the basic SCR encounter model is a binomial encounter model in which distance is a covariate. As such, it is similar to classical distance sampling models (Buckland et al., 2001). Both have distance as a covariate but, in classical distance sampling problems, the focus is on the distance between the observer and the animal at an instant in time, not the distance between a trap and an animal's home range center. As a practical matter, in distance sampling, "distance" is *observed* for those individuals that appear in the sample. Conversely, in SCR problems, it is only imperfectly observed (we have partial information in the form of trap observations). Clearly, it is preferable to observe distance if possible, but distance sampling requires field methods that are not practical in many situations, e.g., when studying carnivores such as bears or large cats. Furthermore, SCR models allow us to relax many of the assumptions made in classical distance sampling, such as perfect detection at distance zero, and SCR models allow for estimates of quantities other than density, such as home range size, and space usage (Chapters 12 and 13).

5.4 The implied model of space usage

We developed the basic SCR model in terms of a latent variable, $\mathbf{s}$, the home range center or activity center. Surely, the encounter probability model, which relates encounter of individuals in specific traps to $\mathbf{s}$, must somehow imply a certain model for home range geometry and size. Here, we explore the nature of that relationship, and we argue that any given detection model implies a model of space usage—i.e., the amount and extent of area used some prescribed percentage of the time. So we might say, for example, 95% of animal movements are within some distance from an individual's activity center.

Indeed, it is natural to interpret the detection model as the composite of two processes: movement of an individual about its home range, i.e., how it uses space within its home range ("space usage"), and detection *conditional on use* in the vicinity of a trapping device. This implies the decomposition of encounter probability according to:

$$\Pr(\text{encounter at } \mathbf{x}|\mathbf{s}) = \Pr(\text{encounter}|\text{usage of } \mathbf{x}, \mathbf{s}) \Pr(\text{usage of } \mathbf{x}|\mathbf{s}).$$

As before, $\mathbf{s}$ is the activity center and, here, $\mathbf{x}$ is any potential trap location on the landscape. We will think of $\mathbf{x}$ as a pixel of some small size. In practice, it might make sense to think about the first component, i.e., Pr(encounter|usage of $\mathbf{x}$, $\mathbf{s}$) as being a constant (e.g., if $\mathbf{x}$ is the area in close proximity to a trap). In that case, the encounter probability model is directly proportional to this model for individual movements about their home range center which determines the use frequency of each location $\mathbf{x}$. This is a sensible heuristic model for what ecologists would call a central place forager, although, as we have stated previously, it may be meaningful as a description of transient space usage as well (that is, the space usage during the period of sampling).

To motivate a specific model for space usage, imagine the area we are interested in consists of some large number of small pixels (i.e., we're thinking of a discrete representation of space), and that we have some kind of perfect observation device (e.g., continuous telemetry) so that we observe every time an individual moves into a pixel. After a long period of time, we observe an enormous sample size of values of $\mathbf{x}$. We tally those up into each pixel, producing the frequency $m(\mathbf{x}, \mathbf{s})$, which is the "true" usage of pixel $\mathbf{x}$ by individual with activity center $\mathbf{s}$. So, then, the usage model should be regarded as a probability mass function for these counts and, naturally, we regard the counts $m(\mathbf{x}, \mathbf{s})$ as a multinomial observation with probabilities $\pi(\mathbf{x}|\mathbf{s})$, and prescribe a suitable model for $\pi(\mathbf{x}|\mathbf{s})$ that describes how use events should accumulate in space. A natural null model for $\pi(\mathbf{x}|\mathbf{s})$ has a decreasing probability of use as $\mathbf{x}$ gets far away from $\mathbf{s}$, i.e., animals spend more time close to their activity centers than far away. We can regard points used by the individual with activity center $\mathbf{s}$ as the realization of a point process with conditional intensity:

$$\pi(\mathbf{x}|\mathbf{s}) = \frac{g(\mathbf{x}, \mathbf{s})}{\sum_x g(\mathbf{x}, \mathbf{s})}, \tag{5.4.1}$$

where $g(\mathbf{x}, \mathbf{s})$ is any positive function. In continuous space, the equivalent representation is:

$$\pi(\mathbf{x}|\mathbf{s}) = \frac{g(\mathbf{x}, \mathbf{s})}{\int g(\mathbf{x}, \mathbf{s})dx}.$$

If we use a negative exponential function, then this produces a standard resource selection function (RSF) model (e.g., Manly et al., 2002, Chapter 8). But, we often use a Gaussian kernel, i.e.,

$$g(\mathbf{x}, \mathbf{s}) = \exp(-d(\mathbf{x}, \mathbf{s})^2/(2\sigma^2))$$

so that contours of the probability of space usage resemble a bivariate normal or Gaussian probability distribution function.

To apply this model of space usage to SCR problems we allow for imperfect detection by introducing a "thinning" mechanism applied to the true counts $m(\mathbf{x}, \mathbf{s})$. This yields, precisely, our Gaussian encounter probability model where the thinning rate is our baseline encounter probability p_0 for each pixel where we place a trap, and $p = 0$ in each pixel where we don't place a trap.

The main take-away point here is that underlying most SCR models is some kind of model of space usage, implied by the specific choice of $g(\mathbf{x}, \mathbf{s})$. Whether or not we have perfect sampling devices, the function we use in the encounter probability model equates to some conditional distribution of points, a utilization distribution, as in Eq. (5.4.1), from which we can compute effective home range area, i.e., the area that contains some percent of the mass of a probability distribution proportional to $g(\mathbf{x}, \mathbf{s})$; e.g., 95% of all space used by an individual with activity center $\mathbf{s}$.

5.4.1 Bivariate normal case

One encounter model that allows direct analytic computation of home range area is the Gaussian encounter probability model

$$p(\mathbf{x}, \mathbf{s}) = p_0 \exp\left(-\frac{1}{2\sigma^2}||\mathbf{x} - \mathbf{s}||^2\right).$$

For this model, encounter probability is proportional to the kernel of a bivariate normal (Gaussian) pdf and so the natural interpretation is that in which movement outcomes (or successive locations of an individual) are draws from a bivariate normal distribution with standard deviation σ. Therefore, this model implies a bivariate normal model of space usage. Under this model we can compute precisely the effective home range area. In particular, if use outcomes are bivariate normal, then $||\mathbf{x} - \mathbf{s}||^2$ has a chi-square distribution with 2 df. The quantity $B(\alpha)$ that encloses $(1 - \alpha)\%$ of all realized distances, i.e., $\Pr(d \leq B(\alpha)) = 1 - \alpha$, is $B(\alpha) = \sigma\sqrt{q(\alpha, 2)}$ where $q(\alpha, 2)$ is the α chi-square critical value on 2 df. For example, to compute $q(.05, 2)$ in **R** we execute the command `qchisq(.95,2)`, which is $q(2, \alpha) = 5.99$. Then, for $\sigma = 1$, $B(\alpha) = 1 \times \sqrt{5.99} = 2.447$. Therefore, 95% of the points used will be within 2.447 (standard deviation) units of the home range center. So, in practice, we can estimate σ by fitting the bivariate normal encounter probability model to some SCR data, and then use the estimated σ to compute the "95% radius," say $r_{.95} = \sigma\sqrt{5.99}$, and convert this to the *95% use area*—the area around $\mathbf{s}$ which contains 95% of the movement outcomes—according to $A_{.95} = \pi r_{.95}^2$.

An alternative bivariate normal model is the bivariate normal hazard rate model:

$$p(\mathbf{x}, \mathbf{s}) = 1 - \exp\left(-\lambda_0 \exp\left(-\frac{1}{2\sigma^2}||\mathbf{x} - \mathbf{s}||^2\right)\right). \qquad (5.4.2)$$

We use λ_0 here because this parameter, the baseline encounter *rate*, can be >1. This arises by assuming the latent "use frequency" $m(\mathbf{x}, \mathbf{s})$ is a Poisson random variable with intensity $\lambda_0 g(\mathbf{x}, \mathbf{s})$. The model is distinct from our Gaussian encounter model $p(\mathbf{x}, \mathbf{s}) = p_0 g(\mathbf{x}, \mathbf{s})$ used previously, although we find that they produce similar results in terms of estimates of density or 95% use area, as long as baseline encounter probability is low. We discuss these two formulations of the bivariate normal model further in Chapter 9.

5.4.2 Calculating space usage

For any encounter model we can compute space usage quantiles empirically by taking a fine grid of points and either simulating movement outcomes with probabilities proportional to $p(\mathbf{x}, \mathbf{s})$ and accumulating area around $\mathbf{s}$, or else we can do this precisely by varying $B(\alpha)$ to find that value within which 95% of all movements are concentrated, i.e., the set of all $\mathbf{x}$ such that $\|\mathbf{x} - \mathbf{s}\| \leq B(\alpha)$. Under any detection model, movement outcomes will occur in proportion to $p(\mathbf{x}, \mathbf{s})$, as long as the probability of encounter is constant, *conditional on use*, and so we can define our space usage distribution according to:

$$\pi(\mathbf{x}|\mathbf{s}) = \frac{p(\mathbf{x}, \mathbf{s})}{\sum_x p(\mathbf{x}, \mathbf{s})}.$$

Given the probabilities $\pi(\mathbf{x}, \mathbf{s})$ for all $\mathbf{x}$, we can find the value of $B(\alpha)$, for any α, such that

$$\left(\sum_{\mathbf{x}:\|\mathbf{x}-\mathbf{s}\| \leq B(\alpha)} \pi(\mathbf{x}, \mathbf{s}) \right) \leq 1 - \alpha$$

(here, we use : to mean "such that"). We have a function called `hra` in the `scrbook` package that computes the home range area for any encounter model and prescribed parameter values. The help file for `hra` has an example of simulating some data. The following commands illustrate this calculation for the bivariate normal model of space usage:

```
##
## Define encounter probability model as R function
##
> pGauss2 <- function(parms,Dmat){
    a0 <- parms[1]
    sigma <- parms[2]
    lp <- parms[1] -(1/(2*parms[2]*parms[2]))*Dmat*Dmat
    p <- 1-exp(-exp(lp))
    p
}

> pGauss1 <- function(parms,Dmat){
    a0 <- parms[1]
    sigma <- parms[2]
    p <- plogis(parms[1])*exp(-(1/(2*parms[2]*parms[2]))*Dmat*Dmat)
    p
}

##
## Execute hra with sigma = .3993
##
> hra(pGauss1,parms=c(-2,.3993),plot=FALSE,xlim=c(0,6),ylim=c(0,6),
      ng=500,tol=.0005)

[1] 0.9784019
radius to achieve 95% of area: 0.9784019
home range area: 3.007353
[1] 3.007353
```

```
## Analytic solution:
##     true sigma that produces area of 3
> sqrt(3/pi)/sqrt(5.99)
[1] 0.3992751
```

What this means is that $B(\alpha) = 0.978$ is the radius that encloses about 95% of all movements under the standard bivariate normal encounter model. Therefore, the area is about $\pi\ 0.978^2 = 3.007$ spatial units. You can change the intercept of the model and find that it has no effect. The true (analytic) value of σ that produces a home range area of 3.0 is 0.3993 which is the value we input to the `hra` function. We can improve on the numerical approximation to home range area (get it closer to 3.0) by increasing the resolution of our spatial grid (increase the `ng` argument) along with the `tol` argument which determines how close to the target value is close enough.

We can also reverse this process, and find, for any detection model, the parameter values that produce a certain $(1 - \alpha)\%$ home range area, which we imagine would be useful for doing simulation studies. The function `hra` will compute the value of the scale parameter that achieves a certain target $(1 - \alpha)\%$ home range area, by simply providing a non-null value of the variable `target.area`. Here we use `target.area = 3.00735` (from above) to obtain a close approximation to the value σ we started with (the `parms` argument is meaningless here):

```
> hra(pGauss1,parms=c(-2,.3993),plot=FALSE,xlim,ylim,ng=500,
     target.area=3.00735,tol=.0005)

Value of parm[2] to achieve 95% home range area of 3.00735: 0.3993674
```

5.4.3 Relevance of understanding space usage

One important reason that we need to be able to deduce "home range area" from a detection model is so that we can compare different models with respect to a common biological currency. Many encounter probability models have some scale parameter, which we might call σ no matter the model, but this relates to 95% area in a different manner under each model. Therefore, we want to be able to convert different models to the same currency. Another reason to understand the relationship between models of encounter probability and space usage is that it opens the door to combining traditional resource selection data from telemetry with spatial capture-recapture data. In Chapter 13 we consider this problem, for the case in which a sample of individuals produces encounter history data suitable for SCR models and, in addition, we have telemetry locations on a sample of individuals. This is achieved by regarding the two sources of data as resulting from the same underlying process of space usage, but telemetry data produce "perfect" observations, like always-on camera traps blanketing a landscape.

5.4.4 Contamination due to behavioral response

Interpretation of encounter probability models as models of animal home range and space usage can be complicated by a number of factors, including whether traps are baited or not. In the case of baited traps, this might lead to a behavioral response (Section 7.2.3), which could affect animal space usage. For example, if traps attract animals from a long distance, it could make typical home ranges appear larger than normal. More likely, in our view, it wouldn't change the typical size of a range but would change how individuals use their range, e.g., by moving from baited trap to baited trap, so that observed movement distances of individuals are typically larger than normal.

In other cases, the reliance on Euclidean distance in models for encounter probability might be unrealistic and can lead to biased estimates of density (Royle et al., 2013a). For example, animals might concentrate their movements along trails, roads, or other landscape features. In this case, models that accommodate other distance metrics can be considered. We present models based on least-cost path in Chapter 12.

5.5 Simulating SCR data

It is always useful to simulate data, because it allows you to understand the system that you're modeling and calibrate your understanding with specific values of the model parameters. That is, you can simulate data using different parameter values until you obtain data that "look right" based on your knowledge of the specific situation that you're interested in. Here we provide a simple script to illustrate how to simulate spatial encounter history data. In this exercise, we simulate data for 100 individuals and a 25-trap array laid out in a 5×5 grid of unit spacing. The specific encounter model is the Gaussian model given above and we used the code below to simulate data used in subsequent analyses. The 100 activity centers were simulated on a state-space defined by an 8×8 square within which the trap array was centered (thus, the trap array is buffered by 2 units). Therefore, the density of individuals in this system is fixed at $100/64$:

```
> set.seed(2013)
# Create 5 x 5 grid of trap locations with unit spacing
> traplocs <- cbind(sort(rep(1:5,5)),rep(1:5,5))
> ntraps <- nrow(traplocs)
# Compute distance matrix:
> Dmat <- e2dist(traplocs,traplocs)

# Define state-space of point process. (i.e., where animals live).
# "buffer" just adds a fixed buffer to the outer extent of the traps.
#
> buffer <- 2
> xlim <- c(min(traplocs[,1] - buffer),max(traplocs[,1] + buffer))
> ylim <- c(min(traplocs[,2] - buffer),max(traplocs[,2] + buffer))

> N <- 100      # population size
> K <-  20      # number nights of effort

> sx <- runif(N,xlim[1],xlim[2])     # simulate activity centers
> sy <- runif(N,ylim[1],ylim[2])
> S <- cbind(sx,sy)
```

```
# Compute distance matrix:
> D <- e2dist(S,traplocs)      # distance of each individual from each trap

> alpha0 <- -2.5         # define parameters of encounter probability
> sigma <- 0.5           # scale parameter of half-normal
> alpha1 <- 1/(2*sigma*sigma) # convert to coefficient on distance

# Compute Probability of encounter:
#
> probcap <- plogis(-2.5)*exp(- alpha1*D*D)

# Generate the encounters of every individual in every trap
> Y <- matrix(NA,nrow=N,ncol=ntraps)
> for(i in 1:nrow(Y)){
    Y[i,] <- rbinom(ntraps,K,probcap[i,])
  }
```

The data matrix produced above has N rows (individuals) and J columns (traps), with each element being the frequency of encounter (out of K) of individuals in traps. We remind the reader that, in presenting **R** or other code snippets throughout the book, we will deviate from our standard variable expressions for some quantities. In particular, we sometimes substitute words for integer variable designations: `nind` (for n), `ntraps` (for J), and `nocc` (for K). In our opinion this leaves less to be inferred by the reader in trying to understand code snippets.

Subsequently, we will generate data using this code packaged in an **R** function called `simSCR0` in the package `scrbook`, which takes a number of arguments including `discard0`, which, if `TRUE`, will return only the encounter histories for captured individuals. A second argument is `array3d`, which, if `TRUE`, returns the three-dimensional encounter history array instead of the aggregated individual and trap-specific encounter frequencies (see below). Finally, we provide a random number seed, `rnd = 2013`, to ensure repeatability of the analysis. We obtain a data set as above using the following command:

```
> data <- simSCR0(discard0=TRUE, array3d=FALSE, rnd=2013)
```

The **R** object `data` is a list, so let's take a look at what's in the list and then harvest some of its elements for further analysis below:

```
> names(data)
[1] "Y"        "traplocs" "xlim"     "ylim"     "N"      "alpha0"     "beta"
[8] "sigma"    "K"

## Grab encounter histories from simulated data list
> Y <- data$Y
## Grab the trap locations
> traplocs <- data$traplocs
```

5.5.1 Formatting and manipulating data sets

Conventional capture-recapture data are easily stored and manipulated as a two-dimensional array, an `nind` × `K` (individuals by sample occasions) matrix, which is maximally informative for any conventional capture-recapture model, but not for spatial capture-recapture models. For SCR models we must preserve the spatial information (trap locations of capture) in the encounter history information. We will routinely analyze data in three standard formats:

(1) The basic two-dimensional data format, which is an `n` × `J` encounter frequency matrix such as that simulated previously. These are the total number of encounters in each trap, summed over the K sample occasions.
(2) The maximally informative three-dimensional array, for which we establish here the convention that it has dimensions `n` × `J` × `K`.
(3) We use a compact format—the "encounter data file" which we describe in Section 5.9.

To simulate data in the most informative format—the "3-d array"—we can use the **R** commands given previously but replace the last four lines with the following:

```
> Y <- array(NA,dim=c(N,ntraps,K))

> for(i in 1:nrow(Y)){
     for(j in 1:ntraps){
        Y[i,j,1:K] <- rbinom(K,1,probcap[i,j])
     }
  }
```

We see that a collection of K binary encounter events are generated for *each* individual and for *each* trap. The probabilities of those Bernoulli trials are computed based on the distance from each individual's home range center and the trap (see calculation above), and those are housed in the matrix `probcap`. Our data simulator function `simSRC0` will return the full 3-d array if `array3d=TRUE` is specified in the function call. To recover the 2-d matrix from the 3-d array, and subset the 3-d array to individuals that were captured, we do this:

```
# Sum over the "sample occasions" dimension (3rd margin of the array)
> Y2d <- apply(Y,c(1,2),sum)

# Compute how many times each individual was captured
> ncaps <- apply(Y2d,1,sum)

# Keep those individuals that were captured
> Y <- Y[ncaps>0, ,]
```

5.6 Fitting model SCR0 in BUGS

If we somehow knew the value of N then we could fit this model directly because, in that case, it is a special kind of logistic regression model, one with a random effect (**s**) that enters into the model in a peculiar fashion. Our aim here is to analyze the known-N problem, using our simulated data, as an incremental step in our progress toward fitting more generally useful models. To begin, we use our simulator to grab a data set and then harvest the elements of the resulting object for further analysis:

```
> data <- simSCR0(discard0=FALSE,rnd=2013)
> y <- data$Y
> traplocs <- data$traplocs

# In this case nind=N because we're doing the known-N problem
#
> nind <- nrow(y)
> X <- data$traplocs
> J <- nrow(X)      # number of traps
> K <- data$K
> xlim <- data$xlim
> ylim <- data$ylim
```

Note that we specify `discard0 = FALSE` so that we have a "complete" data set, i.e., one with the all-zero encounter histories corresponding to uncaptured individuals. Now, within an **R** session, we can create the **BUGS** model file and fit the model using the following commands:

```
cat("
model{
alpha0 ~ dnorm(0,.1)
logit(p0) <- alpha0
alpha1 ~ dnorm(0,.1)
sigma <- sqrt(1/(2*alpha1))
for(i in 1:N){    # note N here -- N is KNOWN in this example
  s[i,1] ~ dunif(xlim[1],xlim[2])
  s[i,2] ~ dunif(ylim[1],ylim[2])
  for(j in 1:J){
   d[i,j] <- pow(pow(s[i,1]-X[j,1],2) + pow(s[i,2]-X[j,2],2),0.5)
   y[i,j] ~ dbin(p[i,j],K)
   p[i,j] <- p0*exp(- alpha1*d[i,j]*d[i,j])
   }
  }
}
",file = "SCR0a.txt")
```

This model describes the Gaussian encounter probability model, but it would be trivial to modify that to various others including the logistic described above. We have to constrain the encounter probability to be in [0, 1], which we do here by defining `alpha0` to be the logit of the intercept parameter `p0`. Note that the distance covariate is computed within the **BUGS** model specification given the matrix of trap locations, `X`, which is provided to **WinBUGS** as data.

Next, we do a number of organizational activities including bundling the data for **WinBUGS**, defining some initial values, the parameters to monitor, and some basic MCMC settings. We choose initial values for the activity centers **s** by generating uniform random numbers in the state-space but, for the observed individuals, we replace those values by each individual's mean trap coordinate of its encounters:

```
### Starting values for activity centers, s
> sst <- cbind(runif(nind,xlim[1],xlim[2]),runif(nind,ylim[1],ylim[2]))
> for(i in 1:nind){
       if(sum(y[i,])==0) next
       sst[i,1] <- mean( X[y[i,]>0,1] )
       sst[i,2] <- mean( X[y[i,]>0,2] )
  }

> data <- list (y=y, X=X, K=K, N=nind, J=J, xlim=xlim, ylim=ylim)
> inits <- function(){
       list (alpha0=rnorm(1,-4,.4), alpha1=runif(1,1,2), s=sst)
  }

> library(R2WinBUGS)
> parameters <- c("alpha0","alpha1","sigma")
> out <- bugs (data, inits, parameters, "SCR0a.txt", n.thin=1, n.chains=3,
              n.burnin=1000,n.iter=2000,debug=TRUE,working.dir=getwd())
```

There is little to say about the preceding operations other than to suggest that you might explore the output and investigate additional analyses by running the `simSCR0` script provided in the **R** package `scrbook`.

For purposes here, we ran 1,000 burn-in and 1,000 post-burn-in iterations, and 3 chains, to obtain 3,000 posterior samples. Because we know N for this particular data set we only have two parameters of the detection model to summarize (`alpha0` and `alpha1`), along with the derived parameter σ, the scale parameter of the Gaussian kernel, i.e., $\sigma = \sqrt{1/(2\alpha_1)}$. When the object `out` is produced we print a summary of the results as follows:

```
> print(out,digits=2)
 Inference for Bugs model at "SCR0a.txt", fit using WinBUGS,
  3 chains, each with 2000 iterations (first 1000 discarded)
  n.sims = 3000 iterations saved
             mean     sd    2.5%     25%     50%     75%   97.5% Rhat  n.eff
alpha0      -2.50   0.22   -2.95   -2.65   -2.48   -2.34   -2.09 1.01    190
alpha1       2.44   0.42    1.64    2.15    2.44    2.72    3.30 1.00    530
sigma        0.46   0.04    0.39    0.43    0.45    0.48    0.55 1.00    530
deviance   292.80  21.16  255.60  277.50  291.90  306.00  339.30 1.01    380

 [...some output deleted...]
```

The data were generated with `alpha0` $= -2.5$ and `alpha1` $= 2$. The estimates look reasonably close to those data-generating values and we probably feel pretty good about the performance of the Bayesian analysis and MCMC algorithm that **WinBUGS** cooked up based on our sample size of 1 data set. It is worth noting that the `Rhat` statistics indicate convergence.

5.7 Unknown *N*

In all real applications N is unknown. We handled this important issue in Chapter 4 using the method of data augmentation (DA), which we apply here to achieve a realistic analysis of model SCR0. As with the basic closed population models considered previously, we formulate the problem by augmenting our observed data set with a number of "all-zero" encounter histories—what we referred to in Chapter 4 as potential individuals. If n is the number of observed individuals, then let $M - n$ be the number of potential individuals in the data set. For the two-dimensional y_{ij} data structure (n individual $\times$ J traps encounter frequencies) we simply add additional rows of all-zero observations to that data set. Because such "individuals" are unobserved, they therefore necessarily have $y_{ij} = 0$ for all j. A data set with 4 traps and 6 individuals, augmented with 4 potential individuals therefore might look like this:

```
      trap1  trap2  trap3  trap4
 [1,]     1      0      0      0
 [2,]     0      2      0      0
 [3,]     0      0      0      1
 [4,]     0      1      0      0
 [5,]     0      0      1      1
 [6,]     1      0      1      0
 [7,]     0      0      0      0
 [8,]     0      0      0      0
 [9,]     0      0      0      0
[10,]     0      0      0      0
```

We typically have more than 4 traps and, if we're fortunate, many more individuals in our data set.

For the augmented data set, we introduce a set of binary latent variables (the data augmentation variables), z_i, and the model is extended to describe $\Pr(z_i = 1)$, which, in the context of this problem, is the probability that an individual in the augmented data set is a member of the population of size N that was exposed to sampling. In other words, if $z_i = 1$ for one of the all-zero encounter histories, this is implied to be a sampling zero whereas observations for which $z_i = 0$ are "structural zeros." Under DA, we also express the binomial observation model *conditional on* z_i as follows:

$$y_{ij}|z_i \sim \text{Binomial}(K, z_i p_{ij}),$$

where we see that the binomial probability evaluates to 0 if $z_i = 0$ (so y_{ij} is a fixed 0 in that case) and evaluates to p_{ij} if $z_i = 1$.

How big does the augmented data set have to be? We discussed this issue in Chapter 4 where we noted that the size of the data set is equivalent to the upper limit of a uniform prior distribution on N. Practically speaking, it should be sufficiently large so that the posterior distribution for N is not truncated. On the other hand, if it is too large then unnecessary calculations are being done. An approach to choosing M by trial-and-error is indicated. Do a short MCMC run and then consider whether you need to increase M. See Chapter 17 for an example of this. Kéry and Schaub (2012, Chapter 6) provide an assessment of choosing M in closed population models. Using data augmentation, N is a derived parameter, computed by $N = \sum_{i=1}^{M} z_i$. Similarly, *density*, D, is also a derived parameter, computed as $D = N/\text{area}(\mathcal{S})$.

5.7.1 Analysis using data augmentation in WinBUGS

We provide a complete **R** script for simulating and organizing a data set, and analyzing it in **WinBUGS**. As before, we begin by obtaining a data set using our `simSCR0` function and then harvesting the required data objects from the resulting data list. Note that we use the `discard0=TRUE` option this time so that we get a "real looking" data set with no all-zero encounter histories:

```
## Simulate the data and extract the required objects
##
> data <- simSCR0(discard0=TRUE,rnd=2013)
> y <- data$Y
> nind <- nrow(y)
> X <- data$traplocs
> K <- data$K
> J <- nrow(X)
> xlim <- data$xlim
> ylim <- data$ylim
```

After harvesting the data we augment the data matrix `y` with $M - n$ all-zero encounter histories, and create starting values for the variables z_i and the activity centers $\mathbf{s}_i$ of which, for each, we require M values. One thing to take care of in using the **BUGS** engines is the starting values for the activity centers. It is usually helpful to start the $\mathbf{s}_i$ for each observed individual at or near the trap(s) it was captured. All of this happens as follows:

```
## Data augmentation
> M <- 200
> y <- rbind(y,matrix(0,nrow=M-nind,ncol=ncol(y)))
> z <- c(rep(1,nind),rep(0,M-nind))

## Starting values for s
> sst <- cbind(runif(M,xlim[1],xlim[2]),runif(M,ylim[1],ylim[2]))
> for(i in 1:nind){
    sst[i,1] <- mean( X[y[i,]>0,1] )
    sst[i,2] <- mean( X[y[i,]>0,2] )
  }
```

Next, we write out the **BUGS** model specification and save it to an external file called `SCR0b.txt`. The model specification now includes M encounter histories including the augmented potential individuals, the data augmentation parameters z_i, and the data augmentation parameter ψ:

```
> cat("
model{
  alpha0 ~ dnorm(0,.1)
  logit(p0) <- alpha0
  alpha1 ~ dnorm(0,.1)
  sigma <- sqrt(1/(2*alpha1))
  psi ~ dunif(0,1)
```

```
for(i in 1:M){
  z[i] ~ dbern(psi)
  s[i,1] ~ dunif(xlim[1],xlim[2])
  s[i,2] ~ dunif(ylim[1],ylim[2])
  for(j in 1:J){
    d[i,j] <- pow(pow(s[i,1]-X[j,1],2) + pow(s[i,2]-X[j,2],2),0.5)
    y[i,j] ~ dbin(p[i,j],K)
    p[i,j] <- z[i]*p0*exp(- alpha1*d[i,j]*d[i,j])
  }
}
N <- sum(z[])
D <- N/64
}
",file = "SCR0b.txt")
```

The remainder of the code for bundling the data, creating initial values and executing **WinBUGS**, looks much the same as before except with more or differently named arguments:

```
> data <- list (y=y, X=X, K=K, M=M, J=J, xlim=xlim, ylim=ylim)
> inits <- function(){
    list (alpha0=rnorm(1,-4,.4), alpha1=runif(1,1,2), s=sst, z=z)
    }

> library(R2WinBUGS)
> parameters <- c("alpha0","alpha1","sigma","N","D")
> out <- bugs (data, inits, parameters, "SCR0b.txt", n.thin=1,
       n.chains=3, n.burnin=1000,n.iter=2000,debug=TRUE,working.dir=getwd())
```

Note the differences in this new **WinBUGS** model with that appearing in the known-N version—there are not many! The loop over individuals goes up to M now, and there is a model component for the DA variables z. The input data has changed slightly too, as the augmented data set has more rows to include excess all-zero encounter histories. This analysis can be run directly using the `SCR0bayes` function once the `scrbook` package is loaded, by issuing the following commands:

```
> library(scrbook)
> data <- simSCR0(discard0=TRUE,rnd=2013)
> out1 <- SCR0bayes(data,M=200,engine="winbugs",ni=2000,nb=1000)
```

Summarizing the output from **WinBUGS** produces:

```
> print(out1,digits=2)
Inference for Bugs model at "SCR0b.txt", fit using Win BUGS,
   3 chains, each with 2000 iterations (first 1000 discarded)
   n.sims = 3000 iterations saved
           mean    sd   2.5%    25%    50%    75%  97.5% Rhat n.eff
alpha0    -2.57  0.23  -3.04  -2.72  -2.56  -2.41  -2.15 1.01   320
alpha1     2.46  0.42   1.63   2.16   2.46   2.73   3.33 1.02   120
sigma      0.46  0.04   0.39   0.43   0.45   0.48   0.55 1.02   120
N        113.62 15.73  86.00 102.00 113.00 124.00 147.00 1.01   260
D          1.78  0.25   1.34   1.59   1.77   1.94   2.30 1.01   260
deviance 302.60 23.67 261.19 285.47 301.50 317.90 354.91 1.00  1400

[...some output deleted...]
```

The Rhat statistic (discussed in Sections 3.5.2 and 17.6.4) for this analysis indicates satisfactory convergence. We see that the estimated parameters (α_0 and α_1) are comparable to the previous results obtained for the known-N case, and also not too different from the data-generating values. The posterior of N overlaps the data-generating value substantially.

5.7.1.1 Use of other BUGS engines: JAGS

There are two other popular **BUGS** engines in widespread use: **OpenBUGS** (Thomas et al., 2006) and **JAGS** (Plummer, 2003). Both of these are easily called from **R**. **OpenBUGS** can be used instead of **WinBUGS** by changing the package option in the bugs call to package="OpenBUGS". **JAGS** can be called using the function jags in package R2jags which has nearly the same arguments as bugs. Or, it can be executed from the **R** package rjags (Plummer, 2011) which has a slightly different implementation that we demonstrate here as we reanalyze the simulated data set in the previous section (note: the same **R** commands are used to generate the data and package the data, inits and parameters to monitor). The software and **R** packages mentioned here, and elsewhere in the book, are listed in Appendix 1. The function jags.model is used to initialize the model and run the MCMC algorithm for an adaptive period during which tuning of the MCMC algorithm might take place. These samples cannot be used for inference. Then the Markov chains are updated using coda.samples to obtain posterior samples for analysis, as follows:

```
> jinit <- jags.model("SCR0b.txt", data=data, inits=inits,
                      n.chains=3, n.adapt=1000)
> jout <- coda.samples(jinit, parameters, n.iter=1000, thin=1)
```

These commands can be executed using the function SCR0bayes provided with the **R** package scrbook by setting the 'engine' argument to "jags".

5.7.2 Implied home range area

Here, to assess implied space usage differences among encounter probability models, we apply the method described in Section 5.4, and use the function hra to compute the effective home range area under different encounter probability models fit to simulated data. We simulated a data set from the Gaussian kernel model as in Section 5.7 and then we fitted four models to it: (1) the true data-generating Gaussian encounter probability model; (2) the "hazard" or complementary log-log link model (Eq. (5.4.2)); (3) the negative exponential model; and (4) the logit model (Eq. (5.2.2)). We modified the function SCR0bayes for this purpose, which you should be able to do with little difficulty. We fit each model to the same simulated data set using **WinBUGS**, based only on 1,000 post-burn-in samples and 3 chains, which produced the posterior summaries given in Table 5.2. The main thing we see is that, while the implied home range area can vary substantially, there are smaller differences in the estimated N and hence D.

Table 5.2 Posterior mean of model parameters for four different encounter probability models fitted to a single simulated data set, and the effective home range area under each detection model computed using the function hra.

Parameter	Gaussian	Cloglog	Exponential	Logit
N	113.62	114.16	119.69	118.29
D	1.78	1.78	1.87	1.85
α_0	−2.57	−2.60	−1.51	−0.47
α_1	2.46	2.56	3.59	3.86
hra	3.85	3.78	5.51	2.64

5.7.3 Realized and expected population size

In Bayesian analysis of the SCR model, we estimate a parameter N, which is the size of the population for the prescribed state-space (presumably the state-space is defined so as to be relevant to where our traps were located, so N can be thought of as the size of the sampled population). In the context of Efford and Fewster (2012) this is the *realized* population size. Conversely, sometimes we see estimates of *expected* population size reported, which are estimates of $\mathbb{E}(N)$, the expected size of some hypothetical, unspecified population. Usually, the distinction between realized and expected population size is not made in SCR models, because almost everyone only cares about actual populations—and their realized population size.

If you do likelihood analysis of SCR models, then the distinction between realized and expected is often discussed by whether the estimator is "conditional-on-N" (realized) or not (expected). The naming arises because in obtaining the MLE of N, its properties are evaluated *conditional* on N—in particular, if the estimator is unbiased then $\mathbb{E}(\hat{N}|N) = N$ and $\text{Var}(\hat{N}|N) = \tilde{\sigma}^2_{\hat{N}}$ is the sampling variance. This does not conform to any concept or quantity that is relevant to Bayesian inference. If we care about N for the population that we sampled it is understood to be a realization of a random variable, but the relevance of "conditional-on-N" is hard to see. Bayesian analysis will provide a prediction of N that is based on the posterior $[N|y, \theta]$—which is certainly *not* conditional on N.

There is a third type of inference objective that is relevant in practice and that is prediction of N for a population that was not sampled—i.e., a "new" population. To elaborate on this, consider a situation in which we are concerned about the tiger populations in two distinct reserves in India. We do a camera trapping study on one of the reserves to estimate N_1 and we think the reserves are similar and homogeneous so we're willing to apply a density estimate based on N_1 to the second reserve. For the second reserve, do we want a prediction of the realized population size, N_2, or do we want an estimate of its expected value? We believe the former is the proper quantity for inference about the population size in the second reserve. An estimate of N_2 should include the uncertainty with which the mean is estimated (from reserve 1) and it should also include "process variation" for making the prediction of the latent variable N_2.

As a practical matter, to do a Bayesian analysis of this, you could just define the state-space to be the union of the two state-spaces, increase M so that the posterior of the total population size is not truncated, and then have MCMC generate a posterior sample of individuals on the joint state-space. You can tally up the ones that are on $\mathcal{S}_2$ as an estimate of N_2. Alternatively, we can define $\mu = \psi M / A_1$ and then simulate posterior samples of $N_2 \sim \text{Binomial}(M, \mu A_2/M)$ for the new state-space area, A_2.

To carry out a classical likelihood analysis of this second type of problem, what should we do? The argument for making a prediction of a new value of N would go something like this: If you obtain an MLE of N, say $\hat{N}$, then the inference procedure tells us the variance of this *conditional* on N, i.e., $Var(\hat{N}|N)$. This is fine, if we care about the specific value of N that generated our data set. However, if we don't care about the specific value in question then we want to "uncondition" on N to introduce a new variance component. Law of total variance says:

$$\text{Var}(\hat{N}) = \mathbb{E}[\text{Var}(\hat{N}|N)] + \text{Var}[\mathbb{E}(\hat{N}|N)].$$

If $\hat{N}$ is unbiased then we say the unconditional variance is

$$\text{Var}(\hat{N}) = \sigma^2_{\hat{N}} + \text{Var}(N).$$

The first variance component is estimation error and the second component is the "process variance." If you do Bayesian analysis, then you don't have to worry too much about how to compute variances properly. You decide if you care about N, or its expected value, or predictions of some "new" N, and you tabulate the correct posterior distribution from your MCMC output.

The considerations for estimating density are the same. Realized density is N/A, where N is the realized population, unless we put an expectation operator around the N, $\mathbb{E}(N)/A$. The formula for obtaining "expected density" is slightly different depending on whether we assume N has a Poisson distribution, or whether we assume a binomial distribution (under data augmentation). In the latter case ψ is related to the point process intensity (see Chapter 11) in the sense that, under the binomial prior:

$$\mathbb{E}(N) = M \times \psi$$

so, what we think of as "density," D, is a derived parameter $D = M\psi/A$. Under the Poisson point process model density is the canonical parameter and the expected population size is:

$$\mathbb{E}(N) = D \times A.$$

In summary, there are three basic inference problems that relate to estimating population size (or density):

(1) What is the value of N for some population that was sampled? This is what Efford and Fewster call "realized N." In general, we want the uncertainty to reflect having to estimate n_0, the part of the population not seen.

(2) We need to estimate N for some population that we didn't sample but it is "similar" to the population that we have information on. In this case, we have to account for both variation in having to estimate parameters of the distribution of N, and we have to account for process variation in N (i.e., due to the stochastic model of N).
(3) In some extremely limited cases we might care about estimating the expected value of N, $\mathbb{E}(N)$.

5.8 The core SCR assumptions

It's always a good idea to sit down and reflect on the meaning of any particular model, its various assumptions, and what they mean in a specific context. From the statistician's point of view, the basic assumption, the omnibus assumption, as in all of statistics, and for every statistical model, is that "the model is correctly specified." Naturally, that precludes everything that isn't explicitly addressed by the model. To point this out to someone seems to cause a lot of anxiety, so we enumerate here what we think are the most important statistical assumptions of the basic SCR0 model:

- *Demographic closure.* The model does not allow for demographic processes. There is no recruitment or entry into the sampled population. There is no mortality or exit from the sampled population.
- *Geographic closure.* We assume no permanent emigration or immigration from the state-space. However, we allow for "temporary" movements around the state-space and variable exposure to encounter as a result. The whole point of SCR models is to accommodate this dynamic. In ordinary capture-recapture models we have to assume geographic closure to interpret N in a meaningful way.
- *Activity centers are randomly distributed.* That is, uniformity and independence of the underlying point process $\mathbf{s}_1, \ldots, \mathbf{s}_N$ (see Section 5.3).
- *Detection is a function of distance.* A detection model that describes how encounter probability declines as a function of distance from an individual's home range center.
- *Independence of encounters* among individuals. Encounter of any individual is independent of encounter of any other individual.
- *Independence of encounters* of the same individual. Encounter of an individual in any trap is independent of its encounter in any other trap, and subsequent sample occasion.

It's easy to get worried and question the whole SCR enterprise just on the grounds that these assumptions combine to form such a simplistic model, one that surely can't describe the complexity of real populations. On this sentiment, a few points are worth making. First, you don't have inherently fewer assumptions by using an ordinary capture-recapture model but, rather, the SCR model relaxes a number of important assumptions compared to the non-spatial counterpart. For one, here, we're not assuming that p is constant for all individuals but rather that p varies substantially as a matter of the spatial juxtaposition of individuals with traps. So maybe the manner in which p varies isn't quite right, but that's not an argument that supports doing less modeling. Fundamentally, a distance-based model for p has some basic biological justification in virtually every capture-recapture study. Secondly, for some of these core assumptions such as uniformity, and independence of individuals and of encounters, we expect a fair amount of robustness to departures. They function primarily to allow us to build a model and an estimation scheme and we don't usually think they represent real populations (of course, no model does!). Third, we can extend these assumptions in many different ways and we do that to varying extents in this book, and more work remains to be done in this regard. Fourth, we can also evaluate the reasonableness of the assumptions formally in some cases using standard methods of assessing model fit (Chapter 8).

Sometimes you see in capture-recapture literature statements like "we assume no marks are lost," "marks are correctly identified," and similar things. We will typically neglect such statements because, in our view, we should separate statistical assumptions about model parameters or aspects of the probability model, from what are essentially logistical or operational assumptions about how we interpret our data, or our ability to conduct the study.

5.9 Wolverine camera trapping study

We provide an illustration of some of the concepts we've introduced previously in this chapter by analyzing camera trapping study of wolverines *Gulo gulo* (Magoun et al., 2011; Royle et al., 2011b). In this study, wolverines were individually identified from their ventral pattern (Fig. 1.1), thus allowing for the application of SCR models. The study took place in SE Alaska (Figure 5.4) where 37 cameras were operational for variable periods of time (min = 5 days, max = 108 days, median = 45 days). A consequence of this is that the number of sampling occasions, K, is variable for each camera. Thus, we must provide a vector of sample sizes as data to **BUGS** and modify the model specification in Section 5.7 accordingly.

5.9.1 Practical data organization

To carry out an analysis of these data, we require the matrix of trap coordinates and the encounter history data. We usually store data in two distinct data files which contain all the information needed for an analysis. These files are:

- The encounter data file (EDF), containing a record of at which traps and when each individual encounter occurred.
- The trap deployment file (TDF), which contains the coordinates of each trap, along with information indicating which sample occasions each trap was operating.

Encounter Data File (EDF)—We store the encounter data in an efficient "flat" file format which is easily manipulated in **R** and easy to create in Excel and other spreadsheets that are widely used for data management. The file structure is a simple matrix with four columns (1) **session ID**, the trap *session* which usually corresponds to a year or a primary period in the context of a Robust Design situation, but it could

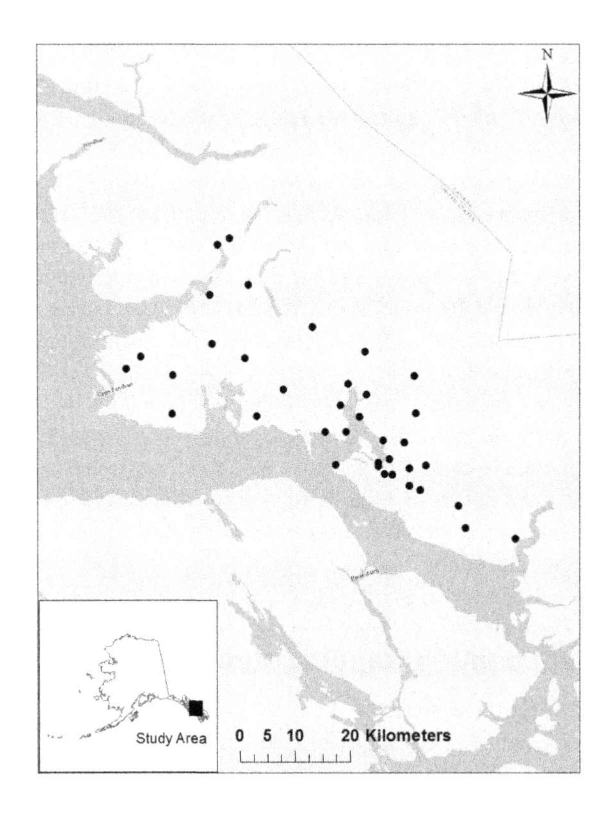

FIGURE 5.4

Wolverine camera trap locations (black dots) from a study that took place in SE Alaska. See Magoun et al. (2011) for details.

also correspond to a distinct spatial unit (see Section 6.5.4 and Chapter 14). For a single-year study (as considered here) this should be an integer that is the same for all records; (2) **individual ID**, the individual identity, being an integer from 1 to n (repeated for multiple captures of the same individual), indicating which individual the record (row) of the matrix belongs to; (3) **occasion ID**, the integer sample occasion which generated the record; and (4) **trap ID**, the trap identity, an integer from 1 to J, the number of traps. The structure of the EDF is the same as used in the `secr` package (Efford, 2011a) and similar to that used in the **SPACECAP** (Gopalaswamy et al., 2012a), and `SCRbayes` (Russell et al., 2012) packages, both of which have a 3-column format (`trapID`, `indID`, `sampID`). We note that the naming of the columns is irrelevant for most analyses we do in this book, although `secr` and other software may have requirements on variable naming.

To illustrate this format, the wolverine data are available in the package `scrbook` by typing:

```
> data(wolverine)
```

which contains a list with elements `wcaps` (the EDF) and `wtraps` (the TDF). We see that `wcaps` has 115 rows, each representing a unique encounter event including the trap identity, the individual identity, and the sample occasion index (`sample`). The first five rows of `wcaps` are:

```
> wolverine$wcaps[1:5,]
     year  individual  day  trap
[1,]    1           2  127     1
[2,]    1           2  128     1
[3,]    1           2  129     1
[4,]    1          18  130     1
[5,]    1           3  106     2
```

The first column here, labeled `year`, is in integer indicating the year or session of the encounter. All these data come from a single year (2008) and so `year` is set to 1. The variable `trap` will have to correspond to the row of a matrix containing the trap coordinates—in this case the TDF file `wtraps`, which we describe further below.

Note that the information provided in this encounter data file `wcaps` does not represent a completely informative summary of the data. For example, if no individuals were captured in a certain trap or during a certain sample occasion, then this compact data format will have no record of this occasion or trap. Thus, we need to know J, the number of traps, *and* K, the number of sample occasions, when reformatting this SCR data format into a 2-d encounter frequency matrix or 3-d array.

For our purposes here, we need to convert the `wcaps` file into the $n \times J$ array of binomial encounter frequencies, although more general models, such as those containing a behavioral response, might require an encounter-history formulation of the model which requires a full 3-d array. To obtain our encounter frequency matrix, we do this the hard way by first converting the encounter data file into a 3-d array and then summarizing to trap totals. We have a handy function, `SCR23darray`, which takes the compact encounter data file, and converts it to a 3-d array, and then we use the **R** function `apply` to summarize over the sample occasion dimension (by convention here, this is the 2nd dimension). To apply this to the wolverine data we do this:

```
> y3d <- SCR23darray(wolverine$wcaps,wolverine$wtraps)
> y <- apply(y3d,c(1,2),sum)
```

See the help file for more information on `SCR23darray`.

Trap Deployment File (TDF)—The other important information needed to fit SCR models is the "trap deployment file" (TDF) which provides additional information not contained in the encounter data file. The traps file has $K + 3$ columns. The first column is assumed to be a trap identifier, columns 2 and 3 are the easting and northing coordinates (assumed to be in a Euclidean coordinate system), and columns 4 to $K + 3$ are binary indicators of whether each trap was operational during each sample occasion. The first 10 rows (out of 37) and 10 columns (out of 167) of the trap deployment file for the wolverine data are shown as follows:

```
> wolverine$wtraps[1:10,1:10]

   Easting  Northing  1  2  3  4  5  6  7  8
1   632538   6316012  0  0  0  0  0  0  0  0
2   634822   6316568  1  1  1  1  1  1  1  1
3   638455   6309781  0  0  0  0  0  0  0  0
4   634649   6320016  0  0  0  0  0  0  0  0
5   637738   6313994  0  0  0  0  0  0  0  0
6   625278   6318386  0  0  0  0  0  0  0  0
7   631690   6325157  0  0  0  0  0  0  0  0
8   632631   6316609  0  0  0  0  0  0  0  0
9   631374   6331273  0  0  0  0  0  0  0  0
10  634068   6328575  0  0  0  0  0  0  0  0
```

This tells us that trap 2 was operated during occasions (days) 1–8, but the other traps were not operational during those periods. To compute the vector of sample sizes K, and extract the trap locations, we do this:

```
> traps <- wolverine$wtraps
> traplocs <- traps[,1:2]
> K <- apply(traps[,3:ncol(traps)],1,sum)
```

This results in a matrix `traplocs`, which contains the coordinates of each trap, and a vector K containing the number of days that each trap was operational. We now have all the information required to fit a basic SCR model in **BUGS**.

Summarizing the data for the wolverine study, we see that 21 unique individuals were captured a total of 115 times. Most individuals were captured 1–6 times, with 4, 1, 4, 3, 1, and 2 individuals captured 1–6 times, respectively. In addition, 1 individual was captured 8 and one captured 14 times, and 2 individuals each were captured 10 and 13 times. The number of unique traps that captured a particular individual ranged from 1 to 6, with 5, 10, 3, 1, 1, and 1 individual captured in each of 1–6 different traps, respectively, for a total of 50 unique wolverine-trap encounters. These numbers might be hard to get your mind around whereas some tabular summary is often more convenient. For that it seems natural to tabulate individuals by trap and total encounter frequencies. For the wolverine data, we reproduce Table 5.1 from Royle et al. (2011b) as Table 5.3. Generally, effective estimation in SCR studies requires a sufficient sample size of spatial recaptures, which are captures of the same individual in multiple traps. We address this in the context of sampling design in Chapter 10. Therefore, it is informative to understand how many unique traps each individual was captured in, and the total number of encounters.

5.9.2 Fitting the model in WinBUGS

Here, we fit the simplest SCR model with the Gaussian encounter probability model, although we revisit these data and fit additional models in later chapters. Model SCR0 is summarized by the following four elements:

(1) $y_{ij}|\mathbf{s}_i \sim \text{Binomial}(K, z_i\, p_{ij})$;
(2) $p_{ij} = p_0 \exp(-\alpha_1 \, \|\mathbf{x}_j - \mathbf{s}_i\|^2)$;

Table 5.3 Individual frequencies of capture for wolverines encountered by camera traps in SE Alaska in 2008. Rows index unique traps of capture for each individual and columns represent total number of captures (e.g., we captured 4 individuals 1 time, necessarily in only 1 trap; we captured 3 individuals 3 times but in 2 different traps).

	Number of Captures									
Number of Traps	**1**	**2**	**3**	**4**	**5**	**6**	**8**	**10**	**13**	**14**
1	4	1	0	0	0	0	0	0	0	0
2	0	0	3	2	0	2	1	2	0	0
3	0	0	1	1	0	0	0	0	0	1
4	0	0	0	0	0	0	0	0	1	0
5	0	0	0	0	1	0	0	0	0	0
6	0	0	0	0	0	0	0	0	1	0

(3) $\mathbf{s}_i \sim \text{Uniform}(\mathcal{S})$;
(4) $z_i \sim \text{Bernoulli}(\psi)$.

We assume customary flat priors on the structural (hyper-) parameters of the model, $\alpha_0 = \text{logit}(p_0), \alpha_1$ and ψ.

It remains to define the state-space $\mathcal{S}$. For this, we nested the trap array (Figure 5.4) in a rectangular state-space extending 20 km beyond the traps in each cardinal direction. We scaled the coordinate system so that a unit distance was equal to 10 km, producing a rectangular state-space of dimension 9.88×10.5 units ($area = 10,374$ km^2) within which the trap array was nested. As a general rule, we recommend scaling the state-space so that it is defined near the origin $(x, y) = (0, 0)$. While the scaling of the coordinate system is theoretically irrelevant, a poorly scaled coordinate system can produce Markov chains that mix poorly. The buffer of the state-space should be large enough so that individuals with activity centers beyond the state-space boundary are not likely to be encountered (Section 5.3.1). To evaluate this, we fit models for various choices of a rectangular state-space based on buffers from 1.0 to 5.0 units (10–50 km). In the **R** package `scrbook` we provide a function `wolvSCR0` which will fit model SCR0. For example, to fit the model in **WinBUGS** using data augmentation with $M = 300$ potential individuals, the state-space buffer of 1 standardized unit, using three Markov chains each of 12,000 total iterations, discarding the first 2,000 as burn-in, we execute the following **R** commands:

```
> library(scrbook)
> data(wolverine)
> traps <- wolverine$wtraps
> y3d <- SCR23darray(wolverine$wcaps,wolverine$wtraps)
> wolv <- wolvSCR0(y3d,traps,nb=2000,ni=12000,buffer=1,M=300)
```

The argument `buffer` determines the buffer size of the state-space in the scaled units (i.e., 10 km). Note that this analysis takes between 1 and 2 h on many machines (in 2013) so we recommend testing it with lower values of M and fewer iterations. The posterior summaries are shown in Table 5.4.

Table 5.4 Posterior summaries of SCR model parameters for the wolverine camera trapping data from SE Alaska, using state-space buffers from 10 up to 50 km. Each analysis was based on 3 chains, 12,000 iterations, 2,000 burn-in, for a total of 30,000 posterior samples.

	σ			N			D		
Buffer	**Mean**	**SD**	**n.eff**	**Mean**	**SD**	**n.eff**	**Mean**	**SD**	**n.eff**
10	0.65	0.06	1,800	39.63	6.70	7,100	5.97	1.00	7,100
15	0.64	0.06	510	48.77	9.19	3,300	5.78	1.09	3,300
20	0.64	0.06	1,200	59.84	11.89	20,000	5.77	1.15	20,000
25	0.64	0.05	3,600	72.40	14.72	2,700	5.79	1.18	2,700
30	0.63	0.05	5,600	86.42	17.98	3,900	5.82	1.21	3,900
35	0.63	0.05	4,500	101.79	21.54	30,000	5.85	1.24	30,000
40	0.64	0.05	410	118.05	26.17	410	5.87	1.30	450
45	0.64	0.05	10,000	134.43	28.68	3,300	5.83	1.24	3,300
50	0.63	0.05	4,700	151.61	31.65	3,400	5.79	1.21	3,400

5.9.3 Summary of the wolverine analysis

We see that the estimated density is roughly consistent as we increase the state-space buffer from 15 to 55 km. We do note that the data augmentation parameter ψ (and, correspondingly, N) increases with the size of the state-space in accordance with the relationship $E(N) = D \times A$. However, density is more or less constant as we increase the size of the state-space beyond a certain point. For the 10 km state-space buffer, we see a slight effect on the posterior distribution of D because the state-space is not sufficiently large. The full results from the analysis based on a 20 km state-space buffer are given in Table 5.5.

Our point estimate of wolverine density from this study, using the posterior mean from the state-space based on the 20 km buffer, is approximately 5.77 individuals/1,000 km^2 with a 95% posterior interval of $[3.86, 8.29]$. Density is estimated imprecisely, which might not be surprising given the low sample size ($n = 21$ individuals!). This seems to be a basic feature of carnivore studies although it should not (in

Table 5.5 Posterior summaries of SCR model parameters for the wolverine camera trapping data from SE Alaska. The model was run with the trap array centered in a state-space with a 20 km rectangular buffer.

Parameter	Mean	SD	2.5%	25%	50%	75%	97.5%	Rhat
N	59.84	11.89	40.00	51.00	59.00	67.00	86.00	1
D	5.77	1.15	3.86	4.92	5.69	6.46	8.29	1
α_1	1.26	0.21	0.87	1.11	1.25	1.40	1.71	1
p_0	0.06	0.01	0.04	0.05	0.06	0.06	0.08	1
σ	0.64	0.06	0.54	0.60	0.63	0.67	0.76	1
ψ	0.20	0.05	0.12	0.17	0.20	0.23	0.30	1

our view) preclude the study of their populations by capture-recapture nor attempts to estimate density or vital rates.

It is worth thinking about this model, and these estimates, computed under a rectangular state-space, roughly centered over the trapping array (Figure 5.4). Does it make sense to define the state-space to include, for example, ocean? What are the possible consequences of this? What can we do about it? There's no reason at all that the state-space has to be a regular polygon—we defined it as such here strictly for convenience and for ease of implementation in **WinBUGS**, where it enables us to specify the prior for the activity centers as uniform priors for each coordinate. While it would be possible to define a more realistic state-space using some general polygon GIS coverage, it might take some effort to implement that in the **BUGS** language, although it is not difficult to devise custom MCMC algorithms to do that (see Chapter 17). Alternatively, we recommend using a discrete representation of the state-space—i.e., approximate $\mathcal{S}$ by a grid of G points. We discuss this in Section 5.10.

5.9.4 Wolverine space usage

The parameter α_1 is related to the home range radius (Section 5.4). For the Gaussian model we interpret the scale parameter σ, related to α_1 by $\alpha_1 = 1/(2\sigma^2)$, as the radius of a bivariate normal model of space usage. In this case $\sigma = 0.64$ standardized units (10 km), corresponds to $0.64 \times 10 = 6.4$ km. It can be argued then that 95% of space used by an individual is within $6.4 \times \sqrt{5.99} = 15.66$ km of the home range center. The effective "home range area" is the then the area of this circle, which is $\pi \times 15.66^2 = 770.4$ km^2. Using our handy function `hra` we do this:

```
hra(pGauss1,parms=c(-2,1/(2*.64*.64)),xlim=c(-1,7),ylim=c(-1,7))

[1] 7.731408
```

which is in units of 100 km^2, or 773.1 km^2. The difference in this case is due to numerical approximation of our all-purpose tool `hra`. This home range size is relatively huge for measured home ranges of wolverines, which range between 100 and 535 km^2 (Whitman et al., 1986).

Royle et al. (2011b) reported estimates for σ in the range 6.3–9.8 km depending on the model, which isn't too different from what we report here.[1] However, these estimates are larger than the typical home range sizes suggested in the literature. One possible explanation is that if a wolverine is using traps as a way to get yummy chicken, so it's moving from trap to trap instead of adhering to "normal" space usage patterns, then the implied home range size might not be biologically meaningful. Thus, interpretation of detection models in terms of home range area depends on

[1] Royle et al. (2011b) expressed the model as $\text{cloglog}(p_{ij}) = \alpha_0 - (1/\sigma^2) * d_{ij}^2$, but the estimates of σ reported in their Table 5.2 are actually based on the model according to $\text{cloglog}(p_{ij}) = \alpha_0 - \frac{1}{2\sigma^2} * d_{ij}^2$, and so the estimates of σ they report in units of km are consistent with what we report here except based on the complementary log-log (Gaussian hazard) model, instead of the Gaussian encounter probability model.

some additional context or assumptions, such as that traps don't effect individual space usage patterns. As such, we caution against direct biological interpretations of home range area based on σ, although SCR models can be extended to handle more general, non-Euclidean, patterns of space usage (Chapters 12 and 13).

We can calibrate the size of the state-space by looking at the estimated home range radius of the species. We should target a buffer of width 2 to $3 \times \sigma$ in order that the probability of encountering an individual is very close to 0 beyond the prescribed state-space. Essentially, by specifying a state-space, we're setting $p = 0$ for individuals beyond the prescribed state-space. For the wolverine data, with σ in the range of 6–9 km, a state-space buffer of 20 km appears to be sufficiently large.

5.10 Using a discrete habitat mask

The SCR model developed previously in this chapter assumes that individual activity centers are distributed uniformly over the prescribed state-space. Clearly, this will not always be a reasonable assumption. In Chapter 11, we develop models that allow explicitly for non-uniformity of the activity centers by modeling covariate effects on density. A simplistic method of affecting the distribution of activity centers, which we address here, is to modify the shape and organization of the state-space explicitly. For example, we might be able to classify the state-space into distinct blocks of habitat and non-habitat. In that case we can remove the non-habitat from the state-space and assume uniformity of the activity centers over the remaining portions judged to be suitable habitat. There are several ways to approach this: We can use a grid of points to represent the state-space, i.e., the set of coordinates $\mathbf{s}_1, \ldots, \mathbf{s}_G$, and assign equal probabilities to each possible value. Alternatively, we can retain the continuous formulation of the state-space but attempt to describe constraints analytically, or we can use polygon clipping methods to enforce constraints on the state-space in the MCMC analysis. We focus here on the formulation of the basic SCR model in terms of a discrete state-space but, in Chapter 17, we demonstrate the latter approach based on using polygon operations to define an irregular state-space. Use of a discrete state-space can be computationally expensive in **WinBUGS**. That said, it isn't too difficult to perform the MCMC calculations in **R** (discussed in Chapter 17). The **R** package `SPACECAP` (Gopalaswamy et al., 2012a) was motivated as a discrete-space implementation of an MCMC algorithm for basic SCR models.

While clipping out non-habitat seems like a good idea, we think investigators should go about this very cautiously. We might prefer to do it when non-habitat represents a clear-cut restriction on the state-space such as a reserve boundary or a lake, ocean, or river. But, having the capability to do this also encourages people to want to define "habitat" vs. "non-habitat" based on their understanding of the system, whereas it is usually unknown whether the animal being studied has the same understanding. Moreover, differentiating the landscape by habitat or habitat quality must affect the geometry and morphology of home ranges (see Chapter 13) much more so than the plausible locations of activity centers. That is, a home range centroid could, in actual fact, occur in a shopping mall parking lot if there is pretty good habitat around the shopping mall, so there is probably no sense precluding it as the location for an activity center. It would generally be better to include some definition of habitat quality in the model for the detection probability (Royle et al., 2013a, b), which we address in Chapters 12 and 13.

5.10.1 Evaluation of coarseness of habitat mask

The coarseness of the state-space should not really have much of an effect on estimates if the grain is sufficiently fine relative to typical animal home range sizes. Why is this? We have two analogies that can help us understand this. First is the relationship to model M_h. As noted in Section 5.3.2 above, we can think about SCR models as a type of finite mixture (Norris and Pollock, 1996; Pledger, 2004) where we are fortunate to be able to obtain direct information about which group individuals belong to (group being location of activity center). In the standard finite-mixture models we typically find that a small number of groups (e.g., 2 or 3 at most) can explain high levels of heterogeneity and are adequate for most data sets of small to moderate sample sizes. We therefore expect a similar effect in SCR models when we discretize the state-space. We can also think about discretizing the state-space as being related to numerical integration where we find (see Chapter 6) that we don't need a very fine grid of support points to evaluate the integral to a reasonable level of accuracy. We demonstrate this here by reanalyzing simulated data using a state-space defined by a different number of support points. We provide an **R** script called `SCR0bayesDss` in the **R** package `scrbook`. For this comparison we generated the actual activity centers as a continuous random variable and thus the discrete state-space is, strictly speaking, an approximation to truth. That said, we regard all state-space specifications as approximations to truth in the sense that they represent a component of the SCR model.

As with our **R** function `SCR0bayes`, the modification `SCR0bayesDss` will use either **WinBUGS** or **JAGS**. In addition, it requires a grid resolution argument (`ng`) which is the dimension of 1 side of a square state-space. To execute this function we do, for example for a 9×9 grid:

```
> library(scrbook)
> data <- simSCR0(discard0=TRUE,rnd=2013)   # Generate data set

# Run with JAGS
> out1 <- SCR0bayesDss(data,ng=9,M=200,engine="jags",ni=2000,nb=1000)

# Run with WinBUGS
> out2 <- SCR0bayesDss(data,ng=9,M=200,engine="winbugs",ni=2000,nb=1000)
```

We fit this model to the same simulated data set for 6×6, 9×9, 12×12, and 15×15 state-space grids. For **WinBUGS**, we used 3 chains of 5,000 total length 1,000 burn-in, which yields 12,000 total posterior samples. Summary results are shown in Table 5.6. The results are broadly consistent except for the 6×6 case. We see that the runtime increases with the size of the state-space grid (not unexpected), such that we imagine it would be impractical to run models with more than a few hundred state-space grid points. We found (not shown here) that the runtime of **JAGS**

Table 5.6 Comparison of the effect of state-space grid coarseness on estimates of N for a simulated data set with true $N = 100$. Posterior summaries and runtime are given. Results obtained using **WinBUGS** run from `R2WinBUGS`.

Grid Size	Mean	SD	NaiveSE	Time-seriesSE	Runtime (s)
6×6	111.670	16.614	0.152	0.682	2,274
9×9	114.230	17.991	0.164	0.833	4,300
12×12	115.981	17.384	0.159	0.763	7,100
15×15	115.379	17.937	0.164	0.832	13,010

is much faster and, furthermore, relatively *constant* as we increase the grid size. We suspect that **WinBUGS** is evaluating the full conditional for each activity center at all possible values of the discrete grid whereas it may be that **JAGS** is evaluating the full conditional at only a subset of values or perhaps using previous calculations more effectively. While this might suggest that one should always use **JAGS** for this analysis, we found in our analysis of the wolverine data (next section) that **JAGS** could be extremely sensitive to starting values, producing MCMC algorithms that often simply do not work for some problems, so be careful when using **JAGS**. The performance of either should improve if we compute the full distance matrix outside of **BUGS** and pass it as data, although we haven't fully evaluated this approach.

FIGURE 5.5

Three habitat mask (state-space) grids used in the comparison of the effect of pixel size on the estimated density surface of wolverines. The three cases are 2-km (left), 4-km (center), and 8-km (right) spacing of state-space points, extending 40 km from the vicinity of the trap array. Larger dots are the camera trap locations.

5.10.2 Analysis of the wolverine camera trapping data

We reanalyzed the wolverine data using discrete state-space grids with points spaced by 2, 4, and 8 km (see Figure 5.5). These were constructed from a 40 km buffered state-space, deleting the points over water (see Royle et al., 2011b). Our interest in doing this was to evaluate the relative influence of grid resolution on estimated density. The coarser grids will be more efficient from a computational standpoint and so we would prefer to use them, but only if there is no strong influence on estimated density. The posterior summaries for the three habitat grids are given in Table 5.7. We see that the density estimates are quite a bit larger than obtained in our analysis (Table 5.4) based on a rectangular, continuous state-space. We also see that there are slight differences depending on the resolution of the state-space grid. Interestingly, the effectiveness of the MCMC algorithms, as measured by effective sample size (`n.eff`), is pretty remarkably different. Furthermore, the finest grid resolution (2 km spacing) took about 6 days to run and thus, it would not be practical for large problems or with many models.

Table 5.7 Posterior summaries for the wolverine camera trapping data, using model SCR0, with a Gaussian hazard encounter probability model, and a discrete habitat mask of three different resolutions: 2, 4, and 8 km. Parameters are p_0 = baseline encounter probability, σ is the scale parameter of the Gaussian kernel, ψ is the data augmentation parameter, N and D are population size and density, respectively. Models fitted using **WinBUGS**, 3 chains, each with 11,000 iterations (first 1,000 discarded) producing 30,000 posterior samples.

	Mean	SD	2.5%	25%	50%	75%	97.5%	Rhat	n.eff
2 km spacing									
N	86.56	16.94	57.00	75.00	85.00	97.00	124.00	1.00	510
D	8.78	1.72	5.78	7.60	8.62	9.83	12.57	1.00	510
p_0	0.05	0.01	0.03	0.04	0.05	0.05	0.06	1.01	320
σ	0.62	0.05	0.54	0.59	0.62	0.65	0.73	1.01	160
ψ	0.43	0.09	0.27	0.37	0.43	0.49	0.63	1.00	560
4 km spacing									
N	89.25	17.44	59.00	77.00	88.00	100.00	127.00	1.00	1100
D	9.01	1.76	5.96	7.77	8.88	10.10	12.82	1.00	1100
p_0	0.05	0.01	0.03	0.04	0.05	0.05	0.07	1.00	2500
σ	0.61	0.04	0.53	0.58	0.61	0.64	0.71	1.00	1600
ψ	0.45	0.09	0.28	0.38	0.44	0.50	0.64	1.00	1300
8 km spacing									
N	83.18	16.14	56.00	72.00	82.00	93.00	119.00	1.00	700
D	8.28	1.61	5.57	7.17	8.16	9.26	11.84	1.00	700
p_0	0.05	0.01	0.03	0.04	0.04	0.05	0.06	1.00	560
σ	0.68	0.05	0.59	0.64	0.67	0.71	0.77	1.01	220
ψ	0.42	0.09	0.26	0.36	0.41	0.47	0.61	1.00	940

5.11 Summarizing density and activity center locations

One of the most useful aspects of SCR models is that they are parameterized in terms of individual locations—i.e., *where* each individual lives—and, thus, we can compute many useful and interesting summaries of the activity centers using output from an MCMC simulation, including maps of density (the number of activity centers per unit area), estimates of N for any well-defined polygon, or estimates of where the activity centers for specific individuals reside. In Bayesian analysis by MCMC, obtaining such summaries entails no added calculations, because we need only post-process the output for the individual activity centers to obtain the desired summaries (Section 3.5.4). We demonstrate that in this section. Note that you have to be sure to retain the MCMC history for the $\mathbf{s}$ variables and the data augmentation variables z in order to do the following analyses.

5.11.1 Constructing density maps

Because SCR models are spatially explicit, it is natural to want to summarize the results of fitting a model by producing a map of density. Using Bayesian analysis by MCMC, it is most easy to make a map of *realized* density. We can do this by tallying up the number of activity centers $\mathbf{s}_i$ in pixels of arbitrary size and then producing a nice multi-color spatial plot of the result. Specifically, let $B(\mathbf{x})$ indicate a pixel centered at $\mathbf{x}$, then

$$N(\mathbf{x}) = \sum_{i=1}^{M} I(\mathbf{s}_i \in B(\mathbf{x}))$$

(here, $I(arg)$ is the indicator function which evaluates to 1 if arg is true, and 0 otherwise) is the population size of pixel $B(\mathbf{x})$, and $D(\mathbf{x}) = N(\mathbf{x})/\|B(\mathbf{x})\|$ is the local density. Note that these $N(\mathbf{x})$ parameters are just "derived parameters" as we normally obtain from posterior output using the appropriate Monte Carlo average (see Chapter 3).

One thing to be careful about, in the context of models in which N is unknown, is that, for each MCMC iteration m, we only tabulate those activity centers which correspond to individuals in the sampled population, i.e., for which the data augmentation variable $z_i = 1$. In this case, we take all of the output for MCMC iterations $m = 1, 2, \ldots,$ `niter` and compute this summary:

$$N(\mathbf{x}, m) = \sum_{i:z_{i,m}=1} I(\mathbf{s}_{i,m} \in B(\mathbf{x})).$$

Thus, $N(\mathbf{x}, 1), N(\mathbf{x}, 2), \ldots,$ is the Markov chain for parameter $N(\mathbf{x})$. In what follows we will provide a set of **R** commands for doing this calculation and making a basic image plot from the MCMC output:

Step 1: Define the center points of each pixel $B(\mathbf{x})$, or point at which local density will be estimated:

```
> xg <- seq(xlim[1],xlim[2], ,50)
> yg <- seq(ylim[1],ylim[2], ,50)
```

Step 2: Extract the MCMC histories for the activity centers and the data augmentation variables. Note that these are each $N \times$ `niter` matrices. Here we do this assuming that **WinBUGS** was run producing the **R** object named `out`:

```
> Sxout <- out$sims.list$s[, ,1]
> Syout <- out$sims.list$s[, ,2]
> z <- out$sims.list$z
```

Step 3: We associate each coordinate with the proper pixel using the **R** command `cut`. Note that we keep only the activity centers for which $z = 1$ (i.e., individuals that belong to the population of size N):

```
> Sxout <- cut(Sxout[z==1], breaks=xg, include.lowest=TRUE)
> Syout <- cut(Syout[z==1], breaks=yg, include.lowest=TRUE)
```

Step 4: Use the `table` command to tally up how many activity centers are in each $B(\mathbf{x})$:

```
> Dn <- table(Sxout, Syout)
```

Step 5: Use the `image` command to display the resulting matrix:

```
> image(xg, yg, Dn/nrow(z), col=terrain.colors(10))
```

We'll apply this analysis to create a density map from the wolverine MCMC output shortly. It is worth emphasizing here that density maps will not usually appear uniform despite that we have assumed that activity centers are uniformly distributed. This is because the observed encounters of individuals provide direct information about the location of the $i = 1, 2, \ldots, n$ activity centers and thus their "estimated" locations will be affected by the observations. In a limiting sense, were we to sample space intensely enough, every individual would be captured a number of times and we would have considerable information about all N point locations. Consequently, the uniform prior would have almost no influence at all on the estimated density surface in this limiting situation. Thus, in practice, the influence of the uniformity assumption decreases as the fraction of the population encountered, and the total number of encounters per individual, increases.

On the non-intuitiveness of `image`—the **R** function `image`, invoked for a matrix M by `image(M)`, might not be very intuitive to some—it plots $M[1, 1]$ in the lower left corner. If you want $M[\,]$ to be plotted "as you look at it" then $M[1, 1]$ should be in the upper left corner. We have a function, `rot` which does that. If you do `image(rot(M))` then it puts the image on the monitor as if it was a map you were looking at. You can always specify the x and y labels explicitly as we did above.

Spatial dot plots—A cruder version of the density map can be made using our "spatial dot map" function `spatial.plot` (in `scrbook`). This function requires, as input, point locations and the value to be displayed. A simplified version of this function is as follows:

```
> spatial.plot <- function(x,y){
    nc <- as.numeric(cut(y,20))
    plot(x,pch=" ")
    points(x,pch=20,col=topo.colors(20)[nc],cex=2)
    image.scale(y,col=topo.colors(20))
  }
#
# To execute the function do this:
#
> spatial.plot(cbind(xg,yg), Dn/nrow(z))
```

5.11.2 Wolverine density map

We return to the wolverine study from SE Alaska (Figure 5.4) and we produce a density map of wolverines from that analysis. We include the function `SCRdensity`, which requires a specific data structure as shown below. In particular, we have to package up the MCMC history for the activity centers and the data augmentation variables z into a list. This also requires that we add those variables to the parameters-to-be-monitored list when we pass things to **BUGS**.

We used the posterior output from the wolverine model fitted previously to compute a relatively coarse version of a density map, using 100 pixels in a 10×10 grid (Figure 5.6 top panel) and using 900 pixels arranged in a 30×30 grid (Figure 5.6 lower panel) for a fine-scale map. The **R** commands for producing such a plot (for a short MCMC run) are as follows:

```
> library(scrbook)
> data(wolverine)
> traps <- wolverine$wtraps
> y3d <- SCR23darray(wolverine$wcaps,wolverine$wtraps)

# This takes 341 seconds on a standard CPU circa 2011
> out <- wolvSCR0(y3d,traps,nb=1000,ni=2000,buffer=1,M=100,keepz=TRUE)

> Sx <- out$sims.list$s[, ,1]
> Sy <- out$sims.list$s[, ,2]
> z <- out$sims.list$z
> obj <- list(Sx=Sx,Sy=Sy,z=z)
> tmp <- SCRdensity(obj,nx=10,ny=10,scalein=100,scaleout=100)
```

In these figures density is expressed in units of individuals per 100 km^2, while the area of the pixels is about 103.7 km^2 and 11.5 km^2, respectively. That calculation is based on:

```
> total.area <- (ylim[2]-ylim[1])*(xlim[2]-xlim[1])*100
> total.area/(10*10)
[1] 103.7427
> total.area/(30*30)
[1] 11.52697
```

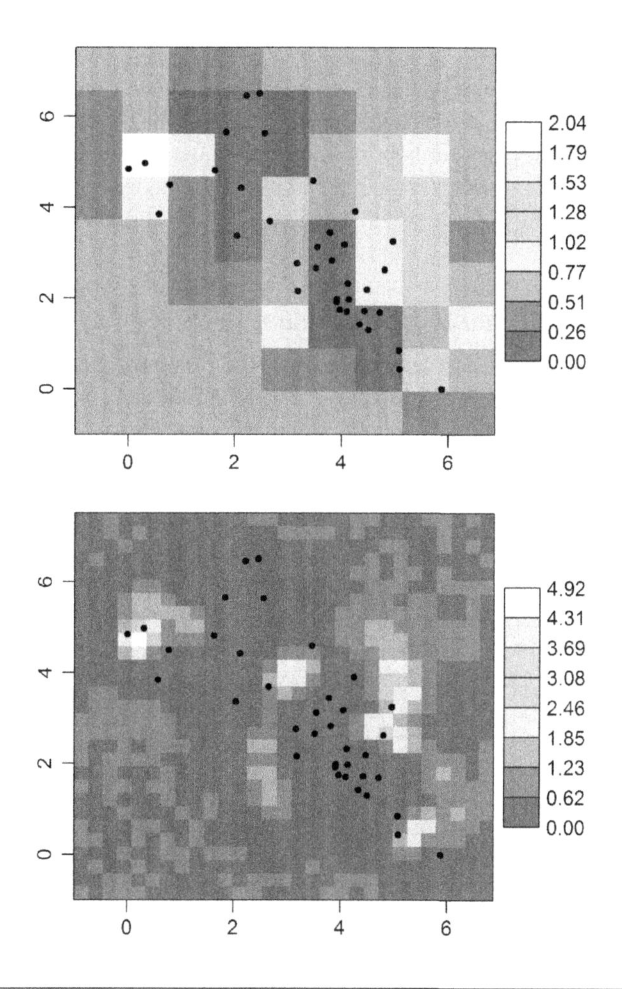

FIGURE 5.6

Density of wolverines (individuals per 100 km^2) in SE Alaska in 2007 based on model SCR0. Map grid cells are about 103.7 km^2 (top panel) and 11.5 km^2 (bottom panel) in area. Dots are the trap locations.

A couple of things are worth noting: First, as we move away from "where the data live" (away from the trap array) we see that the density approaches the mean density. This is a property of the estimator as long as the detection function decreases sufficiently rapidly as a function of distance. Relatedly, it is also a property of statistical smoothers such as splines, kernel smoothers, and regression smoothers—predictions tend toward the global mean as the influence of data diminishes. Another way to think of it is that it is a consequence of the prior, which imposes uniformity, and as you get far away from the data, the predictions tend to the expected constant density under the prior. Another thing to note about this map is that density is not 0 over water (although the coastline is not shown). This might be perplexing to some who are fairly certain that wolverines do not like water. However, there is nothing about this model that recognizes water from non-water and so the model predicts over water *as if* it were habitat similar to that within which the array is nested. But, all of this is OK as far as estimating density goes and, furthermore, we can compute valid estimates of N over any well-defined region, which presumably wouldn't include water if we so wished. Alternatively, areas covered by water could be masked out, which we discussed in Section 5.10.2.

5.11.3 Predicting where an individual lives

The density maps in the previous section show the expected number of individuals per unit area. A closely related problem is that of producing a map of the probable location of a specific individual's activity center. For any observed encounter history, we can easily generate a posterior distribution of $\mathbf{s}_i$ for individual i. In addition, for an individual that is *not* captured, we can use the MCMC output to produce a corresponding plot of where such an individual might live, say $\mathbf{s}_{n+1}$. Obviously, all such uncaptured individuals (for $i = n + 1, \ldots, N$) should have the same posterior distribution. To illustrate, we show the posterior distribution of $\mathbf{s}_1$, the activity center for the individual labeled 1 in the data set, in Figure 5.7. This individual was captured a single time at trap 30, which is circled in Figure 5.7. We see that the posterior

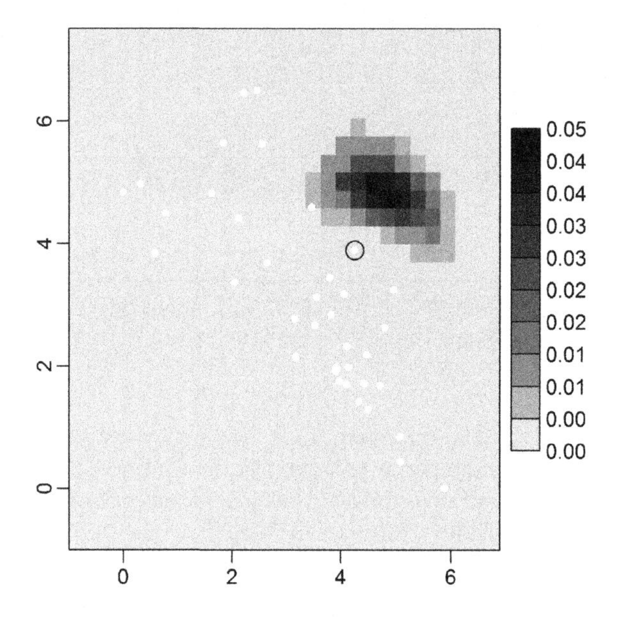

FIGURE 5.7

Posterior probability distribution of $\mathbf{s}_1$, the activity center for individual 1 in the wolverine data set. This individual was captured a single time in one trap (trap 30) which is circled. White dots are trap locations.

distribution is affected by traps of capture *and* traps of non-capture in fairly intuitive ways. In particular, because there are other traps in close proximity to trap 30 in which individual 1 was *not* captured, the model pushes its activity center away from the trap array. The help file for `SCRdensity` shows how to calculate Figure 5.7.

5.12 Effective sample area

One of the key issues in using ordinary capture recapture models that we've brought up over and over again is the issue that the area sampled by a trapping array is unknown—in other words, the N that is estimated by capture-recapture models does not have an explicit region of space associated with it. Classically, this has been addressed in the ad hoc way of prescribing an area that contains the trap array, usually by adding a buffer of some width, which is not estimated as part of the capture-recapture model. In SCR models we avoid the problem of not having an explicit linkage between N and "area," by prescribing explicitly the area within which the underlying point process is defined—the state-space of the point process.

However, it is possible to provide a characterization of effective sampled (or sample) area (ESA) under any SCR model. This is directly analogous to the calculation of "effective strip width" in distance sampling (Buckland et al., 2001; Borchers et al., 2002). The conceptual definition of ESA follows from equating density to "apparent density"—ESA is the magic number that satisfies that equivalence:

$$D = N/A = n/ESA.$$

In other words, the ratio of N to the area of the state-space should be equal to the ratio of the observed sample size n to this number ESA. Both of these should equal density. So, to compute ESA for a model, we substitute $\mathbb{E}(n)$ for n into the above equation, and solve for ESA, to get:

$$ESA = \mathbb{E}(n)/D.$$

Our following development assumes that D is constant, but these calculations can be generalized to allow for D to vary spatially. Imagine our habitat mask for the wolverine data, or the bins we just used to produce a density map, then we can write $\mathbb{E}(n)$ according to

$$\mathbb{E}(n) = \sum_{s} \text{Pr}(\text{encounter}|\mathbf{s})\mathbb{E}(N(\mathbf{s})),$$

where if we prefer to think of this more conceptually, we could replace the summation with an integration (which, in practice, we would just replace with a summation, and so we just begin there). In this expression, note that $\mathbb{E}(N(\mathbf{s}))$ is the expected population size at pixel $\mathbf{s}$ which is the density times the area of the pixel, i.e., $\mathbb{E}(N(\mathbf{s})) = D \times a$. Therefore,

$$\mathbb{E}(n) = D \times a \times \sum_{s} \text{Pr}(\text{encounter}|\mathbf{s}),$$

and (plugging this into the expression above for ESA)

$$ESA = \frac{D \times a \times \sum_s \Pr(\text{encounter}|\mathbf{s})}{D}.$$

We see that D cancels and we have ESA $= a \times \sum_s \Pr(\text{encounter}|\mathbf{s})$. So what you have to do here is substitute in $\Pr(\text{encounter}|\mathbf{s})$ and just sum them up over all pixels. For the Bernoulli model SCR0

$$\Pr(\text{encounter}|\mathbf{s}) = 1 - (1 - p(\mathbf{s}))^K$$

with slight modifications when encounter probability depends on covariates. Thus,

$$ESA = a \sum_s (1 - (1 - p(\mathbf{s}))^K). \tag{5.12.1}$$

Clearly, the calculation of ESA is affected by the use of a habitat mask, because the summation in Eq. (5.12.1) only occurs over pixels that define the state-space.

For the wolverine camera trapping data, we used the 2×2 km habitat mask and the posterior means of p_0 and σ (see Section 5.10.2) to compute the probability of encounter for each $\mathbf{s}$ of the mask points. The result is shown graphically in Figure 5.8. The ESA is the sum of the values plotted in that figure multiplied by 4, the area of each

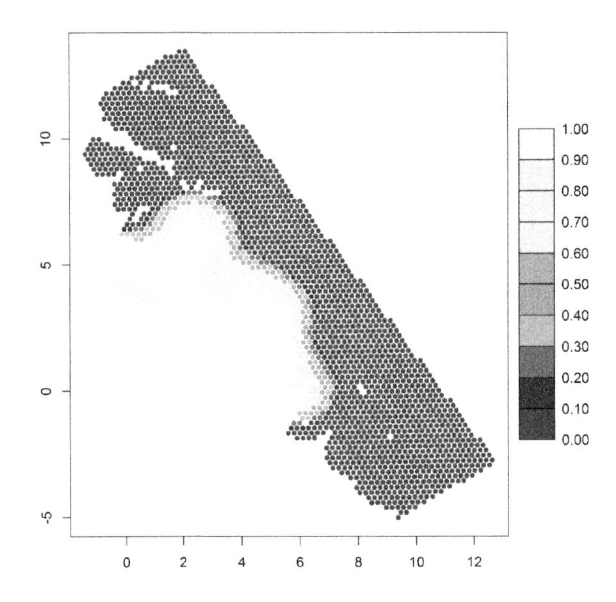

FIGURE 5.8

Probability of encounter used in computing effective sampled area for the wolverine camera trapping array, using the parameter estimates (posterior means) for the 2×2 km habitat mask.

pixel. For the wolverine study, the result is 2507.152 km^2. We note that the probability of encounter declines rapidly to 0 as we move away from the periphery of the camera trap array, indicating the state-space constructed from a 40 km buffered trap array was indeed sufficient for the analysis of these data. And, grid cells contribute less to ESA as their distance to the array increases. An **R** script for producing this figure is in the `wolvESA` function of the `scrbook` package.

5.13 Summary and outlook

In this chapter, we introduced the simplest SCR model—"model SCR0"—which is an ordinary capture-recapture model like model M_0, but augmented with a set of latent individual effects, $\mathbf{s}_i$, which relate encounter probability to some sense of individual location using a covariate, "distance," from $\mathbf{s}_i$ to each trap location. Thus, individuals in close proximity to a trap will have a higher probability of encounter, and *vice versa*. The explicit modeling of individual locations and distance in this fashion resolves classical problems related to estimating density using non-spatial models: unknown sample area for density estimation, and heterogeneous encounter probability due to variable exposure to traps.

SCR models are closely related to classical individual covariate models ("model M_x," as introduced in Chapter 4), but with imperfect information about the individual covariate. Therefore, they are also not too dissimilar from standard GLMMs used throughout statistics and, as a result, we find that they are easy to analyze using standard MCMC methods encased in black boxes such as **WinBUGS** or **JAGS**. We will also see that they are easy to analyze using likelihood methods, which we address in Chapter 6.

Formal consideration of the collection of individual locations $(\mathbf{s}_1, \ldots, \mathbf{s}_N)$ is fundamental to all models considered in this book. In statistical terminology, we think of this collection of points as a realization of a point process. Because SCR models formally link individual encounter history data to an underlying point process, we can obtain formal inferences about the point process. For example, we showed how to produce a density map (Figure 5.6), or even a probability map for an individual's home range center (Figure 5.7). We can also use SCR models as the basis for doing more traditional point process analyses, such as testing for "complete spatial randomness" (CSR) (see Chapter 8), and computing other point process summaries (Illian et al., 2008).

Part of the promise and ongoing challenge of SCR models is to develop models that reflect interesting biological processes, for example interactions among points or temporal dynamics in point locations. In this chapter we considered the simplest possible point process model in which points are independent and uniformly ("randomly") distributed over space. Despite the simplicity of this model, it should suffice in many applications of SCR models, although we do address generalizations in later chapters. Moreover, even though the *prior* distribution on the point locations is uniform, the realized pattern may deviate markedly from uniformity as the observed encounter data provide information to impart deviations from uniformity. Thus, estimated density maps will typically appear distinctly non-uniform (as we saw in the wolverine example). In applications of the basic SCR model, we find that this simple *a priori* model can effectively reflect or adapt to complex realizations of the underlying point process. For example, if individuals are highly territorial then the data should indicate this in the form of individuals not being encountered in the same trap, and the resulting posterior distribution of point locations should therefore reflect non-independence. Obviously, the complexity of posterior estimates of the point pattern will depend on the quantity of data, both number of individuals and captures per individual. Because the point process is such an integral component of SCR models, the state-space of the point process plays an important role in developing SCR models. As we emphasized in this chapter, the state-space is part of the model. And, under certain circumstances, it can have an influence on parameter estimates and other inferences, such as model selection (Chapter 8).

One concept we introduced in this chapter, which has not been discussed much in the literature on SCR models, is the manner in which the encounter probability model relates to a model of space usage by individuals. The standard SCR models of encounter probability can all be motivated as simplistic models of space usage and movement, in which individuals make random use decisions from a probability distribution proportional to the encounter probability model. This simple formulation suggests a direct extension to produce more realistic models, which we discuss in Chapter 13. We consider a number of other important extensions of the basic SCR model in later chapters. For example, we consider models that include covariates that vary by individual, trap, or over time (Chapter 7), spatial covariates on density (Chapter 11), and open populations (Chapter 16), and methods for model assessment and selection (Chapter 8) among other topics. We also consider technical details of maximum likelihood (Chapter 6) and Bayesian (Chapter 17) estimation, so that the interested reader can develop or extend methods to suit their own needs.

CHAPTER

Likelihood Analysis of Spatial Capture-Recapture Models

So far, we have mainly focused on Bayesian analysis of spatial capture-recapture models. And, in the previous chapter we learned how to fit some basic spatial capture-recapture models in **BUGS** engines including **WinBUGS** and **JAGS**. Despite our focus on Bayesian analysis, it is instructive to develop the basic concepts and ideas behind likelihood methods and frequentist inference for SCR models. We recognized earlier (Chapter 5) that SCR models are versions of binomial (or other) GLMs, but with random effects. Throughout statistics, such models are routinely analyzed by likelihood, methods. In particular, likelihood analysis is based on the integrated or marginal likelihood, in which the random effects are removed, by integration, from the conditional-on-**s** likelihood (**s** being the individual activity center). This has been the approach taken by Borchers and Efford (2008), Dawson and Efford (2009), and related papers. Therefore, in this chapter, we provide some conceptual and technical foundation for likelihood-based analysis of spatial capture-recapture models.

We will show here that it is straightforward to compute the maximum likelihood estimates (MLE) for SCR models by integrated likelihood. We develop the MLE framework using **R**, and we also provide a basic introduction to the **R** package `secr` (Efford, 2011a) which does likelihood analysis of SCR models (see also the standalone program **DENSITY** (Efford et al., 2004)). To set the context for likelihood analysis of SCR models, we first analyze the SCR model when N is known because, in that case, analysis proceeds as in standard GLMMs. We generalize the model to allow for unknown N using both conventional ideas based on the "full likelihood" (e.g., Borchers et al., 2002) and also using a formulation based on data augmentation. We obtain the MLEs for the SCR model from the wolverine camera trapping study (Magoun et al., 2011) analyzed in previous chapters to compare/contrast the results.

6.1 MLE for SCR with known N

We noted in Chapter 5 that, with N known, the basic SCR model is a type of binomial model with a random effect. For such models we can obtain maximum likelihood estimators of model parameters based on integrated likelihood. The integrated likelihood is based on the marginal distribution of the data, y, in which the random effects are removed by integration from the conditional-on-**s** distribution of the observations.

Spatial Capture-Recapture. http://dx.doi.org/10.1016/B978-0-12-405939-9.00006-2

See Chapter 2 for a review of marginal, conditional, and joint distributions. Conceptually, any SCR model begins with a specification of the conditional-on-$\mathbf{s}$ model $[y|\mathbf{s}, \boldsymbol{\alpha}]$ and a "prior distribution" for $\mathbf{s}$, say $[\mathbf{s}]$. Then, the marginal distribution of the data y is

$$[y|\boldsymbol{\alpha}] = \int_{\mathcal{S}} [y|\mathbf{s}, \boldsymbol{\alpha}][\mathbf{s}]d\mathbf{s}.$$

When viewed as a function of $\boldsymbol{\alpha}$ for purposes of estimation, the marginal distribution $[y|\boldsymbol{\alpha}]$ is often referred to as the *integrated likelihood.*

It is worth analyzing the simplest SCR model with known N in order to understand the underlying mechanics and basic concepts. These are directly relevant to the manner in which many capture-recapture models are classically analyzed, such as model M_h, and individual covariate models (see Chapter 4).

To develop the integrated likelihood for SCR models, we first identify the conditional-on-$\mathbf{s}$ likelihood. We work with the Bernoulli encounter model here (SCR0) although the basic approach is the same for all other SCR models. The observation model for each encounter observation y_{ij}, for individual i and trap j, specified conditional-on-$\mathbf{s}_i$, is

$$y_{ij}|\mathbf{s}_i \sim \text{Binomial}(K, p_{\alpha}(\mathbf{x}_j, \mathbf{s}_i)), \tag{6.1.1}$$

where we have indicated the dependence of encounter probability, p_{ij}, on $\mathbf{s}$ and parameters $\boldsymbol{\alpha}$ explicitly. For example, p_{ij} might be the Gaussian model given by

$$p_{ij} = \text{logit}^{-1}(\alpha_0)\exp(-\alpha_1\|\mathbf{x}_j - \mathbf{s}_i\|^2),$$

where $\alpha_1 = 1/(2\sigma^2)$. The joint distribution of the data for individual i is the product of J such terms (i.e., contributions from each of J traps).

$$[\mathbf{y}_i|\mathbf{s}_i, \boldsymbol{\alpha}] = \prod_{j=1}^{J} \text{Binomial}(K, p_{\alpha}(\mathbf{x}_j, \mathbf{s}_i)).$$

We note this assumes that encounter of individual i in each trap is independent of encounter in every other trap, conditional on $\mathbf{s}_i$. This is the fundamental property of the basic model SCR0. The marginal likelihood is computed by removing $\mathbf{s}_i$, by integration from the conditional-on-$\mathbf{s}$ likelihood, so we compute:

$$[\mathbf{y}_i|\boldsymbol{\alpha}] = \int_{\mathcal{S}} [\mathbf{y}_i|\mathbf{s}_i, \boldsymbol{\alpha}][\mathbf{s}_i]d\mathbf{s}_i.$$

In most SCR models, $[\mathbf{s}] = 1/A(\mathcal{S})$ where $A(\mathcal{S})$ is the area of the prescribed state-space $\mathcal{S}$ (but see Chapter 11 for alternative specifications of $[\mathbf{s}]$). The joint likelihood for all N individuals, assuming independence of encounters among individuals, is the product of N such terms:

$$\mathcal{L}(\boldsymbol{\alpha}|\mathbf{y}_1, \mathbf{y}_2, \ldots, \mathbf{y}_N) = \prod_{i=1}^{N} [\mathbf{y}_i|\boldsymbol{\alpha}].$$

We emphasize that two independence assumptions are explicit in this development: independence of trap-specific encounters within individuals and independence among individuals. In particular, these assumptions are only valid when individuals are not physically restrained or removed upon capture, and when traps do not "fill up."

The key operation for computing the likelihood is solving a two-dimensional integration problem. There are some general-purpose **R** packages that implement a number of multi-dimensional integration routines including `adapt` (Genz et al., 2007) and `R2cuba` (Hahn et al., 2010). In practice, we won't rely on these extraneous **R** packages (except see Chapter 11 for an application of `R2cuba`), but instead will use perhaps less efficient methods in which we replace the integral with a summation over an equal-area mesh of points on the state-space $\mathcal{S}$ and explicitly evaluate the integrand at each point. We invoke the rectangular rule for integration here,[1] in which we evaluate the integrand on a regular grid of points of equal area and compute the average of the integrand over that grid of points. Let $u = 1, 2, \ldots, nG$ index a grid of nG points, $\mathbf{s}_u$, where the area of grid cells is constant. In this case, the integrand, i.e., the marginal pmf of $\mathbf{y}_i$, is approximated by

$$[\mathbf{y}_i|\boldsymbol{\alpha}] = \frac{1}{nG}\sum_{u=1}^{nG}[\mathbf{y}_i|\mathbf{s}_u, \boldsymbol{\alpha}]. \tag{6.1.2}$$

This is a specific case of the general expression that could be used for approximating the integral for any arbitrary distribution $[\mathbf{s}]$. The general case is

$$[\mathbf{y}_i|\boldsymbol{\alpha}] = \frac{A(\mathcal{S})}{nG}\sum_{u=1}^{nG}[\mathbf{y}_i|\mathbf{s}_u, \boldsymbol{\alpha}][\mathbf{s}_u].$$

Under the uniformity assumption, $[\mathbf{s}] = 1/A(\mathcal{S})$ and thus the grid-cell area cancels in the above expression to yield Eq. (6.1.2). The rectangular rule for integration can be seen as an application of the Law of Total Probability for a discrete random variable $\mathbf{s}$, having nG unique values with equal probabilities $1/nG$.

6.1.1 Implementation (simulated data)

Here, we will illustrate how to carry out this integration and optimization based on the integrated likelihood using simulated data (see Section 5.5). Using `simSCR0`, we simulate data for 100 individuals and an array of 25 traps laid out in a 5×5 grid of traps having unit spacing. The specific encounter model is the Gaussian model. The 100 activity centers were simulated on a state-space defined by an 8×8 square within which the trap array was centered (thus the trap array is buffered by 2 units). In the following **R** commands we generate the data and harvest the required data objects:

[1] e.g., http://en.wikipedia.org/wiki/Rectangle_method.

```
  ## simulate a complete data set
> data <- simSCR0(discard0=FALSE,rnd=2013)
  ## extract the objects that we need for analysis
> y <- data$Y
> traplocs <- data$traplocs
> nind <- nrow(y)    ## in this case nind=N
> J <- nrow(traplocs)
> K <- data$K
> xlim <- data$xlim
> ylim <- data$ylim
```

Now, we need to define the integration grid, say `G`, which we do with the following set of **R** commands (here, `delta` is the grid spacing):

```
> delta <- .2
> xg <- seq(xlim[1]+delta/2,xlim[2]-delta/2,by=delta)
> yg <- seq(ylim[1]+delta/2,ylim[2]-delta/2,by=delta)
> npix <- length(xg)          # valid for square state-space only
> G <- cbind(rep(xg,npix),sort(rep(yg,npix)))
> nG <- nrow(G)
```

In this case, the integration grid is set up as a grid with spacing $\delta = 0.2$, which produces a 40×40 grid of points for evaluating the integrand with the state-space buffer set at 2. We note that the integration grid is set up here to correspond exactly to the state-space used in simulating the data. However, in practice, we wouldn't know this, and our estimate of N (for the unknown case, see below) could be sensitive to choice of the extent of the integration grid. As we've discussed previously, density, which is N standardized by the area of the state-space, will not be so sensitive in most cases.

We are now ready to compute the conditional-on-**s** likelihood and carry out the marginalization described by Eq. (6.1.2). We need to do this by defining an **R** function that computes the likelihood for the integration grid, as a function of the data objects `y` and `traplocs`, which were created above. However, it is a bit untidy to store the grid information in your workspace, and define the likelihood function in a way that depends on these things that exist in your workspace. Therefore, we build the **R** function so that it computes the integration grid *within* the function, thereby avoiding potential problems if our trapping grid locations change, or if we want to modify the state-space buffer. We therefore define the function, called `intlik1`, to which we pass the data objects and other information necessary to compute the marginal likelihood. This function is available in the `scrbook` package (use `?intlik1` at the **R** prompt). The code is reproduced here:

```
intlik1 <- function(parm,y=y,X=traplocs, delta=.2, ssbuffer=2){

  Xl <- min(X[,1]) - ssbuffer    ## These lines of code are setting up the
  Xu <- max(X[,1]) + ssbuffer    ## support for the integration which is
  Yu <- max(X[,2]) + ssbuffer    ## the same as the state-space of "s"
  Yl <- min(X[,2]) - ssbuffer
  xg <- seq(Xl+delta/2,Xu-delta/2,,length=npix)
  yg <- seq(Yl+delta/2,Yu-delta/2,,length=npix)
  npix<- length(xg)
```

```
    G <- cbind(rep(xg,npix),sort(rep(yg,npix)))
    nG <- nrow(G)
    D <- e2dist(X,G)

    alpha0 <- parm[1]
    alpha1 <- exp(parm[2])  # alpha1 restricted to be positive here

    probcap <- plogis(alpha0)*exp(-alpha1*D*D)
    Pm <- matrix(NA,nrow=nrow(probcap),ncol=ncol(probcap))
                    # Frequency of all-zero encounter histories
    n0 <- sum(apply(y,1,sum)==0)
                    # Encounter histories with at least 1 detection
    ymat <- y[apply(y,1,sum)>0,]
    ymat <- rbind(ymat,rep(0,ncol(ymat)))
    lik.marg <- rep(NA,nrow(ymat))

    for(i in 1:nrow(ymat)){
        ## Nextline: log conditional likelihood for ALL possible values of s
        Pm[1:length(Pm)] <- dbinom(rep(ymat[i,],nG),K,probcap[1:length(Pm)],
                                   log=TRUE)
        ## Next line: sum the log conditional likelihoods, exp() result
        ##   same as taking the product
        lik.cond <- exp(colSums(Pm))
        ## Take the average value == computing marginal
        lik.marg[i] <- sum( lik.cond*(1/nG))
    }
    ## n0 = number of all-0 encounter histories
    nv <- c(rep(1,length(lik.marg)-1),n0)
    return( -1*( sum (nv*log(lik.marg)) ) )
}
```

We emphasize that this and subsequent functions are not meant to be general-purpose routines for solving all of your SCR problems but, rather, they are meant for illustrative purposes—so you can see how the integrated likelihood is constructed and how we connect it to data and other information that is needed.

The function `intlik1` accepts as input the encounter history matrix, `y`, the trap locations, `X`, and the state-space buffer. This allows us to vary the state-space buffer and easily evaluate the sensitivity of the MLE to the size of the state-space. Note that we have a peculiar handling of the encounter history matrix `y`. In particular, we remove the all-zero encounter histories from the matrix and tack on a single all-zero encounter history as the last row, which then gets weighted by the number of such encounter histories (`n0`). This is a bit long-winded and strictly unnecessary when N is known, but we did it this way because the extension to the unknown-N case is now transparent (as we demonstrate in the following section). The matrix `Pm` holds the log-likelihood contributions of each encounter frequency for each possible state-space location of the individual. The log contributions are summed up and the result exponentiated on the next line, producing `lik.cond`, the conditional-on-**s** likelihood (Eq. (6.1.1)). The marginal likelihood (`lik.marg`) sums up the conditional elements weighted by the probabilities [**s**] (Eq. (6.1.2)).

This is a fairly primitive function that doesn't allow much flexibility in the data structure. For example, it assumes that K, the number of replicates, is constant for each trap. Further, it assumes that the state-space is a square. We generalize this to some extent later in this chapter.

Here is the **R** command for maximizing the likelihood using `nlm` (the function `optim` could also be used) and saving the results into an object called `frog`. The output is a list shown below for the simulated data set

```
> starts <- c(-2,2)
> frog <- nlm(intlik1,starts,y=y,X=traplocs,delta=.1,ssbuffer=2,hessian=TRUE)
> frog

$minimum
[1] 297.1896

$estimate
[1] −2.504824 2.373343

$gradient
[1] −2.069654e-05 1.968754e-05

$hessian
          [,1]      [,2]
[1,]  48.67898 -19.25750
[2,] -19.25750  13.34114

$code
[1] 1

$iterations
[1] 11
```

Details about this output can be found on the help page for `nlm`. We note briefly that `frog$minimum` is the negative log-likelihood value at the MLEs, which are stored in the `frog$estimate` component of the list. The order of the parameters is as they are defined in the likelihood function so, in this case, the first element (value = −2.504824) is the logit transform of p_0 and the second element (value = 2.373343) is the value of α_1, the "coefficient" on distance-squared. The Hessian is the observed Fisher information matrix, which can be inverted to obtain the variance-covariance matrix using the command:

```
> solve(frog$hessian)
```

It is worth drawing attention to the fact that the estimates are slightly different than the Bayesian estimates reported in Section 5.6. There are several reasons for this. First, Bayesian inference is based on the posterior distribution and it is not generally the case that the MLE should correspond to any particular value of the posterior distribution. If the prior distributions in a Bayesian analysis are uniform, then the (multivariate) mode of the posterior is the MLE, but note Bayesians almost always report posterior *means* so that there will typically be some discrepancy. Secondly, we have implemented an approximation to the integral here and there might be a slight bit of error induced by that. We will evaluate that shortly. Third, the Bayesian analysis by MCMC is itself subject to some amount of Monte Carlo error, which the analyst should always be aware of in practical situations. All of these different explanations are likely responsible for some of the discrepancy. Despite these, we see general consistency between the two estimates.

In summary, for the basic SCR model, computing the integrated likelihood is a simple task when N is known. Even for N unknown it is not too difficult, and we will do that shortly.

6.2 MLE when N is unknown

Here, we build on the previous introduction to integrated likelihood and consider the case in which N is unknown. We will see that adapting the analysis based on the known-N model is straightforward for the more general problem. The main distinction is that we don't observe the all-zero encounter history so we have to make sure we compute the probability for that encounter history. We do that by including a row of zeros in the encounter history matrix and treating the number of such all-zero encounter histories (that is, the number of individuals *not* encountered) as an unknown parameter of the model. Call that unknown quantity n_0, so that $N = n_0 + n$, where n is the number of unique individuals encountered. We will usually parameterize the likelihood in terms of n_0, because optimization over a parameter space in which $\log(n_0)$ is unconstrained is preferred to a parameter space in which N must be constrained so that $N \geq n$. With n_0 unknown, we have to be sure to include a combinatorial term in the likelihood to account for the fact that, of the n observed individuals, there are $\binom{N}{n}$ ways to realize a sample of size n. The combinatorial term involves the unknown n_0 and thus it must be included in the likelihood. In evaluating the log-likelihood, we have to compute terms such as the log-factorial, $\log(N!) = \log((n_0 + n)!)$. We do this in **R** by making use of the log-gamma function (`lgamma`) and the identity

$$\log(N!) = \text{lgamma}(N + 1).$$

Therefore, to compute the likelihood, we require the following three components:
(1) The marginal probability of each $\mathbf{y}_i$ as before:

$$[\mathbf{y}_i|\boldsymbol{\alpha}] = \int_{\mathcal{S}} [\mathbf{y}_i|\mathbf{s}_i, \boldsymbol{\alpha}][\mathbf{s}_i] d\mathbf{s}_i.$$

(2) The probability of an all-0 encounter history:

$$\pi_0 = [\mathbf{y} = \mathbf{0}|\boldsymbol{\alpha}] = \int_{\mathcal{S}} \text{Binomial}(\mathbf{0}|\mathbf{s}_i, \boldsymbol{\alpha})[\mathbf{s}_i] d\mathbf{s}_i.$$

(3) The combinatorial term: $\binom{N}{n}$.

Then, the marginal likelihood has this form:

$$\mathcal{L}(\boldsymbol{\alpha}, n_0|\mathbf{y}) = \frac{N!}{n!n_0!}\left\{\prod_{i=1}^{n}[\mathbf{y}_i|\boldsymbol{\alpha}]\right\}\pi_0^{n_0}. \quad (6.2.1)$$

This is discussed in Borchers and Efford (2008, p. 379) as the conditional-on-N form of the likelihood—we also call it the "binomial form" of the likelihood because of its appearance.

Operationally, things proceed much as before: We compute the marginal probability of each observed $\mathbf{y}_i$, i.e., by removing the latent $\mathbf{s}_i$ by integration. In addition, we compute the marginal probability of the "all-zero" encounter history, and make sure to weight it n_0 times. We accomplish this by "padding" the data set with a single encounter history having $y_{n+1,j} = 0$ for all traps $j = 1, 2, \ldots, J$. Then we make sure to include the combinatorial term in the likelihood or log-likelihood computation. We demonstrate this by analyzing a simulated data set. To set some things up in our workspace we do this:

```
## Obtain a simulated data set
> data <- simSCR0(discard0=TRUE, rnd=2013)

## Extract the items we need for analysis
> y <- data$Y
> nind <- nrow(y)
> traplocs <- data$traplocs
> J <- nrow(traplocs)
> K <- data$K
```

Recall that these data are simulated by default with $N = 100$, on an 8×8 unit state-space representing the trap locations buffered by two units, although you can modify the simulation script easily.

As before, the likelihood is defined as an **R** function, `intlik2`, which takes as an argument the unknown parameters of the model and additional arguments as prescribed: the encounter history matrix `y`, the trap locations `traplocs`, the spacing of the integration grid (argument `delta`), and the state-space buffer. Here is the new likelihood function:

```
intlik2 <- function(parm,y=y,X=traplocs,delta=.3,ssbuffer=2){

  Xl <- min(X[,1]) - ssbuffer
  Xu <- max(X[,1]) + ssbuffer
  Yu <- max(X[,2]) + ssbuffer
  Yl <- min(X[,2]) - ssbuffer

  xg <- seq(Xl+delta/2,Xu-delta/2,delta)
  yg <- seq(Yl+delta/2,Yu-delta/2,delta)
  npix.x <- length(xg)
  npix.y <- plength(yg)
  area <- (Xu-Xl)*(Yu-Yl)/((npix.x)*(npix.y))
  G <- cbind(rep(xg,npix.y),sort(rep(yg,npix.x)))
  nG <- nrow(G)
```

```
  D <- e2dist(X,G)
  # extract the parameters from the input vector
  alpha0 <- parm[1]
  alpha1 <- exp(parm[2])
  n0 <- exp(parm[3])  # note parm[3] lives on the real line
  probcap <- plogis(alpha0)*exp(-alpha1*D*D)
  Pm <- matrix(NA,nrow=nrow(probcap),ncol=ncol(probcap))
  ymat <- rbind(y,rep(0,ncol(y)))

  lik.marg <- rep(NA,nrow(ymat))
  for(i in 1:nrow(ymat)){
     Pm[1:length(Pm)] <- (dbinom(rep(ymat[i,],nG),K,probcap[1:length(Pm)],
                           log=TRUE))
     lik.cond <- exp(colSums(Pm))
     lik.marg[i] <- sum(lik.cond*(1/nG) )
  }
  nv <- c(rep(1,length(lik.marg)-1),n0)
  ## part1 here is the combinatorial term.
  ## math: log(factorial(N)) = lgamma(N+1)
  part1 <- lgamma(nrow(y)+n0+1) - lgamma(n0+1)
  part2 <- sum(nv*log(lik.marg))
  return( -1*(part1+part2))
}
```

To execute this function for the data we created with `simSCR0`, we execute the following commands (saving the results in our friend `frog`). This results in the usual output, including the parameter estimates, the gradient, and the numerical Hessian, which is useful for obtaining asymptotic standard errors (see below):

```
> starts <- c(-2.5,0,4)
> frog <- nlm(intlik2,starts,hessian=TRUE,y=y,X=traplocs,delta=.2,ssbuffer=2)

Warning message:
In nlm(intlik2, starts, hessian = TRUE, y = y, X = traplocs, delta = 0.2,  :
  NA/Inf replaced by maximum positive value

> frog
$minimum
[1] 113.5004

$estimate
[1] -2.538333 0.902807 4.232810

[... additional output deleted ...]
```

Executing `nlm` here usually produces one or more **R** warnings due to numerical calculations happening on extremely small or large numbers (calculation of p near the edge of the state-space), and they also happen if a poor parameterization is used, which produces evaluations of the objective function beyond the boundary of the parameter space (e.g., $n_0 < 0$). Such numerical warnings can often be minimized or avoided altogether by picking judicious starting values of parameters or properly transforming or scaling the parameters but, most of the time they can be ignored. You will see from the `nlm` output that the algorithm performed satisfactorily in minimizing the objective function. The estimate of population size, $\hat{N}$, for the state-space (using the default state-space buffer) is

```
> Nhat <- nrow(y) + exp(4.2328)     ### This is n + MLE of n0
> Nhat
[1] 110.9099
```

This estimate differs from the data-generating value ($N = 100$), as we might expect for a single realization. We usually will present an estimate of uncertainty associated with this MLE, which we can obtain by inverting the Hessian. Note that $\text{Var}(\hat{N}) = n + \text{Var}(\hat{n}_0)$. Since we have parameterized the model in terms of $\log(n_0)$, we use the delta method[2] described in Williams et al. (2002, Appendix F4) (see also Ver Hoef, 2012) to obtain the variance on the scale of n_0 as follows:

```
> (exp(4.2328)^2)*solve(frog$hessian)[3,3]
[1] 260.2033

> sqrt(260)
[1] 16.12452
```

Therefore, the asymptotic "Wald-type" confidence interval for N is $110.91 \pm 1.96 \times 16.125 = (79.305, 142.515)$. To report this in terms of density, we scale appropriately by the area of the prescribed state-space, which is 64 square units in area. Our MLE of D is $\hat{D} = 110.91/64 = 1.733$ individuals per square unit. To get the standard error for $\hat{D}$ we need to divide the SE for $\hat{N}$ by the area of the state-space, and so $\text{SE}(\hat{D}) = (1/64) \times 16.12452 = 0.252$.

6.2.1 Integrated likelihood under data augmentation

The analysis developed in the previous sections is based on the likelihood in which N (or n_0) is an explicit parameter. This is usually called the "full likelihood" or sometimes "unconditional likelihood" (Borchers et al., 2002), because it is the likelihood for all individuals in the population, not just those that have been captured, i.e., it is not the likelihood that is *conditional on capture*. It is also possible to express an alternative unconditional likelihood using data augmentation, replacing the parameter N with ψ as we described in Section 4.2 (also see Royle and Dorazio, 2008, Section 7.1.6). While it is possible to carry out likelihood analysis of models under data augmentation, we primarily advocate data augmentation for Bayesian analysis.

6.2.2 Extensions

So far, we have only considered basic SCR models with no additional covariates. However, in practice, we are interested in covariate effects including "behavioral response," sex-specificity of parameters, and potentially others. Some of these can be added directly to the likelihood if the covariate is fixed and known for all individuals, captured or not. An example is a behavioral response, which amounts to having a

[2]We found a good set of notes on the delta approximation on Dr. David Patterson's ST549 notes: http://www.math.umt.edu/patterson/549/Delta.pdf.

covariate $C_{ik} = 1$ if individual i was captured prior to occasion k and $C_{ik} = 0$ otherwise. For uncaptured individuals, $C_{ik} = 0$ for all k. Royle et al. (2011b) called this a global behavioral response, because the covariate is defined for all traps, no matter the trap in which an individual was captured. We also define a *local* behavioral response, that occurs at the level of the trap, i.e., $C_{ijk} = 1$ if individual i was captured in trap j prior to occasion k, etc. Trap-specific covariates such as trap type or status, or time-specific covariates such as date, are easily accommodated as well. As an example, Kéry et al. (2010) develop a model for the European wildcat (*Felis silvestris*) in which traps are either baited or not (a trap-specific covariate with only two values), and encounter probability varies over time in the form of a quadratic seasonal response. We consider models with behavioral response or fixed covariates in Chapter 7. The integrated likelihood routines we provided above can be modified directly for such cases, and we leave that to you to investigate.

Sex specificity is more difficult to deal with, since sex is not known for uncaptured individuals (and sometimes not even for all captured individuals). To analyze such models, we do Bayesian analysis of the joint likelihood using data augmentation (Gardner et al., 2010b; Russell et al., 2012), discussed further in Chapter 7. For such covariates (i.e., that are not fixed and known for all individuals), it is somewhat more challenging to do MLE based on the joint or full likelihood as we have developed it above. Instead, it is more conventional to use what is colloquially referred to as the "Huggins-Alho"-type model, which is one of the approaches taken in the software package `secr` (Efford, 2011a). We introduce the `secr` package in Section 6.5.

6.3 Classical model selection and assessment

In most analyses, one is interested in choosing from among various potential models, or ranking models, or otherwise assessing the relative merits of a set of models. A good thing about classical analysis based on likelihood is that we can apply Akaike Information Criterion (AIC) methods (Burnham and Anderson, 2002; Borchers and Efford, 2008, Section 6.5) without difficulty. AIC is convenient for assessing the relative merits of these different models, although if there are only a few models, it is not objectionable to use hypothesis tests or confidence intervals to determine importance of effects. A second model selection context has to do with choosing among various detection models (e.g., Gaussian, hazard rate, exponential, etc.), although, as a general rule, we don't recommend this application of model selection. This is because there is hardly ever (if at all) a rational subject-matter based reason motivating specific distance functions. As a result, we believe that doing too much model selection can lead to overfitting and thus potentially overstatement of precision. This is the main reason that we haven't loaded you down with a basket of models for detection probability so far, although we discuss many possibilities in Chapter 7.

Goodness-of-fit or model-checking—For many standard capture-recapture models, it is possible to identify goodness-of-fit statistics based on the multinomial likelihood (Cooch and White, 2006, Chapter 5), and evaluate model adequacy using formal statistical tests. Similar strategies can be applied to SCR models using expected cell frequencies based on the marginal distribution of the observations. Also, because computing MLEs is somewhat more efficient in many cases compared to Bayesian analysis, it is sometimes feasible to use bootstrap methods. At the present time, there are few applications of goodness-of-fit testing for SCR models based on likelihood inference (but see Borchers and Efford, 2008, Section 6.5). We discuss the use of Bayesian p-values for assessing model fit in Chapter 8. A potential practical problem in trying to evaluate goodness-of-fit is that, in realistic sample sizes, fit tests may lack the power to detect departures from the model under consideration and so they may not be generally useful in practice.

6.4 Likelihood analysis of the wolverine camera trapping data

Here, we compute the MLEs for the wolverine data using an expanded version of the function we developed in the previous section. To accommodate that each trap might be operational a variable number of nights, we provided an additional argument to the likelihood function (allowing for a vector $\mathbf{K} = (K_1, \ldots, K_J)$), which requires also a modification to the construction of the likelihood. The more general function (`intlik3`) is given in the **R** package `scrbook`. In addition, this modification accommodates the state-space as a rectangle, and we included a line of code to compute the state-space area, which we apply below for computing density. To use this function to obtain the MLEs for the wolverine camera trap study, we execute the following commands (note: these commands will execute if you type `example(intlik3)`):

```
> library(scrbook)
> data(wolverine)

> traps <- wolverine$wtraps
> traplocs <- traps[,2:3]/10000
> K.wolv <- apply(traps[,4:ncol(traps)],1,sum)

> y3d <- SCR23darray(wolverine$wcaps,traps)
> y2d <- apply(y3d,c(1,2),sum)

> starts <- c(-1.5,0,3)

> wolv <- nlm(intlik3,starts,hessian=TRUE,y=y2d,K=K.wolv,X=traplocs,
              delta=.2,ssbuffer=2)

> wolv
$minimum
[1] 220.4313

$estimate
[1] -2.8176120 0.2269395 3.5836875

[.... output deleted ....]
```

Of course we're interested in obtaining an estimate of population size for the prescribed state-space, or density, and associated measures of uncertainty which we do using the delta method. To do all of that we need to manipulate the output of `nlm`, since we have our estimate in terms of $\log(n_0)$. We execute the following commands:

```
> wolv <- nlm(intlik3,starts,hessian=TRUE,y=y2d,K=K.wolv,X=traplocs,delta=.2,
             ssbuffer=2)
> Nhat <- nrow(y2d)+exp(wolv$estimate[3])
> area <- attr(intlik3(starts,y=y2d,K=K.wolv,X=traplocs,delta=.2,ssbuffer=2),
               "SSarea")
> Dhat <- Nhat/area

> Dhat
[1] 0.5494947

> SE <- (1/area)*exp(wolv$estimate[3])*sqrt(solve(wolv$hessian)[3,3])

> SE
[1] 0.1087073
```

Our estimate of density is 0.55 individuals per "standardized unit" which is 100 km^2, because we divided UTM coordinates by 10,000. So this is about 5.5 individuals per 1,000 km^2, with a SE of around 1.09 individuals. This compares closely with 5.77 reported in Section 5.9 based on Bayesian analysis of the model.

6.4.1 Sensitivity to integration grid and state-space buffer

The effect of approximating the integral by a discrete mesh of points is that it induces some numerical error that increases as the coarseness of the mesh increases. To evaluate the effect (or sensitivity) of the integration spacing, we obtained the MLEs for a state-space buffer of 2 (standardized units) and for integration grids with spacing $\delta = .3, .2, .1, .05$. The MLEs for these four cases including the relative runtime are given in Table 6.1. We see that the results change only slightly as the integration grid changes. Conversely, the runtime on the platform of the day for the four cases increases rapidly. These runtimes could be regarded in relative terms, across platforms, for gaging the decrease in speed as the fineness of the integration grid increases.

Table 6.1 Runtime and MLEs for different integration grid resolutions for the wolverine camera trapping data.

		Estimates		
δ	Runtime(s)	$\hat{\alpha}_0$	$\hat{\alpha}_1$	$\widehat{\log(n_0)}$
0.30	9.9	−2.820	1.258	3.570
0.20	32.3	−2.818	1.255	3.584
0.10	115.1	−2.817	1.255	3.599
0.05	407.3	−2.817	1.255	3.607

Table 6.2 Results of the effect of the state-space buffer on the MLE. Given are the state-space buffer, area of the state-space (area), the MLE of N ($\hat{N}$) for the prescribed state-space, and the corresponding MLE of density ($\hat{D}$).

Buffer	Area	$\hat{N}$	$\hat{D}$
1.0	66.982	37.733	0.563
1.5	84.362	46.210	0.548
2.0	103.743	57.006	0.549
2.5	125.123	69.036	0.552
3.0	148.503	82.175	0.553
3.5	173.884	96.440	0.555
4.0	201.264	111.835	0.556

We studied the effect of the state-space buffer on the MLEs, using a fixed $\delta = .2$ for all analyses. We used state-space buffers of 1–4 units stepped by .5. As we can see (Table 6.2), the estimates of D stabilize rapidly and the incremental difference is small for buffers larger than 1.5 units.

6.4.2 Using a habitat mask (restricted state-space)

In Section 5.10 we used a discrete representation of the state-space in order to have control over its extent and shape. This makes it easy to do things like clip out non-habitat, or create a *habitat mask* that defines suitable habitat. Clearly, that formulation of the model is relevant to the calculation of the marginal likelihood in the sense that the discrete state-space is equivalent to the integration grid. Thus, for example, we could easily compute the MLE of parameters under some model with a restricted state-space merely by creating the required state-space at whatever grid resolution is desired, and then inputting that state-space into the likelihood functions above, instead of computing it within the function. We can create an explicit state-space grid for integration from arbitrary polygons or GIS shapefiles, which we demonstrate here. Our approach is to create the integration grid (state-space grid) outside of the likelihood evaluation, and then determine which points of the grid lie in the polygon defined by the shapefile using functions in the **R** packages `sp` and `maptools`. For each point in the state-space grid (object `G`, in the code below which is assumed to exist in your workspace), we determine whether it is inside the polygon,[3] identifying such points with a value of `mask=1`, and `mask=0` for points that are *not* in the polygon. We load the shapefile by applying the `readShapeSpatial` function.

[3]We perform this check using the `over` function. This function takes as its second argument an object of the class "SpatialPolygons" or "SpatialPolygonsDataFrame," which can hold additional information for each polygon, and the output value of the function differs slightly for these two classes: if using a "SpatialPolygons" object, the function returns a vector of length equal to the number of points (e.g., in the example above), but if using a "SpatialPolygonsDataFrame" it returns a data frame (e.g., see Section 17.7 in Chapter 17). If you use the `over` function, make sure you know the class of your second argument so that when processing the function output you index it correctly.

We have saved the result into an **R** data object called SSp, which is in the scrbook package. Here are the **R** commands for doing this (see also the helpfile ?intlik4):

```
> library(maptools)
> library(sp)
> library(scrbook)

#### If we have the .shp file in place, we would use this command:
#### SSp <- readShapeSpatial('Sim_Polygon.shp')
#### The object SSp is in data(fakeshapefile)
> data(fakeshapefile)
> Pcoord <- SpatialPoints(G)
> PinPoly <- over(Pcoord,SSp)  ### determine if each point is in polygon
> mask <- as.numeric(!is.na(PinPoly[,1]))  ## convert to binary 0/1
> G <- G[mask==1,]
```

We created the function intlik4 to accept the integration grid as an explicit argument; this function is also available in the package scrbook.

We apply this modified function to the wolverine camera trapping study. Royle et al. (2011b) created 2, 4, and 8-km state-space grids so as to remove "non-habitat" (mostly ocean, bays, and large lakes). To set up the wolverine data and fit the model with the habitat mask using maximum likelihood we execute the following commands (data loaded as in Section 6.4):

```
> starts <- c(-1.5,0,3)
> wolv <- nlm(intlik4, starts, y=y2d, K=K.wolv, X=traplocs, G=G)

> wolv

$minimum
[1] 225.8355

$estimate
[1] -2.9955424 0.2350885 4.1104757

[... some output deleted ...]
```

As before, we convert the parameter estimates to estimates of total population size for the prescribed state-space, and then obtain an estimate of density (per 1,000 km^2) using the area computed as the number of pixels in the state-space grid, G, multiplied by the area per grid cell. In the present case (the calculation above) we used a state-space grid with 2 km $\times$ 2 km pixels. Finally, we compute a standard error using the delta approximation:

```
> area <- nrow(G)*4
# Nhat = n (observed) + MLE of n0 (not observed)
> Nhat <- 21 + exp(wolv$estimate[3])
> SE <- exp(wolv$estimate[3])*sqrt(solve(wolv$hessian)[3,3])
> D <- (Nhat/(nrow(G)*area))*1000
> SE.D <- (SE/(nrow(G)*area))*1000
```

We did this for each 2 km, 4 km, and 8 km state-space grid, which produced the estimates summarized in Table 6.3. These estimates compare with the 8.6 (2-km grid) and 8.2 (8-km grid) reported in Royle et al. (2011b) based on a clipped state-space as described in Section 5.10.

Table 6.3 Maximum likelihood estimates (MLEs) and asymptotic standard errors (SE) for the wolverine camera trapping data using 2, 4, and 8 km state-space grids and a habitat mask.

Grid	α_0	α_1	$\log(n_0)$	N	SE	D(1000)	SE
2	−3.00	1.27	4.11	81.98	16.31	8.31	1.65
4	−2.99	1.34	4.16	84.88	16.76	8.57	1.69
8	−3.05	1.08	4.06	78.89	15.31	7.85	1.52

6.5 DENSITY and the R package `secr`

DENSITY is a software program developed by Efford (2004) for fitting spatial capture-recapture models based mostly on classical maximum likelihood estimation and related inference methods. Efford (2011a) has also released an **R** package called `secr`, that contains much of the functionality of **DENSITY**, but also incorporates new models and features. Here, we briefly introduce the `secr` package, which we prefer to use over **DENSITY**, because it allows us to remain in the **R** environment for data processing and summarization. We provide a brief introduction to `secr` and some of its capabilities here, and we also use it for doing some analysis in other parts of this book (e.g., Chapter 7). We believe that `secr` will be sufficient for many (if not most) of the SCR problems that one might encounter. It provides a flexible analysis platform, with a large number of summary features, and "publication ready" output. Its user interface is clean and intuitive to **R** users, and it has been stable, efficient, and reliable in the (fairly extensive) evaluations that we have done.

To install and run models in `secr`, you must download the package and load it in **R**.

```
> install.packages("secr")
> library(secr)
```

`secr` allows the user to simulate data and fit a suite of models with various detection functions and covariate responses. It also contains a number of helpful constructor functions for creating objects of the proper class that are recognized by other `secr` functions. The `secr` help manual can be accessed with the command:

```
> RShowDoc("secr-manual", package = "secr")
```

We note that `secr` has many capabilities that we will not cover or do so only sparingly. We encourage you to read through the manual, the extensive documentation, and the vignettes, in order to get a better understanding of what the package is capable of.

The main model-fitting function in `secr` is called `secr.fit`, which makes use of the standard **R** model specification framework with tildes. As an example, the equivalent of the basic model SCR0 is fitted as follows:

```
> secr.fit(capturedata, model = list(D ~ 1, g0 ~ 1, sigma ~ 1),
           buffer = 20000)
```

where capturedata is the object created by secr containing the encounter history data and the trap information, and the model expression g0~1 indicates the intercept-only (i.e., constant) model. Note that we use p_0 for the baseline encounter probability parameter, which is g_0 in secr notation. A number of possible models for encounter probability can be fitted including both pre-defined variables (e.g., t and b corresponding to "time" and "behavior"), and user-defined covariates of several kinds. For example, to include a global behavioral response, this would be written as g0~b. The discussion of this (global versus local trap-specific behavioral response) and other covariates is developed more in Chapter 7. We can also model covariates on density in secr, which we discuss in Chapter 11. It is important to note that secr requires the buffer distance to be defined in meters and density will be returned as number of animals per hectare. Thus to make comparisons between secr and output from other programs, we will often have to convert the density to the same units.

Before we can fit a model, the data must first be packaged properly for secr. We require data files that contain two types of information: trap layout (location and identification information for each trap), which is equivalent to the trap deployment file (TDF) described in Section 5.9, and the capture data file containing sampling *session*, animal identification, trap occasion, and trap location, equivalent in information content to the encounter data file (EDF). Sample session can be thought of as primary period identifier in a robust design-like framework—it could represent a yearly sample or multiple sample periods within a year, each of them producing data on a closed population. We discuss "multi-session" models in more detail in Section 6.5.4, and Chapter 14.

There are three important constructor functions that help bundle your data for use in secr: read.traps, make.capthist, and read.mask. We provide a brief description of each here, and apply them to our wolverine camera trapping data in Section 6.5.2:

1. read.traps: This function points to an external file *or* **R** data object containing the trap coordinates, and other information, and requires specification of the type of encounter devices (described in Section 6.5.1). A typical application of this function looks like the following, invoking the data= option when there is an existing **R** object containing the trap information:

```
> trapfile <- read.traps(data=traps, detector="proximity")
```

2. make.capthist: This function takes the EDF and combines it with trap information, and the number of sampling occasions. A typical application looks like this:

```
> capturedata <- make.capthist(enc.data, trapfile, fmt="trapID",
                 noccasions=165)
```

   See ?make.capthist for definition of distinct file formats. Specifying fmt = trapID is equivalent to our EDF format.
3. read.mask: If there is a habitat mask available (as described in Section 6.4.2), then this function will organize it so that secr.fit knows what to do with it. The

function accepts either an external file name (see ?read.mask for details of the structure) or a $nG \times 2$ **R** object, say mask.coords, containing the coordinates of the mask. A typical application looks like the following:

```
> grid <- read.mask(data=mask.coords)
```

These constructor functions produce output that can then be used in the fitting of models using secr.fit.

6.5.1 Encounter device types and detection models

The secr package requires that you specify the type of encounter device. Instead of describing models by their statistical distribution (Bernoulli, Poisson, etc.), secr uses certain operational classifications of detector types including "proximity," "multi," "single," "polygon," and "signal." For camera trapping/hair snares we might consider "proximity" detectors or "count" detectors. The "proximity" detector type allows, at most, one detection of each individual at a particular detector on any occasion (i.e., it is equivalent to what we call the Bernoulli or binomial encounter process model, or model SCR0). The "count" detector designation allows repeat encounters of each individual at a particular detector on any occasion. There are other detector types that one can select such as: "polygon" detector type, which allows for a trap to be a sampled polygon (Royle and Young, 2008), which we discussed further in Chapter 15, and "signal" detector, which allows for traps that have a strength indicator, e.g., acoustic arrays (Dawson and Efford, 2009). The detector types "single" and "multi" refer to traps that retain individuals, thus precluding the ability for animals to be captured in other traps during the sampling occasion. The "single" type indicates trap that can only catch one animal at a time (single-catch traps), while "multi" indicates traps that may catch more than one animal at a time (multi-catch). These are both variations of the multinomial encounter models described in Chapter 9.

As with all SCR models, secr fits an encounter probability model ("detection function" in secr terminology) relating the probability of encounter to the distance of a detector from an individual activity center. secr allows the user to specify one of a variety of detection functions including the commonly used half-normal ("Gaussian"), hazard rate ("Gaussian hazard"), and (negative) exponential models. There are 12 different functions as of version 2.3.1 (see Table 7.1). The different detection functions are defined in the secr manual and can be found by calling the help function for the detection function:

```
> ?detectfn
```

Most of the detection functions available in secr contain some kind of a scale parameter that is usually labeled σ. The units of this parameter default to meters in the secr output. We caution that the meaning of this parameter depends on the specific detection model being used, and it should not be directly compared as a measure of home range size across models. Instead, as we noted in Section 5.4, most encounter probability models imply a model of space usage and fitted encounter models should be converted to a common currency such as implied home range area.

6.5.2 Analysis using the secr package

To demonstrate the use of the secr package, we will show how to do the same analysis of the wolverine data as in Section 5.9. To use the secr package, the data need to be formatted in a similar manner to that analysis. For example, in Section 5.9 we introduced a standard data format for the encounter data file (EDF) and trap deployment file (TDF). The EDF shares the same format as that used by the secr package, with 1 row for every encounter observation and 4 columns representing trap session ("Session"), individual identity ("ID"), sample occasion ("Occasion"), and trap identity ("trapID"). For a standard closed population study that takes place during a single season, the "Session" column is all 1's, to indicate a single primary sampling occasion. In addition to providing the encounter data file (EDF), we must tell secr information about the traps, which is formatted as a data frame with column labels "trapID," "x," and "y," the last two being the coordinates of each trap, with additional columns representing the operational state of each trap during each occasion (1 = operational, 0 = not).

We now walk through an application of secr to the wolverine camera trapping data. To read in the trap locations and other related information, we make use of the constructor function read.traps, which also requires that we specify the detector type. The detector type is important because it determines the likelihood that secr will use to fit the model. Here, we have selected "proximity," which corresponds to the Bernoulli encounter model in which individuals are captured at most once in each trap during each sampling occasion:

```
> library(secr)
> library(scrbook)
> data(wolverine)

> traps <- as.matrix(wolverine$wtraps)
> dimnames(traps) <- list(NULL,c("trapID","x","y",paste("day",1:165,sep="")))
> traps1 <- as.data.frame(traps[,1:3])
> trapfile1 <- read.traps(data=traps1,detector="proximity")
```

Here, we note that trap coordinates are extracted from the wolverine data but we do *not* scale them. This is because secr defaults to coordinate scaling of meters, which is the extant scaling of the wolverine trap coordinates. Note that we add appropriate column labels to the "traps" data frame. An important aspect of the wolverine study is that while the camera traps were operated over a 165-day period, each trap was operational during only a portion of that period. We need to provide the trap operation information, which is contained in the columns to the right of the trap coordinates in our standard trap deployment file (TDF). Unfortunately, this is less easy to do in secr,[4] which requires an external file with a single long string of 1's and 0's, indicating the days in which each trap was operational (1) or not (0). The read.traps function will not allow for this information on trap operation if the data exists as an **R** object—instead, we can create this external file and then read it back in with read.traps using these commands:

[4]as of v. 2.3.1.

```
> hold <- rep(NA,nrow(traps))
> for(i in 1:nrow(traps)){
+     hold[i] <- paste(traps[i,4:ncol(traps)],collapse="")
+ }
> traps1 <- cbind(traps[,1:3],"usage"=hold)

> write.table(traps1, "traps.txt", row.names=FALSE, col.names=FALSE)
> trapfile2 <- read.traps("traps.txt",detector="proximity")
```

These operations can be accomplished using the function `scr2secr`, which is provided in the **R** package `scrbook`.

After reading in the trap data, we now need to create the encounter matrix using the `make.capthist` command, where we provide the capture histories in EDF format, which is the existing format of the data input file `wcaps`. In creating the capture history, we also provide the trapfile created previously, the format (e.g., here EDF format is `fmt= "trapID"`), and finally, the number of occasions.

```
# Grab the encounter data file and format it:
#
wolv.dat <- wolverine$wcaps
dimnames(wolv.dat) <- list(NULL,c("Session","ID","Occasion","trapID"))
wolv.dat <- as.data.frame(wolv.dat)
wolvcapt2 <- make.capthist(wolv.dat,trapfile2,fmt="trapID",noccasions=165)
```

We next set up a habitat mask using the 2×2 km grid we used previously in the analysis of the wolverine data and then pass the relevant objects to `secr.fit` as follows:

```
# Grab the habitat mask (2 x 2 km) and format it:
#
> gr2 <- (as.matrix(wolverine$grid2))
> dimnames(gr2) <- list(NULL,c("x","y"))
> gr2 <- read.mask(data=gr2)
#
# To fit the model we use secr.fit:
#
wolv.secr2 <- secr.fit(wolvcapt2,model=list(D ~ 1, g0 ~ 1, sigma ~ 1),
                       buffer=20000,mask=gr2)
```

We are using the "proximity detector" (model SCR0), which is the default model, so we do not need to make any specifications in the command line, except to provide the buffer size (in meters). To specify different models, you can change the default model `D~1, g0~1, sigma~1`. We provide all of these commands and additional analyses in the `scrbook` package with the function called `secr_wolverine`. Printing the output object produces the following (slightly edited):

```
> wolv.secr2

secr 2.3.1, 15:52:45 29 Aug 2012

Detector type      proximity
Detector number    37
Average spacing    4415.693 m
x-range            593498 652294 m
y-range            6296796 6361803 m
```

```
N animals        :  21
N detections     :  115
N occasions      :  165
Mask area        :  987828.1 ha

Model            :  D ~ 1 g0 ~ 1 sigma ~ 1
Fixed (real)     :  none
Detection fn     :  halfnormal
Distribution     :  poisson
N parameters     :  3
Log likelihood   :  -602.9207
AIC              :  1211.841
AICc             :  1213.253

Beta parameters (coefficients)

            beta     SE.beta        lcl        ucl
D      -9.390124  0.22636698  -9.833795  -8.946452
g0     -2.995611  0.16891982  -3.326688  -2.664535
sigma   8.745547  0.07664648   8.595323   8.895772

Variance-covariance matrix of beta parameters
                    D              g0          sigma
D       0.0512420110   -0.0004113326   -0.003945371
g0     -0.0004113326    0.0285339045   -0.006269477
sigma  -0.0039453711   -0.0062694767    0.005874683

Fitted (real) parameters evaluated at base levels of covariates
        link      estimate    SE.estimate           lcl           ucl
D        log  8.354513e-05   1.915674e-05  5.360894e-05  1.301982e-04
g0     logit  4.762453e-02   7.661601e-03  3.466689e-02  6.509881e-02
sigma    log  6.282651e+03   4.822512e+02  5.406315e+03  7.301037e+03
```

The object returned by `secr.fit` provides extensive default output when printed. Much of this is basic descriptive information about the model, the traps, or the encounter data. We focus here on the parameter estimates. Under the fitted (real) parameters, we find D, the density, given in units of individuals/hectare (1 hectare $= 10{,}000\ \text{m}^2$). To convert this into individuals/1,000 km^2, we multiply by 100,000, thus our density estimate is 8.35 individuals/1,000 km^2. The parameter σ is given in units of meters, and so this corresponds to 6.283 km. Both of these estimates are very similar to those obtained in our likelihood analysis summarized in Table 6.3 which, for the 2×2 km grid, returned $\hat{D} = 8.31$ with a SE of $(100{,}000 \times 1.915674e - 05) = 1.9156$ and, accounting for the scale difference (1 unit = 10,000 m in the previous analysis), $\hat{\sigma} = \sqrt{1/(2\hat{\alpha}_1)} \times 10{,}000 = 6.289$ km. The difference in the MLE between Table 6.3 and those produced by `secr` could be due to subtle differences in internal tuning of optimization algorithms, starting values, or other numerical settings. In addition, the likelihood used by `secr` is based on a Poisson prior for N (see Section 6.5.3).

6.5.3 Likelihood analysis in the `secr` package

The `secr` package does likelihood analysis of SCR models for most classes of models as developed by Borchers and Efford (2008). Their formulation deviates slightly from the binomial form we presented in Section 6.2 above (though Borchers and Efford (2008) also mention the binomial form). Specifically, the likelihood that `secr` implements is that based on removing N from the likelihood by integrating the binomial likelihood (Eq. (6.2.1)) over a Poisson prior for N—what we will call the *Poisson-integrated likelihood* as opposed to the conditional-on-N (*binomial-form*) considered previously.

To develop the Poisson-integrated likelihood we compute the marginal probability of each $\mathbf{y}_i$ and the probability of an all-zero encounter history, π_0, as before, to arrive at the marginal likelihood in the binomial-form:

$$\mathcal{L}(\boldsymbol{\alpha}, n_0|\mathbf{y}) = \frac{N!}{n!n_0!}\left\{\prod_i^n [\mathbf{y}_i|\boldsymbol{\alpha}]\right\}\pi_0^{n_0}.$$

Now, Borchers and Efford (2008) assume that $N \sim \text{Poisson}(\Lambda)$ and they do a further level of marginalization over this prior distribution:

$$\sum_{n_0=0}^{\infty} \frac{N!}{n!n_0!}\left\{\prod_i^n [\mathbf{y}_i|\boldsymbol{\alpha}]\right\}\pi_0^{n_0}\frac{\exp(-\Lambda)\Lambda^N}{N!}.$$

In Chapter 11 we write $\Lambda = \mu\|\mathcal{S}\|$ where $\|\mathcal{S}\|$ is the area of the state-space and μ is the density ("intensity") of the point process. Carrying out the summation above produces exactly this marginal likelihood:

$$\mathcal{L}_2(\boldsymbol{\alpha}, \Lambda|\mathbf{y}) = \left\{\prod_i^n [\mathbf{y}_i|\boldsymbol{\alpha}]\right\}\Lambda^n \exp(-\Lambda(1-\pi_0)),$$

which is Eq. (6.1.2) of Borchers and Efford (2008), except for notational differences. It also resembles the binomial form of the likelihood in Eq. (6.2.1) with $\Lambda^n \exp(-\Lambda\pi_0)$ replacing the combinatorial term and the $\pi_0^{n_0}$ term. We emphasize there are two marginalizations going on here: (1) the integration to remove the latent variables $\mathbf{s}$; and (2) summation to remove the parameter N. We provide a function for computing this in the `scrbook` package, called `intlik3Poisson`. The help file for that function shows how to conduct a small simulation study to compare the MLE under the Poisson-integrated likelihood with that from the binomial form.

The essential distinction between our MLE and that of Borchers and Efford (2008) as implemented in `secr` is whether you keep N in the model or remove it by integration over a Poisson prior. If you have prescribed a state-space explicitly with a sufficiently large buffer, then we imagine there should be hardly any difference at all between the MLEs obtained by either the Poisson-integrated likelihood or the binomial form of the likelihood, which retains N as an explicit parameter. There is a subtle distinction in the sense that under the binomial form, we estimate the realized population size N for the state-space whereas, for the Poisson-integrated form, we estimate the *prior* expected value which would apply to a hypothetical new study of a similar population (see Section 5.7.3).

Both models (likelihoods) assume **s** is uniformly distributed over space, but for the binomial model we make no additional assumption about N whereas we assume N is Poisson using the formulation in secr from Borchers and Efford (2008). Using data augmentation we could do a similar kind of integration but integrate N over a binomial (M, ψ) prior—which we referred to as the binomial-integrated likelihood in Section 4.2.4.

6.5.4 Multi-session models in secr

In practice we will often deal with SCR data that have some meaningful stratification or group structure. For example, we might conduct mist netting of birds on K consecutive days, repeated, say, T times during a year, or perhaps over T years. Or we might collect data from R distinct trapping grids. In these cases, we have T or R groups, which we might reasonably regard as being samples of independent populations. While the groups might be distinct sites, years, or periods within years, they could also be other biological groups such as sex or age. Conveniently, secr fits a specific model for stratified populations—referred to as *multi-session* models. These models build on the Poisson assumption which underlies the integrated likelihood used in secr (as described in Section 6.5.3). To understand the technical framework underlying multi-session models, let N_g be the population size of group g and *assume*

$$N_g \sim \text{Poisson}(\Lambda_g).$$

Naturally, we model group-specific covariates on Λ_g:

$$\log(\Lambda_g) = \beta_0 + \beta_1 C_g,$$

where C_g is some group-specific covariate such as a categorical index to the group, or a trend variable, or a spatial covariate, such as treatment effect or habitat structure, if the groups represent spatial units. Under this model, we can marginalize *all* N_g parameters out of the likelihood to concentrate the likelihood on the parameters β_0 and β_1 precisely as discussed in Section 6.5.3. This Poisson hierarchical model is the basis of the multi-session models in secr.

To implement a multi-session model (or stratified population model) in secr, we provide the relevant stratification information in the "Session" variable of the input encounter data file (EDF). If "Session" has multiple values, then a "multi-session" object is created by default and session-specific variables can be described in the model. For example, if the session has two values for males and females then we can estimate sex-specific densities, and baseline encounter probability p_0 (g_0 in secr) by just doing this:

```
> out <- secr.fit(capdata, model=list(D ~ session, g0 ~ session, sigma ~ 1),
          buffer=20000)
```

See Chapter 8 for the **R** code to set this up. More detailed analysis is given in Section 8.1 where we fit a number of different models to the wolverine camera trapping data and apply methods of model selection to obtain model-averaged estimates of density.

We can also easily implement stratified population models in the various **BUGS** engines using data augmentation (Converse and Royle, 2012; Royle and Converse, In review), which we discuss in Chapter 14.

6.5.5 Some additional capabilities of `secr`

The `secr` package has capabilities to do a complete analysis of SCR data, including model fitting, selection, and many summary analyses. In the previous sections, we've given a basic overview, and we do more in later chapters of this book. Here we mention a few of these other capabilities that you should know about as you use `secr`. Of course, you should skim through the associated documentation (`?secr`) to see more of what is available.

6.5.5.1 *Alternative observation models*

`secr` fits a wide range of alternative observation models besides the Bernoulli encounter model, including multinomial encounter models for "multi-catch" and "single catch" traps, models for sound attenuation from acoustic detection devices, and many others. We discuss many of these other methods in Chapter 9 and elsewhere in the book.

6.5.5.2 *Summary statistics*

`secr` provides a useful default summary of the data, but it also has summary statistics about animal movement including mean-maximum distance moved (the function `MMDM`). For example, see the help page `?MMDM`, which lists a number of other summary functions for `capthist` objects:

```
> moves(capthist)
> dbar(capthist)
> RPSV(capthist)
> MMDM(capthist, min.recapt = 1, full = FALSE)
> ARL(capthist, min.recapt = 1, plt = FALSE, full = FALSE)
```

The function `moves` returns the observed distances moved, `dbar` returns the average distance moved, `RPSV` produces a measure of dispersion about the home range center, and `ARL` gives the *Asymptotic Range Length* which is the asymptote of an exponential model fit to the observed range length vs. the number of detections of each individual (Jett and Nichols, 1987).

6.5.5.3 *State-space buffer*

`secr` will produce a warning if the state-space buffer is chosen too small. For example, in fitting the wolverine data as in Section 6.5.2 but with a 1,000 m buffer, we see the following warning message:

```
Warning message:
In secr.fit(wolvcapt2, model=list(D ~ 1, g0 ~ 1, sigma ~ 1), buffer=1000):
  predicted relative bias exceeds 0.01 with buffer = 1000
```

This should cause you to contemplate increasing the state-space buffer, if that is a reasonable thing to do in the specific application.

6.5.5.4 Model selection and averaging

`secr` does likelihood ratio tests to compare nested models using the function `LR.test`. You can create model selection tables based on AIC or AICc, using the function `AIC`, and obtain model-averaged parameter estimates using the function `model.average` (See Chapter 8 for examples).

6.5.5.5 Population closure test

`secr` has a population closure test with the function `closure.test`, which implements the tests of Stanley and Burnham (1999) or Otis et al. (1978). The function is used like this: `closure.test(object, SB = FALSE)`. Here, `object` is a capthist object and `SB` is a logical variable that, if `TRUE`, produces the Stanley and Burnham (1999) test.

6.5.5.6 Density mapping and effective sample area

`secr` produces likelihood versions of the various summaries of posterior density and effective sample area that we discussed in Chapter 5. For example, while `secr` reports estimates of the expected value of N or density directly in the summary output from fitting a model, you can use the function `region.N` to produce estimates of N for any given region. In addition, `secr` has functions for creating maps of detection contours for individuals traps, or for the entire trap array. See the functions `pdot.contour`, and `fxi.contour` for computing the two-dimensional pdf of the locations of one or more individual activity centers (as in Section 5.11.3). In the context of likelihood analysis, estimation of a random effect **s** is based on a plug-in application of Bayes' Rule. When **s** has a uniform distribution, and we use a discrete evaluation of the integral, it can be computed simply by renormalizing the likelihood:

$$[\mathbf{s}|\mathbf{y}, \theta] = \frac{[\mathbf{y}|\mathbf{s}, \theta]}{\sum_s [\mathbf{y}|\mathbf{s}, \theta]}.$$

Any of the `intlik` functions given previously in this chapter can be easily modified to return the posterior distribution of **s** for any, or all, individuals, or an individual that is not encountered.

Effective sample area (see Section 5.12) can be calculated in `secr` using the functions `esa` and `esa.plot`).

6.5.5.7 Covariate models

`secr` has many capabilities for modeling covariates. It has a number of built-in models that allow certain covariates on encounter probability, which we cover to a large extent in Chapters 7 and 8. `secr` also allows covariates to be built into the density model (see Chapter 11). It has some built-in response surface models, allowing for the fitting of linear or quadratic response surfaces. This is done by modifying the density model in `secr.fit`. For example, `D` $\sim$ 1 is a constant density surface, and `D` $\sim$ `x` + `y` fits a linear response surface, etc. See the manual `secr-densitysurfaces.pdf` for details.

There are a number of ways to incorporate your own "custom" covariates into a model (as opposed to pre-specified models). One way is to use the `addCovariates` function and supply it a `mask` or `traps` object along with some "spatialdata." Or, if you have covariates at each trap location then it will extrapolate to all points on the habitat mask. There's also a method by which the user can create a function of geographic coordinates, `userDfn`, which seems to provide additional flexibility, although we haven't used this method. There is a handy function `predictDsurface` for producing density maps under the specified model for density.

6.6 Summary and outlook

In this chapter, we discussed basic concepts related to classical analysis of SCR models based on likelihood methods. Analysis is based on the so-called integrated or marginal likelihood, in which the individual activity centers (random effects) are removed from the conditional-on-**s** likelihood by integration. We showed how to construct the integrated likelihood and fit some simple models in the **R** programming language. In addition, likelihood analysis for some broad classes of SCR models can be accomplished using the **R** package `secr` (Efford, 2011a), which we provided a brief introduction to. In later chapters we provide more detailed analyses of SCR data using likelihood methods and the `secr` package.

Why or why not use likelihood inference exclusively? For certain specific models, it is more computationally efficient to produce MLEs (for an example see Chapter 12). And, likelihood analysis makes it easy to do model selection by AIC and compute standard errors or confidence intervals. However, **BUGS** is extremely flexible in terms of describing models and we can devise models in the **BUGS** language easily that we cannot fit in `secr`. For example, in Chapter 16 we consider open population models, which are straightforward to develop in **BUGS** but, so far, there is no available platform for doing MLE of such models. We can also fit models in **BUGS** that accommodate missing covariates in complete generality (e.g., unobserved sex of individuals), and we can adopt SCR models to include auxiliary data types. For example, we might have camera trapping and genetic data and we can describe the models directly in **BUGS** and fit a joint model (Gopalaswamy et al., 2012b). To do maximum likelihood estimation, we have to write a custom new piece of code for each model[5] or hope someone has done it for us. You should have some capability to develop your own MLE routines with the tools we provided in this chapter.

[5]Although we may be able to handle multiple survey methods together in `secr` using the multi-session models.

CHAPTER

Modeling Variation in Encounter Probability

7

In previous chapters we showed how to fit basic spatial capture-recapture models using Bayesian analysis (in **WinBUGS** or **JAGS**; Chapter 5) or by classical likelihood methods (Chapter 6 or using `secr`). We mostly focused on a specific observation model, the Bernoulli or binomial model, for devices such as "proximity detectors" (although we extend this model to Poisson and multinomial-type observation models in Chapter 9). We have not, however, described a general framework for modeling covariates of individuals, traps, or over time that may influence encounter probability. In practice, investigators are concerned with explicit factors or covariates that might influence variation in parameters. Such covariates include time (e.g., day-of-year, or season), behavior (e.g., if there is an effect of trapping on subsequent capture probabilities), sex of the individual, and trap type (e.g., various camera types, or different constructions for hair snares). Traditionally, in the non-spatial capture-recapture literature such models were called "model M_t," "model M_h," or "model M_b," identifying models that account for variation in detection probability as a function of time, "individual heterogeneity" or "behavior," where behavior describes whether or not an individual had been previously captured. In SCR models, more complex covariate models are possible because we might also have trap-specific covariates, or covariates that vary spatially over the landscape, and because we generally have more than one parameter describing the detection function: Most encounter probability functions include a baseline encounter rate (λ_0) or probability (p_0) parameter, and a scale parameter (σ), which takes on different interpretations depending on the specific encounter probability function under consideration.

In this chapter, we generalize the basic SCR model to accommodate both alternative detection functions as well as many different kinds of covariates. We focus on the binomial observation model used throughout Chapters 5 and 6 and the Gaussian encounter model (also called the "half-normal" model in the distance sampling literature), but the extension to other observation models is straightforward (and other encounter probability models with different functions of distance are considered in Section 7.1). Specifically, we consider three distinct types of covariates—those which are fixed, partially observed, or completely unobserved (latent). Fixed covariates are those that are fully observed; for example, the dates of all sampling occasions. Partially observed individual covariates are those which are not known for all observations; for example, the sex of an individual cannot always be determined from photos taken

Spatial Capture-Recapture. http://dx.doi.org/10.1016/B978-0-12-405939-9.00007-4

during camera trapping. Even if we are able to observe the sex of all individuals sampled, we cannot know it for those individuals never observed during the study. And finally, unobserved covariates are those which we cannot observe at all, for example, the home range size of individuals, or unstructured random "individual effects."

We will see that models containing these different types of covariates are relatively easy to describe in **WinBUGS** or **JAGS**, and therefore to analyze using Bayesian analysis of the joint likelihood based on data augmentation, thus providing a coherent and flexible framework for inference for all classes of SCR models. Throughout the chapter, we will continue to develop the analysis of the black bear study introduced in Chapter 4, using the software **JAGS**. We also consider the likelihood analysis of many of these models; to do so, we continue to use the **R** package `secr`, and we introduce some ideas of model comparison using AIC (Section 7.4 at the end of the chapter). There are other types of covariates that we do *not* cover in this chapter; for example, covariates that vary across the landscape might affect density (consider these covariates in Chapter 11). Alternatively, landscape covariates might affect the way individuals use space and we develop more realistic models of encounter probability in which covariates affect space usage in Chapter 12.

7.1 Encounter probability models

In Chapter 5, we developed a basic spatial capture-recapture model using a standard encounter probability function based on the kernel of a normal (Gaussian) probability distribution:

$$p_{ij} = p_0 \exp(-\alpha_1 ||\mathbf{x}_j - \mathbf{s}_i||^2),$$

where $||\mathbf{x}_j - \mathbf{s}_i||$ is the distance between trap location, $\mathbf{x}_j$ and individual activity center, $\mathbf{s}_i$ and $\alpha_1 = 1/(2\sigma^2)$. We argued (see Section 5.4) that one can view this model as corresponding to an explicit model of space usage—namely, that individual locations are draws from a bivariate normal distribution. Other detection models are possible, including a logit model of the form:

$$\text{logit}(p_{ij}) = \alpha_0 + \alpha_1 ||\mathbf{x}_j - \mathbf{s}_i||. \tag{7.1}$$

However, there's nothing preventing us from constructing a myriad of other models for encounter probability as a function of distance. The most commonly used detection probability models are also those used in the distance sampling literature: the half-normal (Gaussian), the hazard, and the negative exponential. The negative exponential model is:

$$p_{ij} = p_0 \exp(-\alpha_1 ||\mathbf{x}_j - \mathbf{s}_i||),$$

where we define $\alpha_1 = 1/\sigma$. We could also use the general power model (Russell et al., 2012):

$$p_{ij} = p_0 \exp(-\alpha_1 ||\mathbf{x}_j - \mathbf{s}_i||^{\theta}),$$

of which the Gaussian and exponential models are special cases. Another model that could be considered is the Gaussian hazard rate model (Hayes and Buckland, 1983):

$$p_{ij} = 1 - \exp(-\lambda_0 \exp(-\alpha_1 ||\mathbf{x}_j - \mathbf{s}_i||^2)),$$

which was previously discussed in Chapter 5.

At this time, the **R** package `secr` allows the user to access 12 different encounter probability models (termed "distance functions" in `secr`), of which some are only used for simulating data (see Table 7.1). These encounter probability models can also be implemented in **R**, **WinBUGS**, **JAGS**, etc. Previous studies have shown that SCR estimates of density are robust to the choice of the encounter probability model (Efford, 2004; Efford et al., 2009a; Russell et al., 2012), but further investigation is warrented as these studies were limited in scope.

Insofar as all these encounter probability models are symmetric and stationary, they are pretty crude descriptions of space usage by real animals. This is not to say they are inadequate descriptions of the data and, as we discuss in Chapters 12 and 13, we can use them as the basis for producing more realistic models of space usage.

As we see (Table 7.1), most distance functions have a scale parameter, σ. It is important to note that σ is not comparable under these different encounter probability models and should not be regarded as "home range radius" in general. While there is

Table 7.1 Basic encounter probability models ("distance functions") available in `secr`. (Table taken from the `secr` help files). Notation deviates from that used in the text. In this table g_0 is the baseline encounter rate or probability parameter used in `secr` which is equivalent to our p_0 or λ_0 depending on context. d is distance defined as we have done throughout, as the distance between the activity center and the trap. One can read more on this specific table by loading the `secr` package and using the `help` command in **R** (`?detectfn`).

	Name	Params	Function
0	half-normal	g_0, σ	$g(d) = g_0 e^{-d^2/(2\sigma^2)}$
1	hazard rate	g_0, σ, z	$g(d) = g_0(1 - e^{-(d/\sigma)^{-z}})$
2	exponential	g_0, σ	$g(d) = g_0 e^{-d/\sigma}$
3	compound half-normal	g_0, σ, z	$g(d) = g_0[1 - \{1 - e^{-d^2/(2\sigma^2)}\}^z]$
4	uniform	g_0, σ	$g(d) = g_0,\ d \le \sigma$; $g(d) = 0$, otherwise
5	w exponential	g_0, σ, w	$g(d) = g_0,\ d < w$; $g(d) = g_0 e^{(-(d-w)/\sigma)}$, otherwise
6	annular normal	g_0, σ, w	$g(d) = g_0 e^{(-(d-w)^2/(2\sigma^2))}$
7	cumulative lognormal	g_0, σ, z	$g(d) = g_0[1 - F(d - \mu)/s)]$
8	cumulative gamma	g_0, σ, z	$g(d) = g_0\{1 - G(d, k, \theta)\}$
9	binary signal strength	b_0, b_1	$g(d) = 1 - F\{-(b_0 + b_1 d)\}$
10	signal strength	β_0, β_1,S	$g(d) = 1 - F[\{c - (\beta_0 + \beta_1 d)\}/S]$
11	signal strength spherical	β_0, β_1,S	$g(d) = 1 - F[\{c - (\beta_0 + \beta_1(d-1) - 10 * log_{10}(d^2))\}/S]$

often a relationship between σ and home range size, that relationship varies depending on the model under consideration. We demonstrate how to fit different encounter probability models in the Bayesian framework here, and then provide information on the likelihood analysis (in `secr`) in a separate section below.

7.1.1 Bayesian analysis with `bear.JAGS`

To demonstrate how to incorporate various types of covariates into models for encounter probability using **JAGS**, we return to the data collected during the Fort Drum bear study. This data set was first introduced in Chapter 4 but to refresh your memory, there were 38 baited hair snares that were operated between June and July 2006. The snares were checked each week for a total for $K = 8$ sample occasions and $n = 47$ individual bears were encountered at least once. The data are provided in the **R** package `scrbook` and an **R** function called `bear.JAGS` allows the user to easily pick which model to analyze. The function `bear.JAGS` will set up the data, write the model, define the MCMC settings (e.g., initial values, number of iterations etc.) and, finally, run the selected model in **JAGS**. In addition to choosing which model to run, the user can also specify the number of chains, iterations, and length of the burn-in phase. Calling the function will provide all the code to implement the models independently as well. In the following sections we will present the model code and output for the most commonly employed models; for all analyses we ran three chains with a burn-in of 500 iterations and 20,000 saved iterations.

7.1.2 Bayesian analysis of encounter probability models

In Panel 7.1, we present the basic SCR model and show how to specify the negative exponential encounter probability model. To call each of these from the function `bear.JAGS` set `model='SCR0'` or `model='SCRexp'` in the function call, respectively. To reduce repetition of the **R** coding, we include the basic code here and then only show modifications when necessary throughout the chapter. All of the **R** code can be found within the `bear.JAGS` function. The function begins by loading the required **R** libraries, as well as the Ft. Drum bear data set. This data set includes a three-dimensional data array (called `bearArray` in our code), with dimensions `nind` × `ntraps` × `nreps` representing the capture histories of `nind` captured individuals at `ntraps` trap locations. In the Bayesian analysis, data augmentation is used to estimate N and therefore the `bearArray` data must be augmented with $M -$ `nind` all zero encounter histories. In models without time dependence, the augmented `bearArray` (called `Yaug` in the code) will be reduced to a two-dimensional array (denoted `y` in the code) that has dimensions `M` × `ntraps`.

```
> library(rjags) # Load the necessary libraries
> library(scrbook)

> data(beardata)    # Attach the bear data for Ft. Drum
> ymat <- beardata$bearArray
> trapmat <- beardata$trapmat
```

```
model{
alpha0 ~ dnorm(0,.1)                        # Prior distributions
logit(p0) <- alpha0
alpha1 <- 1/(2*sigma*sigma)
sigma ~ dunif(0, 15)
psi ~ dunif(0,1)

   for(i in 1:M){
    z[i] ~ dbern(psi)
    s[i,1] ~ dunif(xlim[1],xlim[2])
    s[i,2] ~ dunif(ylim[1],ylim[2])
   for(j in 1:J){
    d[i,j] <- pow(pow(s[i,1]-X[j,1],2) + pow(s[i,2]-X[j,2],2),0.5)
    y[i,j] ~ dbin(p[i,j],K)
    p[i,j] <- z[i]*p0*exp(- alpha1*d[i,j]*d[i,j])  # Gaussian model
   #p[i,j] <- z[i]*p0*exp(- alpha1*d[i,j])         # exponential model
 }
}
N <- sum(z[])
D <- N/area
}
```

PANEL 7.1

JAGS model specification for a basic SCR model with Gaussian encounter probability function and the alternative exponential encounter probability function.

```
> nind <- dim(beardata$bearArray)[1]
> K <- dim(beardata$bearArray)[3]
> ntraps <- dim(beardata$bearArray)[2]
> M <- 650
> nz <- M-nind

# Create augmented array
> Yaug <- array(0, dim=c(M,ntraps,K))
> Yaug[1:nind, ,] <- ymat
> y <- apply(Yaug,1:2, sum)
```

The function `bear.JAGS` also establishes the upper and lower limits on the state space by centering the trap array coordinates (which are imported with the `beardata` and saved in the code above as `trapmat`) and then buffering by 20 km.

Applying the SCR model with Gaussian encounter probability model provides an estimate (posterior mean) of $D = 0.167$ bears per km^2 and with the negative exponential encounter probability model the posterior mean is the same, $D = 0.167$. In distance sampling, the use of different encounter probability models often results in very different estimates of density (especially when using the negative exponential

Table 7.2 Posterior summaries of SCR model parameters under the Gaussian and exponential encounter probability models, for the Fort Drum black bear data.

Parameter	Mean	SD	2.5%	97.5%
Gaussian				
N	500.63	66.652	371.000	628.000
D	0.17	0.022	0.122	0.207
p_0	0.11	0.014	0.081	0.135
σ	1.99	0.131	1.762	2.275
ψ	0.77	0.104	0.566	0.966
Exponential				
N	512.06	65.771	382.000	634.000
D	0.17	0.022	0.130	0.210
p_0	0.34	0.056	0.246	0.465
σ	1.12	0.095	0.951	1.323
ψ	0.79	0.102	0.584	0.974

model). There are two main reasons why the different models may have less of an impact on the density estimates under the SCR models. First, we can estimate the baseline encounter probability parameter (p_0). In most distance sampling models, detection at distance 0 is set to 1. In Table 7.2, the posterior mean of p_0 is 0.11 under the Gaussian model and 0.34 under the negative exponential model. The larger baseline encounter probability under the negative exponential model reduces the impact of the quick decline in detection as a function of distance. Secondly, the detection probability function here is governing "movement" of individuals (which we have more information on than in distance sampling), not the whole detection process, so the shape of the detection probability function does not impact the density estimation as much.

In all analyses it is important to check that the size of the augmented data set (M) is sufficiently large and does not impact the estimate of N. Here, the 97.5% percentile for N is 628 (Table 7.2), thus not reaching the $M = 650$ value. We could also increase M and compare the posterior of N under the different scenarios as another check that the data augmentation is sufficient.

The estimate (posterior mean) of σ under the negative exponential model is 1.12, which is distinct from our estimate of σ under the Gaussian model, $\sigma = 1.996$. The interpretation of σ in the two models is really quite different. In the normal model it can be interpreted as the standard deviation of a bivariate normal movement model whereas the manner in which σ relates to "area used" for the negative exponential model has nothing to do with a bivariate normal model of movement. This highlights that it is important for the user to know what detection probability function is used and what the interpretation of σ might be in relation to the home range size (Section 5.4).

We now move onto incorporating covariates into the model using the **JAGS** language. For the rest of this chapter, we will stick to the Gaussian encounter probability model shown in Panel 7.1.

7.2 Modeling covariate effects

The basic strategy for modeling covariate effects is to include them on the baseline encounter rate or probability parameter, p_0 (or λ_0), or the scale parameter of the encounter model, σ, or in some cases, both parameters.

Broadly speaking, we recognize (here) three types of covariates. Fixed covariates are fully observable and might vary by trap alone (e.g., type of trap, baited or not, disturbance regime, even habitat), sample occasion (e.g., day of season or weather conditions), or both (e.g., behavior, weather—if over a large region). Another class of covariates are those which vary at the level of the individual (and possibly also over time). As a technical matter, and as noted before, these are different from fixed covariates because we cannot see all of the individuals and the covariates are almost always incompletely observed (if at all). The lone exception is the effect of previous capture, used to model a behavioral response to capture, which is known for all individuals, captured or not (i.e., an animal never captured/observed has never been captured before). And finally, we have completely unobserved covariates such as heterogeneity in home range size. We consider heterogeneity in a separate section below since there are a suite of models for describing latent heterogeneity.

To develop covariate models, we assume a standard sampling design in which an array of J traps are operated for K sample occasions, producing encounter histories for n individuals. For the null model, there are no time-varying covariates that influence encounter, there are no explicit individual-specific covariates, and there are no covariates that influence density. For fixed effects, we can easily incorporate these into the encounter probability model, just as we would do in any standard GLM or GLMM, on some suitable scale for the baseline encounter probability, $p_{0,ijk}$. For example,

$$\begin{aligned}\text{logit}(p_{0,ijk}) &= \alpha_0 + \alpha_2 C_{ijk},\\ p_{ijk} &= p_{0,ijk}\exp(-\alpha_1 ||\mathbf{x}_j - \mathbf{s}_i||^2),\end{aligned}$$

where C_{ijk} is some covariate that varies (potentially) by individual (i), trap (j), and occasions (k), and α_2 is the coefficient to be estimated. How we define specific covariates (e.g., trap-specific versus individual-specific) will influence exactly how we include them in the model. Table 7.3 shows examples of covariates by type—trap,

Table 7.3 Examples of different types of covariates in SCR models.

Covariate Type	Examples
individual	sex, age, home range
trap	baited/not, habitat (see also Chapter 13)
time	season, shedding, weather
individual × time	global behavioral response
trap × time	trap failures
individual × trap × time	local behavioral response

individual, and time—and also gives examples of some combined types. These are the types of covariates we will specifically address in this chapter, demonstrating how to analyze the different types in the following sections.

7.2.1 Date and time

Researchers may be interested in modeling the effect of date or chronological time on encounter probability. For example, in a long-term hair snare study, we may expect that seasonal shedding (Wegan et al., 2012) will influence encounter probabilities directly. Or, we may expect behaviors such as denning, mating, etc., to influence the encounter of certain species at certain times of year (Kéry et al., 2011). There are two common ways to incorporate date or time information into a model for encounter probability. For cases with a small number of sampling occasions we can fit a time-specific intercept (analogous to "model M_t" in classical capture-recapture (Otis et al., 1978)). In this model, there are K sampling occasion-specific parameters to reflect potential variation in sampling effort or other factors that might vary across samples. Alternatively, we can model parametric functions of date or time such as polynomial or sinusoidal functions.

In the first case, we allow each sampling occasion, k, to have its own baseline encounter probability, e.g.,

$$\text{logit}(p_{0,k}) = \alpha_{0,k},$$

so that

$$p_{ijk} = p_{0,k} \exp(-\alpha_1 ||\mathbf{x}_j - \mathbf{s}_i||^2).$$

This description of the model includes k occasion-specific baseline encounter probabilities. Thus, if there are four sampling occasions, then there are four different baseline encounter probabilities. We imagine that complete time specificity of p_0 (i.e., one distinct value for each sample occasion) would be most useful in situations where there are just a few sampling occasions (if there are many, this formulation will dramatically increase the number of parameters to be estimated) or when we do not expect systematic patterns over time (e.g., explainable by a polynomial function or time-varying covariates).

To implement this in **JAGS**, α_0 has to be estimated for each time period k either using an index vector or dummy variables (as described in Chapter 2 and Section 4.3) and this can be done by only changing only a few lines in Panel 7.1:

```
alpha0[k] ~ dnorm(0,.1)
logit(p0[k]) <- alpha0[k]
.......
.......
y[i,j,k] ~ dbin(p[i,j,k],K)
p[i,j,k] <- z[i]*p0[k]*exp(- alpha1*d[i,j]*d[i,j])
```

Since the model contains a parameter for each time period, the encounter histories must be time-dependent. Thus, a three-dimensional data array (called `bearArray` in

Table 7.4 Posterior summaries of parameters from an SCR model with time-dependent baseline encounter probability for the Ft. Drum black bear data set.

Parameter	Mean	SD	2.5%	97.5%
N	509.24	66.13	381.00	632.00
D	0.17	0.02	0.13	0.21
$p_0(t=1)$	0.06	0.02	0.03	0.10
$p_0(t=2)$	0.05	0.02	0.02	0.09
$p_0(t=3)$	0.15	0.03	0.09	0.22
$p_0(t=4)$	0.14	0.03	0.09	0.21
$p_0(t=5)$	0.15	0.03	0.09	0.22
$p_0(t=6)$	0.12	0.03	0.07	0.19
$p_0(t=7)$	0.15	0.03	0.09	0.22
$p_0(t=8)$	0.08	0.02	0.04	0.13
σ	1.96	0.12	1.73	2.22
ψ	0.78	0.10	0.58	0.97

our code), with dimensions `nind` × `ntraps` × `nreps`, is required (recall that we use the three-dimensional augmented array called `Yaug` with dimensions `M` × `ntraps` × `nreps` for the Bayesian analysis). In addition to using the three-dimensional data array, the initial values must be updated so that there are K values generated for α_0. And finally, this means that another nested *for loop* is needed in the code to account for the K sample occasions. A side note: the computation time will increase quite a bit; this model for the bear data may take up to 15 h or more on your machine to obtain a sufficient posterior sample.

Running this model with the function `bear.JAGS` by setting `model='SCRt'` returns estimates of density similar to those from the model without covariates (see Table 7.4), but now we have a characterization of variation in encounter probability over time. Encounter probability seems to increase for the first few time periods before stabilizing around 0.14, dropping off again at the end of the study. The differences in encounter probability from the first time periods to the others might actually be due to something like a behavioral response (see below) or possibly seasonal differences in the efficiency of the sampling technique. Researchers have found that hair snares are more effective at different times of the year (even within season) due to shedding (Wegan et al., 2012). In this particular example, our density estimates (posterior means) are similar to the base model, likely because the differences in encounter probability between occasion were not that large. In a longer term study or in one with greater variation in the encounter probability, the implication of such differences might have a bigger impact on the estimates of density and σ.

The occasion-specific intercepts (baseline encounter probability) model might not be the most appropriate for all scenarios and could require the estimation of many parameters if we had many sampling occasions; for example, the wolverine study from Chapter 5.9, where there were 165 daily sampling occasions. Particularly in such a case as the wolverine study, variation in the encounter process over time is to be expected. Instead of fitting a model with K baseline encounter probabilities, we can include date as a linear (or quadratic, etc.) effect. An example can be found in Kéry et al. (2011) who incorporated a day-of-year covariate, both as a linear and a quadratic effect, into their SCR model of European wildcats; the data had been collected over a year-long period and cat behavior was expected to vary seasonally, thus influencing the probability of encounter. In these cases, we would specifically incorporate day-of-year (variable "`Date`"), a numeric covariate, as:

$$\begin{aligned} \text{logit}(p_{0,ijk}) &= \alpha_0 + \alpha_2 \texttt{Date}_k, \\ p_{ijk} &= p_{0,ijk} \exp(-\alpha_1 ||\mathbf{x}_j - \mathbf{s}_i||^2), \end{aligned}$$

or a quadratic effect of day-of-year:

$$\begin{aligned} \text{logit}(p_{0,ijk}) &= \alpha_0 + \alpha_2 \texttt{Date}_k + \alpha_3 \texttt{Date}_k^2, \\ p_{ijk} &= p_{0,ijk} \exp(-\alpha_1 ||\mathbf{x}_j - \mathbf{s}_i||^2). \end{aligned}$$

7.2.2 Trap-specific covariates

In some studies it makes sense to model encounter probability as a function of local or trap-specific covariates. These can be one of two types: genuine trap covariates that describe the trap or encounter site, such as whether a trap is baited or not, or how many traps were set at a sampling location, or what kind of bait was used, etc., or local covariates that describe the likelihood that an animal would use the habitat in the vicinity of the trap (see Chapter 13 for more on this situation). We imagine that these covariates, of either type, should affect baseline encounter probability. For example, Sollmann et al. (2011) found a large difference in the encounter probability of jaguars due to traps being located on roads, which the animals were using to travel along, as opposed to traps placed off of roads. In this case, the trap type is a binary variable—on/off road (another binary variable could be baited/non-baited). We can write this as:

$$\begin{aligned} \text{logit}(p_{0,j}) &= \alpha_{0,type_j}, \\ p_{ijk} &= p_{0,j} \exp(-\alpha_1 ||\mathbf{x}_j - \mathbf{s}_i||^2). \end{aligned}$$

Here, we use an index variable, "type," an integer value for the trap-specific covariate. Thus for our example of on/off road, we would have $\texttt{type}_j = 1$ if trap j is on a road and $\texttt{type}_j = 2$ otherwise, and we would estimate two separate α_0 parameters—one for on-road and one for off-road cameras. An alternative way to express the 2-category model, using dummy variables, requires that we specify our "type" vector as $\texttt{Type}_j = 0$ if trap j is on a road and $\texttt{Type}_j = 1$ otherwise, and write the model as

$$\text{logit}(p_{0,ijk}) = \alpha_0 + \alpha_2 \texttt{Type}_j.$$

Now, α_0 is the baseline encounter probability (on the logit scale) for traps on a road ($\texttt{Type}_j = 0$) and α_2 is the effect on baseline encounter probability of a trap being of `Type` = 1. This general setup also allows for more than two categories, say if four different camera models were used in a study, we would use a set of three binary dummy variables to allow for estimation of the different encounter rates (i.e., the intercept). While these models are equivalent, and should yield identical results, sometimes one parameterization might work better than the other in **WinBUGS** or **JAGS** (Kéry, 2010).

7.2.3 Behavior or trap response by individual

One of the most basic encounter models is that which accommodates a change in encounter probability as a result of initial encounter. This is colloquially referred to as "trap happiness" or "trap shyness," or in other words, a behavioral response of individuals to being captured (Otis et al., 1978). If a trap is baited with a food source, an individual might come back for more. On the other hand, if being captured is traumatic then an individual might learn to avoid traps. Both of these types of responses can occur in most species depending on the type of encounter mechanisms being employed. Moreover, behavioral response can be either global (Gardner et al., 2010b) or local (Royle et al., 2011b). The local response is a trap-specific response while a global response suggests that initial capture provides a net increase or decrease in subsequent probabilities of capture (across all traps). A behavioral response does not need to be enduring (i.e., persist for the entire study after the individual has been captured/observed for the first time) but can be ephemeral, if, for example, an animal only responds to recapture on the occasion immediately after initial capture (Yang and Chao, 2005; Royle, 2008). While we will focus the examples in this chapter on enduring behavioral effects, extending such a model to the case of an ephemeral response should not pose any difficulties.

To describe these behavioral models we need to create a binary matrix that indicates if an individual has been captured previously. For the global behavioral response, define the $n \times K$ matrix, $\mathbf{C}$, where $C_{ik} = 1$ if individual i was captured at least once prior to session k, otherwise $C_{ik} = 0$

$$\begin{aligned} \text{logit}(p_{0,ik}) &= \alpha_0 + \alpha_2 C_{ik}, \\ p_{ijk} &= p_{0,ik} \exp(-\alpha_1 ||\mathbf{x}_j - \mathbf{s}_i||^2). \end{aligned}$$

For the local behavioral response, which is trap specific, we create an array, C_{ijk}, that indicates if an individual i has been previously captured in trap j at time k. (For the augmented individuals, the entries are all 0 since the animals were never captured.) We then include this in the model in the exact same form as above (with the sole difference that both C and p are now also indexed by k):

$$\begin{aligned} \text{logit}(p_{0,ijk}) &= \alpha_0 + \alpha_2 C_{ijk}, \\ p_{ijk} &= p_{0,ijk} \exp(-\alpha_1 ||\mathbf{x}_j - \mathbf{s}_i||^2). \end{aligned}$$

Since the behavioral response is occasion specific, to implement either the local or global response model in **JAGS**, we will have to use the three-dimensional array

Table 7.5 Posterior summaries of parameter estimates from the SCR model with a global behavioral response in encounter for the Fort Drum black bear data set.

Parameter	Mean	SD	2.5%	97.5%
N	577.56	54.30	452.00	648.00
D	0.19	0.02	0.15	0.21
α_0	−2.81	0.24	−2.91	−2.36
α_2	0.90	0.23	0.45	1.35
σ	2.00	0.13	1.77	2.28
ψ	0.88	0.08	0.69	0.99

of the augmented capture histories (`M` × `ntraps` × `nreps`) as we did for the time-varying encounter probability model above. The code must loop over each sampling occasion, but otherwise, the model varies only a little from the basic SCR model shown in Panel 7.1. Here is the specification of the occasion-specific (k) loop:

```
for(k in 1:K){
  logit(p0[i,j,k]) <- alpha0 + alpha2*C[i,j,k]
  y[i,j,k] ~ dbin(p[i,j,k],1)
  p[i,j,k] <- z[i]*p0[i,j,k]*exp(-alpha1*d[i,j]*d[i,j]).
}
```

Despite only minor changes to the **BUGS** code, this model can require quite a bit of time and computational effort. Implementing the behavioral models with the function `bear.JAGS` by setting `model='SCRb'` or `model='SCRB'` for the local or global model, respectively, returns the results shown in Table 7.5. There is a strong global behavioral response suggested by the posterior mean of $\alpha_2 = 0.90$. The estimate of N and subsequently D are larger than under the model without a behavioral response; here we estimate the posterior mean of $N = 577.56$, whereas in the SCR0 model, we estimated the posterior mean as $N = 500$. This makes sense given the large estimate of α_2, which suggests that bears are trap happy. In situations where animals are trap happy, the null model overestimates encounter probability (i.e., the bears that are never observed have a lower encounter probability than those that have been captured in the study) and thereby reduces the estimate of N. We do not include the results here, but the estimates were similar under the local behavioral response model.

7.2.4 Individual covariates

Individual covariates are those which are measured (or are measurable) on individuals, so we get to observe them only for the captured individuals. Sex is a simple example of an individual covariate, but one of the most commonly used in capture-recapture studies. The sex of an individual can influence many aspects of its ecology and behavior, including for example, the frequency of movement, seasonal behavior, and its home range size. This is common in studies of carnivores where females often have smaller home ranges than males (Gardner et al., 2010b; Sollmann et al., 2011). Additionally, we may find differences in the baseline encounter probability between males and females because females may move around less frequently, or possibly because they are less likely to use landscape structures that researchers may target with sampling devices in order to increase sample size, such as roads (e.g., Salom-Pérez et al., 2007). Therefore, we can imagine that sex may impact both the baseline encounter probability α_0 and the typical home range size, so that α_1 might also be sex-specific also. The fully sex-specific model is:

$$\begin{aligned} \text{logit}(p_{0,i}) &= \alpha_{0,sex_i}, \\ p_{ijk} &= p_{0,i} \exp(-\alpha_{1,sex_i} ||\mathbf{x}_j - \mathbf{s}_i||^2), \end{aligned}$$

where sex_i is a variable indicating the sex of each individual (1 = male, 2 = female). While we might know the sex of all individuals observed in the study, we will never know the sex of individuals that are not observed (Gardner et al., 2010b). It is also possible that we may not be able to determine the sex of individuals that are observed during the study. For example, photographic captures do not necessarily result in pictures that allow the sex to be absolutely determined, thus sometimes resulting in missing values of this covariate for animals captured in the study. We deal with this slightly differently depending on the inference framework that we adopt (Bayesian or likelihood). Here we demonstrate the Bayesian implementation and we discuss the likelihood approach using `secr` below in Section 7.4.2. Before proceeding with that, we note that it would be possible also to model covariates directly on the parameter σ (or its logarithm), e.g., $\log(\sigma_i) = \theta_1 + \theta_2 \texttt{sex}_i$ (see Section 8.1). One or the other (or perhaps *some* other) parameterization may yield a better performing MCMC algorithm or provide a more natural or preferred interpretation. In the context of Bayesian analysis, given that priors are not invariant to transformation of the parameters, this may be a consideration in choosing the particular parameterization.

Specifying a fully sex-specific model for **JAGS** is similar to the time-specific model shown above. We need to use an index or dummy variable to let α_0 and/or α_1 be defined separately for males and females. The main difference in this specification is that we do not observe sex for the augmented individuals. Therefore, we have missing observations of the covariate for those individuals. As a result, sex is regarded as a random variable and so the missing values can be estimated along with the other structural parameters of the model.

Because we are regarding sex as a random variable, we have to specify a distribution for it. With only two possible outcomes, it is natural to suppose that $\texttt{Sex}_i \sim \text{Bernoulli}(\psi_{sex})$ where the parameter ψ_{sex} is the sex ratio of the population. We assume our default non-informative prior for this parameter: $\psi_{sex} \sim \text{Uniform}(0, 1)$. The model specification in Panel 7.2 demonstrates how to incorporate a partially observed covariate (i.e., "sex"). It is important to note that in the previous equation, sex_i is a vector with two categories indicating the sex of each individual (e.g., 1 = male, 2 = female). This corresponds directly to having a binary indicator of sex (e.g., $\texttt{Sex}_i = 1$ if individual i is female, and 0 otherwise). In the Bayesian formulation

```
model{

psi ~ dunif(0,1)                               # Prior distributions
psi.sex ~ dunif(0,1)
for(t in 1:2){
  alpha0[t] ~ dnorm(0,.1)
  logit(p0[t]) <- alpha0[t]
  alpha1[t] <- 1/(2*sigma[t]*sigma[t])
  sigma[t] ~ dunif(0, 15)
}

for(i in 1:M){
  z[i] ~ dbern(psi)
  Sex[i] ~ dbern(psi.sex)                       # Sex is binary
  Sex2[i] <- Sex[i] + 1                         # Convert to categorical
  s[i,1] ~ dunif(xlim[1],xlim[2])
  s[i,2] ~ dunif(ylim[1],ylim[2])

  for(j in 1:J){
    d[i,j] <- pow(pow(s[i,1]-X[j,1],2) + pow(s[i,2]-X[j,2],2),0.5)
    y[i,j] ~ dbin(p[i,j],K)
    p[i,j] <- z[i]*p0[Sex2[i]]*exp(-alpha1[Sex2[i]]*d[i,j]*d[i,j])
   }
 }
N <- sum(z[])
D <- N/area
}
```

PANEL 7.2

JAGS model specification for an SCR model with sex-specific encounter probability parameters.

of the model, we use both the binary indicator (`Sex`) and a categorical indicator (`Sex2` = `Sex` + 1). The former (termed `Sex` in Panel 7.2) allows us to specify the Bernoulli distribution for the random variable, and the latter (termed `Sex2`) allows us to use the dummy or indicator variable specification in the model.

In both **JAGS** or **BUGS**, missing data are indicated by `NA` in the data objects passed to the program through the `bugs` or `jags` functions in **R**. To set up the data, we need to create a vector of length M with the first n elements being 0 if individual i is a female, or 1 if i is a male (for the Fort Drum black bear data the function `bear.JAGS` extracts this information automatically from the `beardata` object), and the subsequent $M - n$ elements being `NA`. It is generally a good idea to provide starting values for the missing data, but we cannot provide starting values for observed data; in this case where one vector (or other object) contains both observed and missing data, initial values for the observed data have to be specified as `NA`. The code snippet

Table 7.6 Posterior summaries of parameter estimates from sex-specific SCR models for the Fort Drum black bear data set.

Parameter	Mean	SD	2.5%	97.5%
N	509.98	66.35	376.00	631.00
D	0.17	0.02	0.12	0.21
$p_{0,female}$	0.14	0.02	0.09	0.19
$p_{0,male}$	0.09	0.02	0.06	0.13
σ_{female}	1.54	0.13	1.31	1.83
σ_{male}	2.68	0.39	2.09	3.62
ψ_{sex}	0.31	0.07	0.19	0.45
ψ	0.78	0.10	0.58	0.97

below shows you how to set up the data including the `Sex` vector and the initial values function (the remainder of the code is identical to what we've shown before).

```
> sex <- beardata$sex #the sex data for captured individual
> Sex <- c(sex-1, rep(NA, nz)) #sex enters as 1/2, this recodes it to 0/1
                              #so we can use Bernoulli distribution

> data <- list(y=y,Sex=Sex, M=M,K=K, J=ntraps, xlim=xlim, ylim=ylim,area=areaX)
> params <- c('psi','p0','N', 'D', 'sigma', 'psi.sex')
> inits <- function() { list(z=c(rep(1,nind), rbinom(nz,1,0.5)),psi=runif(1),
          s=cbind(runif(M, xlim[1],xlim[2]), runif(M,ylim[1],ylim[2])),
          psi.sex=runif(1),Sex=c(rep(NA, nind), rbinom(nz,1,0.5)),
          sigma=runif(2,2,3),alpha0=runif(2)) }
```

The **BUGS** model specification is shown in Panel 7.2. Our estimate of density under the fully sex-specific model is still very similar to the previous models (Table 7.6), and while the baseline detection was not very different between males and females, we can see that they had very different σ estimates (note that the BCIs do not overlap). As usual, you can reproduce this analysis by calling the function `bear.JAGS` and set `model='SCRsex'`.

7.3 Individual heterogeneity

Here we consider SCR models with individual heterogeneity. Capture-recapture models with individual heterogeneity in detection probability, so-called model M_h, have a long history in classical capture-recapture models and they have special relevance to SCR (Section 4.4). While the advent of SCR models may appear to have rendered the use of classical model M_h obsolete (because one major source of heterogeneity, namely exposure to the trap array, is being accounted for explicitly), we may still wish to consider heterogeneity models for other biological reasons. It is reasonable to expect in real populations that there exists heterogeneity in home range size and so we think that α_1 could exhibit heterogeneity among individuals. As we noted previously, it may be advantageous or desirable in some cases to model heterogeneity directly in terms of the scale parameter of the encounter probability function, σ, or some other transformation of the "distance coefficient," perhaps even 95% home range area.

In this section, we describe a class of spatial capture-recapture models to allow for individual heterogeneity in encounter probability. In particular, one class of models we propose explicitly admits individual heterogeneity in home range *size*. In addition, we consider a standard representation for heterogeneity in which an additive individual-specific random effect is included in the linear predictor for baseline encounter probability.

7.3.1 Models of heterogeneity

An obvious extension to the SCR model is to include an additive individual effect, analogous to classical "model M_h." We'll call this model "SCRh":

$$\begin{aligned} \text{logit}(p_{0,i}) &= \alpha_0 + \eta_i, \\ p_{ijk} &= p_{0,i} \exp(-\alpha_1 ||\mathbf{x}_j - \mathbf{s}_i||^2), \end{aligned}$$

where η_i is an individual random effect having distribution $[\eta|\sigma_p]$. A popular class of models arises by assuming $\eta_i \sim \text{Normal}(0, \sigma_p^2)$ (Coull and Agresti, 1999; Dorazio and Royle, 2003). We show how to implement this specific SCRh model in Panel 7.3, and this model can be used to analyze the Ft. Drum bear data by calling the function `bear.JAGS` and setting `model='SCRh'`. While we show one possible implementation here, many other random effects distributions are possible. A popular one is the finite-mixture of point masses (Norris and Pollock, 1996; Pledger, 2004) which we demonstrate how to fit using `secr` in Section 7.4.3.

7.3.2 Heterogeneity induced by variation in home range size

An alternative heterogeneity model, one that has more of a direct biological motivation and interpretation, describes heterogeneity in home range size among individuals. To model heterogeneity in home range area, we can assume a distribution for a transformation of the scale parameter of the encounter probability model such as σ^2, or $\log(\sigma^2)$, etc. We call this "model SCRah" (Ah here for area-induced heterogeneity).

Consider the following log-normal model for the individual scale parameter of the Gaussian encounter probability model, σ_i^2:

$$\log(\sigma_i^2) \sim \text{Normal}(\mu_{hra}, \tau_{hra}^2),$$

then the 95% home range area has a scaled log-normal distribution with mean

$$E(\sigma_i^2) = 6\pi \exp(\mu_{hra} + \tau_{hra}^2/2).$$

The variance is slightly more complicated, but you can look up the variance of a log-normal distribution and combine it with the 95% home range area calculation in Section 5.4 to work out the implied variance of home range area under this model. We show two examples of the implied *population* distribution of home range area

```
model{

alpha0 ~ dnorm(0,.1)                          # Prior distributions
alpha1 <- 1/(2*sigma*sigma)
sigma ~ dunif(0, 15)
psi ~ dunif(0,1)
tau_p ~ dgamma(.001,.001)

for(i in 1:M){
  eta[i] ~ dnorm(0, tau_p)                    # Individual level variables
    z[i] ~ dbern(psi)
  s[i,1] ~ dunif(xlim[1],xlim[2])
  s[i,2] ~ dunif(ylim[1],ylim[2])

   for(j in 1:J){                             # The "likelihood" etc..
   d[i,j] <- pow(pow(s[i,1]-X[j,1],2) + pow(s[i,2]-X[j,2],2),0.5)
   y[i,j] ~ dbin(p[i,j],K)
   logit(p0[i,j]) <- alpha0 + eta[i]
   p[i,j] <- z[i]*p0[i,j]*exp(- alpha1*d[i,j]*d[i,j])
  }
 }
N <- sum(z[])                                  # N, D are derived
D <- N/area
}
```

PANEL 7.3

JAGS model specification for the SCRh model with Gaussian encounter probability model and additive normal random effect.

under this log-normal model, corresponding to a mean home range area of about 6.9 units squared (Figure 7.1). The left panel shows a standard deviation in home range area of 2.88 and the right panel shows a standard deviation in home range area of 0.70. The two cases were generated by tweeking the μ_{hra} and τ^2_{hra} parameters of the log-normal distribution to achieve a constant expected value of home range area, but different standard deviations.

7.4 Likelihood analysis in `secr`

Previously, in Chapter 6, we introduced the **R** package `secr` and described the likelihood-based inference approach taken by that package (see Section 6.5.3). Here we discuss how to implement some standard covariate models in `secr` and provide an example of model selection using AIC. As we saw in Chapter 6, `secr` uses the standard **R** model specification syntax, defining the dependent and independent variable

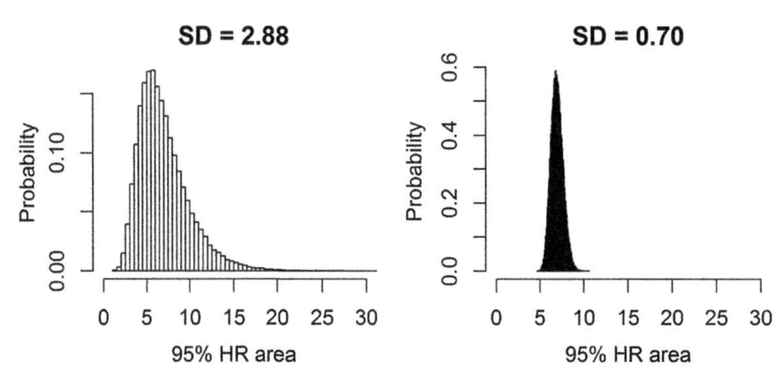

FIGURE 7.1

Population distribution of home range area for a model in which $\log(\sigma^2)$ has a normal distribution with mean μ_{hra} and variance τ^2_{hra}. The parameters were chosen to yield a constant expected value of about 6.9 units squared, but to produce two different levels of heterogeneity: A population standard deviation of 2.88 (left panel) and 0.70 (right panel).

relationship using tildes (e.g., `y ~ x`). Thus, in `secr` we might have `g0 ~ b` or `sigma ~ t`; when left unspecified or set to 1 (e.g., `g0 ~ 1`), this will default to a model with no covariates (i.e., constant parameter values). A number of default model formulas for the baseline and scale parameter of the encounter probability model are available in `secr`. Additionally, `secr` allows us to specify covariates on density (we cover this in Chapter 11), which are set for example as `D ~ habitat`.

To demonstrate models with various types of covariates using `secr`, we continue using the Fort Drum black bear data. We include in the `scrbook` package a function called `secr.bear` that will format the data (see Chapter 6 for the `secr` data format) and then fit and compare eight models (details shown in Panel 7.4). We have described all of these models in the previous sections, so we only briefly comment here on how to fit certain models in `secr` and compare them using AIC, and give a few helpful notes.

7.4.1 Notes for fitting standard models

In the `secr` package, the encounter probability model is called the "detection function" and it is specified by changing the "`detectfn`" option (an integer code) within the `secr.fit` command. Table 7.1 shows some of the possible encounter probability models that `secr` allows; the default is that based on the kernel of a bivariate normal probability distribution function (hence we call this the Gaussian model, but it is referred to as "half-normal" in `secr`) and the (negative) exponential is `detectfn = 2`. See model 2 in Panel 7.4 for how to fit the exponential model to the Fort Drum bear data set.

The `secr` package easily fits a range of SCR equivalents of standard capture-recapture models. The package has pre-defined versions of the classic model M_t where

```
1. null model with a bivariate normal encounter probability  model
bear_0=secr.fit(bear.cap, model=list(D ~ 1, g0 ~ 1, sigma ~ 1))

2. null model with an exponential encounter probability model
bear_0exp=secr.fit(bear.cap, model=list(D ~ 1, g0 ~ 1, sigma ~ 1),
                   detectfn=2)

3. model with fixed time effects
bear_t=secr.fit(bear.cap, model=list(D ~ 1, g0 ~ t, sigma ~ 1))

4. global behavioral model
bear_B=secr.fit(bear.cap, model=list(D ~ 1, g0 ~ b, sigma ~ 1))

5. trap-specific behavioral response
bear_b=secr.fit(bear.cap, model=list(D ~ 1, g0 ~ bk, sigma ~ 1))

6. global behavior model with fixed time effects
bear_bt=secr.fit(bear.cap, model=list(D ~ 1, g0 ~ b+t, sigma ~ 1))

7. sex-specific model
bear_sex=secr.fit(bear.cap, model=list(D ~ session, g0 ~ session,
                  sigma ~ session))

8. heterogeniety model
bear_h2=secr.fit(bear.cap, model=list(D ~ 1, g0 ~ h2, sigma ~ h2))
```

PANEL 7.4

Models called from secr.bear function. All models use buffer = 20,000.

each occasion has its own encounter probability, as well as a linear trend in baseline encounter probability over occasions (in a spatial modeling framework σ could also be an occasion specific parameter, but having encounter probability change with time seems like the more common case). For the classical time-effects type of model with K distinct parameters secr uses "t" to denote this in the model specification formula (see model 3 in Panel 7.4); whereas, for a linear trend over occasions secr uses "T".

The global trap response model (what we called model M_B), or a local trap-specific behavioral response (model M_b) can be fitted in secr using formulae with "b" for the global response model and "bk" for the local trap response model (see models 4 and 5 in Panel 7.4; note that to fit the trap-specific behavioral response model you need version 2.3.1 or newer of secr).

7.4.2 Sex effects

Incorporating sex effects into models with secr can be done a few different ways, but there are not pre-defined models for this. A limitation of fitting models with sex effects in `secr` is that it does not accommodate missing values of the sex variable. Thus, in all cases, individuals that are of unknown sex must be removed from the data set (recall that in a Bayesian framework we can keep these individuals in the data set by specifying a distribution for the individual covariate "sex"). In `secr`, the easiest way to include sex effects is to code sex as a "session" variable using the multi-session models (see Section 6.5.4 for a description of the multi-session models), providing two sessions, one representing males and one for females (see model 7 in Panel 7.4). This method provides two separate density estimates, which can then be combined into a total density.

7.4.3 Individual heterogeneity

To incorporate heterogeneity, `secr` fits a set of finite mixture models (Norris and Pollock, 1996; Pledger, 2004). These are expensive in terms of parameters but they have been widely adopted because they are easy to analyze using likelihood methods, as the marginal distribution of the data is just a sum of a small number of components. Using `secr`, individual heterogeneity can be incorporated into the encounter probability model using default models for either a 2- or 3-component finite mixture model using the "`h2`" or "`h3`" model terms. The 2-part mixture is shown in model 8 of Panel 7.4 and the 3-part mixture can easily be fit by substituting `h3` for `h2`. We only showed the SCRh logit-normal mixture in the version above (see Section 7.3.1), but finite-mixture models can also be fit in **JAGS** or **BUGS**.

7.4.4 Model selection in `secr` using AIC

One practical advantage to using the `secr` package, or likelihood inference in general, is the convenience of automatic model selection using AIC (Burnham and Anderson, 2002). The `secr` package has a number of convenient functions for computing AIC and producing model selection tables, or doing model-averaging (as described in Chapter 8). Running the function `secr.bear`, which calls all of the models we have described, will return, in addition to all model results, an AIC table with all of the summarized results including the AIC values, delta AIC, and model weights (see Table 7.7 or reproduce results in **R** using: `out<- secr.bear(); out$AIC.tab`).

It is important to note that AIC is not comparable between a multi-session model and a model that is not a multi-session model. Therefore, to compare the sex-specific model (which uses "sessions") with all the other models including the null, time, and behavioral models, we coded the data set initially as a multi-session design when first entering it to `secr`. This results in all the model outputs listing separate parameter estimates for each session, even the null model with no covariates; however, the estimates are the same for both "sessions" in all but the sex-specific model (in other words, we don't specify any effect of session on parameters, except in the sex specific model).

The results from this AIC analysis are straightforward to interpret; the model with a local trap response of encounter probability, "bk," has a model weight of 1 and

Table 7.7 Log-likelihood, AIC, AICc, δAICc, and AIC weight for several models run in secr for the Fort Drum black bear data set.

Model	logLik	AIC	AICc	δAICc	AICwt
bear.b	−641.7215	1291.443	1292.395	0.000	1
bear.h2	−653.8382	1319.676	1321.776	29.381	0
bear.0exp	−663.9152	1333.830	1334.389	41.994	0
bear.B	−677.6175	1363.235	1364.187	71.792	0
bear.bt	−668.3044	1358.609	1366.152	73.757	0
bear.sex	−677.7151	1367.430	1369.530	77.135	0
bear.t	−674.4134	1368.827	1374.938	82.543	0
bear.0	−686.2455	1378.491	1379.049	86.654	0

thus, according to AIC, 100% support compared with the other models in this model set. The 2-part finite-mixture model for g_0 and σ has the second lowest AIC, but considering the large δAICc compared to the local trap response model we would probably not consider it any further.

7.5 Summary and outlook

There are endless covariates and encounter probability models that can be defined and our goal in this chapter was to introduce basic types of covariate models and demonstrate how to implement them in **BUGS** and `secr`. Essentially, SCRs are GLMMs and therefore we develop covariate models in much the same way, using a suitable transformation (link function) of the parameter(s). In SCR models, we typically have two parameters of the encounter probability model for which we might specify covariate models—the baseline encounter probability (or rate) parameter, and a scale parameter that is related in many cases to the home range size of the species. A few examples of different covariate models are given in Table 7.3. We can also consider covariates by their classification as fixed, partially observed, or unobserved (see Table 7.8). This classification of covariate types can be important because the MLE and Bayesian approaches to dealing with partially and unobserved covariates are often different. This was seen above in how the covariate `Sex` was handled in the two frameworks.

Table 7.8 Examples of different covariate classifications.

Covariate Class	Examples
Fixed	baited, weather, habitat
Partially observed	sex, age,
Unobserved	home range size, ind. effects

While the move to spatially explicit models in capture-recapture studies has largely rendered the basic non-spatial CR models obsolete, they remain useful for categorizing the *spatial* extensions of these standard CR models. The extended models include the standard M_0, M_t, M_b, and M_h, but also new models that allow for trap-specific information such as "baited/not-baited" or "on/off road." In addition, in Chapters 11–13, we explore models for explaining variation in encounter probability and density based on spatial covariates that describe variation in landscape or habitat conditions.

CHAPTER

Model Selection and Assessment

8

Our purpose in life is to analyze models. By that, we mean one or more of the following basic four tasks: (1) estimate parameters, (2) make predictions of unobserved random variables, (3) evaluate the relative merits of different models or choosing a best model (model selection), and (4) check whether a specific model appears to provide a reasonable description of the data or not (model checking, assessment, or "goodness-of-fit"). In previous chapters we addressed the problems of estimation of model parameters, and making predictions of latent variables, **s** or **z**, or functions of these variables such as density or population size. In this chapter, we focus on the last two of these basic inference tasks: model selection (which model or models should be favored), and model assessment (do the data appear to be consistent with a particular model).

We review basic strategies of model selection using both likelihood methods (as implemented in the `secr` package) and Bayesian analysis. Specifically, we review a number of standard methods of model selection that apply to "variable selection" problems, when our set of models consists of distinct covariate effects and they represent constraints of some larger model. For classical analysis based on likelihood, model selection by Akaike Information Criterion (AIC) is the standard approach (Burnham and Anderson, 2002). For Bayesian analysis we rely on a number of different methods. We demonstrate the use of the deviance information criterion (DIC) (Spiegelhalter et al., 2002), although it has deficiencies when applied to hierarchical models in some cases (Millar, 2009). We use the Kuo and Mallick indicator variable selection approach (Kuo and Mallick, 1998) which produces direct statements of posterior model probabilities which we think are useful, and leads directly to model-averaged estimates of density. There is a good review paper recently by O'Hara and Sillanpää (2009) that discusses these and many other related ideas for variable selection. In addition we also recommend Link and Barker (2010, Chapter 7) for general information on model selection and assessment.

To check model adequacy in a Bayesian framework, or whether a specific model provides a satisfactory description of our data set, we rely exclusively on the Bayesian p-value framework (Gelman et al., 1996). For assessing fit of SCR models, part of the challenge is coming up with good measures of model fit, and there does not appear much definitive guidance in the literature on this point. Following Royle et al. (2011a), we break the problem up into two components which we attack separately: (1) Conditional on the underlying point process, does the encounter model fit? (2) Do the uniformity and independence assumptions appear adequate for the point process

Spatial Capture-Recapture. http://dx.doi.org/10.1016/B978-0-12-405939-9.00008-6

model of activity centers? The latter component of model fit has a considerable precedence in the ecological literature as it is analogous to the classical problem of testing "complete spatial randomness" (Cressie, 1991; Illian et al., 2008).

We apply some of these methods to the wolverine camera trapping data first introduced in Chapter 5 to investigate sex specificity of model parameters and whether there is a behavioral response to encounter. We note that individuals are drawn to the camera trap devices by bait and therefore it stands to reason that once an individual discovers a trap, it might be more likely to return subsequently, a response termed "trap happiness". We evaluate whether certain models for encounter probability appear to be adequate descriptions of the data, and we evaluate the uniformity assumption for the underlying point process.

8.1 Model selection by AIC

Using classical analysis based on likelihood, model selection is easily accomplished using AIC (Burnham and Anderson, 2002) which we demonstrate below. The AIC of a model is simply twice the negative log-likelihood evaluated at the MLE, penalized by the number of parameters (k) in the model:

$$\text{AIC} = -2\log L(\hat{\boldsymbol{\theta}}|\mathbf{y}) + 2k.$$

Models with small values of AIC are preferred. It is common to use a modified ("corrected") AIC referred to as AIC_c for small sample sizes which is

$$\text{AIC}_c = -2\log L(\hat{\boldsymbol{\theta}}|\mathbf{y}) + \frac{2k(k+1)}{n-k-1},$$

where n is the sample size. Two important problems with the use of AIC and AIC_c are that they don't apply directly to hierarchical models that contain random effects, unless they are computed directly from the marginal likelihood (for SCR models we can do this, see Chapter 6). Moreover, it is not clear what should be the effective sample size n in the calculation of AIC_c, as there can be covariates that affect individuals, that vary over time, or space. We do not offer strict guidelines as to when to use a small sample size adjustment.

The **R** package `secr` computes and outputs AIC automatically for each model fitted and it provides some capabilities for producing a model selection table (function `AIC`) and also doing model-averaging (function `model.average`), which we recommend for obtaining estimates of density based on multiple models.

8.1.1 AIC analysis of the wolverine data

We provide an example of model selection for the wolverine camera trapping data using `secr`. We consider a model set with distinct models to accommodate various types of sex specificity of model parameters:

Model 0: Model SCR0 with constant density and constant encounter model parameters;

Model 1: Model SCR0 with constant encounter probability parameter values shared by both male and female wolverines but with sex-specific density only;
Model 2: Sex-specific density, sex-specific p_0 but constant σ;
Model 3: Sex-specific density, sex-specific σ but constant p_0;
Model 4: Sex-specific density, sex-specific p_0 and sex-specific σ.

To model sex-specific abundance (density), we use the multi-session models fitted by `secr` (introduced in Section 6.5.4), which allow one to model session-specific effects on density, baseline encounter probability, p_0 (labeled g_0 in `secr`), and the scale parameter σ of the encounter probability model. Using this formulation, we define the "Session" variable to be a *categorical* sex code having value 1 or 2 (demonstrated below) and thus *session*-specific parameters represent *sex*-specific parameters. For example, if we model session-specific density, D, then this corresponds to Model 1 in our list above. We note that "Model 0" in our list corresponds to a model where all of the encounter histories have the same session ID. This model is one of constant density, which implies that the population sex ratio is fixed at 0.5, i.e., $\psi_{sex} = 0.5$.

Although `secr` also uses the logit/log-linear predictors as the default for modeling covariates on baseline encounter probability and the scale parameter, respectively, `secr` does something different with the multi-session models. It reports estimates in a *session mean* parameterization (equivalent to, in **BUGS**, using an index variable instead of a set of dummy variables), and not the *session effect* (i.e., deviation from the intercept) which arises from the use of dummy variables. We show this **BUGS** model description in Section 8.2.2.

To fit these models using `secr`, we load the wolverine data and do a slight bit of formatting to prepare the data objects for analysis by `secr`. The key difference from our analysis in Chapter 6 is that, here, we use the wolverine sex information (`wolverine$wsex`) which is a binary 0/1 variable (1 = male) and we add 1 so that we can define a categorical "Session" variable (having values 1 or 2). We also have a function `scr2secr` which converts a standard trap deployment file (TDF) matrix into a `secr` object of class "`traps`." The **R** commands are as follows (contained in the help file `?secr_wolverine`):

```
> library(secr)
> library(scrbook)
> data(wolverine)
> traps <- as.matrix(wolverine$wtraps)

## Name variables as required by secr
> dimnames(traps) <- list(NULL,c("trapID","x","y",paste("day",1:165,sep="")))
## Convert trap information to a secr "traps" object
> trapfile <- scr2secr(scrtraps=traps,type="proximity")

## Grab the wolverine state-space grid (2km here)
> gr <- as.matrix(wolverine$grid2)
> dimnames(gr) <- list(NULL,c("x","y"))
> gr2 <- read.mask(data=gr)
```

```
## Grab the encounter data, and re-name variables
> wolv.dat <- wolverine$wcaps
> dimnames(wolv.dat) <- list(NULL,c("Session","ID","Occasion","trapID"))

## Convert binary 0/1 sex variable to categorical 1/2 for "session"
> wolv.dat[,1] <- wolverine$wsex[wolv.dat[,2]]+1
> wolv.dat <- as.data.frame(wolv.dat)

## Convert to capthist object
> wolvcapt <- make.capthist(wolv.dat,trapfile,fmt="trapID",noccasions=165)
```

Once the data have been prepared in this way, we use the `secr` function `secr.fit` to fit the different models, and then the function `AIC` to package the models together and summarize them in the form of an AIC table, with rows of the table ordered from best to worst. The function `model.average` performs AIC-based model-averaging of the parameters specified by the `realnames` variable (below this is demonstrated for the parameter density, D). Because this function defaults to averaging by AIC$_c$, we slightly modified this function (called `model.average2`) to do model-averaging by either AIC or AIC$_c$ as specified by the user. The model fitting commands look like this for Model 0 and Model 1:

```
> model0 <- secr.fit(wolvcapt, model=list(D~1, g0~1, sigma~1),
           buffer=20000)
> model1 <- secr.fit(wolvcapt, model=list(D~session, g0~1, sigma~1),
           buffer=20000)
```

Next we use the function `AIC`, passing the fit objects from all five models, and that produces the following output (abbreviated horizontally to fit on the page):

```
> AIC(model0,model1,model2,model3,model4)
            model         ... npar   logLik       AIC       AICc    dAICc AICwt
model0 D~1 g0~1 sigma~1 ...   3   -627.2603 1260.521 1261.932 0.000 0.5831
model2         ..         ...   5   -624.9051 1259.810 1263.810 1.878 0.2280
model1         ..         ...   4   -627.2365 1262.473 1264.973 3.041 0.1275
model4         ..         ...   6   -624.6632 1261.326 1267.326 5.394 0.0393
model3         ..         ...   5   -627.2358 1264.472 1268.472 6.540 0.0222
```

By default, the `AIC` functions orders the models by AIC$_c$. Model-averaging the results is done as follows:

```
> model.average(model0,model1,model2,model3,model4,realnames="D")
              estimate    SE.estimate            lcl           ucl
session=1  2.707190e-05  7.913577e-06  1.544474e-05  4.745224e-05
session=2  2.927423e-05  8.270402e-06  1.700631e-05  5.039193e-05
```

As usual, estimates and standard errors of the individual model parameters can be obtained from the `secr.fit` summary output of any of the `modelX` objects shown above. The default output of estimated density is in individuals per ha, so we have to scale this up to something more reasonable. To get into units of per 1000 km^2, we need to multiply by 100 to get to units of km^2 and then multiply by 1000.

Table 8.1 Model selection results for the wolverine models of sex specificity, with/without a habitat mask. Fitting was done using `secr` with a half-normal (Gaussian) encounter probability model. Models are ordered by *AIC*. Density, *D*, is reported in units of individuals per 1,000 km^2. Model abbreviations indicate which parameters are sex-specific in order $D/p_0/\sigma$.

				Female			Male		
Model	**npar**	**AIC**	**AICc**	**D**	p_0	σ	**D**	p_0	σ
No Habitat Mask									
2: sex/sex/1	5	1259.8	1263.8	2.45	0.08	6435.51	3.16	0.04	6435.51
0: 1/1/1	3	1260.5	1261.9	2.83	0.06	6298.66	2.83	0.06	6298.66
4: sex/sex/sex	6	1261.3	1267.3	2.59	0.08	6080.70	2.99	0.04	6833.16
1: sex/1/1	4	1262.5	1265.0	2.69	0.06	6298.69	2.96	0.06	6298.69
3: sex/1/sex	5	1264.5	1268.5	2.70	0.06	6280.49	2.95	0.06	6319.03
With Habitat Mask									
2: sex/sex/1	5	1268.1	1272.1	3.64	0.07	6382.88	4.73	0.03	6382.88
4: sex/sex/sex	6	1268.7	1274.7	3.87	0.07	5859.40	4.41	0.03	7039.09
0: 1/1/1	3	1271.2	1272.6	4.18	0.05	6282.62	4.18	0.05	6282.62
1: sex/1/1	4	1273.1	1275.6	3.98	0.05	6282.65	4.38	0.05	6282.65
3: sex/1/sex	5	1275.1	1279.1	3.93	0.05	6357.26	4.41	0.05	6220.22

This produces an estimated density of about 2.71 for `session=1` (females) and 2.93 for `session = 2` (males). We can use the generic **R** function `predict` applied to the `secr.fit` output to obtain specific information about the MLEs on the natural scale.

We don't necessarily agree with the use of AIC$_c$ here and think its better to use AIC, in general. This is because, as noted previously, it is not clear what the effective sample size is for most capture-recapture problems. While we have 21 individuals in the data set, most of the model structure has to do with encounter probability samples and for that there are hundreds of observations. We do note that the AIC and AIC$_c$ results are not entirely consistent. By looking at the best model by AIC (Table 8.1), we find that the model with sex-specific density and sex-specific baseline encounter probability, p_0, is preferred (Model 2). This is just slightly better than the null model (Model 0) with no sex effects at all which has an implied fixed sex ratio of $\psi_{sex} = 0.50$.

We fit the same models but now using a modified state-space which excludes the ocean (this is called a habitat mask in `secr`). Results are shown in Table 8.1 (bottom half of the table). We see AIC values are smaller for the model without the mask. In general, it is probably acceptable to compare these different fits (with and without habitat mask) by AIC because we recognize the mask as having the effect of modifying the random effects distribution (i.e., of the activity centers, **s**) and the results should be sensitive to choice of the distribution for **s**. That said, we tend to prefer the mask model because it makes sense to exclude the areas of open water from the state-space of **s**. For females the model-averaged density is 3.88 individuals per 1,000 km^2 and for males the model-averaged density estimate is 4.46 individuals per 1,000 km^2 as we see here:

```
> model.average(model0b,model1b,model2b,model3b,model4b,realnames="D")
              estimate    SE.estimate          lcl          ucl
session=1  3.876615e-05  1.189102e-05  2.153795e-05  6.977518e-05
session=2  4.459658e-05  1.323696e-05  2.523280e-05  7.882022e-05
```

This is quite a bit higher than that based on the rectangular state-space (i.e., not specifying a habitat mask). This is not surprising given that **the state-space is part of the model** and the specific state-space modification we made here, which reduces the area from the rectangular state-space, should be extremely important from a biological standpoint.

8.2 Bayesian model selection

Model selection is somewhat less straightforward as a Bayesian, and there is no canned all-purpose method like AIC. As such we recommend a pragmatic approach, in general, for all problems, based on a number of basic considerations:

1. For a small number of fixed effects we think it is reasonable to adopt a conventional "hypothesis testing" approach—i.e., if the posterior for a parameter overlaps zero substantially, then it is probably reasonable to discard that effect from the model.
2. Calculation of posterior model probabilities: In some cases we can implement methods which allow calculation of posterior model probabilities. One such idea is the indicator variable selection method from Kuo and Mallick (1998). For this, we introduce a latent variable $w \sim \text{Bern}(.5)$ and expand the model to include the variable w as follows:

$$\text{logit}(p_{ijk}) = \alpha_0 + w \times \alpha_1 \times C_{ijk}.$$

   The importance of the covariate C is then measured by the posterior probability that $w = 1$.
3. The Deviance Information Criterion (DIC): Bayesian model selection is now routinely carried out using DIC (Spiegelhalter et al., 2002), although its effectiveness in hierarchical models depends very much on the manner in which it is constructed (Millar, 2009). We recommend using it if it leads to sensible results, but we think it should be calibrated to the extent possible for specific classes of models. This has not yet been done in the literature for SCR models, to our knowledge.
4. Logical argument: For something like sex specificity of certain parameters, it seems to make sense to leave an extra parameter in the model no matter what because, biologically, we should expect a difference (e.g., home range size). In some cases failure to apply logical argument leads to meaningless tests of gratuitous hypotheses (Johnson, 1999).

In all modeling activities, as in life itself, the use of logical argument should not be underutilized.

8.2.1 Model selection by DIC

The availability of AIC makes the use of likelihood methods convenient for problems where likelihood estimation is achievable. For Bayesian analysis, DIC seemed like a general-purpose equivalent, at least for a brief period of time after its invention. However, there seem to be many variations of DIC, and a consistent version is not always reported across computing platforms. Even statisticians don't have general agreement on practical issues related to the use of DIC (Millar, 2009). Despite this, it is still widely reported. We think DIC is probably reasonable for certain classes of models that contain only fixed effects, or for which the latent variable structure is the same across models so that only the fixed effects are varied (this covers many SCR model selection problems). However, it would be useful to see some calibration of DIC for some standardized model selection problems.

Model deviance is defined as negative twice the log-likelihood; i.e., for a given model with parameters θ: $\text{Dev}(\theta) = -2\log L(\theta|\mathbf{y})$. The DIC is defined as the posterior mean of the deviance, $\overline{\text{Dev}}(\theta)$, plus a measure of model complexity, p_D:

$$\text{DIC} = \overline{\text{Dev}}(\theta) + p_D.$$

The standard definition of p_D is

$$p_D = \overline{\text{Dev}}(\theta) - \text{Dev}(\bar{\theta}),$$

where the second term is the deviance evaluated at the posterior mean of the model parameter(s), $\bar{\theta}$. The p_D here is interpreted as the effective number of parameters in the model. Gelman et al. (2004) suggest a different version of p_D based on one-half the posterior variance of the deviance:

$$p_V = \text{Var}(\text{Dev}(\theta)|\mathbf{y})/2.$$

This is what is produced from **WinBUGS** and **JAGS** if they are run from `R2WinBUGS` or `R2jags`, respectively. It is less easy to get DIC summaries from `rjags`, so we used `R2jags` in our analyses below.

8.2.2 DIC analysis of the wolverine data

We repeated the analysis of the wolverine models with sex specificity, but this time doing a Bayesian analysis paralleling the likelihood analysis we did above with `secr`, using the logit/log parameterization of the model parameters. To do so in **BUGS**, we used dummy variables. Thus, we can express models allowing for sex specificity using a dummy variable `Sex` and new parameters ($\alpha_{sex}, \beta_{sex}$) which represent the *effect* of covariate `Sex`:

$$\text{logit}(p_{0,i}) = \alpha_0 + \alpha_{sex}\texttt{Sex}_i$$

and

$$\log(\sigma_i) = \log(\sigma_0) + \beta_{sex}\texttt{Sex}_i.$$

In these expressions, the sex variable $\texttt{Sex}_i$ is a binary variable where $\texttt{Sex}_i = 0$ corresponds to female, and $\texttt{Sex}_i = 1$ corresponds to male.

Unlike the multi-session model in secr, we carry out the analysis of the sex-specific model here by putting all of the data into a single data set, and explicitly accounting for the covariate "sex" in the model by assigning it a Bernoulli prior distribution with ψ_{sex} being the proportion of males in the population. In this case, we produce "Model 0" above, the model with no sex effect on density, by setting the population proportion of males at one-half: $\psi_{sex} = 0.5$ (see also Section 7.2.4). As usual, handling of missing values of the sex variable is done seamlessly which might be a practical advantage of Bayesian analysis in situations where sex is difficult to record in the field which may lead to individuals of unknown sex (i.e., missing values).

The **BUGS** model specification for the most complex model, Model 4, is shown in Panel 8.1. This model has a sex-specific intercept, scale parameter, σ, and density. We provide an **R** script named wolvSCR0ms in the scrbook package which fits each model. The function uses **JAGS** by default for the fitting, with the R2jags

```
alpha.sex ~ dunif(-3,3)              ## Prior distributions
beta.sex  ~ dunif(-3,3)
sigma0 ~ dunif(0,50)
alpha0 ~ dnorm(0,.1)
psi ~ dunif(0,1)                     ## Data augmentation parameter
psi.sex  ~ dunif(0,1)                ## Probability of ''male''

for(i in 1:M){                       ## DA loop
  wsex[i] ~ dbern(psi.sex)           ## Latent sex state (male = 1)
  z[i] ~ dbern(psi)                  ## DA variables, activity centers, etc..
  s[i,1] ~ dunif(Xl,Xu)
  s[i,2] ~ dunif(Yl,Yu)
  logit(p0[i]) <- alpha0 + alpha.sex*wsex[i]
  log(sigma.vec[i]) <- log(sigma0) + beta.sex*wsex[i]
  alpha1[i] <- 1/(2*sigma.vec[i]*sigma.vec[i])
  for(j in 1:ntraps){
    mu[i,j] <- z[i]*p[i,j]
    y[i,j] ~ dbin(mu[i,j],K[j])
    dd[i,j] <- pow(s[i,1] - traplocs[j,1],2)  + pow(s[i,2] - traplocs[j,2],2)
    p[i,j]  <-  p0[i]*exp( - alpha1[i]*dd[i,j] )
   }
 }
```

PANEL 8.1

Part of the **BUGS** specification for complete sex specificity of model parameters. This is a simplified version of the model contained in the wolvSCR0ms script, which allows users to run any of the models presented in Table 8.2.

Table 8.2 DIC results for the five models of sex specificity fitted to the wolverine camera trapping data, using the function `wolvSCR0ms`. Results are based on three chains of length 61,000, with a burn-in of 1,000, yielding 180,000 posterior samples.

Model	Meandev	p_D	DIC	Rank
Model 0	441.01	77.09	518.10	1
Model 1	441.78	77.50	519.28	3
Model 2	440.12	78.44	518.56	2
Model 3	443.31	79.47	522.79	5
Model 4	441.24	80.07	521.32	4

package. The template of this function is the model specification in Panel 8.1, which gets modified depending on the model we wish to fit using a command line option `model`. For example, `model = 1` fits the model with constant parameter values for males and females, but sex-specific population sizes (`model = 0` constrains the male probability parameter, ψ_{sex}, to be 0.5). The **R** function fits each of the five models using a binary indicator variable to turn "on" or "off" each effect. Here is how we obtain the MCMC output for each of the five models:

```
> wolv0 <- wolvSCR0ms(nb=1000,ni=21000,buffer=2,M=200,model=0)
> wolv1 <- wolvSCR0ms(nb=1000,ni=21000,buffer=2,M=200,model=1)
> wolv2 <- wolvSCR0ms(nb=1000,ni=21000,buffer=2,M=200,model=2)
> wolv3 <- wolvSCR0ms(nb=1000,ni=21000,buffer=2,M=200,model=3)
> wolv4 <- wolvSCR0ms(nb=1000,ni=21000,buffer=2,M=200,model=4)
```

We fitted the five models to the wolverine data and summarize the DIC computation results in Table 8.2. The model rank has Model 0, Model 2, Model 1, Model 4, and Model 3. Interestingly, this is the same order as the models based on AIC_c which we found above (see Table 8.1). The posterior mean and SD of model parameters for each model are given in Table 8.3.

8.2.3 Bayesian model-averaging with indicator variables

A convenient way to deal with model selection and averaging problems when doing Bayesian analysis by MCMC is to use the method of model indicator variables (Kuo and Mallick, 1998). Using this approach, we expand the model to include a set of prescribed models as specific reductions of a larger model. This has been demonstrated in some specific capture-recapture models in Royle and Dorazio (2008, Section 3.4.3) and in the context of SCR by Tobler et al. (2012). A useful aspect of this method is that model-averaged parameters are produced by default. We emphasize the need to be careful of reporting model-averaged parameters that don't have a common interpretation in the different models because they may be meaningless (e.g., like averaging apples and oranges). For example, if a regression parameter is in a specific

Table 8.3 Posterior summaries of model parameters for models with varying sex specificity of model parameters ranging from Model 0 = no sex specificity to Model 4 = fully sex-specific (see text). Models are based on the Gaussian encounter probability model, each with 21,000 iterations, 1,000 burn-in, 3 chains for a total of 60,000 posterior samples.

Parameter	Model 0		Model 1		Model 2		Model 3		Model 4	
	Mean	*SD*	*Mean*	*SD*	*Mean*	*SD*	*Mean*	*SD*	*Mean*	*SD*
N	60.02	11.91	60.24	11.93	59.37	11.97	59.67	11.97	58.77	11.75
D	5.79	1.15	5.81	1.15	5.72	1.15	5.75	1.15	5.66	1.13
α_0	−2.81	0.18	−2.82	0.17	−2.44	0.25	−2.82	0.18	−2.43	0.25
α_{sex}	0.00	1.73	0.00	1.73	−0.75	0.34	0.00	1.73	−0.79	0.36
σ_0	0.64	0.06	0.64	0.05	0.66	0.06	0.65	0.08	0.63	0.09
β_{sex}	0.00	1.73	−0.01	1.73	0.01	1.74	−0.01	0.17	0.10	0.18
ψ_{sex}	0.50	0.29	0.52	0.10	0.56	0.10	0.52	0.11	0.54	0.11
ψ	0.30	0.07	0.30	0.07	0.30	0.07	0.30	0.07	0.30	0.07
deviance	441.01	12.42	441.78	12.45	440.12	12.53	443.31	12.61	441.24	12.66
	pD = 77.1		pD = 77.5		pD = 78.4		pD = 79.5		pD = 80.1	
	DIC = 518.1		DIC = 519.3		DIC = 518.6		DIC = 522.8		DIC = 521.3	

model, then the posterior is informed by the data and a specific MCMC draw is from the appropriate posterior distribution. On the other hand, if the regression parameter is not in the model then the MCMC draw is obtained directly from the prior distribution, and so we need to think carefully about whether it makes sense to report an average of such a thing (in the vast majority of cases the answer is no). But some parameters like N or density, D, do have a consistent interpretation and we support producing model-averaged results of those parameters.

To implement the Kuo and Mallick approach, we expand the model to include the latent indicator variables, say w_m, for variable m in the model, such that

$$w_m = \begin{cases} 1 & \text{linear predictor includes covariate } m, \\ 0 & \text{linear predictor does not include covariate } m. \end{cases}$$

We assume that the indicator variables w_m are mutually independent with

$$w_m \sim \text{Bernoulli}(0.5)$$

for each variable $m = 1, 2, \ldots,$ in the model. For example, with two variables, the expanded model has the linear predictor:

$$\text{logit}(p_{ijk}) = \alpha_0 + \alpha_1 w_1 C_{1,i} + \alpha_2 w_2 C_{2,ijk},$$

where, let's suppose, $C_{1,i}$ is an individual covariate such as sex, and $C_{2,ijk}$ is a behavioral response covariate which is individual, trap, and occasion-specific.

We can assume a parallel model specification on the parameter σ which is liable to vary by individual-level covariates such as sex:

$$\log(\sigma_i) = \beta_0 + \beta_1 w_3 C_{1,i}.$$

Using this indicator variable formulation of the model selection problem we can characterize unique models by the sequence of w variables. In this case, each unique sequence (w_1, w_2, w_3) represents a model, and we can tabulate the posterior frequencies of each model by postprocessing the MCMC histories of (w_1, w_2, w_3), as we demonstrate shortly. This method then evaluates all possible combinations of covariates or 2^m models.

Conceptually, analysis of this expanded model within the data augmentation framework does not pose any additional difficulty. One broader, technical consideration is that posterior model probabilities are well known to be sensitive to priors on parameters (Aitkin, 1991; Link and Barker, 2006; Tenan et al., 2013). See also Royle and Dorazio (2008, Section 3.4.3) and Link and Barker (2010, Section 7.2.5). What might normally be viewed as vague or non-informative priors are not usually innocuous or uninformative when evaluating posterior model probabilities. The use of AIC seems to avoid this problem largely by imposing a specific and perhaps undesirable prior that is a function of the sample size (Kadane and Lazar, 2004). One possible solution is to compute posterior model probabilities under a model in which the prior for parameters is fixed at the posterior distribution under the full model (Aitkin, 1991). The Gibbs Variable Selection (GVS) method (Dellaportas et al., 2000) resolves this by prescribing distinct priors depending on whether a variable is in the model or not (Tenan et al., 2013). At a minimum, one should evaluate the sensitivity of posterior model probabilities to different prior specifications.

8.2.3.1 Analysis of the wolverine data

The **R** script `wolvSCR0ms` in the package `scrbook` provides the model indicator variable implementation for the fully sex-specific SCR model. It is run by setting `model=5` in the function call. We note again that it is not very useful to report most parameter estimates from this model because their marginal posterior is a mixture from the prior (when a value of the indicator variable of 0 is sampled) and draws informed by the data (i.e., from the posterior, when a 1 is drawn for the indicator variable w). On the other hand, the parameters N and density D should be reported and they represent marginal posteriors over all models in the model set. In effect, model-averaging is done as part of the MCMC sampling. The variable "mod" in the expressions below contains the two binary indicator variables (w above) which multiply the "sex" term in each of the p_0 and σ model components:

$$\text{logit}(p_{0,i}) = \alpha_0 + \texttt{mod}[1]\alpha_{sex}\texttt{sex}_i$$

and

$$\log(\sigma_i) = \log(\sigma_0) + \texttt{mod}[2]\beta_{sex}\texttt{sex}_i.$$

The third element of mod determines whether the ψ_{sex} parameter is estimated or fixed at $\psi_{sex} = 0.5$ which is accomplished with the line of **BUGS** code as follows:

```
sex.ratio <- psi.sex*mod[3] +.5*(1-mod[3]).
```

The MCMC output for "mod" was post-processed to obtain the model weights using the following **R** commands:

```
> mod <- wolv5$BUGSoutput$sims.list$mod
> mod <- paste(mod[,1],mod[,2],mod[,3],sep="")
>
> table(mod)
mod
  000     001     010     011     100     101     110     111
17181    4935    1057     296   25211    8337    2275     708
> round(table(mod)/length(mod), 3)
mod
  000     001     010     011     100     101     110     111
0.286   0.082   0.018   0.005   0.420   0.139   0.038   0.012
```

This results in a comparison of all eight possible models (based on $m = 3$ covariates) instead of just the five models we originally proposed. We see that the best model is that labeled (1,0,0) which, according to our construction above, has mod[1]=1, mod[2]=0, and mod[3]=0. This is the model having sex-specific baseline encounter probability p_0, and $\psi_{sex} = 0.5$. This model has posterior model probability 0.420. The model with no sex specificity at all (i.e., model (0,0,0)) has posterior probability 0.286, and the remaining posterior mass is distributed over the other six models. We could arrive at a qualitatively similar conclusion about which effects should be in the model using a more ad hoc approach based on looking at the posterior mass for each parameter under the full model (Model 4; see Table 8.3, in part). Considering the sex-specific intercept, it appears to be very important as its posterior mass is mostly away from zero. On the other hand, the coefficient on $\log(\sigma)$ is concentrated around 0, and the estimated ψ_{sex} (probability that an individual is a male) is 0.54 with a large posterior standard deviation. We might therefore be inclined to discard the sex effect on $\log(\sigma)$ based on classical thinking-like-a-hypothesis-testing-person and settle for the model with a sex-specific intercept in the encounter probability model. This is consistent with our indicator variable approach, which found that model $(1, 0, 0)$ has posterior probability of 0.420.

We can obtain model-averaged estimates from the indicator variable approach, which produces direct model-averaged estimates of N and D:

```
  mu.vect sd.vect   2.5%    25%    50%    75%  97.5%  Rhat n.eff
D   5.695   1.133  3.759  4.916  5.591  6.362  8.193 1.002  3600
N  59.077  11.758 39.000 51.000 58.000 66.000 85.000 1.002  3600
```

The model-averaged estimate (posterior mean) of density is $\hat{D} = 5.695$ which is hardly different from our model-specific estimates (Table 8.3) and, in particular, from Model 2 which has only a sex-specific intercept.

8.2.4 Choosing among detection functions

Another approach to implementing model selection is to introduce a categorical "model identity" variable which is itself a parameter of the model. Using this approach, then each distinct model is associated with a unique set of covariates or other set of model features. This is convenient especially when we cannot specify the linear predictor as some general model that reduces to various alternative submodels simply by switching binary variables on or off. In the context of SCR models, choosing among different encounter probability models would be an example. For this case we do something like this `mod ~ dcat(probs[])` where `probs` is a vector with elements $1/(\#models)$, and the encounter probability matrix is filled-in depending on the value of `mod`. In particular, instead of a two-dimensional array `p[i,j]`, we build `p[i,j,m]` for each of $m = 1, 2, \ldots, M$ models. An example in the **BUGS** language with three distinct models is:

```
 mod   ~ dcat(probs[])
##
## Using a double loop construction fill-in p[i,j,m] for each model:
##
 p[i,j,1] <- p0[1]*exp( - alpha1[1]*dist2[i,j] )
 p[i,j,2] <- 1-exp(-p0[2]*exp( - alpha1[2]*dist2[i,j] ) )
 logit(p[i,j,3]) <- p0[3] - alpha1[3]*dist2[i,j]

 mu[i,j] <- z[i]*p[i,j,mod]
 y[i,j] ~ dbin(mu[i,j],K[j])
```

As before the posterior probabilities can be highly sensitive to priors on the different model parameters and sometimes mixing is really poor and, in general, we've experienced mixed success trying to carry out model selection using this construction. We do provide a template **R/JAGS** script (`wolvSCR0ms2`) in the `scrbook` package which has an example of choosing among three different encounter probability models: The Gaussian encounter probability, Gaussian hazard, and logistic model with the square of distance (defined in Section 7.1). The key things to note are that there are three intercepts and three different "`alpha1`" parameters (the coefficient on distance). The parameters should not be regarded as equivalent across the models, so it is important to have them separately defined (and estimated) for each model. In our analysis we used a vague normal prior (precision = 0.1) for the intercept parameter (either log or logit scale of baseline encounter probability p_0) and a Uniform(0,5) prior for one-half the inverse of the coefficient on distance squared. In the **BUGS** model specification the priors look like this:

```
for(i in 1:3){
 alpha0[i] ~ dnorm(0,.1)
 sigma[i] ~ dunif(0,5)
 alpha1[i] <- 1/(2*sigma[i]*sigma[i])
}
```

Then, we create a probability of encounter for each individual, trap *and* model so that the holder object "`p`" in the model description is a three-dimensional array (sometimes this would have to be a four or five-dimensional array in more complex models with time effects, etc.), so that construction of the encounter probability models looks like this:

```
p[i,j,1] <- p0[1]*exp( - alpha1[1]*dist2[i,j] )
p[i,j,2] <- 1-exp(-p0[2]*exp( - alpha1[2]*dist2[i,j] ) )
logit(p[i,j,3]) <- p0[3] - alpha1[3]*dist2[i,j]
```

where

```
logit(p0[1]) <- alpha0[1]
log(p0[2]) <- alpha0[2]
p0[3]  <- alpha0[3]
```

You can experiment with the `wolvSCR0ms2` script to investigate the importance of different models of encounter probability and whether they have an affect on the inferences.

8.3 Evaluating goodness-of-fit

In practical settings, we estimate parameters of a desirable model, or maybe fit a bunch of models and report estimates from all of them or a model-averaged summary of density. An important question is: Is our model worth anything? In other words, does the model appear to be an adequate description of our data? Formal assessment of model adequacy or goodness-of-fit is a challenging problem and there are no all-purpose algorithms for doing this in either frequentist or Bayesian paradigms. Moreover, there are some philosophical challenges to evaluating model fit, such as, if we do model-averaging then should all of the models have to fit? Or should the averaged model have to fit? What do we do if none of the models fit? We don't know the answers to these questions and we won't try to answer them. Instead, we will provide what guidance we can on taking the first steps to evaluating fit, of a single model, as if it were a cherished family heirloom of great importance. We suggest that if you have a model that you really like, a single model, then it is a sensible thing to check that the model provides a good fit to your data. If it does not, we do not think the model is useless but just that some thought should be put into why the model doesn't fit so that, perhaps, some remediation might happen as future data are collected. After all, you may have spent 2, 3, or many more years of your life collecting that data set, perhaps thousands of hours, and therefore it seems a reasonable proposition to expect to do some estimation and analysis of the model regardless of model fit. You can still learn something from a model that does not pass some technical test of model fit.

For either Bayesian or classical inference, the basic strategy to assessing model fit is to come up with a fit statistic that depends on the parameters and the data set, which we denote by $T(\mathbf{y}, \theta)$, and then compute this for the observed data set, and compare its value to that computed for perfect data sets simulated under the correct model. In the case of classical inference, we will often rely on the standard practice of parametric bootstrapping (Dixon, 2002), where we simulate data sets conditional on the MLE $\hat{\theta}$ and compare realizations of the fit statistic with what we've observed. The **R** package `unmarked` (Fiske and Chandler, 2011) contains generic bootstrapping methods for certain hierarchical models, including distance sampling (e.g., see Sillett et al., 2012, for an application). In simple cases, using classical inference methods, it is sometimes possible to identify a test statistic of theoretical merit, perhaps with a known asymptotic distribution. For examples from capture-recapture see Burnham et al. (1987), Lebreton et al. (1992), and Chapter 5 of Cooch and White (2006). For Bayesian analysis we use the Bayesian p-value method (Gelman et al., 1996) (we introduced the Bayesian p-value in Section 3.9.1). Using this approach, data sets are simulated based on a posterior sample of the model parameters, θ, and the value of a fit statistic for the simulated data sets, usually based on the discrepancy of the observed data from its expected values, is compared to that for the actual data. In most cases, whether Bayesian or frequentist, the main idea for assessing model fit is the same: We compare data sets from the model we're interested in with the data set we have in hand. If they appear to be consistent with one another, then our faith in the model increases, at least to some extent, and we say "the model fits."

To date, there have been few suggestions on how to evaluate fit of SCR models from a classical perspective (but see Borchers and Efford, 2008, Section 6.5). For Bayesian analysis of SCR models, there has not been a definitive or general proposal for a fit statistic or even a class of fit statistics, although a few specialized implementations of Bayesian p-values have been provided (Gardner et al., 2010a; Royle et al., 2011a; Gopalaswamy et al., 2012a,b; Russell et al., 2012). While we universally adopt the Bayesian p-value approach, and suggest some fit statistics in the following text, we caution that there is no general expectation to support how well they should do. As such, one might consider doing some kind of custom evaluation or calibration when using such methods, to gage the power of the test (ability to reject under specific departures from the model).

8.4 The two components of model fit

For most SCR models, there are at least two distinct components of model fit, and we propose to evaluate these two distinct components individually. First, we can ask, are the data consistent with the *observation* model, conditional on the underlying point process? We can evaluate this based on the encounter frequencies of individuals *conditional* on (posterior samples of) the underlying point process $\mathbf{s}_1, \ldots, \mathbf{s}_N$. We discuss some potential fit statistics for addressing this in Section 8.4.2. Second, we can evaluate whether the data appear consistent with the *state* process model (i.e., the "uniformity" assumption of the point process). For the simple model of independence and uniformity, this is similar to the assumption of *complete spatial randomness* (CSR) which we consider in Section 8.4.1 below. Actually, this is not strictly the assumption of CSR because of the binomial assumption on N under data augmentation, so we instead use the term *spatial randomness*.

8.4.1 Testing uniformity or spatial randomness

Historically, especially in ecology, there has been an extraordinary amount of interest in whether a realization of a point process indicates "complete spatial randomness," i.e., that the points are distributed uniformly and independently in space. Two good references for such things are Cressie (1991, Chapter 8) and Illian et al. (2008).[1] In the context of animal capture-recapture studies, the spatial randomness hypothesis is manifestly false, purely on biological grounds. Typically individuals will be clustered, or more regular (for territorial species), than expected under spatial randomness and heterogeneous habitat will generate the appearance of clustering even if individuals are distributed independently of one another. While we recommend modeling spatial structure explicitly when possible (Chapters 11–13), the uniformity assumption may be an adequate description of data sets in some situations. Further, we find that it is generally flexible enough to reflect non-uniform patterns in the data, because we do observe some direct information about some of the point locations.

The basic technical framework for evaluating the spatial randomness hypothesis is based on counts of activity centers in cells or bins. For that we use any standard goodness-of-fit test statistic, based on griding (i.e., binning) the state-space of the point process into $g = 1, 2, \ldots, G$ cells or bins, and we tabulate $N_g \equiv N(\mathbf{x}_g)$ the number of activity centers in bin g, centered at coordinate $\mathbf{x}_g$. Specifically, let $B(\mathbf{x})$ indicate a bin centered at coordinate $\mathbf{x}$, then[2] $N(\mathbf{x}) = \sum_{i=1}^{N} I(\mathbf{s}_i \in B(\mathbf{x}))$ is the population size of bin $B(\mathbf{x})$. In Section 5.11.1, we used the summaries $N(\mathbf{x})$ for producing density maps from MCMC output. Here, we use them for constructing a fit statistic. We have used the Freeman-Tukey statistic of the form:

$$T(\mathbf{N}, \theta) = \sum_{g} \left(\sqrt{N_g} - \sqrt{\mathbb{E}(N_g)}\right)^2,$$

where $\mathbb{E}(N_g)$ is estimated by the mean bin count. An alternative assessment of fit is based on the following statistic: Conditional on N, the total number of activity centers in the state-space $\mathcal{S}$, the bin counts N_g should have a binomial distribution. It will usually suffice to approximate the binomial cell counts by Poisson cell counts, in which case we can use the classical "index-of-dispersion" test (Illian et al., 2008, p. 87), based on the variance-to-mean ratio:

$$ID = (G - 1) \times s^2 / \bar{N},$$

where s^2 is the sample variance of the bin counts and $\bar{N}$ is the sample mean. When the point process realization is *observed*, as in classical point pattern modeling (but not in SCR), this statistic has approximately a Chi-square distribution on $(G - 1)$

[1]We also like Tony Smith's lecture notes (Univ. of Penn. ESE 502), which can be found at, `http://www.seas.upenn.edu/~ese502/NOTEBOOK/Part_I/3_Testing_Spatial_Randomness.pdf`, accessed January 24, 2013.

[2]$I(arg)$ is the indicator function which evaluates to 1 if arg is true, otherwise 0.

degrees of freedom under the spatial randomness hypothesis. If $s^2/\bar{N} > 1$, clustering is suggested whereas, $s^2/\bar{N} < 1$ suggests the point process is too regular.

Whatever statistic we choose as our basis for assessing spatial randomness, the important technical issue is that we don't observe the point process and so the standard statistics for evaluating spatial randomness cannot be computed directly. However, using Bayesian analysis, we do have a posterior sample of the underlying point process and so we suggest computing the posterior distribution of any statistic in a Bayesian p-value framework. For a given posterior draw of all model parameters, we can obtain a posterior sample of $N(\mathbf{x})$ by taking all the output for MCMC iterations $m = 1, 2, \ldots,$ and computing:

$$N(\mathbf{x})^{(m)} = \sum_{z_i^{(m)}=1} I\left(\mathbf{s}_i^{(m)} \in B(\mathbf{x})\right).$$

Thus, $N(\mathbf{x})^{(1)}, N(\mathbf{x})^{(2)}, \ldots,$ is the Markov chain for the derived parameter $N(\mathbf{x})$.

In addition to computing the bin counts for the model based on our observed data at each iteration of the MCMC algorithm, simultaneously we generate a realization of the activity centers $\mathbf{s}_i$ under the spatial randomness model, and we obtain bin counts using these simulated activity centers, $\tilde{N}(\mathbf{x})$. For each of the posterior samples—that based on the real data, and that based on the simulated activity centers, we compute the fit statistic. The fit statistic based on the bin counts for the actual data is:

$$T(\mathbf{N}, \theta) = \sum_{x} \left(\sqrt{N(\mathbf{x})} - \sqrt{\bar{N}(\mathbf{x})}\right)^2,$$

whereas the fit statistic based on a simulated realizations of points under the spatial randomness hypothesis is:

$$T(\tilde{\mathbf{N}}, \theta) = \sum_{x} \left(\sqrt{\tilde{N}(\mathbf{x})} - \sqrt{\bar{N}(\mathbf{x})}\right)^2.$$

And we compute the Bayesian p-value by tallying up the proportion of times that $T(\tilde{\mathbf{N}}, \theta)$ is larger than $T(\mathbf{N}, \theta)$, as an estimate of: $p = \Pr(T(\tilde{\mathbf{N}}, \theta) > T(\mathbf{N}, \theta))$. The **R** function `SCRgof` in our package `scrbook` will do this, given the output from **JAGS** (Section 8.4.3).

8.4.1.1 ***Sensitivity to bin size***

Evaluating fit of point process models based on bin counts is sensitive to the number of bins (Illian et al., 2008, pp. 87–88). This is related to the classical problem of fit testing in binary regression because, in a point process model, as the bins get small, the bin counts go to 0 or 1 and standard fit statistics (e.g., based on deviance or Pearson residuals) are known not to be very useful. There is some good discussion of this in McCullagh and Nelder (1989, Section 4.4.5). What it boils down to is, using the example of the Pearson residual statistic considered by McCullagh and Nelder (1989), the fit statistic is exactly a deterministic function of the sample size only, which clearly should not be regarded as useful for model fit. This is why, in order to do a check of model fit when you have a binary response, one must always aggregate the data in some fashion. In the context of testing spatial randomness, computing the test statistic we described above has us chop up the region $\mathcal{S}$ into bins, and tally up N_g, the frequency of activity centers in each bin g. Suppose we choose the bin size to be extremely small such that $\mathbb{E}(N_g)$ tends to N/G (N being the number of activity centers) and N_g tends to a binary outcome. Therefore the fit statistic has N components that have value $N_g = 1$, and it has $G - N$ components that have value $N_g = 0$. Therefore, the fit statistic resembles:

$$T(\mathbf{N}, \theta) = \sum_{g \ni N_g = 1}^{N} (1 - \sqrt{N/G})^2 + \sum_{g \ni N_g = 0}^{G-N} (N/G)^2 = N(1 + (G - N)/G)$$

(here $\ni$ means "such that"). If G is huge relative to N, then we see that $T(\mathbf{N}, \theta)$ tends to about $2N$, which does not provide any meaningful assessment of model fit. So, in the limit in which the bin counts become binary, the fit statistic loses all its variability to the specific model used and is just a deterministic function of N. As a practical matter, it probably makes sense to restrict the number of bins to *fewer* than the number of observed individuals in the sample size. In typical SCR applications this will therefore result, usually, in very large (and few) bins, and presumably not much power.

There are some extensions that help resolve the issue of sensitivity to bin size. We can construct fit statistics based not just on quadrat counts but also the neighboring quadrat counts—this is the Greig-Smith method (Greig-Smith, 1964). In addition, there are a myriad of "distance methods" for evaluating point process models, and we believe that many of these can (and will) be adapted to SCR models. Again the main feature is that the point process on which inference is focused is completely latent in SCR models—so this makes the fit assessment slightly different than in classical point analysis.

8.4.1.2 Sensitivity to state-space extent

An issue that we have not investigated is that any model assessment that applies to a *latent* point process is probably sensitive to the size of the state-space. As the size of the state-space increases then the bin counts (far away from the data) *are* independent binomial counts with constant density, and so we can overwhelm the fit statistic with extraneous "data" simulated from the posterior, which is equal to the prior as we move away from the data, and therefore uninformed by the observed data that live in the vicinity of the trap array. Therefore we recommend computing these goodness-of-fit statistics in the vicinity of the trap array only. Perhaps, as an ad hoc rule of thumb, less than the average trap spacing from the rectangle enclosing the trap array. For example, if the average trap spacing is, say, 10 km, then the bins used to obtain the observed and predicted activity centers should not extend any further from the traps than 5 km. This should be a matter of future research.

8.4.2 Assessing fit of the observation model

In evaluating the spatial randomness hypothesis, we were able to draw on well-established ideas from point process modeling. On the other hand, it is less clear how to approach goodness-of-fit evaluation of the observation model. For many SCR problems, we have a three-dimensional data array of *binary* observations, y_{ijk} for individual i, trap j, and sample occasion k. As discussed in the previous section, we need to construct fit statistics based on observed and expected frequencies that are aggregated in some fashion. We recommend focusing on summary statistics that represent aggregated versions of y_{ijk} over 1 or 2 of the dimensions. We describe three such fit statistics below. We recognize that, depending on the model, some information about model fit will be lost by summarizing the data in this way. For example if there is a behavioral response and we aggregate over time to focus on the individual and trap level summaries then some information about lack-of-fit due to temporal structure in the data is lost.

Fit statistic 1: individual x trap frequencies. We summarize the data by individual-and trap-specific counts aggregated over all sample occasions. Using standard "dot notation" to represent summed quantities, we express that as: $y_{ij.} = \sum_{k=1}^{K} y_{ijk}$. Conditional on $\mathbf{s}_i$, the expected value under any encounter model is:

$$\mathbb{E}(y_{ij.}) = p_{ij}K$$

(or K_j if the traps are operational for variable periods). If there is time-varying structure to the model, then expected values would have to be computed according to $\mathbb{E}(y_{ij.}) = \sum_k p_{ijk}$. Then we can define a fit statistic from the Freeman-Tukey residuals according to:

$$T_1(\mathbf{y}, \theta) = \sum_i \sum_j \left(\sqrt{y_{ij.}} - \sqrt{\mathbb{E}(y_{ij.})}\right)^2,$$

where we use θ here to represent the collection of all parameters in the model. This statistic is conditional on $\mathbf{s}$ as well as on the data augmentation variables $\mathbf{z}$ and we compute it for *each* iteration of both the MCMC algorithm for both the observed data set and for a new data set simulated from the posterior distribution, say $\tilde{\mathbf{y}}$.

We could also use a similar fit statistic derived from summarizing over traps to obtain an `nind` $\times K$ matrix of count statistics. We imagine that either summary of the data will probably be too disaggregated (have mostly values of 0) in most practical settings to have much power.

Fit statistic 2: Individual encounter frequencies. SCR models represent a type of model for heterogeneous encounter probability, like model M_h, but with an explicit factor (space) that explains part of the heterogeneity. For model M_h, the individual encounter frequencies are the sufficient statistic for model parameters. Therefore, it makes intuitive sense to provide some kind of omnibus fit assessment of the core heuristic that the SCR model is adequately explaining the heterogeneity using a model M_h-like statistic based on individual encounter frequencies. So, we build a fit statistic based on the individual total encounters (Russell et al., 2012), $y_{i..} = \sum_j \sum_k y_{ijk}$. In addition, the expected value is a similar summary over traps and occasions: $\mathbb{E}(y_{i..}) = \sum_j \sum_k p_{ijk}$. Then, define statistic T_2 according to:

$$T_2(\mathbf{y}, \theta) = \sum_i \left(\sqrt{y_{i..}} - \sqrt{\mathbb{E}(y_{i..})} \right)^2.$$

We imagine this test statistic should provide an omnibus test of extra-binomial variation and should therefore capture some effect of variable exposure to encounter of individuals, although we have not carried out any evaluations of power under specific alternatives. Obviously, in using this statistic, we lose information on departures from the model that might be trap- or time-specific.

Fit statistic 3: Trap frequencies. We construct an analogous statistic based on aggregating over individuals and replicates to form trap-encounter frequencies: $y_{.j.} = \sum_i \sum_k y_{ijk}$ (Gopalaswamy et al., 2012b) and the expected value is a similar summary over individuals and occasions: $\mathbb{E}(y_{.j.}) = \sum_i \sum_k p_{ijk}$. Then statistic T_3 is:

$$T_3(\mathbf{y}, \theta) = \sum_j \left(\sqrt{y_{.j.}} - \sqrt{\mathbb{E}(y_{.j.})} \right)^2$$

This seems like a sensible fit statistic because we can think of SCR models as spatial models for counts (Chandler and Royle, 2013). Therefore, we should seek models that provide good predictions of the observable spatial data, which are the trap totals. In this context, it might even make sense to pursue cross-validation-based methods for model selection. Cross-validation is a standard method of evaluating spatial models such as in kriging or spline smoothers, so we could as well develop such ideas based on the trap-specific frequencies.

8.4.3 Does the SCR model fit the wolverine data?

We use the ideas described in the previous section to evaluate goodness-of-fit of the SCR model to the wolverine camera trapping data. We consider first whether the simple model of spatial randomness of the activity centers is adequate. We think that the encounter model shouldn't have a large effect on whether the spatial randomness assumption is adequate or not, so we fit "Model 0" (in which parameters are *not* sex-specific) using an **R** script provided in the function `wolvSCR0gof` which will default to fitting the model in **JAGS**. This is the same script as `wolvSCR0ms` except that it saves the MCMC output for the activity centers **s** and the data augmentation variables z, which are required in order to compute the Bayesian p-value test of spatial randomness.

The MCMC output is processed with the **R** function `SCRgof` which computes the test of spatial randomness based on bin counts, using the Bayesian p-value calculation. The function `SCRgof` requires a few things as inputs: (1) the output from a **BUGS** run (in particular, the activity center coordinates and the data augmentation variables); (2) the number of bins to use for computing activity center counts; (3) the trap locations, and (4) the buffer around the trap array to use for making the bins and computing the bin counts. This buffer could be that used in defining the state-space for the model fitting, but we think it should be relatively tighter to the trap array than the state-space used in model fitting. For the wolverine analysis, where we're using 10-km grid cells (1 unit = 10 km) and a 20-km buffer for model fitting, we'll use a state-space buffer of 0.4 units (4 km) for computing the fit statistic. The **R** commands to fit the model and obtain the goodness-of-fit result are as follows:

```
> wolv1 <- wolvSCR0gof(nb=1000,ni=6000,buffer=2,M=200,model=0)

> bugsout <- wolv1$BUGSoutput$sims.list

> traplocs <- wolverine$wtraps[,2:3]
> traplocs[,1] <- traplocs[,1] - min(traplocs[,1])
> traplocs[,2] <- traplocs[,2] - min(traplocs[,2])
> traplocs <- traplocs/10000

> set.seed(2013)   # set seed so Bayesian p-value is the same each time

> SCRgof(bugsout,5,5,traplocs=traplocs,buffer=.4)

Cluster index observed: 1.099822
Cluster index simulated: 1.000453
P-value index of dispersion: 0.408
P-value2 freeman-tukey: 0.6842667
```

The output produced by `SCRgof` is the index of dispersion based on the ratio of the variance to the mean (see above), which is computed as the posterior mean index of dispersion for the latent point process, and the average value for simulated data. If this value is >1 then clustering is suggested, which we see a (very) minor amount of evidence for here. Two Bayesian p-values are produced: the first is based on the cluster index, and the second is based on the Freeman-Tukey statistic calculated as described in Section 8.4.1. Because our p-values aren't close to 0 or 1, we judge that the model of spatial randomness provides an adequate description of the data. You can verify that a similar result is obtained if we use the model with fully sex-specific parameters (Model 4).

Next, we did a Bayesian p-value analysis of the observation component of the model, using the three fit statistics described in Section 8.4.2. These statistics can be calculated as part of the **BUGS** model specification or by postprocessing the MCMC output returned from a **BUGS** run. The **R** script `wolvSCR0gof` contains the relevant calculations. For example, to compute fit statistic T_1, we have to add some commands to the **BUGS** model specification such as the following (note: this is only a fraction of the model specification):

```
......
for(j in 1:ntraps){
  mu[i,j] <- w[i]*p[i,j]

  y[i,j]~dbin(mu[i,j],K[j])
```

```
    ynew[i,j]~dbin(mu[i,j],K[j])

    err[i,j] <- pow(pow(y[i,j],.5) - pow(K[j]*mu[i,j],.5),2)
    errnew[i,j] <- pow(pow(ynew[i,j],.5) - pow(K[j]*mu[i,j],.5),2)
}

T1obs <- sum(err[,])
T1new <- sum(errnew[,])
.......
```

Similar calculations are carried out to obtain the posterior samples of test statistics 2 (individual totals) and 3 (trap totals). For the wolverine data, the Bayesian p-value calculations produce:

```
> mean(wolv1$BUGSoutput$sims.list$T1new>wolv1$BUGSoutput$sims.list$T1obs)
[1] 0

> mean(wolv1$BUGSoutput$sims.list$T2new>wolv1$BUGSoutput$sims.list$T2obs)
[1] 0.17

> mean(wolv1$BUGSoutput$sims.list$T3new>wolv1$BUGSoutput$sims.list$T3obs)
[1] 0.02066667
```

Based on statistic T_2, we might conclude that the model is adequate for explaining individual heterogeneity although the other two statistics suggest a general lack-of-fit of the observation model. A similar result is obtained using the fully sex-specific model. We note that one individual was captured eight times in one trap, which is pretty extreme under a model which assumes independent Bernoulli trials. We surmise that the trap counts simply are not well explained by this model.

In attempt to resolve this problem, we extended the model to include a local (trap-specific) behavioral response (following Royle et al., 2011b) which can be fitted using the sample **R** script `wolvSCRMb`. To fit a model using **WinBUGS**, and then compute the Bayesian p-values we do this:

```
> wolv.Mb <- wolvSCRMb(nb=1000,ni=6000,buffer=2,M=200)

> mean(wolv.Mb$sims.list$T1new>wolv.Mb$sims.list$T1obs)
[1] 0.9666667

> mean(wolv.Mb$sims.list$T2new>wolv.Mb$sims.list$T2obs)
[1] 0.3644667

> mean(wolv.Mb$sims.list$T3new>wolv.Mb$sims.list$T3obs)
[1] 0.4990667
```

This model seems to fit better, and so we might prefer reporting estimates from it, which we do in Table 8.4. (the behavioral response parameter is labeled α_2 in the table). Estimated density is about 1 individual higher per 1,000 km^2 compared with the various models that lack a behavioral response. It might be useful to try these fit assessment exercises using the habitat mask as described in Section 5.10. That takes an extremely long time to run in **BUGS** though, especially for the behavioral response model.

Table 8.4 Posterior summary statistics for local (trap-specific) behavioral response model M_b fitted to the wolverine camera trapping data using **WinBUGS**. The parameter α_2 is the local (trap-specific) behavioral response parameter. $T_x()$ are the posterior summaries of fit statistics $x = 1,2,3$ used in the Bayesian p-value analysis (see text for definitions). Results are based on three chains, each with 6,000 iterations (first 1,000 discarded) for a total of 15,000 posterior samples.

Parameter	Mean	SD	2.5%	50%	97.5%	Rhat	n.eff
N	71.32	19.07	42.00	69.00	114.02	1.00	2100
D	6.87	1.84	4.05	6.65	10.99	1.00	2100
σ	0.88	0.13	0.68	0.86	1.17	1.00	730
p_0	0.01	0.00	0.01	0.01	0.02	1.01	530
α_1	0.69	0.19	0.37	0.67	1.10	1.00	730
α_2	2.50	0.27	1.99	2.50	3.04	1.00	700
ψ	0.36	0.10	0.20	0.35	0.58	1.00	2600
T_1^{obs}	54.71	6.12	43.69	54.39	67.47	1.00	3900
T_1^{new}	64.73	7.62	50.93	64.39	80.96	1.00	3900
T_2^{obs}	13.93	4.07	7.25	13.53	23.04	1.00	5700
T_2^{new}	12.65	3.35	6.93	12.36	20.07	1.00	2000
T_3^{obs}	12.80	1.74	9.80	12.64	16.61	1.00	2400
T_3^{new}	12.94	3.05	7.77	12.67	19.58	1.00	15000

8.5 Quantifying lack-of-fit and remediation

Molinari-Jobin et al. (2013) used a strategy for assessing model fit in dynamic occupancy models similar to that which we suggested above. They constructed a fit statistic based on aggregating the data over replicate samples (k), to obtain the total detections per site i and year j. They used a Bayesian p-value analysis based on a Chi-squared test statistic (also see Kéry and Schaub, 2012, Chapter 12). Their analysis suggested a model that didn't fit, and, so they computed the "lack-of-fit ratio" (see Kéry and Schaub, 2012, Section 12.3)—the ratio of the fit statistic computed for the observed data to that of the replicate (simulated) data sets. They interpret this analogous to the overdispersion coefficient in generalized linear models (McCullagh and Nelder, 1989), usually called the c-hat statistic in capture-recapture literature (see Cooch and White, 2006, Chapter 5). Molinari-Jobin et al. (2013) reported the lack-of-fit ratio for their model to be 1.14 which suggests a minor lack-of-fit, compared to perfect data having a value of 1, because the posterior standard deviations will be too small by a factor of $\sqrt{1.14} = 1.07$. In classical capture-recapture applications of goodness-of-fit assessment, inference for non-fitting models is dealt with by inflating the resulting SEs (of the non-fitting model), by the square root of c-hat. We believe that these ideas related to quantifying lack-of-fit and understanding its effect could also be applied to SCR models, although we have not yet explored this.

8.6 Summary and outlook

In this chapter, we offered some general strategies for model selection and model checking, or assessment of model fit. We think the strategies we outlined for model selection are fairly standard and can be effectively applied to many SCR modeling problems. Some technical issues of Bayesian analysis need to be addressed (in general) before Bayesian methods are more generally useful and accessible. For one thing, Bayesian model selection based on the indicator variable approach of Kuo and Mallick (1998) can be tediously slow even for small data sets, and so improved computation will improve our ability to do Bayesian model selection in practical situations. Also, and most importantly, sensitivity to prior distributions is an important issue. Further research and practice might identify preferred prior configurations for SCR that provide a good calibration in relevant model selection problems. Finally, we believe that cross-validation should prove to be a useful method in model assessment and selection, as SCR models are a form of spatial model of counts, and so it is natural to pick models that predict the observable spatial counts (i.e., at trap locations) well. Bayesian model selection is an active area of research and development, and we expect more general and efficient methods will be developed, and even exist now for application. For example, Tenan et al. (2013) present an evaluation of two additional methods, including the product space method (Carlin and Chib, 1995) and Gibbs variable selection (GVS) (Dellaportas et al., 2000), which could provide improvement over the methods we've used in SCR applications. These ideas merit consideration.

For Bayesian model assessment, or goodness-of-fit checking, we suggested a framework based on independent testing of the spatial model of independence and uniformity, and testing fit of the observation model conditional on the underlying point process. These ideas are based on mostly ad hoc attempts in a number of published applications (e.g., Royle et al., 2009a, 2011a; Gopalaswamy et al., 2012b; Russell et al., 2012). While we think this general strategy should be fruitful, we know of no studies on the power to detect various model departures, and so the ideas should be viewed as experimental. We have not addressed assessment of model fit for SCR models using likelihood methods, although we imagine that standard bootstrapping ideas should be effective, perhaps based on the fit statistics (or similar ones) we suggested here for computing Bayesian p-values.

Clearly there is much research to be done on assessment of model fit in SCR models. Other approaches from spatial point process modeling should be pursued including nearest-neighbor methods or distance-based methods. In addition, studies to evaluate the power to detect relevant departures from the standard assumptions, and the robustness of inferences about N or density, need to be conducted. If the spatial randomness model appears inadequate, it is possible to fit models that allow for a non-uniform distribution of points (see Chapter 11) and even point process models that allow for interactions among points (Reich et al., in review). On the other hand, we expect that most of these Bayesian p-value tests will have low power in typical data sets consisting of a few to a few dozen individuals. As such, failure to detect a lack-of-fit may not be that meaningful. But, on the other hand, it may not make a difference in terms of density estimates either. We think inference about density should be relatively insensitive to departures from spatial randomness, because we get to observe direct information on some component of the population. For those activity centers, the assumed model of the point process should exert little influence on the placement of the activity centers. Conversely, as is the case with classical closed population models (Otis et al., 1978; Dorazio and Royle, 2003; Link, 2003), inferences may be somewhat more sensitive to bad-fitting models for the observation process.

CHAPTER

Alternative Observation Models

9

In previous chapters we considered various models of *encounter probability*, both in terms of parametric functions of distance and also in terms of covariate models (Chapter 7 and elsewhere). However, we have so far only considered a specific probability model for the observations (we'll call this the "observation model")—the Bernoulli observation model which, in `secr` (Efford, 2011a), is the *proximity detector* model. This assumes that individual and trap-specific encounters are independent Bernoulli trials.

In this chapter, we focus on developing additional observation models. The observation model could be thought of as being determined by the type of device—or the type of "detector" using the terminology of `secr` (Efford, 2011a). We consider models that apply when observations are not binary and, in some cases, that do not require independence of the observations. For example, if sampling devices can detect an individual some arbitrary number of times during an interval, then it is natural to consider observation models for encounter frequencies, such as the Poisson model. Another type of encounter device is the "multi-catch" device (Efford et al., 2009a), which is a physical device that can capture and hold an arbitrary number of individuals. A typical example is a mist net for birds (Borchers and Efford, 2008). It is natural to regard observations from these kinds of studies as independent multinomial observations. A related type of device that produces *dependent* multinomial observations are the so-called *single-catch* traps (Efford, 2004; Efford et al., 2009a). The canonical example are small-mammal live traps which catch and hold a single individual. Competition among individuals for traps induces a complex dependence structure among individual encounters. To date, no formal inference framework has been devised for this method although it stands to reason that the independent multinomial model should be a good approximation in some situations (Efford et al., 2009a). We analyze a number of examples of these different observation models using **JAGS** and also the **R** package `secr` (Efford, 2011a).

9.1 Poisson observation model

The models we analyze in Chapter 5 assumed binary observations—i.e., standard encounter history data—so that individuals are captured at most one time in a trap on

Spatial Capture-Recapture. http://dx.doi.org/10.1016/B978-0-12-405939-9.00009-8

any given sample occasion. This makes sense for many types of DNA sampling (e.g., based on hair snares) because distinct visits to sampled locations or devices within a sampling occasion cannot be differentiated. However, for some encounter devices, or methods, the potential number of encounters is *not* fixed, and so it is possible to encounter an individual some arbitrary number of times during any particular sampling episode. That is, we might observe encounter frequencies $y_{ijk} > 1$ for individual i, trap j, and sampling interval k. As an example, if a camera device is functioning properly it may be programed to take photos every few seconds if triggered. For a second example, suppose we are searching a quadrat or length of trail for scat, we may find multiple samples from the same individual. Therefore, we seek observation models that accommodate such encounter frequency data. In general, any discrete probability mass function could be used for this purpose, including the standard models for count data used throughout ecology, the Poisson, and negative binomial. Here we focus on using the Poisson model only although other count frequency models have been considered for SCR models (Efford et al., 2009b).

Let y_{ijk} be the frequency of encounter for individual i in trap j during occasion k, then assume:

$$y_{ijk} \sim \text{Poisson}(\lambda_{ij}),$$

where the expected encounter frequency λ_{ij} depends on both individual and trap (we could, of course, also make λ time-dependent). As we did in the binary model of Chapter 5, we now seek to model the expected value of the observation (which was p_{ij} in Chapter 5) as a function of the individual activity center $\mathbf{s}_i$. We assume that

$$\lambda_{ij} = \lambda_0 g(\mathbf{x}_j, \mathbf{s}_i),$$

where $g(\mathbf{x}, \mathbf{s})$ is any positive valued function, such as the negative exponential or the bivariate Gaussian kernel, and λ_0 is the baseline encounter rate—the expected number of encounters if a trap is placed precisely at an individual's home-range center (note: in `secr` the notation for this is g_0). Then, $\lambda_0 g(\mathbf{x}_j, \mathbf{s}_i)$ is the expected encounter rate in trap $\mathbf{x}_j$ for an individual having activity center $\mathbf{s}_i$. Note that

$$\log(\lambda_{ij}) = \log(\lambda_0) + \log(g(\mathbf{x}_j, \mathbf{s}_i)).$$

Equating $\alpha_0 \equiv \log(\lambda_0)$, and, if $g(\mathbf{x}, \mathbf{s}) \equiv \exp(-d(\mathbf{x}, \mathbf{s})^2/(2\sigma^2))$ (i.e., the Gaussian encounter probability model), then:

$$\log(\lambda_{ij}) = \alpha_0 - \alpha_1 d(\mathbf{x}_j, \mathbf{s}_i)^2, \qquad (9.1.1)$$

where $\alpha_1 = 1/(2\sigma^2)$, which is the same linear predictor as we have seen for the Bernoulli model in Chapter 5. This Poisson SCR model is therefore a type of Poisson generalized linear mixed model (GLMM).

We can accommodate covariates at the level of the individual-, trap-, or sample-occasion by including them on the baseline encounter rate parameter λ_0. For example, if C_j is some covariate that depends on trap only, then we express the relationship between λ_0 and C_j as:

$$\log(\lambda_{0,ijk}) = \alpha_0 + \alpha_2 C_j$$

and, therefore, covariates on the logarithm of baseline encounter probability appear also as linear effects on λ_{ij}. In general, covariates might also affect the coefficient on the distance term (α_1) (e.g., sex of individual). We don't get into too much discussion of general covariate models here, because the same principles apply as we discussed in Chapters 7 and 8.

For models in which we do not have covariates that vary across the sample occasions k, we can aggregate the observed data by the property of compound additivity of the Poisson distribution (if x and y are *iid* Poisson with mean λ then $x + y$ is Poisson with mean 2λ). Therefore,

$$y_{ij} = \left(\sum_{k=1}^{K} y_{ijk} \right) = \text{Poisson}(K\lambda_0 g(\mathbf{x}_j, \mathbf{s}_i)).$$

We see that K and λ_0 serve the same role as affecting the base encounter rate. Since the observation model is the same, probabilistically speaking, for all values of K, evidently we need only $K = 1$ "survey" from which to estimate model parameters (Efford et al., 2009b). We know this intuitively, as sampling by multiple traps serves as replication in SCR models. This has great practical relevance to the conduct of capture-recapture studies and the use of SCR models. For example, if individuality is obtained by genetic information from scat sampling, one should only have to carry out a single spatial sampling of the study area. However, one must be certain that sufficient spatial recaptures will be obtained so that effective parameter estimation is possible.

9.1.1 Poisson model of space usage

It is natural to interpret the Poisson encounter model as a model of space usage resulting from movement of individuals about their home range (Section 5.4). Imagine we have perfect samplers in every pixel of the landscape so that whenever an individual moves from one pixel to another, we can record it. Let m_{ij} be the number of times individual i was recorded in pixel j (i.e., it selected or used pixel j). Then, we might think of the Poisson model for the observed *use* frequencies:

$$m_{ij} \sim \text{Poisson}(\lambda_0 g(\mathbf{x}_j, \mathbf{s}_i)),$$

where λ_0 is related to the baseline movement rate of the animal (how often it moves). This model of space usage gives rise to the standard resource selection function (RSF) models (see Chapter 13). But now suppose our samplers are not perfect but, rather, record only a fraction of the resulting visits. A sensible model is

$$y_{ij} | m_{ij} \sim \text{Binomial}(m_{ij}, p).$$

The marginal distribution of y_{ij} is:

$$y_{ij} \sim \text{Poisson}(p_0 g(\mathbf{x}_j, \mathbf{s}_i)),$$

where p_0 is a composite of the movement rate and conditional detection probability p. Therefore, we see that encounters accumulate in proportion to the frequency of outcomes of an individual using space (or "selecting resources").

We introduced an interpretation of SCR models in terms of movement and space usage in Section 5.4, and it is one of the main underlying concepts of SCR models that is not present in ordinary capture-recapture models. As we noted there, the underlying model of space usage is only as complex as the encounter probability model, which has been, so far in this book, only symmetric and stationary (does not vary in space). We generalize this model of space usage in Chapter 13.

9.1.2 Poisson relationship to the Bernoulli model

There is a sense in which the Poisson and Bernoulli models can be viewed as consistent with one another. Note that under the Poisson model, the relationship between the expected count and the probability of "at least 1 detection" is given by

$$\Pr(y > 0) = 1 - \exp(-\lambda), \tag{9.1.2}$$

where $\mathbb{E}(y) = \lambda$. Therefore, if we equate the event "encountered" with the event that the individual was captured at least once time under the Poisson model, i.e., $y > 0$, then it would be natural to set $p_{ij} = \Pr(y > 0)$ according to Eq. (9.1.2). That is, we can use Eq. (9.1.2) as the model for encounter probability for binary observations. This is the "hazard rate" model in distance sampling.

In fact, as λ gets small, the Poisson model is a close approximation to the Bernoulli model in the sense that outcomes concentrate on $\{0, 1\}$, i.e., $\Pr(y \in \{0, 1\}) \to 1$ as $\lambda \to 0$. Indeed, under the Poisson model, $\Pr(y > 0) \to \lambda$ for small values of λ. This phenomenon is shown in Figure 9.1 where the left panel shows a plot of $\lambda_{ij} = \lambda_0 g(\mathbf{x}_j, \mathbf{s}_i)$ vs. distance, and superimposed on that is a plot of $p_{ij} = 1 - \exp(-\lambda_{ij})$ vs. distance, for values $\lambda_0 = 0.1$ and $\sigma = 1$, and the right

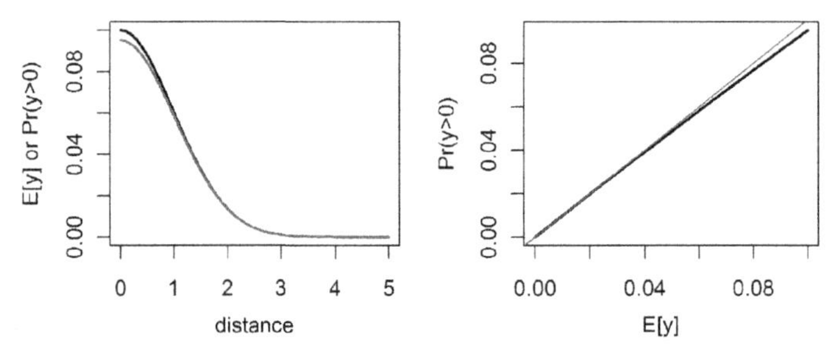

FIGURE 9.1

Poisson approximation to the binomial model. As the Poisson mean approaches 0, then $\Pr(y > 0)$ under the Poisson model approaches λ and therefore $y \sim \text{Poisson}(\lambda)$ is well approximated by a Bernoulli model with parameter $p = \lambda$.

panel shows a plot of $\Pr(y > 0)$ vs. $\mathbb{E}(y)$. We see that the two quantities are practically indistinguishable. To evaluate the closeness of the approximation, you can execute the following **R** commands, which we used to produce Figure 9.1:

```
> x <- seq(0.001,5, ,200)
> lam0 <- .1
> sigma <- 1
> lam <- lam0*exp(-x*x/(2*sigma*sigma))

> par(mfrow=c(1,2))
> p1 <- 1-exp(-lam)
> plot(x, lam, ylab="E[y] or Pr(y>0)",xlab="distance",type="l",lwd=2)
> lines(x,p1,lwd=2,col="red")
> plot(lam, p1, xlab="E[y]",ylab="Pr(y>0)",type="l",lwd=2)
> abline(0,1,col="red")
```

To summarize, if y is Poisson, then, as λ gets small,

$$\begin{aligned} \Pr(y > 0) &\approx \mathbb{E}(y), \\ 1 - \exp(-\lambda_0 g(\mathbf{x}, \mathbf{s})) &\approx \lambda_0 g(\mathbf{x}, \mathbf{s}). \end{aligned} \tag{9.1.3}$$

What all of this suggests is that if we have very few observations > 1 in our SCR data set, then we won't lose much information by using the Bernoulli model. On the other hand, the Poisson model may have some advantages in terms of analytic or numerical tractability in some cases. Further, this approximation explains the close correspondence we have found between these two versions of the Gaussian encounter probability model (Section 5.4). Namely, the Gaussian hazard model and the Gaussian encounter probability model are close approximations because $1 - \exp(-\lambda) \approx \lambda$ if λ is small.

Even in such cases where the Poisson and Bernoulli models are not quite equivalent, we might choose to truncate individual encounter frequencies to binary observations anyhow (transforming counts to 0/1 is called "quantizing"). We might do this intentionally in some cases, such as when the distinct encounter events are highly dependent as often happens in camera trap studies when the same individual moves back and forth in front of a camera during a short period of time. But sometimes, truncation is a feature of the sampling. For example, in the case of bear hair snares, the number of encounters might be well approximated by a Poisson distribution but we cannot determine unique visits and so only get to observe the binary event "$y > 0$." In this case, we might choose to model the encounter probability for the binary encounter using Eq. (9.1.3). This is equivalent to the complementary log-log link model from Chapter 5:

$$\text{cloglog}(p_{ij}) = \log(\lambda_0) + \log(g(\mathbf{x}, \mathbf{s})),$$

where $\text{cloglog}(u) = \log(-\log(1 - u))$.

9.1.3 A cautionary note on modeling encounter frequencies

Other models for counts might be appropriate. For example, ecologists are especially fond of the negative binomial model for count data because it accommodates overdispersion (White and Bennetts, 1996; Kéry et al., 2005; Ver Hoef and Boveng, 2007) but other models for excess-Poisson variation are possible. For example, we might add a normally distributed random effect to the linear predictor for λ (Coull and Agresti, 1999).

As a general rule, we favor the Bernoulli observation model even if our sampling scheme produces encounter frequencies. The main reason is that, with frequency data, we are forced to confront a model choice problem (i.e., Poisson, negative binomial, log-normal mixture) that is wholly unrelated to the fundamental space usage process that underlies the genesis of SCR data. Repeated encounters over short time intervals are not likely to be the result of independent encounter events. E.g., an individual moving back and forth in front of a camera yields a cluster of observations that is not informative about the underlying process of space usage. Similarly, in scat surveys dogs are used to locate scats which are processed in the laboratory for individuality (Kohn et al., 1999; MacKay et al., 2008; Thompson et al., 2012). The process of local scat deposition is not strictly the outcome of movement or space usage but rather the outcome of complex behavioral considerations as well as dependence in detection of scat by dogs. For example, dogs find (or smell) one scat and then are more likely to find one or more nearby ones, if present, or they get into a den or latrine area and find many scats. The additional assumption required to model variation in observed frequencies (i.e., conditional on location) provides no information about space usage and density, and we feel that the model selection issue should therefore be avoided.

9.1.4 Analysis of the Poisson SCR model in BUGS

We consider the simplest possible model here in which we have no covariates that vary over sample occasions $k = 1, 2, \ldots, K$ so that we work with the aggregated individual- and trap-specific encounters:

$$y_{ij} \sim \left(\sum_{k=1}^{K} y_{ijk}\right) \sim \text{Poisson}(K\lambda_0 g(\mathbf{x}_j, \mathbf{s}_i))$$

and we consider the bivariate normal form of $g(\mathbf{x}, \mathbf{s})$:

$$g(\mathbf{x}, \mathbf{s}) = \exp(-d(\mathbf{x}, \mathbf{s})^2/(2\sigma^2))$$

so that

$$\log(\lambda_{ij}) = \alpha_0 - \alpha_1 d(\mathbf{x}_j, \mathbf{s}_i)^2,$$

where $\alpha_0 = \log(\lambda_0)$ and $\alpha_1 = 1/(2\sigma^2)$.

As usual, we approach Bayesian analysis of these models using data augmentation (Section 4.2). Under data augmentation, we introduce a collection of all-zero encounter histories to bring the total size of the data set up to M, and a corresponding set of data augmentation variables $z_i \sim \text{Bernouli}(\psi)$. Then the observation model is specified conditional on z according to:

$$y_{ij} \sim \text{Poisson}(z_i K \lambda_{ij}),$$

which evaluates to a point mass at $y = 0$ if $z = 0$. In other words, the observation model under data augmentation is a zero-inflated Poisson model which is easily analyzed by Bayesian methods, e.g., in one of the **BUGS** dialects or, alternatively, using likelihood methods where the same principles as in Chapter 6 apply.

9.1.5 Simulating data and fitting the model

Simulating a sample SCR data set under the Poisson model requires only a couple of minor modifications to the procedure we used in Chapter 5 (see the function `simSCR0`). In particular, we modify the block of code that defines the model to be that of $\mathbb{E}(y)$ and not $\Pr(y = 1)$, and we change the random variable generator from `rbinom` to `rpois`:

```
##
## S =activity centers and traplocs defined as in simSCR0()
##
## Compute distance between activity centers and traps:
> D <- e2dist(S,traplocs)

## Define parameter values:
> alpha0 <- -2.5
> sigma <- 0.5
> alpha1 <- 1/(2*sigma*sigma)

## Encounter probability model:
> muy <- exp(alpha0)*exp(-alpha1*D*D)

## Now generate the encounters of every individual in every trap
> Y <-matrix(NA,nrow=N,ncol=ntraps)
> for(i in 1:nrow(Y)){
  Y[i,] <- rpois(ntraps,K*muy[i,])
  }
```

We modified our simulation code from Chapter 5 to simulate Poisson encounter frequencies for each trap, and we use this new code to generate, and subsequently analyze, an ideal data set using **BUGS**. The Poisson simulator function `simPoissonSCR` is available in the `scrbook` package (it can produce 3-d encounter history data too, although we don't do that here). Here is an example of simulating a data set, harvesting the required data objects, and doing the data augmentation:

```
## Simulate data and extract data elements
##
> data <- simPoissonSCR(discard0=TRUE,rnd=2013)
> y <- data$Y
> nind <- nrow(y)
> X <- data$traplocs
> K <- data$K
```

```
> J <- nrow(X)
> xlim <- data$xlim
> ylim <- data$ylim

## Data augmentation
> M <- 200
> y <- rbind(y,matrix(0,nrow=M-nind,ncol=ncol(y)))
> z <- c(rep(1,nind),rep(0,M-nind))
```

The process for fitting the model in **WinBUGS** or **JAGS** is identical to what we've done previously in Chapter 5. In particular, we set up some starting values, package the data and inits, identify the parameters to be monitored, and then send everything off to our MCMC engine. Here are the commands for fitting the Poisson observation model (these commands are also shown in the help file for `simPoissonSCR`):

```
## Starting values for activity centers
##
> sst <- X[sample(1:J,M,replace=TRUE),]
> for(i in 1:nind){
  if(sum(y[i,])==0) next
  sst[i,1] <- mean( X[y[i,]>0,1] )
  sst[i,2] <- mean( X[y[i,]>0,2] )
  }
## Dithered a little bit from trap locations
> sst <- sst + runif(nrow(sst)*2,0,1)/8
> data <- list (y=y,X=X,K=K,M=M,J=J,xlim=xlim,ylim=ylim)
> inits <- function(){
   list (alpha0=rnorm(1,-2,.4),alpha1=runif(1,1,2),s=sst,z=z,psi=.5)
   }
> parameters <- c("alpha0","alpha1","N","D")
```

Next, we write the **BUGS** model to an external file:

```
> cat("
model{
 alpha0 ~ dnorm(0,.1)
 alpha1 ~ dnorm(0,.1)
 psi ~ dunif(0,1)

 for (i in 1:M){
   z[i] ~ dbern(psi)
   s[i,1] ~ dunif(xlim[1],xlim[2])
   s[i,2] ~ dunif(ylim[1],ylim[2])
   for(j in 1:J){
      d[i,j] <- pow(pow(s[i,1]-X[j,1],2) + pow(s[i,2]-X[j,2],2),0.5)
      y[i,j] ~ dpois(lam[i,j])
      lam[i,j] <- z[i]*K*exp(alpha0)*exp(- alpha1*d[i,j]*d[i,j])
   }
 }
 N <- sum(z[])
 D <- N/64
}
",file = "SCR-Poisson.txt")
```

To fit the model we execute bugs in the usual way:

```
> library(R2WinBUGS)
> out1 <- bugs (data, inits, parameters, "SCR-Poisson.txt", n.thin=1,
                n.chains=3,n.burnin=1000,n.iter=2000,working.dir=getwd(),
                debug=TRUE)
```

Or, using **JAGS**, we would do something like this:

```
> library(rjags)
> jm <- jags.model("SCR-Poisson.txt", data=data, inits=inits,
                   n.chains=3, n.adapt=1000)
> out2 <- coda.samples(jm, parameters, n.iter=1000, thin=1)
```

Summarizing the output from the **WinBUGS** run produces the following:

```
> print(out1,digits=2)
Inference for Bugs model at "SCR-Poisson.txt", fit using WinBUGS,
 3 chains, each with 2000 iterations (first 1000 discarded)
 n.sims = 3000 iterations saved
           mean    sd   2.5%    25%    50%    75%  97.5% Rhat n.eff
alpha0    -2.57  0.19  -2.95  -2.69  -2.57  -2.44  -2.19 1.00  2600
alpha1     2.34  0.36   1.69   2.08   2.32   2.57   3.12 1.00  3000
N        114.13 15.25  87.97 103.00 113.00 124.00 147.00 1.01   370
D          1.78  0.24   1.37   1.61   1.77   1.94   2.30 1.01   370
deviance 329.95 21.92 290.00 314.20 329.50 344.40 375.80 1.00  1700
...
[..some output deleted..]
...
```

We see that the estimates are close to the data generating values: $\alpha_0 = 2.5$, $\alpha_1 = 2.0$, and $N = 100$, as we should expect.

9.1.6 Analysis of the wolverine study data

We reanalyzed the data from the wolverine camera trapping study that were first introduced in Section 5.9. We modified the **R** script from the function wolvSCR0 to fit the Poisson model (see the help file for wolvSCR0pois). Executing this function produces the results shown in Table 9.1. The results are almost indistinguishable from the Bernoulli model fitted previously, where we had a posterior mean for N of 59.84 and that for σ was 0.64. You can edit the script wolvSCR0pois to obtain more posterior samples, or modify the model in some way.

9.1.7 Count detector models in the secr package

The **R** package secr can fit Poisson or negative binomial encounter frequency models. The formatting of data and structure of the analysis proceeds in a similar fashion to the Bernoulli model described in Section 6.5, except that we specify the detector="count" option when the traps object is created. The setup proceeds as follows:

Table 9.1 Results of fitting the SCR model with Poisson observation model to the wolverine camera trapping data. Posterior summaries were obtained using **WinBUGS** with 3 chains, each with 6,000 iterations, discarding the first 1,000 as burn-in, to yield a total of 15,000 posterior samples.

Parameter	Mean	SD	2.5%	50%	97.5%	Rhat	n.eff
N	60.12	11.91	40.00	59.00	87.00	1.00	630
D	5.80	1.15	3.86	5.69	8.39	1.00	630
$\log(p_0)$	−2.89	0.17	−3.22	−2.89	−2.57	1.00	5000
λ_0	0.06	0.01	0.04	0.06	0.08	1.00	5000
σ	0.64	0.06	0.54	0.64	0.76	1.00	730
ψ	0.30	0.07	0.19	0.30	0.45	1.00	650

```
> library(secr)
> library(scrbook)
> data(wolverine)

> traps <- as.matrix(wolverine$wtraps)
> dimnames(traps) <- list(NULL,c("trapID","x","y",
+                         paste("day",1:165,sep="")))
> traps1 <- as.data.frame(traps[,1:3])
> trapfile1 <- read.traps(data=traps1,detector="count")
```

You can proceed with analysis of these data and compare/contrast with the Bayesian analysis given above, or the results of the Bernoulli model fitted in Chapter 6.

9.2 Independent multinomial observations

Several types of encounter devices yield multinomial observations in which an individual can be caught in a single trap during a particular encounter occasion, but traps might catch any number of individuals. Mist netting is the canonical example of such a "multi-catch" device (Efford et al., 2009a). Also, some kinds of bird or mammal cage-traps, hold multiple animals, as do pit-fall traps, which are commonly used for many species of herptiles. Another type of sample method that might be viewed (in some cases) as a multi-catch device are area-searches for example, of reptiles where we think of a small polygon as the "trap"—we could get multiple individuals in the same plot but not, in the same sample occasion, at different plots. The key features of this independent multinomial or multi-catch model are: (1) capture of an individual in a trap is *not* independent of its capture in other traps, because initial capture precludes capture in any other trap and (2) individuals behave independently of one another, so whether a trap captures some individual doesn't have an affect on whether it captures another. A situation in which the second assumption is violated arises under "single-catch" trap systems, which we address in Section 20.1.5.

In this case we assume the observation $\mathbf{y}_{ik}$ for individual i during sample occasion k is a multinomial observation that consists of a sequence of 0's and a single 1 indicating the trap of capture, or "not captured". For the "not captured" event we define an additional element of the observation vector, by convention element $J+1$. As an example, if we capture an individual in trap 2 during some occasion of a study involving $J = 6$ traps. Then, the multinomial observation has length $J + 1 = 7$, and the observation is $\mathbf{y}_i = (0, 1, 0, 0, 0, 0, 0)$. An individual not captured at all would have the observation vector $(0, 0, 0, 0, 0, 0, 1)$. If we sample for five occasions in all and the individual is also caught in trap 4 during occasion 3, but otherwise uncaptured, then the five encounter observations for that individual are as follows:

```
occasion |-------trap ------| "not captured"
          1   2   3  4   5  6       7
          ----------------------------------------
    1     0   1   0  0   0  0       0
    2     0   0   0  0   0  0       1
    3     0   0   0  1   0  0       0
    4     0   0   0  0   0  0       1
    5     0   0   0  0   0  0       1
```

Statistically we regard the *rows* of this data matrix as *independent* multinomial trials.

Analogous to our previous Bernoulli and Poisson models, we seek to construct the multinomial cell probabilities for each individual as a function of *where* that individual lives, through its center of activity $\mathbf{s}$. Thus, we suppose that

$$\mathbf{y}_{ik}|\mathbf{s}_i \sim \text{Multinomial}(1, \boldsymbol{\pi}(\mathbf{s}_i)), \tag{9.2.1}$$

where $\boldsymbol{\pi}(\mathbf{s}_i)$ is a vector of length $J+1$, where $\pi_{i,J+1}$, the last cell, corresponds to the probability of the event "not captured". Now we have to construct these cell probabilities in some meaningful way that depends on each individual's $\mathbf{s}$. We use the standard multinomial logit with distance as a covariate:

$$\pi_{ij} = \frac{\exp(\alpha_0 - \alpha_1 d_{ij})}{1 + \sum_j \exp(\alpha_0 - \alpha_1 d_{ij})}$$

for $j = 1, 2, \ldots, J$, and for $J + 1$, i.e., "not captured",

$$\pi_{i,J+1} = \frac{\exp(0)}{1 + \sum_j \exp(\alpha_0 - \alpha_1 d_{ij})}$$

or, more commonly, we use a squared-distance term, d_{ij}^2, to correspond to our Gaussian kernel model for encounter probability. Whatever function of distance we use in the construction of multinomial probabilities will have a direct correspondence to the standard encounter probability models we used in the Bernoulli or Poisson models as well (see Section 5.4).

It is convenient to express these multinomial models short-hand as follows, e.g., for the Gaussian encounter probability model:

$$\text{mlogit}(\pi_{ij}) = \alpha_0 - \alpha_1 d_{ij}^2.$$

In this way we can refer to models with covariates in a more concise way. For example, a model with a trap-specific covariate, say C_j, is:

$$\text{mlogit}(\pi_{ij}) = \alpha_0 - \alpha_1 d_{ij}^2 + \alpha_2 C_j,$$

or we could include occasion-specific covariates too, such as behavioral response.

A statistically equivalent distribution to the multinomial is the *categorical* distribution. If $\mathbf{y}$ is a multinomial trial with probabilities $\boldsymbol{\pi}$ then the *position* of the non-zero element of $\mathbf{y}$ is a categorical random variable with probabilities $\boldsymbol{\pi}$. We express this for SCR models as

$$\mathbf{y}|\mathbf{s} \sim \text{Categorical}(\boldsymbol{\pi}(\mathbf{s})).$$

In the SCR context, the categorical version of the multinomial trial corresponds to the *trap of capture*. Using our example above with 6 traps we could as well say y_{ik} is a categorical random variable with possible outcomes $(1, 2, 3, 4, 5, 6, 7)$, where outcome $y = 7$ corresponds to "not captured." Therefore, for our illustration in the previous table, $y_{i1} = 2$, $y_{i2} = 7$, $y_{i3} = 4$, and so on.

For simulating and fitting data in the **BUGS** engines we will typically use the categorical representation of the model because it is somewhat more convenient. We have found that fitting multinomial models in **WinBUGS** is less efficient than **JAGS** (Royle and Converse, in review), which we use in the subsequent examples involving multinomial observation models.

9.2.1 Multinomial resource selection models

The multinomial probabilities in Eq. (9.2.2) look similar to the multinomial resource selection function (RSF) model for telemetry data (Manly et al., 2002; Lele and Keim, 2006). This suggests how we might model landscape or habitat covariates using such methods—i.e., by including them as explicit covariates in a larger multinomial model for "use"—which, if we take the product of use with encounter, produces a model for the observable encounter data. This leads naturally to the development of models that integrate RSF data from telemetry studies with SCR data (Royle et al., 2012a), which is the topic of Chapter 13.

9.2.2 Simulating data and analysis using JAGS

We're going to show the nugget of a simulation function, which is implemented in the function `simMnSCR` found in the **R** package `scrbook`. The first lines of the following **R** code make use of some things that you need to define, but we omit them here (e.g., `xlim`, `ylim` are the boundaries of the state-space, `N` is the population size, etc.):

```
##
## Simulate random activity centers:
```

```
##        (first define N, xlim, ylim, etc.)
##
> S <- cbind(runif(N,xlim[1],xlim[2]),runif(N,ylim[1],ylim[2]))
## Distance from each individual to each trap
> D <- e2dist(S,traplocs)

## Set paramter values
> sigma <- 0.5
> alpha0 <- -1
> alpha1 <- -1/(2*sigma*sigma)

## make an empty data matrix and fill it up with data
> Ycat <- matrix(NA,nrow=N,ncol=K)
>  for(i in 1:N){
    for(k in 1:K){
    lp <- alpha0 + alpha1*D[i,]*D[i,]
    cp <- exp(c(lp,0))
    cp <- cp/sum(cp)
    Ycat[i,k] <- sample(1:(ntraps+1),1,prob=cp)
   }
  }
```

We save the data in the matrix `Ycat` to clarify that it is the categorical observation representing "trap of capture". The matrix `Ycat` here has N rows and, to do an analysis that mimics a real situation, we would have to discard the uncaptured individuals. The function `simMnSCR` will also simulate data that includes a behavioral response, which will be the typical situation in small-mammal trapping problems (see Converse and Royle, 2012, for details).

Here we use our function `simMnSCR` to simulate a data set with $K = 7$ occasions. We'll run the model using **JAGS** which we have found is much more effective for this class of models. We get the data set up for analysis by augmenting the size of the data set to $M = 200$. In addition, we choose starting values for $\mathbf{s}$ and the data augmentation variables z. For starting values of $\mathbf{s}$ we cheat a little bit here and use the true values for the observed individuals and then augment the $M \times 2$ matrix $\mathbf{S}$ with $M - n$ randomly selected activity centers. In practice, we recommend using the average encounter location of observed individuals as starting values for their activity centers. Our function `spiderplot` returns the mean observed location of the `nind` encountered individuals. The parameters input to `simMnSCR` are the intercept α_0, $\sigma = \sqrt{1/(2\alpha_1)}$ for the Gaussian encounter probability model, and α_2 is the behavioral response parameter. The data simulation and setup proceeds as follows:

```
> set.seed(2013)
> parms <- list(N=100,alpha0= -.40, sigma=0.5, alpha2= 0)
> data <- simMnSCR(parms, K=7, ssbuff=2)
> nind <- nrow(data$Ycat)

> M <- 200
> Ycat <- rbind(data$Ycat,matrix(nrow(data$X)+1,nrow=(M-nind),ncol=data$K))
```

```
> Sst <- rbind(data$S,cbind(runif(M-nind,data$xlim[1],data$xlim[2]),
                            runif(M-nind,data$ylim[1],data$ylim[2])))
> zst <- c(rep(1,160),rep(0,40))
```

The model specification is not much more complicated than the binomial or Poisson models given previously. The main consideration is that we define the cell probabilities for each trap $j = 1, 2, \ldots, J$ and then define the last cell probability, $J+1$, for "not captured", to be the complement of the sum of the others. The code is shown in Panel 9.1. In the last lines of code we specify N and density, D, as derived parameters.

To fit the model, we need to package everything up (initial values, parameters, data) and send it off to **JAGS** to build an MCMC simulator for us (these commands are shown in the help file for `simMnSCR`). In addition to the usual data objects, we also pass the limits of the rectangular state-space (`ylim`, `xlim`, both 1×2 vectors) and the

```
model{
psi ~ dunif(0,1)
alpha0 ~ dnorm(0,10)
sigma ~dunif(0,10)
alpha1 <-  1/(2*sigma*sigma)

for(i in 1:M){
  z[i] ~ dbern(psi)
  S[i,1] ~ dunif(xlim[1],xlim[2])
  S[i,2] ~ dunif(ylim[1],ylim[2])
  for(j in 1:ntraps){
    #distance from capture to the center of the home range
    d[i,j] <- pow(pow(S[i,1]-X[j,1],2) + pow(S[i,2]-X[j,2],2),1)
  }
  for(k in 1:K){
    for(j in 1:ntraps){
      lp[i,k,j] <- exp(alpha0 - alpha1*d[i,j])*z[i]
      cp[i,k,j] <- lp[i,k,j]/(1+sum(lp[i,k,]))
    }
    cp[i,k,ntraps+1] <- 1-sum(cp[i,k,1:ntraps])  # last cell = not captured
    Ycat[i,k] ~ dcat(cp[i,k,])
  }
}

N <- sum(z[1:M])
A <- ((xlim[2]-xlim[1])*trap.space)*((ylim[2]-ylim[1])*trap.space)
D <- N/A
}
```

PANEL 9.1

BUGS model specification for the independent multinomial observation model. For data simulation and model fitting see the help file `?simMnSCR` in the **R** package `scrbook`.

scale of the standardized units, called `trap.space` here because we typically will define the trap coordinates to be an integer grid. If the trap spacing is 10 m and we want units of density in terms of individuals per m^2, then we input `trap.space=10`. The analysis is carried out as follows:

```
> inits <- function(){  list (z=zst,sigma=runif(1,.5,1), S=Sst) }

# Parameters to monitor
> parameters <- c("psi","alpha0","alpha1","sigma","N","D")

# Bundle the data. Note this reuses "data"
> data <- list (X=data$X,K=data$K, trap.space=1,Ycat=Ycat,M=M,
                ntraps=nrow(data$X),ylim=data$ylim,xlim=data$xlim)

> library(R2jags)
> out <- jags (data, inits, parameters, "model.txt", n.thin=1,
               n.chains=3, n.burnin=1000, n.iter=2000)
```

The posterior summaries are provided in the following **R** output (recall that $N = 100$, $\alpha_0 = -.40$, and $\sigma = 0.5$):

```
> out
Inference for Bugs model at "model.txt", fit using jags,
 3 chains, each with 2000 iterations (first 1000 discarded)
 n.sims = 3000 iterations saved
         mu.vect sd.vect    2.5%     25%     50%     75%   97.5% Rhat n.eff
D          1.873   0.189   1.531   1.750   1.859   2.000   2.250 1.006  1300
N        119.867  12.107  98.000 112.000 119.000 128.000 144.000 1.006  1300
alpha0    -0.435   0.151  -0.738  -0.535  -0.439  -0.331  -0.146 1.004   580
alpha1     2.195   0.286   1.658   2.004   2.180   2.372   2.785 1.003  2400
psi        0.599   0.069   0.465   0.552   0.599   0.645   0.739 1.006  1400
sigma      0.480   0.032   0.424   0.459   0.479   0.500   0.549 1.003  2400
deviance 892.164  21.988 850.922 877.417 891.561 906.246 937.728 1.003   950

[... output deleted ...]
```

We see that the posterior means of each parameter are very close to the data generating values, considering the high degree of posterior uncertainty.

9.2.3 Multinomial relationship to the Poisson

The multinomial is related to the Poisson encounter rate model by a conditioning argument. Let y_{ij} be the number of encounters for individual i in trap j. If $y_{ij} \sim$ Poisson (λ_{ij}), then, conditional on the *total* number of captures (i.e., across all traps), $y_i = \sum_j y_{ij}$, the trap encounter frequencies are multinomial with probabilities

$$\pi_{ij} = \frac{\lambda_{ij}}{\sum_j \lambda_{ij}}$$

for $j = 1, 2, \ldots, J$. Or equivalently the *trap of capture* is categorical with probabilities π_{ij} as given above. Under the Gaussian kernel model, these probabilities are:

$$\pi_{ij} = \frac{\exp(-\alpha_1 d(\mathbf{x}_j, \mathbf{s}_j)^2)}{\sum_j \exp(-\alpha_1 d(\mathbf{x}_j, \mathbf{s}_i)^2)}, \tag{9.2.2}$$

where, we note, the intercept α_0 has been canceled from both the numerator and denominator. This makes sense because, here, these probabilities describe the trap-specific capture probabilities *conditional on capture*. Therefore, the model is not completely specified, absent a model for the "overall" probability of encounter or the expected frequency of captures, say ϕ_i. Depending on how we specify a model for this quantity ϕ_i, we can reconcile it directly with the Poisson model. Let y_i be the total number of encounters for individual i and suppose y_i has a Poisson distribution with mean ϕ_i. Then, marginalizing Eq. (9.2.1) over the Poisson distribution for y_i produces the original set of iid Poisson frequencies with probabilities:

$$\lambda_{ij} = \phi_i \pi_{ij}$$

for $j = 1, 2, \ldots, J$. In particular, if we suppose that $\phi_i = \sum_j \exp(\alpha_0 - \alpha_1 d(\mathbf{x}, \mathbf{s})^2)$ then the marginal distribution of y_{ij} is Poisson with mean $\exp(\alpha_0 - \alpha_1 d(\mathbf{x}, \mathbf{s})^2)$, equivalent to Eq. (9.1.1).

In summary, the Poisson and multinomial models are equivalent in how they model the distribution of captures among traps. It stands to reason that, if the encounter rate of individuals is low, we could use the Poisson and multinomial models interchangeably. In fact, based on our discussion in Section 9.1.2 above we could use any of the binomial/Poisson/multinomial models with little ill effect when encounter rate is low.

9.2.4 Avian mist-netting example

We analyze data from a mist-netting study of ovenbirds (*Seiurus aurocapilla*), conducted at the Patuxent Wildlife Research Center, Laurel MD, by D.K. Dawson and M.G. Efford. The data from this study are available in the `secr` package and have been analyzed previously by Efford et al. (2004), see also Borchers and Efford (2008). Forty-four mist-nets spaced 30 m apart on the perimeter of a 600-m × 100-m rectangle were operated on 9 or 10 non-consecutive days in late May and June for 5 years from 2005 to 2009. The ovenbird data can be loaded as follows:

```
> library(secr)
> data(ovenbird)
```

The data set consists of adult ovenbirds caught during sampling in each of the 5 years (one ovenbird was killed in 2009 indicated by a negative net number in the encounter data file). As with most mist-netting studies, nets are checked multiple times during a day (e.g., every hour during a morning session). However, for this data set, the within-day recaptures are not included so each bird has at most a single capture per day. Therefore, the multinomial model (detector type "multi" in `secr`) is appropriate. Although several individuals were captured in more than 1 year, this information is not used in the models presently offered in `secr`, but we do make use of it in the development of open models in Chapter 16.

9.2.4.1 Multiple sample sessions

Up to this point we have only dealt with a basic closed population sampling situation consisting of repeated sample occasions on a single population of individuals using a single array of traps. In practice, many studies produce repeated samples over longer periods of time over which demographic closure isn't valid, or at different locations where the populations are completely distinct. We adopt the `secr` terminology of *session* for such replication by groups in time or space, and the models are *multi-session* models, although we think of such models as being relevant to any stratified population (see Chapter 14). We introduced `secr`'s multi-session models in Section 6.5.4. In the case of the ovenbird data, sampling was carried out in multiple years, with a number of sample occasions within each year (9 or 10), a type of data structure commonly referred to as "the robust design" (Pollock, 1982). In this context, it stands to reason that there is recruitment and mortality happening across years. In Chapter 16 we model these processes explicitly, but here, we provide an analysis of the data that does not require explicit models for recruitment and survival, regarding the yearly populations as independent strata, and fitting a multi-session model.

When the sessions represent explicit time periods, the multi-session model of `secr` can be thought of as a type of open population model. In particular, a special case of open models arises when we assume N_t (time-specific population sizes) are independent from one time period or session to the next—this can be thought of as a "random temporary emigration" model of the Kendall et al. (1997) variety, and this is similar to the multi-session model implemented in `secr`. In particular, by assuming that N_t is Poisson with mean Λ_t, one can model variation in abundance among sessions based on the Poisson-integrated likelihood in which parameters of Λ_t appear directly in the likelihood as we noted in Section 6.5.4.[1] We provide an analysis (below) of the ovenbird data here using the multi-session models in `secr` with a multinomial observation model. We formalize the multi-session model approach from a Bayesian perspective using data augmentation in Chapter 14 (Converse and Royle, 2012).

A third way to develop models for stratified or grouped populations, not based on multi-session models, but that is convenient in **BUGS**, is to regard the data from each session as an independent data set with its own N_t parameter, and do T distinct data augmentations. Because each N_t is regarded as a free parameter, independent of the other parameters, we'll call this the non-parametric multi-session model to distinguish it from the multi-session model which assumes the N_t are related to one another by having been generated from a common Poisson distribution. By augmenting each year separately in the same **BUGS** model specification, we avoid making explicit model assumptions about the N_t parameters.

Below, and elsewhere in the book, we demonstrate the three approaches described in the preceding paragraphs to analyzing grouped/stratified data using the ovenbird data: (1) In the following section, we provide the non-parametric multi-session model

[1] We do not know of `secr` documentation that states this (or contradicts it). We think this is what is being done, based partially on conversations or emails with M.G. Efford, D.L. Borchers, the various publications on `secr`, and our own thinking about it.

with unconstrained N_t; (2) we demonstrate the Poisson multi-session models from `secr` both here (following section) and in Chapter 14 from a Bayesian standpoint; (3) later, in Chapter 16, we provide a fully dynamic "spatial Jolly-Seber" model and apply it to the ovenbird data.

9.2.4.2 Analysis using JAGS

The ovenbird data are provided as a multi-session `capthist` object `ovenCH`, which, by regarding years as independent strata or sessions, allows for the fitting of the multi-session model. For doing a Bayesian analysis in one of the **BUGS** engines (we use **JAGS** here) there are a number of ways to structure the data and describe the model. We can analyze either a 2-d data set with all years (data augmented) "stacked" into a data set of dimension $(5M) \times 10$ (5 years, M = size of the augmented data set, $K = 10$ replicate sample occasions). Or, we could produce a 3-d array ($M \times J \times K$). We adopted the former approach, analyzing the data as a 2-d array and creating an additional categorical variable for "year" to indicate which stratum (year) each record goes with. We used $M = 100$ per year for a total size of the augmented data of 500. The **BUGS** model specification is shown in Panel 9.2.

Data on individual sex is included with `secr`, but we provide an analysis (model specification shown in Panel 9.2) of a single model for all adults, constant σ across years, constant p_0, and year-specific values of N_t (and hence D_t). For the habitat mask, we used a rectangular state-space created by buffering the mist net grid by 200 meters. There was a single loss-on-capture which we accounted for by fixing $p = 0$ for all subsequent encounters of that individual (indicated by the binary variable `dead`, as shown in Panel 9.2). We have an **R** script in the `scrbook` package called `SCRovenbird`, so you can see how to set up the data and run the model. Executing the script `SCRovenbird` produces the posterior summaries given in Table 9.2. Here, density is in units of birds per ha. The posterior mean of σ is about 76 m, and there is considerable variability in density over the 5-year period, with density peaking at 1.2 birds/ha in year 3, although there is considerable posterior uncertainty. The $\hat{R}$ statistics suggest satisfactory mixing and convergence of the Markov chains.

9.2.4.3 Analysis using `secr`

Included with the ovenbird data are a number of models fitted as examples. Those include:

```
ovenbird.model.1  fitted secr model -- null
ovenbird.model.1b fitted secr model -- g0 net shyness
ovenbird.model.1T fitted secr model -- g0 time trend within years
ovenbird.model.h2 fitted secr model -- g0 finite mixture
ovenbird.model.D  fitted secr model -- trend in density across years
```

The model fit objects provided in `secr` are based on the use of the habitat mask. However, to make the analyses consistent with our previous analysis in **JAGS**, we refit all of the models here without the habitat mask. The re-analysis proceeds as

```
model{
 alpha0 ~ dnorm(0,.1)
 sigma ~ dunif(0,200)
 alpha1 <- 1/(2*sigma*sigma)

 A <- ((xlim[2]-xlim[1]))*((ylim[2]-ylim[1]))
 for(t in 1:5){
   N[t] <- inprod(z[1:bigM],yrdummy[,t])
   D[t] <- (N[t]/A)*10000  # Put in units of per ha
   psi[t] ~ dunif(0,1)
 }

 for(i in 1:bigM){  # bigM = total size of jointly augmented data set
   z[i] ~ dbern(psi[year[i]])
   S[i,1] ~ dunif(xlim[1],xlim[2])
   S[i,2] ~ dunif(ylim[1],ylim[2])

 for(j in 1:ntraps){  # X = trap locations, S = activity centers
    d2[i,j] <- pow(pow(S[i,1]-X[j,1],2) + pow(S[i,2]-X[j,2],2),1)
  }
 for(k in 1:K){
   Ycat[i,k] ~ dcat(cp[i,k,])
   for(j in 1:ntraps){
     lp[i,k,j] <- exp(alpha0 - alpha1*d2[i,j])*z[i]*(1-dead[i,k])
     cp[i,k,j] <- lp[i,k,j]/(1+sum(lp[i,k,1:ntraps]))
   }
   cp[i,k,ntraps+1] <- 1-sum(cp[i,k,1:ntraps])  # Last cell = not captured
  }
 }
}
```

PANEL 9.2

BUGS model specification for the non-parametric multi-session model in which each N_t is independent of each other. The implied prior (by data augmentation) is that $N_t \sim$ Uniform(0,100). To fit this model to the ovenbird data, see `?SCRovenbird` in the **R** package `scrbook`.

follows, changing the "trend in density across years" model to allow for year-specific density:

```
## Fit constant-density model
> ovenbird.model.1 <- secr.fit(ovenCH)
## Fit net avoidance model
> ovenbird.model.1b <- secr.fit(ovenCH, model =    list (g0 ~ b))
```

Table 9.2 Posterior summary statistics for the ovenbird mist-netting data based on the independent multinomial ("multi-catch") encounter process model. Parameters ψ, N, and D are indexed by year. MCMC was done using `jags` with 3 chains, each with 11,000 iterations, discarding the first 1,000, for a total of 30,000 posterior samples. Note the parameter α_1 is 0 to 3 decimal places, but $\sigma = (\sqrt{1/2\alpha_1})$.

Parameter	Mean	SD	2.5%	50%	97.5%	Rhat	n.eff
D[1]	0.983	0.211	0.636	0.966	1.455	1.002	1,900
D[2]	1.023	0.209	0.673	1.003	1.492	1.001	7,100
D[3]	1.208	0.238	0.807	1.186	1.749	1.004	740
D[4]	0.896	0.195	0.575	0.880	1.333	1.002	3,000
D[5]	0.753	0.177	0.465	0.734	1.149	1.001	4,000
α_0	−3.479	0.160	−3.797	−3.477	−3.171	1.005	490
α_1	0.000	0.000	0.000	0.000	0.000	1.003	1,100
σ	76.214	6.125	65.569	75.758	89.360	1.003	1,100
N[1]	80.423	17.283	52.000	79.000	119.000	1.002	1,900
N[2]	83.685	17.077	55.000	82.000	122.000	1.001	7,100
N[3]	98.822	19.483	66.000	97.000	143.000	1.004	740
N[4]	73.288	15.962	47.000	72.000	109.000	1.002	3,000
N[5]	61.589	14.468	38.000	60.000	94.000	1.001	4,000
ψ[1]	0.403	0.092	0.246	0.395	0.606	1.002	1,600
ψ[2]	0.419	0.091	0.260	0.412	0.620	1.001	6,400
ψ[3]	0.494	0.102	0.315	0.486	0.723	1.004	760
ψ[4]	0.368	0.086	0.221	0.361	0.555	1.002	3,200
ψ[5]	0.310	0.079	0.178	0.302	0.485	1.002	3,500

```
## Fit model with time trend in detection
> ovenbird.model.1T <- secr.fit(ovenCH, model =    list (g0 ~ T))
## Fit model with 2-class mixture for g0
> ovenbird.model.h2 <- secr.fit(ovenCH, model =    list (g0 ~ h2))
## Fit a model with session (year)-specific Density
> ovenbird.model.DT <- secr.fit(ovenCH, model =    list (D ~ session))
```

All of these models can be fitted easily in **JAGS** but the model we fitted previously is roughly equivalent to the last model, `ovenbird.model.DT`, because we allowed for year-specific population sizes (and hence density). So, we'll compare our results from **JAGS** to that model. The `secr` output is extensive and so we do not reproduce it completely here. By default, it summarizes the trap information for each year, encounter information, and then output for each year. Here is an abbreviated version for `ovenbird.model.DT`:

```
> print(ovenbird.model.DT,digits=2)

secr.fit( capthist = ovenCH, model = list(D ~ session), buffer = 300 )
secr 2.3.1, 14:46:52 23 Jan 2013
```

```
$'2005'
Object class      traps
Detector type     multi
Detector number   44
Average spacing   30.27273 m
x-range           −50 49 m
y-range           −285 285 m

[... deleted ...]

            2005 2006 2007 2008 2009
Occasions      9   10   10   10   10
Detections    35   42   52   30   33
Animals       20   22   26   19   16
Detectors     44   44   44   44   44

Model                 : D~session g0~1 sigma~1
Fixed (real)          : none
Detection fn          : halfnormal
Distribution          : poisson
N parameters          : 7
Log likelihood        : -1119.845
AIC                   : 2253.689
AICc                  : 2254.868

[... deleted ...]
```

To do model selection we use the handy helper-function AIC as follows (output edited to fit on the page):

```
AIC (ovenbird.model.1, ovenbird.model.1b, ovenbird.model.1T,
   ovenbird.model.h2, ovenbird.model.DT)

                    model detectfn npar  logLik     AIC     AICc   dAICc
ovenbird.model.1T [edited output]  4  −1111.85 2231.70  2232.10   0.00
ovenbird.model.1b         ....     4  −1117.61 2243.22  2243.63  11.52
ovenbird.model.h2         ....     3  −1121.16 2248.32  2248.57  16.46
ovenbird.model.1          ....     5  −1119.76 2249.52  2250.14  18.03
ovenbird.model.DT         ....     7  −1119.84 2253.68  2254.86  22.75
```

We see that our DT model is way down at the bottom of the list. Instead, the model with a time trend (within-season) in detection probability is preferred, followed by a behavioral response. We encourage you to adapt the **JAGS** model specification for such models which is easily done (see Chapter 7 for many examples). We provide the summary results for the model having D ~ session as follows:

```
> print(ovenbird.model.DT,digits=2)

secr.fit(capthist = ovenCH, model = list(D ~ session), buffer = 300)
secr 2.3.1, 14:46:52 23 Jan 2013

[...deleted...]
```

```
Fitted (real) parameters evaluated at base levels of covariates

 session = 2005
        link estimate SE.estimate    lcl    ucl
D        log    0.920       0.228  0.571  1.484
g0     logit    0.028       0.004  0.021  0.037
sigma    log   78.566       6.379 67.025 92.095

 session = 2006
        link estimate SE.estimate    lcl    ucl
D        log    0.963       0.238  0.598  1.553
g0     logit    0.028       0.004  0.021  0.037
sigma    log   78.566       6.379 67.025 92.095

 session = 2007
        link estimate SE.estimate    lcl    ucl
D        log    1.139       0.282  0.706  1.836
g0     logit    0.028       0.004  0.021  0.037
sigma    log   78.566       6.379 67.025 92.095

 session = 2008
        link estimate SE.estimate    lcl    ucl
D        log    0.832       0.206  0.516  1.341
g0     logit    0.028       0.004  0.021  0.037
sigma    log   78.566       6.379 67.025 92.095

 session = 2009
        link estimate SE.estimate    lcl    ucl
D        log    0.701       0.173  0.435  1.130
g0     logit    0.028       0.004  0.021  0.037
sigma    log   78.566       6.379 67.025 92.095
```

The point estimates (MLEs) of density are uniformly lower than the Bayesian estimates (posterior means) shown in Table 9.2. We expect some difference in this direction due to small-sample skew of the posterior. In addition, there may be slight differences due to the fact that `secr` multi-session model assumes that the N_t have a Poisson prior, but the implementation in **JAGS** using data augmentation is based on a binomial prior. The estimated σ is very similar between the **JAGS** analysis and `secr`.

9.3 Single-catch traps

The classical animal trapping experiment is based on a physical trap that captures a single animal and holds that individual until subsequent handling by a biologist. This type of observation model—the "single-catch" trap—was the original situation considered in the context of spatial capture-recapture by Efford (2004). Nowadays, capture-recapture data are more often obtained by other methods (DNA from hair snares, or scat sampling, camera traps, etc.) but nevertheless the single-catch traps are still widely used in small-mammal studies (Converse et al., 2006b; Converse and Royle, 2012) and other situations.

The single-catch model is basically a multinomial model but one in which the number of available traps is reduced as each individual is captured. As such, the constraints on the joint likelihood for the sample of n encounter histories are very complicated. As a result, at the time of this writing, there has not been a formal development of either likelihood or Bayesian analysis of this model, and applications of SCR models to single-catch systems have used the independent multinomial model as an approximation (see below).

Nevertheless, we can make some progress to describing the basic observation model formally. In particular, if we imagine that all of the individuals captured queued up at the beginning of the capture session to draw a number indicating their order of capture, then there is a nice conditional structure resulting from a "removal process" operating on the traps. The first individual captured has the multinomial observation model:

$$\mathbf{y}_1 \sim \text{Multinomial}(\boldsymbol{\pi}_1),$$

whereas the second individual captured also has a multinomial encounter probability model but with the trap that captured the first individual removed. We might express this as:

$$\mathbf{y}_2 \sim \text{Multinomial}(\boldsymbol{\pi}_2),$$

where

$$\pi_{2j} = \frac{(1 - y_{1j}) \times \exp(\alpha_0 - \alpha_1 d_{ij}^2)}{\sum_j (1 - y_{1j}) \times \exp(\alpha_0 - \alpha_1 d_{ij}^2)}$$

and so on for $i = 3, 4, \ldots, n$. In a certain way, this model is a type of local behavioral response model but where the response is to other individuals being captured. Evidently, the **order of capture** is relevant to the construction of these multinomial cell probabilities. More generally, the *time* of capture of an individual in any trapping interval will affect the encounter probability of subsequently captured individuals, but we think that order of capture might lead to a practical approximation to the single-catch process (this is how we simulate the data in our function `simScSCR`). In the simulation of single-catch data, we randomly ordered the population of individuals for each sample occasion, and then cycled through them, turning off each trap if an individual was captured in it.

9.3.1 Inference for single-catch systems

For the single-catch model, we argued that the observations have a multinomial type of observation model, but the multinomial observations have a unique conditional dependence structure among them owing to the "removal" of traps as they fill up with individuals. Thus, competition for single-catch traps renders the independence assumptions for the independent multinomial model invalid. However, as Efford et al. (2009a) noted, we expect "bias to be small when trap saturation (the proportion of traps occupied) is low. Trap saturation will be higher when population density is high..."

relative to trap density, or when net encounter probability is high. Efford et al. (2009a) did a limited simulation study and found essentially no effective bias and concluded that estimators of density from the misspecified independent multinomial model are robust to the mild dependence induced when trap saturation is low. Naturally then, we expect that the Poisson model could also be an effective approximation under the same set of circumstances.

In the **R** package `scrbook` we provide a function for simulating data from a single-catch system (function `simScSCR`) and fitting the misspecified model (`example(simScSCR)`) in **JAGS** so that you can evaluate the effectiveness of this misspecified model for situations that interest you.

9.3.2 Analysis of Efford's possum trapping data

We provide an analysis here of data from a study of brushtail possums in New Zealand. The data are available with the **R** package `secr` (Efford et al., 2009a); see the help file `?possum` after loading the `secr` package. Originally, the data were analyzed by Efford et al. (2005), and a detailed description of the data set is available in the help file, from which we summarize:

> *Brushtail possums (Trichosurus vulpecula) are an unwanted invasive species in New Zealand. Although most abundant in forests, where they occasionally exceed densities of 15/ha, possums live wherever there are palatable food plants and shelter.*

To load the possum data, execute the following commands:

```
> library(secr)
> data(possum)
```

The study area encompasses approximately 300 ha, and 180 live traps were organized in five distinct grids, shown in Figure 9.2. Each square arrangement of traps consisted of 36 traps with a spacing of 20 m. Thus the squares are 180 m on a side. Individuals were captured, tagged, and released over 5 days during April, 2002. A noteworthy aspect of this study is that it involves replicated grids selected in some fashion from within a prescribed region. From an analysis standpoint, we could adopt the use of the multi-session models, which we used previously to analyze the ovenbird data. This would be useful if we had covariates at the trapping grid level that we wanted to model. Alternatively, we could pool the data from all of the grids and analyze them jointly as if they were based on a single trapping grid (with 180 traps), which is clearly a reasonable approach in this case.

The data file `possumCH` contains 112 encounter histories, and we analyze those here although the last eight of those are recaptures treated as new individuals.[2] The encounter process is not strictly a single-catch multinomial process because, as noted in the `possum` help file, "One female possum was twice captured at two sites on one

[2]M. Efford, personal communication

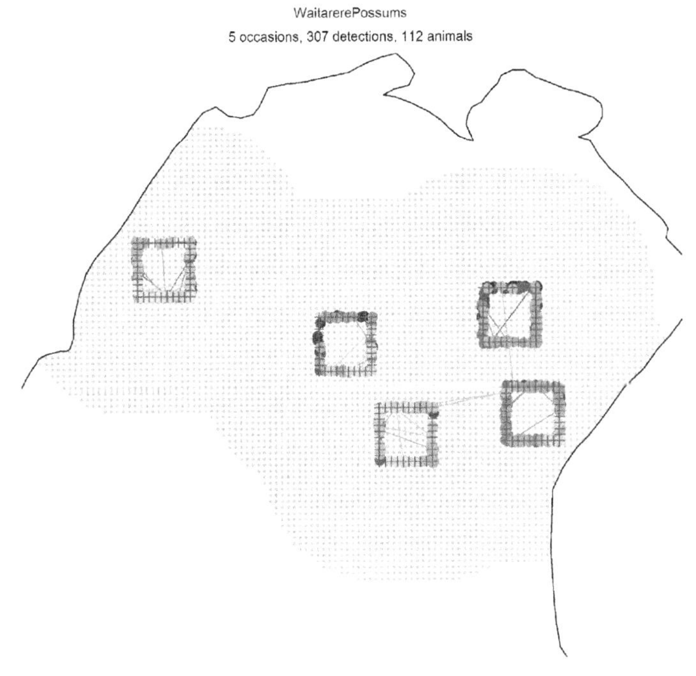

FIGURE 9.2

Trapping grids used in possum study from Efford et al. (2005), data are contained in the **R** package `secr` (Efford, 2011a), refer to the help file `?possum` for additional details of this study.

day, having entered a second trap after being released; one record in each pair was selected arbitrarily and discarded", which is a similar situation to what might happen in bird mist-net studies, as a bird might fly into a net upon release from another. By discarding the two extra-capture events, we can satisfactorily view these data as single-catch data, for which `secr` uses the independent multinomial likelihood (M. Efford, pers. comm.). If multiple same-session captures were common, then it might be worth developing a model describing the number of captures of individual i during sample occasion k, in order to make use of all captures.

For our Bayesian analysis here, we used a rectangular state-space, which doesn't account for any geographic boundaries of the survey region, but we note that a customized habitat mask is included in `secr` and it could be used in a Bayesian analysis. Whether or not we use the mask is probably immaterial as long as we understand the predictions of N or D over the water don't mean anything biological and we probably wouldn't report such predictions. The **JAGS** model specification is based on that of the ovenbird analysis given previously, so we don't reproduce the model here. The

Table 9.3 Results of fitting the independent multinomial observation model to the possum trapping data. Strictly speaking, the trapping device is a "single-catch" trap, and the model represents an intentional misspecification. Density is reported in individuals per ha (Dha). Posterior summaries were obtained using **JAGS** with 3 chains, each with 2,000 iterations, discarding the first 1,000 as burn-in, to yield a total of 3,000 posterior samples.

Parameter	Mean	SD	2.5%	50%	97.5%	Rhat	n.eff
N	235.407	17.435	204.000	235.000	270.000	1.009	340
Dha	1.549	0.115	1.343	1.547	1.777	1.009	340
α_0	−0.935	0.167	−1.270	−0.934	−0.605	1.007	870
α_1	0.000	0.000	0.000	0.000	0.000	1.001	2,800
σ	52.020	2.675	47.067	51.933	57.585	1.001	2,800
ψ	0.783	0.062	0.666	0.782	0.903	1.008	340

R/JAGS script is called `SCRpossum`, which is in the `scrbook` package. The results are summarized in Table 9.3.

The estimated density (posterior mean) is about 1.53 possums/ha. To obtain the `secr` results for the equivalent null model, we execute the following command.

```
> secr.fit(capthist = possumCH, trace = F)
```

which produces (edited) summary output:

```
[... some output deleted ...]

 Fitted (real) parameters evaluated at base levels of covariates
        link   estimate    SE.estimate   lcl          ucl
 D      log     1.6988930  0.17352645    1.3913904    2.0743547
 g0     logit   0.1968542  0.02256272    0.1563319    0.2448321
 sigma  log    51.4689114  2.59981905   46.6204139   56.8216500

 [... some output deleted ...]
```

As we've discussed previously, there are many reasons for why there might be differences between Bayesian and likelihood estimates. For now we just observe that the estimated density is certainly in the ballpark (compared to those in Table 9.3), and so, too, is the estimated σ.

9.4 Acoustic sampling

The last decade has seen an explosion of technology that benefits the study of animal populations. This includes DNA sampling methods that allow for identification from hair or scat, camera trapping and identification software that allow efficient sampling of many mammals, and the resulting statistical technology that helps us to make sense of such data (Borchers and Efford, 2008; Royle and Young, 2008; Efford et al., 2009b; Gopalaswamy et al., 2012b; Sollmann et al., 2013a; Chandler and Royle, 2013). One other extremely promising technology is that of acoustic sampling using microphones or recording devices. That is, instead of having cameras record encounters, or humans pick up scat, we can establish an array of (usually) electronic recording devices which, instead of establishing a visual identity of individuals, record a vocal expression of each individual. In this context, Efford et al. (2009b) referred to audio recorders as "signal strength proximity detectors" to distinguish them from other types of proximity detections, including camera traps, which are *visual* proximity detector. Using audio records, the spatial pattern of the *signal strength* at the different audio recorders or microphones can be used for inference about density (Dawson and Efford, 2009; Efford et al., 2009b) in the same way as the spatial pattern of detections is used in the types of SCR models we have discussed so far. The basic technical formulation of these models comes from Efford et al. (2009b), and it was applied to a field study of birds by Dawson and Efford (2009). In that study, recording devices were organized in groups of 4 (in a square pattern), with an array of 5×15 such clusters, separated by 100 m (300 total recorder locations). This data set, called `signalCH`, is provided with the `secr` package along with some sample analyses and help files. The document `secr-sound.pdf` (Efford and Dawson (2010) that comes with the `secr` package), can be accessed directly from the main help file (`?secr`).

Our development here mostly follows Efford et al. (2009b), but we change some notation to be consistent with our previous material. Let $S(\mathbf{x}, \mathbf{u})$ be the strength of a signal emanating from location $\mathbf{u}$, as recorded by a device at location $\mathbf{x}$. Just as ordinary SCR models represent a model of *encounter frequency* as a function of distance, the acoustic SCR model is a model of sound attenuation as a function of distance. In particular, the acoustic models assumes that S (or a suitable transformation) declines with distance d from the origin of the sound to the recording device. In the context of spatial sampling of animals, the origin is the actual location of some individual animal, and the recording device is something we nailed to a tree, or mounted on a post. For example, a model of sound attenuation used by Dawson and Efford (2009) is the following:

$$S(\mathbf{x}, \mathbf{u}) = \alpha_0 + \alpha_1 d(\mathbf{x}, \mathbf{u}) + \epsilon, \tag{9.4.1}$$

where $\epsilon \sim \text{Normal}(0, \sigma_s^2)$. In many standard situations, S will be measured in decibels, which can be any value on the real line. In this model, the parameters α_0, α_1, and σ_s^2 are to be estimated. We abbreviate the set of parameters by $\boldsymbol{\theta}$.

The basic structure of an acoustic SCR study is not really much different from ordinary SCR studies. Just as ordinary SCR models require that individuals be encountered at >1 trap, these acoustic models require that individuals be heard at >1 recorder. Therefore, the acoustic signals (calls or vocalizations) must be reconcilable and, in fact, reconciled successfully by the investigator. In practice, this would require associating signals that occur at the same instant with the same individual (or making a decision one way or the other). Further, if individuals are actively moving during the sample period (that recorders are functioning) then individuals might be double-counted, thereby biasing estimates of density. In general, the models produce an estimate of density of *sources*, and how that is interpreted depends on whether individuals are stationary or mobile, and other things. In particular, if multiple survey occasions are used (e.g., on different days), then modeling movement of individuals would be essential in order to interpret estimates of density meaningfully. Models that allow some movement should be possible (see Chapters 15 and 16).

9.4.1 The signal strength model

We assert that an individual is detected if S exceeds a threshold, c. The reason for introducing this threshold c is that sound recorders will always record some background sound, and so effective use of the acoustic SCR models requires specification of the threshold of measured signal below which the record is censored (non-detection occurs) and below which the recorded sound is assumed to be background noise. So we assert that an individual is detected if $S > c$, which occurs with probability $\Pr(S > c)$, the encounter probability. To expand on and formalize this, let S_{ij} be the observed value of S for animal i at detector j. The encounter probability is $\Pr(S_{ij} > c)$, which is $\Pr(S_{ij} > c) = 1 - \Pr(S_{ij} < c)$. Expressed in terms of the cumulative distribution function,

$$1 - \Pr\left(\frac{(S_{ij} - \mathbb{E}(S))}{\sigma_s} < \frac{(c - \mathbb{E}(S))}{\sigma_s}\right).$$

Calculation of this requires evaluation of the CDF of a standard normal variate say, $\eta = (S_{ij} - \mathbb{E}(S))/\sigma_s$, being less than $\gamma(\boldsymbol{\theta}) = (c - \mathbb{E}(S))/\sigma_s$, which is a function of all the parameters $\alpha_0, \alpha_1, \sigma_s^2$ and the individual location $\mathbf{u}$ and detector location $\mathbf{x}$. We'll identify it by $\gamma(\boldsymbol{\theta}, \mathbf{x}, \mathbf{u})$ when we need to be explicit about those things. We can compute $\Pr(S_{ij} > c) = 1 - \Pr(\eta < \gamma(\boldsymbol{\theta}, \mathbf{x}, \mathbf{u}))$ easily using any software package including **R** which has a standard function, `pnorm`, for computing the normal cdf. To be more concise, we'll use the $\Phi()$ to represent the normal cdf. Therefore, an individual is encountered whenever $S_{ij} > c$, which happens with probability $\Pr(S_{ij} > c) = 1 - \Phi(\gamma(\boldsymbol{\theta}, \mathbf{x}, \mathbf{u}))$.

Naturally, this quantity should depend on *where* an individual is located at the time of recording—what we call its instantaneous location, $\mathbf{u}$, to distinguish it from its home range center $\mathbf{s}$ and the detector location. The probability of detection is therefore

$$p_{ij} = p(\alpha_0, \alpha_1, \sigma | \mathbf{x}_j, \mathbf{u}_i) = 1 - \Phi(\gamma(\boldsymbol{\theta}, \mathbf{x}, \mathbf{u})),$$

where $\mathbf{u}_i$ is the instantaneous location of individual i and $\mathbf{x}_j$ is the location of trap j. We'll suppose here that the random variables $\mathbf{u}_i$ have state-space $\mathcal{U}$.[3]

The observations from an acoustic survey are the signal strength measurements, and the likelihood of the observed signal strength from individual i at detection device j can be specified by noting that the likelihood is the normal pdf for the observed signal *if* the signal strength is $> c$ and, otherwise, the contribution to the likelihood

[3]We use $\mathcal{U}$ here to avoid confusion with definition of signal strength, S. However, $\mathcal{U}$ is the same state-space as $\mathcal{S}$ in the rest of the book.

is $\Phi(\gamma(\boldsymbol{\theta}, \mathbf{x}, \mathbf{u}))$ (see Eq. 8 of Efford et al. (2009b)):

$$\Pr(S_{ij}|\mathbf{u}_i) = \Phi(\gamma(\boldsymbol{\theta}, \mathbf{x}, \mathbf{u}))^{1-I(S_{ij}>c)}\text{Normal}(S_{ij}; \alpha_0, \alpha_1, \sigma_s, \mathbf{x}_j, \mathbf{u}_i)^{I(S_{ij}>c)}.$$

We can use this as the basis for constructing the likelihood as we did in Chapter 6, which involves the number of individuals not encountered, n_0. The probability that an individual is *not* captured is equal to the probability that its signal strength doesn't exceed c at any microphone. The probability of not being captured at a microphone $\mathbf{x}_j$ is:

$$1 - p_{\mathbf{u},j} = \Phi(\gamma(\boldsymbol{\theta}, \mathbf{x}_j, \mathbf{u})).$$

Therefore, the probability of not being captured at any microphone is:

$$\Pr(\text{all } S_{\mathbf{u},j} < c|\mathbf{u}) = \prod_{j=1}^{J}(1 - p_{\mathbf{u},j}) = \prod_{j=1}^{J} \Phi(\gamma(\boldsymbol{\theta}, \mathbf{x}_j, \mathbf{u})),$$

and the marginal probability of not being captured is

$$\pi_0 = [\text{all } S_{\mathbf{u},j} < c|\boldsymbol{\theta}] = \int_{\mathcal{U}} \left\{ \prod_{j=1}^{J} \Phi(\gamma(\boldsymbol{\theta}, \mathbf{x}_j, \mathbf{u})) \right\} d\mathbf{u},$$

which is used to construct the likelihood as we did in Chapter 6 (see Eq. 6.2.1).

9.4.2 Implementation in `secr`

Fitting acoustic encounter models in `secr` is no more difficult than other SCR models. There is a handy manual (`secr-sound.pdf`) with examples (Efford and Dawson, 2010) that come with the `secr` package. The basic process is that `make.capthist` will make a `capthist` object from a three-dimensional encounter array—which is a binary array indicating whether each individual was detected or not at each recorder/microphone. In the case of signal strength data, `secr` handles the case where # occasions = 1, i.e., the recorders obtained data for a single sample occasion. The "signal" attribute of the `capthist` object contains the signal strength in decibels. The best way to include the signal attribute is to use `make.capthist` in the usual way, providing it with the encounter data and trap data and, in addition, the variable "cutval" (which is c in our notation above) and then provide the signal strength data as an extra column of the `capthist` object. See `?make.capthist` for details.

9.4.3 Implementation in BUGS

We don't know of any Bayesian applications of acoustic SCR models, although we imagine that implementation of such models in the **BUGS** engines should be achievable. Instead of developing a Bayesian version of this model here, we leave it to you to explore simulating data and devising a Bayesian implementation of the acoustic model in one of the **BUGS** engines. Note that for a single occasion, you can simulate the data using the two-stage model (having both **s** and **u**) or you can simulate **u** uniformly without dealing with **s** in the model. The kernel of the **BUGS** model specification should resemble the following snippet:

```
model {
  # Ignoring loops and data augmentation
  u[i,1] ~ dunif(xlim[1], xlim[2])
  u[i,2] ~ dunif(ylim[1], ylim[2])
  mu[i,j] <- alpha0 + alpha1*d[i,j]
  ###
  ### JAGS has this T() truncation feature
  S[i,j] ~ dnorm(mu[i,j], 1/sigma^2)T(c,Inf)
  ###
  gamma[i,j] <- (c - mu[i,j])/sigma
  p[i,j] <- 1 - pnorm(gamma[i,j], 0, 1) # JAGS has pnorm() function
  y[i,j] ~ dbern(p[i,j])
}
```

9.4.4 Other types of acoustic data

Efford and Dawson (2010) noted that various other types of acoustic data might arise for which SCR-like models would be useful.[4] For example, we could measure the *time of arrival* of a vocal cue of some sort at multiple recorders to estimate the number and origin of N queues. Another example is that where we measure *direction* to a queue from multiple devices and do, effectively, a type of statistical triangulation to the multiple but unknown number of sources. This has direct relevance to types of double or multiple-observer sampling that are used in field studies of birds. Normally, two observers stand in close proximity and record birds, reconciling their detections after data collection. An SCR-based formulation of the double-observer method has two observers (or more) standing some distance apart, e.g., 50 or 100 m, and marking individual birds on a map (or at least a direction) and recording a time of detection. The SCR/double-observer method could be applied to such data.

9.5 Summary and outlook

In this chapter we extended SCR models to accommodate alternative models for the observation process, including Poisson and multinomial models. Along with the binomial model described in Chapter 5, this sequence of models will accommodate a substantial majority of contemporary spatial capture-recapture problems, including the four main types of encounter data: binary encounters, multinomial trials from "multi-catch" and "single-catch" (Efford, 2004, 2011a; Royle and Gardner, 2011) trap systems, and Poisson encounter frequency data from devices that can record multiple encounters per occasion of the same individual at a device. We summarize the standard

[4]Some of the following is also related to material presented by D.L. Borchers at the ISEC 2012 conference in Oslo, Norway.

Table 9.4 Different observation models, where we discuss them in this book, and the corresponding `secr` terminology.

observation Model	Where in This Book?	`secr` Name
Bernoulli	Chapter 5	`proximity`
Poisson	Section 9.1	`count`
Multinomial (ind)	Section 9.2	`multi-catch`
Multinomial (dep)	Section 9.3	`single-catch`
Acoustic	Section 9.4	`signal`
Search-encounter	Chapter 15	`polygon` (in part)

observation models and the corresponding `secr` terminology in Table 9.4. What we refer to as search-encounter (or area-search) models (see Chapter 15) are distinct from most of the other classes in that the observation location can also be random (in contrast to traps, where the location is fixed by design). This auxiliary data is informative about an intermediate process related to movement (Royle and Young, 2008).

There is a need for other types of encounter models that arise in practice. We identify a few of them here, although we neglect a detailed development of them at the present time or, in some cases, put that off until later chapters: (1) Removal systems—sometimes traps kill individuals and SCR models can handle that. This can be viewed as a kind of open model, with mortality only, and we handle such models (in part) in Chapter 16; (2) There are models for which only specific summary statistics are observable (Chandler and Royle, 2013; Sollmann et al., 2013a) which we cover in Chapters 18 and 19; (3) We can have multiple observation methods working together as in Gopalaswamy et al. (2012b).

There remains much research to be done to formalize models for certain observation systems. For example, while we think one will usually be able to analyze single-catch systems using the multi-catch model, or even the Bernoulli model if encounter probability is sufficiently low, a formalization of the single-catch model would be a useful development and, we believe, it should be achievable using one or another of the **BUGS** engines. In addition, classical "trapping webs" (Anderson et al., 1983; Wilson and Anderson, 1985a; Jett and Nichols, 1987; Parmenter and MacMahon, 1989; Link and Barker, 1994) have been around for quite some time and it seems like they are amenable to formulation as a type of SCR model although we have not pursued that development simply because trapping webs are rarely used in practice.

CHAPTER

Sampling Design

10

Statistical design is recognized as an important component of animal population studies (Morrison et al., 2008; Williams et al., 2002). Many biologists have been in a situation where some problem with their data could be traced back to a flaw in study design. Commonly, design is thought of in terms of number of samples to take, when to sample, methods of capture, desired sample size (of individuals), power of tests, and related considerations. In the context of spatial sampling problems, where populations of mobile animals are sampled by an array of traps or devices, there are a number of critical design elements. Two of the most important ones are the spacing and configuration of traps (or sampling devices) within the array. For traditional capture-recapture (CR), conceptual and heuristic design considerations have been addressed by a number of authors (e.g., Nichols and Karanth, 2002, Chapter 11), but little formal analysis focused on spatial design of arrays has been carried out. Bondrup-Nielsen (1983) investigated the effect of trapping grid size (relative to animal home range area) on capture-recapture density estimates using a simulation study, and some authors have addressed trap spacing and configuration by sensitivity "re-analysis" (deleting traps and reanalyzing; Wegge et al., 2004; Tobler et al., 2008). The scarcity of simulation-based studies looking at study design issues is surprising, as it seems natural to evaluate prescribed designs by Monte Carlo simulation in terms of their accuracy and precision. In the past few years, however, a growing number of simulation studies addressing questions of study design in the context of spatial capture-recapture (SCR) have been published (e.g., Marques et al., 2011; Sollmann et al., 2012; Efford and Fewster, 2012; Efford, 2011b), the results of which we will discuss throughout this chapter.

In this chapter we recommend a general framework for evaluating design choices for SCR studies using Monte Carlo simulation of specific design scenarios based on trade-offs between available effort, funding, logistics, and other practical considerations—what we call *scenario analysis*. Many study design-related issues can be addressed with preliminary field studies that will give you an idea of how much data you can expect to collect with a unit of effort (a camera trap day or a point count survey, for example). But it is also always useful to perform scenario analysis based on simulation before conducting the actual field survey not only to evaluate the design in terms of its ability to generate useful estimates, but also so that you have an expectation of what the data will look like as they are being collected. This gives you the ability to recognize some pathologies and possibly intervene to resolve issues before they render a whole study worthless. Suppose you design a study to

Spatial Capture-Recapture. http://dx.doi.org/10.1016/B978-0-12-405939-9.00010-4

place 40 camera traps based on expectations of parameter values you obtained from a careful review of the literature, and simulation studies suggest that you should get 3–5 captures of individuals per night of sampling. In the field you find that you're realizing 0 or 1 captures per night and therefore you have the ability to sit down and immediately question your initial assumptions and possibly take some remedial action in order to salvage your project, your PhD thesis or even your career. Simulation evaluation of design *a priori* is therefore a critical element of any field study.

While we recommend scenario analysis as a general tool to understand your *expected data* before carrying out a spatial capture-recapture study, it is possible to develop some heuristics and even analytic results related to the broader problem of model-based spatial design (Müller, 2007) using an explicit objective function based on the inference objective. We outline an approach in this chapter where we identify a variance criterion, namely, the variance of an estimator of N for the prescribed state-space. We show that this depends on the configuration of trap locations, and we provide a framework for optimizing the variance criterion over the design space (the collection of all possible designs of a given size). While there is much work to be done on developing this idea, we believe that it provides a general solution to any type of design problem where the set of candidate trap locations is well defined.

10.1 General considerations

Many biologists have experience with the design of natural resource surveys from a classical perspective (Thompson, 2002; Cochran, 2007), a key feature of which involves sampling space. That is, we identify a sample frame comprised of spatial units and we sample those units randomly (or by some other method, such as generalized random tessellation stratified (GRTS) sampling (Stevens and Olsen, 2004) and measure some attribute. The resulting inference applies to the attribute of the sample frame. There are some distinct aspects of the design of SCR studies which many people struggle with in their attempts to reconcile SCR design with classical survey design problems. We discuss some of these here.

10.1.1 Model-based not design-based

Inference in classical finite-population sampling is usually justified by "design-based" arguments. This means that properties of estimators (bias, variance) are induced by the method of sampling, and they are evaluated by averaging over realizations of the *sample*. The sample is random, but the attribute being observed is not, for the specific sample that is chosen. For example, imagine we have a landscape gridded off into 900 1 km × 1 km grid cells, from which we draw a sample of 100 to measure an attribute such as "percent developed" which we aim to use in a habitat model. In the classical design-based view, the attribute (percent developed) is a static quantity for each of the 900 grid cells and theory tells us that, by taking a random sample, we can expect to obtain estimators (e.g., of the population mean of all 900 grid cells) with good statistical properties, where the expectation is over realizations of the sample of 100 grid cells. That is, if we repeatedly draw samples of size 100 then, over many such samples, the estimator may be unbiased. Conversely, in the SCR modeling framework, properties of our estimators are distinctly model based. We evaluate estimators (usually) for a *fixed* sample of spatial locations, averaged over realizations of the underlying process and data we might generate. Although sometimes we might condition on the data for purposes of inference (if we have our Bayesian hat on), the probability model for the data is fundamental to inference, and the spatial sample of trap locations is always fixed. It therefore makes sense to seek a trap configuration that we expect to produce precise estimates of SCR model parameters, and not judge our design by averaging over many randomly (or otherwise chosen) trap configurations.

10.1.2 Sampling space or sampling individuals?

A fundamental question in any sampling problem is what is the sample frame—or the population we are hoping to extrapolate our sample to. In the context of capture-recapture studies, it is tempting to think of the sample frame as being spatial (the space within "the study area," tiled into quadrats perhaps). Clearly SCR models involve a type of spatial sampling—we have to identify spatial locations for traps, or arrays of traps. However, unlike conventional natural resource sampling, the attribute we measure is *not* directly relevant to the *sample location*, such as where we place a trap and, therefore, it may not be sensible to think of the sample frame as being comprised of spatial units. On the other hand, capture-recapture studies clearly obtain a sample of *individuals* and SCR models are models of *individual* encounter and space use. Therefore, it is more natural to think of the sample frame as a list of N individuals, determined by the definition of the state-space, or a subset of the state-space, i.e., the study area, but the number N is unknown. The purpose of the SCR study is to draw a sample of these N individuals and learn about an individual attribute—namely, where that individual lives. *That* is the sampling context of SCR models. SCR models link the observed data (encounter histories) to this individual attribute via a model (with parameters) which we need to fit. Once we fit that model, we usually use it to make a prediction or estimate the attribute for individuals that did not appear in the sample.

Spatial sampling in SCR studies is important, but only as a device for accumulating individuals in the sample from which we can learn about their inclusion probability. That is, we're not interested in any sample unit attribute directly but, rather, we use spatial units as a means for sampling individuals and obtaining individual-level encounter histories. It makes sense in this context that we should want to choose a set of spatial sample units that provides an adequate sample size of individuals, perhaps as many as possible. The key technical consideration as it relates to spatial sampling and SCR, is that arbitrary selection of sample units has a side-effect that it induces unequal probabilities of inclusion of the sample units (i.e., an individual exposed to more traps is more likely to be included into the sample than an individual exposed to few traps) and so we must also learn about these unequal probabilities of sample inclusion.

The fact that SCR sampling induces unequal probabilities of sampling is consistent with the classical sampling idea of Horvitz-Thompson estimation which has motivated capture-recapture models similar to SCR (Huggins, 1989; Alho, 1990). In the Horvitz-Thompson framework, the sample inclusion probabilities are usually fixed and known. However, in all real animal sampling problems they are unknown because we never know precisely where each individual lives and therefore cannot characterize its encounter probability. Therefore, we have to estimate the sample inclusion probabilities using a model. SCR models achieve this effect formally, using a model that accounts for the organization of individual activity centers and trap locations. This notion of Horvitz-Thompson estimation suggests that perhaps we should consider designing SCR studies based on the Horvitz-Thompson variance estimator as a design criterion. We discuss this a little bit later in this chapter.

10.1.3 Focal population vs. state-space

In SCR models we make a distinction between the focal population—the population of individuals we care about—and those of the state-space, which we are required to prescribe in order to fit SCR models. These are not the same thing. The geographic scope of the population of inference is the region within which animals live that you care about in your study—let's call this "the study area." This is often prescribed for political reasons or legal reasons (e.g., a National Park). To initiate a study, or perhaps motivating the study, you have to draw a line on a map to delineate a study area, although often it is difficult to draw this line, and where you draw it is not so much a statistical/SCR issue. On the other hand, you need to prescribe the state-space to define and fit an SCR model. This is the region that contains individuals that you *might* capture. This is different from the study area in most cases. To design a study, you need a well-defined study area, but the state-space will also be relevant to efficient distribution of traps, and other considerations.

It is helpful to think about this distinction operationally. We define our study area a priori. As a conceptual device, we might think of this as the area that, given an infinite amount of resources, we might wall off so that we can study a real closed population. This "study area" should exist independent of any model or estimator of some population quantity, i.e., the subject-matter context should determine what the study area is. Given a well-defined study area, we use some method to arrange data collecting devices (i.e., traps) within this study area. The method of arrangement can be completely arbitrary but, naturally, we want to choose arrangements that are better in terms of obtaining statistical information from the data we end up collecting.

Lets face it—it's quite a nuisance that animals move around and this makes the idea of a spatial study area kind of meaningless in terms of management in most cases. Wherever you draw a line on a map, there will be animals who live mostly beyond that line that will sometimes be subjected to observation in your study. One of the benefits of SCR models is that they formalize the exposure and contribution of these individuals to your study. That is a good thing. Thus, you can probably be a bit imprecise or practical in your definition of "the study area" and not worry too much.

With these general concepts of spatial sampling and the sampling of individuals in mind, we can now turn our attention to more specific aspects of study design in SCR surveys, namely the spatial arrangement of detectors. We discuss some general concepts, and then focus on a couple of specific case studies that apply to the Bernoulli observation model or passive detection devices. The general concepts are surely relevant to other SCR models, and we suspect that the specific case studies are relevant as well.

10.2 Study design for (spatial) capture-recapture

The importance of adequate trap spacing and overall configuration of the trapping array has long been discussed in the capture-recapture literature. A heuristic based on recognizing the importance of typical home range sizes (Dice, 1938, 1941) and the need to obtain information about home range size from the trap array is that traps should be spaced such that the array of available traps exposes as many individuals as possible but, at the same time, individuals should be available for capture in multiple traps. Thus, good designs should generate a high sample size n (i.e., the number of individuals captured) and a large number of spatial recaptures. These two considerations form a trade-off in building designs. On one hand, having a lot of traps very close together should produce the most spatial recaptures but produce very few unique individuals captured (assuming that studies are limited in the total number of sampling devices they can deploy). On the other hand, spreading the traps out as much as possible, in a nearly systematic or regular design, should yield the most unique individuals, but probably few spatial recaptures. We will formalize this trade-off later, when we consider model-based design of SCR studies.

Traditional CR models require that all individuals in the study area have a probability >0 of being captured, which means that the trap array must not contain "holes" large enough to contain an animal's entire home range (Otis et al., 1978). The reason why "holes" cause a problem in non-spatial models is that they induce heterogeneity in capture probability. If an animal's home range lies completely or partially in a hole, then it will have a different probability of being captured than an individual whose home range is peppered with traps. As a consequence, trap spacing is recommended to be on the same order as the radius of a typical home range (e.g., Dillon and Kelly (2007)). For example, imagine a camera trap study implemented in South America with the objective to survey populations of both jaguars (*Panthera onca*) and ocelots (*Leopardus pardalis*). Ocelots have much smaller home ranges and therefore require closer trap spacing than the large, wide-ranging jaguars. The "no holes" assumption entails some strong restrictions with respect to study design. Although we need not cover an area systematically with traps, there has to be some consistent coverage of the entire area of interest. Often, this is achieved by dividing the study area into grid cells, the size of which approximates an average home range (or possibly the smallest home range recorded for the study species in the study area or a similar area; e.g., Wallace et al. (2003)), and then place (at least) one trap within each cell. In many field situations, especially when dealing with large mammals and accordingly large study areas, achieving this consistent coverage can be extremely challenging or even impossible. Depending on local environmental conditions, parts of the study area can be virtually inaccessible to humans, because of dense vegetation cover, or unsuitable for setting up detectors, because of flooding. Even when accessible, setting up traps in difficult habitat conditions can consume disproportional amounts of time, manpower, and other resources. Regardless of whether the trap spacing does result in holes or not, the problem of spatial heterogeneity in capture probability will still exist because individuals with home ranges near the borders of the trap array will have a different probability of being captured than individuals that spend all their time within the trap array.

Where approaches such as mean maximum distance moved (MMDM) are used in combination with traditional CR models to obtain density estimates (see Chapter 4), trap spacing also has a major effect on movement estimates, since it determines the resolution of the information on individual movement (Parmenter et al., 2003; Wilson and Anderson, 1985a). If trap spacing is too wide, there is little or no information on animal movement because most animals will only be captured at one trap (Dillon and Kelly, 2007). In addition, only a trapping grid that is large relative to individual movement can capture the full extent of such movements, and researchers have suggested that the grid size should be at least four times that of individual home ranges to avoid positive bias in estimates of density (Bondrup-Nielsen, 1983). This recommendation originated in small mammal trapping, and it should be relatively easy to follow when dealing with species covering home ranges <1 ha. However, translated to large mammal research, this can entail having to cover several thousands of square kilometers—a logistical and financial challenge that few projects could realistically tackle.

Though closely related, the requirements in terms of spatial study design for SCR models differ distinctly from those for traditional CR. For one, holes in the study area are of no concern in SCR studies. As a practical matter, some animals within the study area might have a vanishingly small probability of being included in the sample, i.e., $p \approx 0$. The nice thing about SCR models is that N is explicitly tied to the state-space, and not the traps that expose animals to encounter. Within an SCR model, extending inference from the sample to individuals that live in these holes represents an extrapolation (prediction of the model outside the range of the data), but one that the model is capable of producing because we have explicit declarations in the model that it applies to any area within the state-space, even to areas where we can't capture individuals because we happened to not put a trap there. Conversely, ordinary capture-recapture models only apply to individuals that have encounter probabilities that are consistent with the model being considered. Presumably, the existence of a hole in the trap array would introduce individuals with $p = 0$, which is not accommodated in those models. This distinction alone allows for completely new and much more flexible study designs in SCR studies, as compared to traditional CR, such as linear designs, "hollow grids" (detectors trace the outline of a square), or small clusters of grid spread out over larger landscapes (Efford et al., 2005, 2009a; Efford and Fewster, 2012).

Whereas traditional CR studies are concerned with the number of individuals and recaptures and with satisfying the model assumption of all individuals having some probability of being captured, in spatial capture-recapture we are looking at an additional level of information: We need spatially dispersed captures and recaptures. It is not enough to recapture an individual—we need to recapture at least some individuals at several traps. Therefore, in general, design of SCR studies boils down to obtaining three bits of information: total unique individuals captured, total number of recaptures, informative about baseline encounter rate, and spatial recaptures, informative about σ.

Most SCR design choices wind up trading these three things against each other to achieve some optimal (or good) mix. So, for example, if we sample each of a very small number of sites a huge number of times then we can get a lot of recaptures but only very few spatial recaptures, and few unique individuals, etc. This need for spatial recaptures may appear as an additional constraint on study design, but actually, SCR studies are much less restricted than traditional CR studies, because of the way animal movement is incorporated into the model: σ is estimated as a specified function of the auxiliary spatial information collected in the survey and the capture frequencies at those locations. This function is able to make a prediction across distances even when these are latent, including distances larger than the extent of the trap array. When there is enough data across at least some range of distances, the model will do well at making predictions at unobserved distances. The key here is that there needs to be "enough data across some range of distances," which induces some constraint on how large our overall trap array must be to provide this range of distances (e.g., Marques et al., 2011; Efford, 2011b). We will review the flexibility of SCR models in terms of trap spacing and trapping grid size in the following section.

10.3 Trap spacing and array size relative to animal movement

Using a simulation study, Sollmann et al. (2012) investigated how trap spacing and array size relative to animal movement influence SCR parameter estimates, and we will summarize their study here. They simulated encounter histories on an 8×8 trap array with regular spacing of 2 units, using a binomial observation model with a Gaussian hazard encounter model, across a range of values for the scale parameter σ^*. We refer to the scale parameter as σ^* here, because Sollmann et al. (2012) use a slightly different parametrization of SCR models, in which σ^* corresponds to $\sigma \times \sqrt{2}$. The trap array was buffered by $3 \times \sigma^*$.

In Section 5.4 we pointed out that under the Gaussian (or half-normal) detection model σ can be converted into an estimate of the 95% home range or "use area" around $\mathbf{s}_i$. Based on this transformation, values for σ^* were chosen so that there was a scenario where the trap array was smaller than a single individual's home range, i.e., trap spacing was small relative to individual movements ($\sigma^* = 5$ units), a scenario where spaces between traps were large enough to contain entire home ranges ($\sigma^* = 0.5$ units), and two intermediate scenarios and where σ^* was smaller ($\sigma^* = 1$ unit) and larger ($\sigma^* = 2.5$ units) than the trap spacing, respectively. N was 100, the baseline trap encounter rate λ_0 was 0.5 (on the cloglog scale) for all four

Table 10.1 Mean, relative root mean squared error (RRMSE) of the mean, mode, 2.5% and 97.5% quantiles, relative bias of mean (RB), and 95% Bayesian credible interval (BCI) coverage for spatial capture-recapture parameters across 100 simulations for four simulation scenarios, defined by the input value of movement parameter σ^*. N = number of individuals in the state-space; λ_0 = baseline trap encounter rate.

Scenario	Mean	RRMSE	Mode	2.5%	97.5%	RB	BCI
$\sigma^* = 1$ ($\sigma = 0.71$)							
N	108.497	0.172	104.099	78.977	143.406	0.085	96
λ_0	0.518	0.248	0.477	0.303	0.752	0.035	94
σ^*	1.008	0.093	0.990	0.857	1.195	0.008	94
$\sigma^* = 2.5$ ($\sigma = 1.77$)							
N	100.267	0.105	98.456	82.086	121.878	0.003	97
λ_0	0.507	0.118	0.500	0.409	0.623	0.014	92
σ^*	2.501	0.046	2.491	2.267	2.690	<0.001	92
$\sigma^* = 5$ ($\sigma = 3.54$)							
N	102.859	0.137	100.756	77.399	130.020	0.029	88
λ_0	0.505	0.075	0.501	0.435	0.580	0.011	93
σ^*	5.023	0.039	5.001	4.687	5.431	0.005	97

scenarios and trap encounters were generated over four occasions. Table 10.1 shows the results as the average over 100 simulations.

All model parameters were estimated with relatively low bias (<10%) and high to moderate precision (relative root mean squared error, RRMSE < 25%) for all scenarios of σ^*, except $\sigma^* = 0.5$ units, under which model parameters were mostly not estimable (therefore excluded from Table 10.1). Data for the latter case mostly differed from the other scenarios in that fewer animals were captured and very few of the captured animals were recorded at more than 1 trap (Table 10.2). For $\sigma^* = 0.5$, abundance (N) was not estimable in 88% of the simulations, and when estimable, was

Table 10.2 Summary statistics of 100 simulated data sets for four simulation scenarios, defined by the input value of movement parameter σ^*. Individual detection histories were simulated on an 8 × 8 trap array with regular trap spacing of 2 units.

Scenario	Inds. Captured	Total Captures	Inds. Recaptured	Inds. Captured at >1 Trap
$\sigma^* = 0.5$	18.29 (3.84)	25.38 (5.86)	5.52 (2.03)	0.72 (0.95)
$\sigma^* = 1.0$	37.70 (13.44)	69.35 (26.05)	19.48 (7.68)	11.87 (5.43)
$\sigma^* = 2.5$	44.19 (4.67)	231.78 (33.98)	36.60 (4.76)	35.21 (4.73)
$\sigma^* = 5.0$	40.51 (5.15)	427.77 (79.09)	33.09 (4.63)	32.60 (4.76)

underestimated by approximately 50%. This shows that a trap spacing that is considerably too large relative to animal movement may be problematic in SCR studies.

Estimates (posterior means) of N were least biased and most precise under the $\sigma^* = 2.5$ scenario, and in general, all parameters were estimated best under the $\sigma^* = 2.5$ or the $\sigma^* = 5$ scenario. All estimates had the highest relative bias and the lowest precision under the $\sigma^* = 1$ scenario. These results clearly demonstrate that SCR models can successfully handle a range of trap spacing to animal movement ratios, and even perform well when using a trap array smaller than an average home range: at $\sigma^* = 5$, the home range of an individual was approximately 235 units^2, while the trapping grid only covered 196 units^2.

An important consideration in this simulation study is that all but the $\sigma^* = 0.5$ units scenarios provided reasonably large amounts of data, including 20+ individuals captured. When dealing with real-life animals, which are often territorial and may have lower trap encounter rates, a very small grid area compared to an individual's home range may result in the capture of very few individuals. In that case, the sparse data will limit the ability of the model to estimate parameters (Marques et al., 2011), which is true of most models.

To further explore the effects of trap spacing and movement on bias and precision of estimates of N, we expanded the simulation study of Sollmann et al. (2012): we considered a regular 7×7 grid, with trap spacing ranging from $0.5 \times \sigma$ to $4 \times \sigma$, with a state-space that had variable size so that the buffer around the traps was constant in units of σ. For each trap spacing scenario we simulated and analyzed 500 data sets and calculated the RRMSE and relative bias for the estimates of N. Figure 10.1 shows the results of this set of simulations. We see that there is clearly an optimal trap spacing, especially in terms of precision, which is highest at a trap spacing of $1.5 - 2.5 \times \sigma$. Efford (2012) reported similar results and highlighted the trade-off between the number of individuals captured and the number of spatial

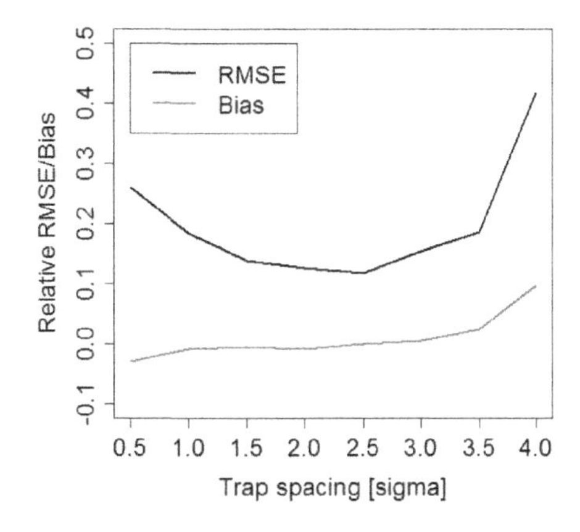

FIGURE 10.1

Relative bias and relative RMSE of estimates of N from an SCR model for a range of trap spacing scenarios.

recaptures—intuitively, the former goes up with an increase in trap spacing, whereas the latter goes down. In summary, in small trap spacing scenarios, the small sample size leads to imprecise estimates, whereas in large trap spacing scenarios, lack of spatial recaptures leads to imprecise and biased estimates.

10.3.1 Black bears from Pictured Rocks National Lakeshore

To see how trap array size influences parameter estimates from spatial capture-recapture models in the real world, Sollmann et al. (2012) also looked at a black bear data set from Pictured Rocks National Lakeshore, Michigan, collected using 123 hair snares distributed over an area of 440 km^2 along the shore of Lake Superior in May-July 2005 (Belant et al., 2005). The SCR model for the bear data included sex-specific encounter rate parameters, and an occasion-specific baseline encounter rate. This was motivated by (a) the lower average number of detections for male bears, (b) the decreasing number of detections over time in the raw data, and (c) the fact that male black bears are known to move over larger areas than females (e.g., Koehler and Pierce, 2003; Gardner et al., 2010b).

To address the impact of a smaller trap array on the parameter estimates, models fitted to the full data set were compared to models fitted to data subsets. The first subset retained only those 50% of the traps closest to the grid center. In the second subset, only the southern 20% of the traps were retained.

Reducing the area of the trap array by 50% created a grid polygon of 144 km^2, which was smaller than an estimated male black bear home range and only 50% larger than a female black bear home range—approximately 260 km^2 and 100 km^2, respectively, when converting estimates of σ^* to home range size. Table 10.3 shows that this did not greatly influence model results, compared to the full data set.

Table 10.3 Posterior summaries of SCR model parameters for black bears, modified from Sollmann et al. (2012).

	Mean (SE)	Mode	2.5%	97.5%
Full data set				
D	10.556 (1.076)	10.448	8.594	12.792
σ^* (males)	7.451 (0.496)	7.323	6.579	8.495
σ^* (females)	2.935 (0.143)	2.939	2.671	3.226
50% of traps				
D	12.648 (1.838)	12.205	9.307	16.713
σ^* (males)	5.354 (0.511)	5.248	4.472	6.473
σ^* (females)	3.318 (0.277)	3.262	2.841	3.910
20% of traps				
D	6.752 (1.611)	5.953	4.000	10.218
σ^* (males)	9.881 (3.572)	7.566	5.121	18.447
σ^* (females)	2.686 (0.391)	2.657	2.121	3.404

Removing 80% of the traps and thereby reducing the area of the trap array to 64 km^2—well below the average black bear home range—had a great effect on sample size (only 25 of the original 83 individuals sampled) and parameter estimates. Particularly, male black bear movement was overestimated and imprecise. The combination of a low baseline trap encounter rate of males and the considerable reduction in sample size led to a low level of information on male movement: 5 of the 12 males were captured at one trap only. Although they moved over smaller areas, owing to their higher trap encounter rate, females were, on average, captured at more traps (3.4 traps per individual, compared to 2.6 for males) so that their movement estimate remained relatively accurate. Overestimated male movements and female trap encounter rates resulted in an underestimate of density of almost 40%. This effect is contrary to what we would expect to see in non-spatial CR models, where a trapping grid that is small relative to animal movement leads to underestimated movement (MMDM) and overestimated density (Bondrup-Nielsen, 1983; Dillon and Kelly, 2007; Maffei and Noss, 2008). While this example again demonstrates the ability of SCR models to deal with a range of trapping grid sizes, it also clearly shows that your study design needs to consider the amount of data you can expect to collect. As an alternative to simulation studies, Efford et al. (2009b) provide a mathematical procedure to determine the expected number of individuals captured and recaptures for a given detector array and set of model parameters.

10.4 Sampling over large areas

Trap spacing is an essential aspect of design of SCR studies. However, it is only the most important aspect if one can uniformly cover a study area with traps. In many practical situations, where the study area is large relative to effort that can be expended, one has to consider other strategies that deviate from a strict focus on trap spacing. There are two general strategies that have been suggested for sampling large areas, which we think are useful in practice, either by themselves or combined: Sampling based on *clusters* of traps and sampling based on moving groups of traps over the landscape.

Karanth and Nichols (2002) describe three strategies for moving traps to achieve coverage of a larger study area, geared toward traditional capture-recapture analysis. Suppose that sampling the entire area of interest requires sampling G sites, then the three strategies are:

1. For every day/sampling occasion, randomly choose x out of your G sites, where x is the number of trapping devices you have at hand. Obviously, this requires that it be relatively easy to move traps around.
2. Move blocks of traps daily that are in proximity to one another. For example, if you divide your total study area into four blocks, sample block 1 for a day, then move traps to block 2 for a day, and so forth, and repeat until each block has been sampled for a sufficient amount of time.

3. If moving blocks of traps daily is too challenging logistically, then you can sample each block for a certain number of days/occasions before moving them to the next block. In this fashion, you only need to move traps to each block once.

In traditional CR we collapse data across traps and assume all individuals in the study area have some probability >0 of being detected. For our data that means that, under scenario (2) the first occasion is defined as the time it takes to sample all 4 blocks once, the second occasion consists of the second round of sampling all blocks, etc. Under scenario (3), we have to combine data from day 1 in each of the blocks to form occasion 1, data from day 2 in each of the blocks forms occasion 2, and so on. Especially scenario 3 makes modeling time-dependent detection difficult, since occasion 1 no longer refers to an actual day or continuous time interval. We do not have that problem in SCR, where accounting for sampling effort at each trap is straightforward, as we first demonstrated for the wolverine example in Section 5.9. Because we are dealing with detection at the trap level, even for design (3) in a spatial framework, we can still look at variation in detection over time. As such, we don't think that one of the above designs is better than the other for SCR models, but rather, all of them may produce adequate SCR data, as long as overall sample size requirements are met.

Efford and Fewster (2012) looked at the performance of different spatial study designs for abundance estimation from traditional and spatial capture-recapture models, including a clustered design, where groups of detectors are spaced throughout the larger region of interest. They found that in a spatial framework this design performed well, although there were indications of a slight positive bias in estimates of N. Such a clustered design enables researchers to increase area coverage without having to increase the number of traps. Efford and Fewster (2012) note that distribution of clusters has to be spatially representative—for example, systematic with a random origin. The issue of spatially representative designs is not limited to SCR and an extensive treatment of the topic can be found in the distance sampling literature (Buckland et al., 2001). Further, the authors stress that, if distances among clusters are large and individuals are unlikely to show up in several clusters, then the method relies on spatial recaptures *within* clusters, meaning that spacing of detectors within clusters has to be appropriate to the movements of the species under study. A clustered type of design is also suggested by Efford et al. (2009b) for acoustic detectors (see Chapter 9.4) with small groups of such detectors (e.g., 2×2) being distributed in a probabilistic fashion across the region of interest.

In practice, employing both of these strategies—clustering and moving traps—might be necessary or advantageous. Sun (2013) used a simulation study to investigate different trap arrangements (Figure 10.2) for a black bear study based on hair snares distributed over a 2625-km^2 study area. She simulated detection data of bears for three trap arrangements including a regular (uniform) coverage of traps, clusters of 4 traps each with a gap between clusters, and a design in which the clusters were moved midway through the study to fill the gap (a sequential or "rotating" design). She found that the precision and accuracy of estimates of N generally decreased when changing

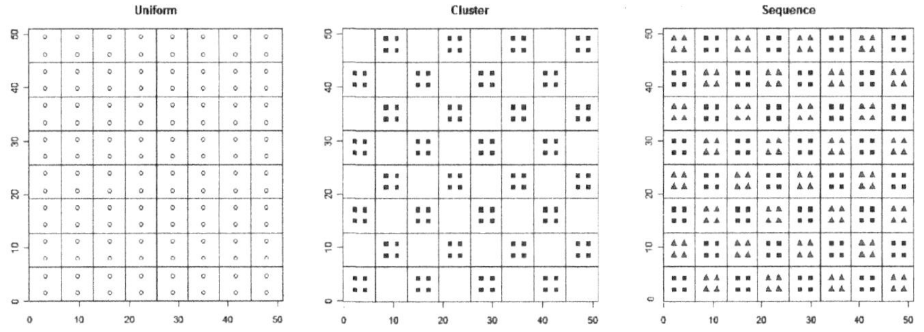

FIGURE 10.2

Three designs evaluated by Sun (2013). The left panel shows uniform coverage of the area with traps (hair snares) equally spaced and static for the duration of the period. The central panel shows clusters of four traps in close proximity, with larger gaps between clusters. The right panel shows a design in which all grid cells are sampled by the cluster of 4 traps, but in a sequential (in time) manner.

from a uniform to a clustered to a rotating design, although the loss of efficiency was relatively small when using the clustered design. The result seems to support that cluster designs can be effective with relatively little loss of efficiency.

Further research on optimal detector configurations, especially for large scale studies, is called for (Efford and Fewster, 2012). More generally, work on formalizing and generalizing these ideas of spatial study design is needed. We believe the model-based spatial design approach that we introduce below is one possible way to tackle that.

10.5 Model-based spatial design

A point we have stressed in previous chapters is that SCR models are basically glorified versions of generalized linear models (GLMs) with a random effect that represents a latent spatial attribute of individuals, the activity center or home range center. This formulation makes analysis of the models readily accessible in freely available software and allows us to adapt and use concepts from this broad class of models to solve problems in spatial capture-recapture. In particular, we can exploit well-established model-based design concepts (Kiefer, 1959; Box and Draper, 1959, 1987; Fedorov, 1972; Sacks et al., 1989; Hardin and Sloane, 1993; Fedorov and Hackl, 1997) to develop a framework for designing spatial trapping arrays for capture-recapture studies. Müller (2007) provides a recent monograph level treatment of the subject that is very accessible.

In the following sections, we adapt these classical methods for constructing optimal designs to obtain the configuration of traps (or sampling devices) in some region (the design space, $\mathcal{X}$), that minimizes some appropriate objective function based on the variance of model parameters, $\boldsymbol{\alpha}$, or N, for a prescribed state-space. In particular, we show that a criterion based on the variance of an estimator of N represents a formal compromise between minimizing the variance of the MLEs of the detection model parameters and obtaining a high expected probability of capture. Intuitively, if our only objective was to minimize the variance of parameter estimates then all of our traps should be in one or a small number of clusters where we can recapture a small number of individuals many times each. Conversely, if our objective was only to maximize the expected probability of encounter then the array should be highly uniform so as to maximize the number of individuals being exposed to capture.

10.5.1 Statement of the design problem

Let $\mathcal{X}$, the *design space*, denote some region within which sampling could occur and let $\mathbf{X} = \mathbf{x}_1, \ldots, \mathbf{x}_J$ denote the *design*, the set of sample locations (e.g., of camera traps), normally we just call these "traps." The design space $\mathcal{X}$ must be prescribed (a priori). Operationally, we could equate $\mathcal{X}$ to the study area itself (which is of management interest) but, in practical cases, there will generally be parts of the study area that we cannot sample. Those areas need to be excluded from $\mathcal{X}$. While $\mathcal{X}$ may be continuous, in practice it will be sufficient to represent $\mathcal{X}$ by a discrete collection of points which is what we do here. This is especially convenient when the geometry of $\mathcal{X}$ is complicated and irregular, which would be the case in most practical applications. The technical problem addressed subsequently is how to choose the locations $\mathbf{X}$ in a manner that produces the "optimal" (lowest variance) for estimating population size or density, or some other quantity of interest.

As usual, we regard the population of N individual activity centers as the outcome of a point process distributed uniformly over the state-space $\mathcal{S}$. The relevance and importance of $\mathcal{S}$ has been established repeatedly in this book, as it defines a population of individuals (i.e., activity centers) and, in practice, it is not usually the same as $\mathcal{X}$ due to the fact that animals move freely over the landscape and the location of traps is typically restricted by policies, ownership, logistics, and other considerations. The objective we pursue here is: Given (1) $\mathcal{X}$, (2) a number of design points, J; (3) the state-space $\mathcal{S}$, (4) an SCR model, and (5) a design criterion $Q(\mathbf{X})$, we want to choose *which* J design points we should select in order to obtain the *optimal* design under the chosen model, where the optimality is with respect to $Q(\mathbf{X})$.

What types of functions make reasonable objective functions, $Q(\mathbf{X})$? We will describe some possible choices for $Q(\mathbf{X})$ below, but it makes sense that they should relate to the variance of estimators of one or more parameters of the SCR model.

We motivate the basic ideas of model-based design with a simple model that proves to be an effective caricature of the SCR model that we'll use shortly. Suppose $\mathbf{s}$ is the activity center of a single individual, and $\mathbf{s}$ is known with certainty. Then, for an array of traps $\mathbf{X}$ we measure a response variable, lets say the strength of an acoustic signal emitted from $\mathbf{s}$, that has a normal distribution. So we have this response variable that has a normal linear model of the form:

$$\mathbf{y} = \mathbf{M}(\mathbf{X}, \mathbf{s})'\boldsymbol{\alpha} + \mathrm{e},$$

where e is random error with $\text{Var}(e) = \delta^2$. In our notation here, $\mathbf{M}(\mathbf{X}, \mathbf{s})$ is some design matrix where, in the context of SCR models, it has 2 columns (for the basic model): A column of 1s, and then a column of distance from each trap $\mathbf{x}_j$ to the activity center $\mathbf{s}$. The design matrix is therefore, for a single individual, a matrix of dimension $J \times 2$.

The inference objective here is to estimate the parameters $\boldsymbol{\alpha}$. The variance-covariance matrix of $\hat{\boldsymbol{\alpha}}$ is, suppressing the dependence on $\mathbf{X}$ for notational convenience,

$$\text{Var}(\hat{\boldsymbol{\alpha}}, \mathbf{X}) = \delta^2(\mathbf{M}(\mathbf{s})'\mathbf{M}(\mathbf{s}))^{-1}.$$

Note that the design points $\mathbf{x}_j$ appear explicitly in the second column of $\mathbf{M}$. Therefore, in considering design for estimation in such models it is natural to choose design points such that the variance of $\hat{\boldsymbol{\alpha}}$ is minimized. Here, $\boldsymbol{\alpha}$ is a vector, and so the "variance" is a matrix (at least 2×2) so we have to work with suitable scalar summaries of that matrix, such as the trace (sum of the diagonals) or the determinant, etc.

For a population of N individuals, if we know *all* N values of $\mathbf{s}$, the design matrix $\mathbf{M}$ has the same basic structure but with N versions stacked up on top of one another, producing a larger $(N \times J) \times 2$ design matrix. The second column of $\mathbf{M}$ contains the information about trap locations (the first column is still just a column of 1s). Therefore, we could, in principle, find the design $\mathbf{X}$ that optimizes some function of the variance-covariance matrix of the model parameters.

The preceding argument assumes we know the activity centers for each individual in the population. When $\mathbf{s}$ is unknown, it might make sense to consider minimizing the expected (spatially averaged) variance:

$$E_{\mathbf{s}}\left\{\text{Var}(\hat{\boldsymbol{\alpha}}, \mathbf{X})\right\} = \frac{\delta^2}{nG}\sum_{s \in \mathcal{S}}(\mathbf{M}'(\mathbf{s})\mathbf{M}(\mathbf{s}))^{-1},$$

where nG is the number of points in the discrete representation of the state-space $\mathcal{S}$. However, this is not the expected variance based on sampling a population of N individuals, just for a single individual having unknown $\mathbf{s}$. Because of the matrix inverse in this expression, it is not sufficient to use a variance criterion that weights this variance by N. As an alternative, we can maximize the expected *information*, the inverse of the variance-covariance matrix, which is more appealing from an analytic point of view. The information matrix for the data based on a single individual, with known $\mathbf{s}$, is: $\mathcal{I}(\hat{\boldsymbol{\alpha}}, \mathbf{X}) = (\mathbf{M}'(\mathbf{s})\mathbf{M}(\mathbf{s}))$. For a population of N individuals, let $\mathbf{M}_i$ be the design matrix for the individual with activity center $\mathbf{s}_i$. Then, the total information for all N individuals is:

$$\mathcal{I}(\hat{\boldsymbol{\alpha}}, \mathbf{X}) = \frac{nG}{\delta^2}\sum_{i=1}^{N}(\mathbf{M}_i'(\mathbf{s}_i)\mathbf{M}_i(\mathbf{s}_i)).$$

The information matrix depends on the design $\mathbf{X}$ through the N individual matrices $\mathbf{M}_1, \ldots, \mathbf{M}_N$. Now, because we don't know each $\mathbf{s}_i$ we must compute the expected total information, over all possible values of $\mathbf{s}_i$, and for *each* of the N $\mathbf{s}_i$ in the population, which is an N-fold summation:

$$E_{\mathbf{s}_1,\ldots,\mathbf{s}_N}\mathcal{I}(\hat{\boldsymbol{\alpha}}, \mathbf{X}) = \frac{nG}{\delta^2}\sum_{i=1}^{N}\sum_{s\in\mathcal{S}}(\mathbf{M}_i'(\mathbf{s}_i)\mathbf{M}_i(\mathbf{s}_i)),$$

which is just N copies of the expected (spatially averaged) information:

$$E_{\mathbf{s}_1,\ldots,\mathbf{s}_N}\mathcal{I}(\hat{\boldsymbol{\alpha}}, \mathbf{X}) = \frac{N * nG}{\delta^2}\sum_{s\in\mathcal{S}}(\mathbf{M}_i'(\mathbf{s}_i)\mathbf{M}_i(\mathbf{s}_i)).$$

It therefore seems sensible to base design of SCR studies on some criterion that is a function of this expected information matrix, e.g., find the design that maximizes the diagonals, or the determinant, or minimizes the trace of the *inverse* (the variance-covariance matrix based on N individuals). This can be done for any number of design points $\mathbf{x}_1, \ldots, \mathbf{x}_J$ using standard exchange algorithms (see Müller, 2007, Chapter 3) and we discuss this below in Section 10.5.5. However, our SCR models are not normal linear models but, rather, more like Poisson or binomial GLMs. We see in the next section that we can adapt these ideas for such models.

10.5.2 Model-based Design for SCR

Following our development of the normal linear model above, suppose for the moment that we know $\mathbf{s}$ for a single individual. In this case, its vector of encounter frequencies in each trap, $\mathbf{y}$, are either binomial or Poisson counts, and the linear predictor has the form:

$$g(\mathbb{E}(\mathbf{y})) = \alpha_0 + \alpha_1 d(\mathbf{x}, \mathbf{s})^2, \tag{10.5.1}$$

for the Gaussian encounter probability model, and where $d(\mathbf{x}, \mathbf{s})$ is the distance between trap location $\mathbf{x}$ and individual activity center $\mathbf{s}$. In vector form, we write this as:

$$g(\mathbb{E}(\mathbf{y})) = \mathbf{M}'\boldsymbol{\alpha},$$

where $\mathbf{M}$ is the $J \times 2$ design matrix where the second column contains the squared pairwise distances between the activity center for individual i and trap j, and thus it depends on both $\mathbf{X}$ (the design) and $\mathbf{s}$.

The asymptotic formula for $\text{Var}(\boldsymbol{\alpha})$ can be specified for any type of GLM. As an example (we make use of this in Section 10.5.1), for the Poisson GLM, the asymptotic variance-covariance matrix of $\hat{\boldsymbol{\alpha}}$, considering a single individual having location $\mathbf{s}$, is[1]

$$\text{Var}(\hat{\boldsymbol{\alpha}}|\mathbf{X}, \mathbf{s}) = (\mathbf{M}(\mathbf{s})'\mathbf{D}(\boldsymbol{\alpha}, \mathbf{s})\mathbf{M}(\mathbf{s}))^{-1}. \tag{10.5.2}$$

This is a function of the design $\mathbf{X}$ as well as $\mathbf{s}$, both of which are balled up in the regression design matrix $\mathbf{M}$, and the matrix $\mathbf{D}$ which is a diagonal matrix having

[1] This is basic GLM theory that derives from the fact that the Poisson is a member of the natural exponential family of distributions, e.g., see McCullagh and Nelder (1989) or Agresti (2002).

elements $\text{Var}(y_j|\mathbf{s}) = \exp(\mathbf{m}'\boldsymbol{\alpha})$ for y_j = the frequency of encounter in trap j and where $\mathbf{m}'$ is the jth row of $\mathbf{M}(\mathbf{s})$. We can compute the expected information under the Poisson model with known N, as in Section 10.5.1.

These ideas are meant to motivate technical concepts related to model-based design, where we know N, and therefore have a convenient variance or information expression to work with. However, in all real capture-recapture applications we won't know N, and so we need to address that issue, which we do in the next section.

10.5.3 An optimal design criterion for SCR

There are a number of appealing directions to pursue for deriving a variance-based criterion upon which to devise designs for spatial capture recapture studies. For one, we could formulate the objective function based on the variance of the Huggins-Alho estimator (Section 4.5) of N. We find that these expressions depend on individual sample inclusion probabilities (if $\mathbf{s}$ is close to traps, the individual has a high probability of being encountered and *vice versa*), and hence the specific trap locations, and parameters of the model. These variance expressions provide a natural design criterion. As an alternative, and slightly simpler formulation, we devise a variance criterion based on the conditional estimator of N having the form

$$\hat{N}_c = \frac{n}{\hat{\bar{p}}},$$

where $\hat{\bar{p}}$ is the MLE of the marginal probability that an individual appears in the sample of n unique individuals, and it depends on the MLE of the parameters of the encounter probability model, $\hat{\boldsymbol{\alpha}}$. We elaborate on the precise form of $\bar{p}$ and the variance of its MLE below. The variance of this estimator is:

$$\text{Var}\left(\frac{n}{\hat{\bar{p}}}\right) = n^2 \times \text{Var}\left(\frac{1}{\hat{\bar{p}}}\right).$$

An important thing to note is that this estimator, and its variance, are *conditional* on the sample size of individuals, n. To "uncondition" on n we use standard rules of expectation and variance, which produces the following expression:

$$\text{Var}(\hat{N}_c(\boldsymbol{\alpha})) = N\bar{p}\{(1-\bar{p}) + N\bar{p}\}\left(\frac{\text{Var}(\hat{\bar{p}})}{\bar{p}^4}\right). \tag{10.5.3}$$

This expression provides an intuitively appealing criterion upon which to devise SCR designs for two reasons: (1) It depends on $\bar{p}$, the marginal probability of encounter. Clearly the variance decreases as $\bar{p}$ increases and, therefore, as the expected sample size of individuals increases, we will obtain a more precise estimate of N. In general, the form of $\bar{p}$ depends on the SCR model being used. We will provide an example below. Obviously, $\bar{p}$ will also depend on the parameter values $\boldsymbol{\alpha}$. (2) The criterion depends on $\text{Var}(\hat{\bar{p}})$. So, designs that estimate $\bar{p}$ well should be preferred. This will also depend on the parameters $\boldsymbol{\alpha}$ and *also* the variance of the MLE, $\hat{\boldsymbol{\alpha}}$. Based on these considerations, we suggest a number of criteria for constructing spatial designs for capture-recapture studies. For convenience we label them $Q_1 - Q_4$:

1. $Q_1 = \text{Trace}(\mathbf{V}_{\hat{\alpha}})$ where $\mathbf{V}_{\hat{\alpha}}$ is the variance-covariance matrix of the MLE of $\boldsymbol{\alpha}$. Designs which minimize this criterion are those which are good for estimating the parameters of the encounter probability model.
2. Q_2 is the variance expression in Eq. (10.5.3). Using this criterion, we should prefer designs that minimize the variance for estimating N.
3. $Q_3 = 1 - \bar{p}$. Designs which minimize this criterion are those which maximize the average capture probability. These should maximize the expected sample size, n.
4. $Q_4 = \text{Var}(\hat{\bar{p}})$. Designs that minimize this criterion provide good estimates of $\bar{p}$.

An important thing to observe is that all of these criteria are functions of the true population size, N, and/or the true parameter values of the encounter probability model. Therefore, to use these criteria in the context of devising SCR designs, we need to have some idea of their values, or be able to obtain estimates from the literature, or a pilot study. Of course, prior information or guesses of parameter values is a typical requirement in the development of sampling designs based on statistical models. However, putting some thought into reasonable values of parameters then gives us a basis for choosing designs that are optimal for our intended estimation objective. We discuss optimizing these or other design criteria over all possible designs in Section 10.5.5.

10.5.4 Too much math for a Sunday afternoon

Here we discuss the calculation of $\bar{p}$ and the variance expressions required to compute the design criteria above, for a particular SCR model.

10.5.4.1 Characterizing $\bar{p}$

In SCR models, an individual with activity center $\mathbf{s}_i$ is captured if it is captured in *any* trap and therefore, under the Bernoulli (passive detector) observation model,

$$\bar{p}(\mathbf{s}_i, \mathbf{X}) = 1 - \prod_{j=1}^{J}(1 - p_{ij}(\mathbf{x}_j, \mathbf{s}_i)),$$

where p_{ij} here is the Gaussian (or other) encounter probability model that depends on distance between traps and activity centers. Using the Gaussian hazard encounter probability model, we have:

$$\bar{p}(\mathbf{s}_i, \mathbf{X}) = 1 - \exp\left(-\lambda_0 \sum_j \exp\left(\alpha_1 \times d(\mathbf{x}_j, \mathbf{s}_i)^2\right)\right),$$

where $\lambda_0 = \exp(\alpha_0)$. Here we emphasize that this is conditional on $\mathbf{s}_i$ and also the design—the trap locations $\mathbf{x}_j$. The *marginal* probability of encounter, i.e., averaging over all possible locations of $\mathbf{s}$, is:

$$\bar{p}(\mathbf{X}) = 1 - \int_{\mathbf{s}} \bar{p}(\mathbf{s}_i, \mathbf{X}) d\mathbf{s}. \tag{10.5.4}$$

This can be calculated directly *given* the design $\mathbf{X}$, and parameters of the model. This is handy because it is used in the variance formulae given subsequently and it is needed to evaluate any of the criteria described above.

10.5.4.2 Characterizing Var $(\hat{\bar{p}})$

Developing an expression for Var $(\hat{\bar{p}})$ depends on the observation model. We work here with the Bernoulli observation model, and the Gaussian hazard encounter probability model (see Huggins (1989) and Alho (1990) for additional context related to computing asymptotic variance expressions under a Bernoulli encounter model). We first express the integral in Eq. (10.5.4) as a summation over a fine mesh of points so that:

$$\bar{p}(\mathbf{X}) = \frac{1}{nG} \sum_{\mathbf{s}} 1 - \bar{p}(\mathbf{s}_i, \mathbf{X}),$$

which, under the Gaussian hazard encounter probability model is:

$$\bar{p}(\mathbf{X}) = \frac{1}{nG} \sum_{\mathbf{s}} \left\{ 1 - \exp\left(- \sum_j \exp\left(\alpha_0 + \alpha_1 d(\mathbf{x}_j, \mathbf{s})^2 \right) \right) \right\}.$$

To obtain the MLE of $\bar{p}(\mathbf{X})$ we plug in the MLE of α_0 and α_1, in this case $\hat{\lambda}_0 = \exp(\hat{\alpha}_0)$ and $\hat{\alpha}_1$. To compute the variance of the MLE of $\bar{p}$, we note that the variance operator can move inside the summation over $\mathbf{s}$, and the subtraction from 1 doesn't affect variance, so we have

$$\text{Var}(\hat{\bar{p}}(\mathbf{X})) = \sum_{\mathbf{s}} \text{Var} \left(\exp\left(- \sum_j \exp\left(\hat{\alpha}_0 + \hat{\alpha}_1 d(\mathbf{x}_j, \mathbf{s})^2 \right) \right) \right).$$

A few applications of the delta method and some arm-waving yield the following expression for the variance of $\hat{\bar{p}}$:

$$\text{Var}(\hat{\bar{p}}(\mathbf{X})) = \sum_{\mathbf{s}} \left(\exp(-\hat{\lambda}_{\mathbf{s}}) \right) \left(\sum_j \hat{\lambda}(\mathbf{x}_j, \mathbf{s})^2 \left(\text{Var}(\hat{\alpha}_0) + d(\mathbf{x}_j, \mathbf{s})^4 \text{Var}(\hat{\alpha}_1) \right) \right),$$

where $\hat{\lambda}(\mathbf{x}, \mathbf{s}) = \exp(\hat{\alpha}_0 + \hat{\alpha}_1 d(\mathbf{x}, \mathbf{s})^2)$ and $\hat{\lambda}_{\mathbf{s}} = \sum_{j=1}^{J} \hat{\lambda}(\mathbf{x}_j, \mathbf{s})$.

10.5.4.3 Characterizing Var ($\hat{\alpha}$)

For a given design $\mathbf{X}$, we just showed how to compute $\text{Var}(\hat{\bar{p}}(\mathbf{X}))$—this is just a calculation involving sums over all points in the state-space and design points—provided we know the variance of the estimator of $\boldsymbol{\alpha}$, $\text{Var}(\hat{\boldsymbol{\alpha}})$, and the true values of the parameters of the encounter probability model. However, it is not so easy to write down the analytic form of the matrix $\text{Var}(\hat{\boldsymbol{\alpha}})$. Some calculus would have to be done on the conditional likelihood (e.g., from Borchers and Efford, 2008) to figure out the asymptotic form of this matrix. For our purposes, we think it might suffice[2] to approximate the matrix, using the analogous result from a Poisson or binomial GLM assuming that N is known, since we have convenient formulas for those (see Eq. (10.5.2) for the Poisson GLM).

The approximate variance given by Eq. (10.5.2) is conditional on the collection of activity centers, $\mathbf{s}_1, \ldots, \mathbf{s}_N$. To resolve this, we take the approach outlined previously (Section 10.5.2) to compute the *expected* total information obtained from a particular realization of N individuals. The expected total (for a population of N individuals) information is N times the expected information:

$$\mathbb{E}(\mathcal{I}(N)) = \frac{N}{nG} \sum_{\mathbf{s}} \mathbf{M}(\mathbf{s})'\mathbf{D}(\mathbf{s})\mathbf{M}(\mathbf{s}).$$

10.5.4.4 Putting it all together

To make use of these mathematical results to evaluate one of the design criteria, we begin by choosing a design, $\mathbf{X}$. Given the design, we compute the expected information quantity, invert it to obtain the variance of $\hat{\boldsymbol{\alpha}}$, and then either use that variance matrix in the calculation of criterion Q_1, or else evaluate some additional quantities along the way to computing the other criteria. We can compute $\bar{p}$ (which is Q_3) for a given design without doing any variance calculations. If we use $\text{Var}(\hat{\boldsymbol{\alpha}})$, along with $\bar{p}$, we can compute $\text{Var}(\hat{\bar{p}})$, which is Q_4. We can combine all of these things together and compute $\text{Var}(\hat{N}_c)$ for a given $\mathbf{X}$, which is Q_2. This results in values of these various criteria for only a single design that we specified. In practice, we would like to search over the whole "design space" and find the optimal (or at least a very good) design of a given size, J. In the next section, we describe a method for efficiently sorting through the design space.

10.5.5 Optimization of the criterion

To compute spatial designs that optimize a given criterion, we need to prescribe values of the model parameters, or at least come up with a ballpark guess of them so that the criterion can be evaluated for any design, i.e., what are the values of $\boldsymbol{\alpha}$ and N we expect in our study? Once we do that, and specify the state-space, $\mathcal{S}$, then we can, in theory, optimize any of our variance criteria over all possible configurations of J traps. For small values of J this problem could possibly be solved using standard

[2]Warning: But we don't know. No warranty is implied.

numerical optimization algorithms as exist in almost every statistical computation environment. However, J will almost always be large enough so as to preclude effective use of such algorithms. This is a common problem in experimental design, and spatial sampling in general, for which sequential exchange or swapping algorithms have been fairly widely adopted (e.g., Wynn, 1970; Fedorov, 1972; Mitchell, 1974; Meyer and Nachtsheim, 1995; Nychka et al., 1997; Royle and Nychka, 1998). The basic idea is to pose the problem as a sequence of one-dimensional optimization problems in which the objective function is optimized over 1 or several coordinates at a time. In the present case, we consider swapping out $\mathbf{x}_j$ for some point in the design space, $\mathcal{X}$, that is nearby $\mathbf{x}_j$ (e.g., a first-order neighbor). Beginning with an initial design, chosen randomly or by some other method, the objective function is evaluated for all possible swaps (4 in the case of first-order neighbors) and whichever point yields the biggest improvement in the criterion under consideration is swapped for the current design point. The algorithm is iterated over all J design points and this continues until convergence is achieved. Such algorithms may yield local optima and optimization for a number of random initial designs can yield incremental improvements. We implemented such a swapping algorithm in **R**, and it is available in the `scrbook` package with the function `SCRdesign`. The algorithm operates on a discrete representation of $\mathcal{S}$ (an arbitrary matrix of coordinates). For each point in the design, **X**, only the nearest neighbors (the number is specified by the user) are considered for swapping into the design during each iteration. For example, to compute `ndesigns = 10` putative optimal designs (each based on a random start) of size $J = 11$, we execute the function as follows:

```
> des <- SCRdesign(S,X,ntraps=11,ndesigns=10,nn=15,sigma=1)
```

where the state-space `S`, and the candidate set `X` are provided as matrices, `nn` is the number of nearest neighbors to inspect for each design point change, and `sigma` is the scale parameter of, in this case, a Gaussian hazard encounter probability model. See the help file `SCRdesign` for examples and analysis of the output.

While swapping algorithms are convenient to implement, and efficient at reducing the criterion in very high-dimensional problems, they do not always yield the global optimum. In practice, as in the examples below, it is advisable to apply the algorithm to a large number of random starting designs (as we showed in the example above). Our experience is that essentially meaningless improvements are realized after searching through a small number of random starting designs.

The design criteria we developed above bear a striking resemblance to design criteria used to construct so-called space-filling designs (Nychka et al., 1997). Such criteria are based on interpoint distances, and space-filling designs seek to optimize some function of distance alone, instead of a variance-based objective function. The benefit of this approach is that one doesn't have to specify the model to produce a design, and space-filling designs have been shown to provide reasonable approximations to designs based on variance criteria under flexible statistical models (Nychka et al., 1997). An interesting fact about model-based SCR design is that the criteria we have outlined above *are* functions of distance alone. This suggests that perhaps other distance-based design criteria might be effective for SCR models.

10.5.6 Illustration

Because the exchange algorithm operates on a discrete version of $\mathcal{S}$, it is trivial to apply to situations in which the state-space is arbitrary in extent and geometry. However, we consider a simplified situation here in order to illustrate the calculation of optimal designs and how they look for an idealized situation.

Consider designing a camera trapping study for a square state-space on $[9, 21] \times [9, 21]$ and with $\mathcal{X}$ being the smaller square $[10, 20] \times [10, 20]$. For this illustration we assumed $\alpha_0 = \log(\lambda_0) = -1.7$ and $\sigma = \sqrt{2}$ so that $\alpha_1 = 1/(2\sigma^2) = 1/4$.

Designs of sizes 11 and 21 were computed using 10 random starting designs. We found the optimal design using each of the four criteria we described above. To refresh your memory Q_1 is the trace of the variance-covariance matrix of $\hat{\boldsymbol{\alpha}}$, Q_2 is the variance of $\hat{N}_c$, Q_3 is $1 - \bar{p}$ (so the design that minimizes this criterion obtains the highest possible $\bar{p}$), and Q_4 is the variance of $\hat{\bar{p}}$. The putative optimal designs (henceforth "best") are shown in Figure 10.3.

The designs are not completely regular but they do have a systematic look to them. For the $J = 11$ designs, the Q_1 design is slightly more compact, with an average closest neighbor distance of 2.59 units vs 3.03 units for the Q_3 design. The two designs are qualitatively similar, providing roughly uniform coverage of the design space $\mathcal{X}$. Conversely, the other two criteria produce designs that are highly clustered. Criterion Q_2, which is optimal for $\hat{N}_c$, produces 2 clusters of traps, 7 traps in one and 4 traps in the other. Finally, designs which are optimal for the criterion Q_4, the variance of estimating $\bar{p}$, produce a single cluster of traps that is roughly centrally located in the design space. This makes sense, because the very dense cluster of traps provides a large number of recaptures near the origin $d = 0$, which, intuitively, provides the

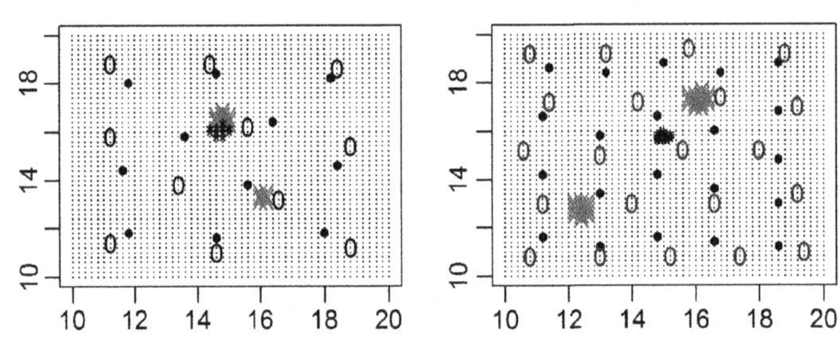

FIGURE 10.3

Best designs for each of 4 design criteria, produced using the exchange algorithm with 15 nearest neighbors and based on 10 random starting designs. The left panel shows the best $J = 11$ designs, and the right panel shows the best $J = 21$ designs. The solid black dots correspond to the best design for Q_1; 0 marks the design for Q_3, tight clusters of "X" for Q_2 and the tightly clustered "*" corresponds to Q_4.

most information about estimating parameter of the encounter probability model. For the $J = 21$ designs, we have an average closest neighbor distance of 1.87 for Q_1 and 2.19 for Q_3 but, qualitatively, the structure of the designs is similar to $J = 11$. The best design for estimating N (minimizing the criterion Q_2) produces 2 clusters, but just with more traps. While these illustrations make sense to us, we're not entirely convinced of the implication that 2 clusters of traps should be optimal with $J = 21$ total traps. However, it is clear that the tight clusters should provide good information about estimating $\bar{p}$ and, by spreading the two clusters out, the expected sample size, n, is maximized.

While the designs for Q_1 and Q_3 are not exactly uniform, they appear very regular, which we should expect given the regularity of both $\mathcal{S}$ and $\mathcal{X}$ in this case. One thing to note is that the trap spacing varies depending on J *even though* σ is the same, so optimal trap spacing should not be viewed as a static thing depending only on the model. Because the designs are not precisely regular, the average closest neighbor distance is not exactly the same as the trap spacing of a regular grid design.

10.5.7 Density covariate models

Many capture-recapture studies will involve one or more landscape or habitat covariates that are thought to affect density D, with the idea of using the methods described in Chapter 11 for modeling and inference. We imagine that it should be possible to extend the model-based framework described previously to accommodate uncertainty due to having to estimate the parameters of the model for density, $\boldsymbol{\beta}$, and this could be included as a feature of the design criterion. In this case, we can think of the captures in a trap being Poisson random variables with mean $\mu(\mathbf{x}, \mathbf{s}) \times D(\mathbf{s})$ and the same arguments as given above can be used to devise design criteria and optimize them. However, in this case we might not only care about estimating N but also (or instead) inference about the parameters $\boldsymbol{\beta}$. Thus, we might choose designs that are good for N or perhaps only good for estimating $\boldsymbol{\beta}$ or perhaps both. Intuitively, we think these two design objectives conflict with one another to some extent. Model-based approaches should favor areas of higher density, for estimating N efficiently, but the design must realize variation in the landscape covariates too (Graves et al., unpublished report).

10.6 Temporal aspects of study design

The spatial configuration of traps is one of the most important aspects of sampling design for capture-recapture studies. Indeed, as we discussed in the previous section, design under SCR models can be thought of as being analogous to classical model-based spatial design, and the concepts and methods from that field can be brought to bear on the design of capture-recapture studies. However, there are other aspects of sampling design that should be considered in capture-recapture studies, including the frequency or length of temporal samples. We touch upon some of these issues here, although without a detailed or formal analysis.

10.6.1 Total sampling duration and population closure

All the models we have discussed so far are *closed population* SCR models, i.e., models that assume that the population under study remains constant during our survey. Traditionally, two different levels of closure have been considered in the capture-recapture literature—demographic and geographic closure (see also Chapter 4). Demographic closure refers to the absence of births and deaths, while geographic closure refers to the absence of immigration and emigration during a study. In traditional capture-recapture, the geographic closure assumption prohibited (in theory, not in field praxis, of course) any movement off the trapping grid. Kendall (1999) explored a range of scenarios of closure violation, focusing on different kinds of movement in and out of the study area, and found that several of these scenarios caused biased abundance estimates from traditional capture-recapture models.

As discussed in Chapter 5, one main objective of SCR models is to relax the geographic closure assumption—the model explicitly allows for movements of individuals about their activity centers, which may have them off the trapping grid at some times, even if the activity center itself is on the grid. SCR models do, however, assume no permanent emigration from, or immigration into, the state-space. The interpretation of demographic closure remains the same in SCR models as it is in traditional CR models.

We have not explored effects of closure violation on SCR abundance and density estimates. Conceptually, we expect estimates to be biased high when births or immigrations happen during a study. For one, the total number of individuals at the study site during the course of the study would be higher than at any particular point in time and correspond to a *cumulative* number of individuals in the study area during the study period. Further, because some individuals are not available for detection for the entire study (they only become available when they are recruited into the state-space) we would expect baseline detection to be underestimated, potentially leading to further positive bias in estimates of abundance. Death or emigrations during a study do not inflate the number of individuals in the state-space, but as animals die and become unavailable for detection, we can again imagine a negative bias in baseline detection and, consequently, some positive bias in N.

To avoid such bias in population estimates, closed population models should typically be applied to short surveys, where short is relative to the life history of the species under study. For example, for small mammals, that might mean a few days, whereas for large, long-lived species with a slow population turnover, several weeks or even a few months can still be considered short enough. In practice, we have no means of guaranteeing a closed population—even if we sample animals for a day, one of the individuals we record may be eaten by a predator later that day, or a dispersing individual may arrive just as we turn our backs. On the other hand, we are faced with the need to collect sufficient data, which, especially for elusive species, pushes us to sample over longer rather than shorter time periods. If we do not have enough sampling devices to cover the entire area of interest at once, study designs based on moving traps (see Section 10.4) can require even longer sampling to accumulate sufficient captures and recaptures. So clearly, in temporal study design we have to strive for a compromise between collecting enough data while still approximating a closed population. For some species we may be able to avoid seasons where violation of demographic closure is particularly likely—for example, migration seasons in migratory birds, or specific breeding seasons (or collective suicide season in lemmings). But for many species such biological seasons might be less clear cut. For example, in warm climates tigers and other large cats can breed year round (Nowak, 1999). As a consequence, guidelines as to what time frame adequately approximates a closed population are generally vague and arm-wavy. Unfortunately, we do not have much more to offer on the subject of how to decide on the length of a study, other than to urge you to think about the biology of your study species *before the study* and choose a time window that seems appropriate for that purpose.

10.6.2 Diagnosing and dealing with lack of closure

Once a field study has been conducted, you may wonder whether the collected data contain any evidence that the closure assumption has been satisfied. Relatively few tests for population closure in traditional capture-recapture have been developed, mostly due to the fact that in the data behavioral variation in detection is indistinguishable from violation of demographic closure (Otis et al., 1978; White et al., 1982). Otis et al., (1978) developed a test for population closure that can handle heterogeneity in detection probability, but does not perform well in the presence of time or behavioral variation in p. Stanley and Burnham (1999) developed a closure test for model M_t (time variation in detection), which works well when there is permanent emigration and a large number of individuals migrate. Both tests are implemented in the program **CloseTest** (Stanley and Richards, 2013).

There are no specific population closure tests for SCR models, for the same reasons that violation of other model assumptions cannot necessarily be distinguished from a lack of population closure. If you are worried that closure might have been violated in your study, one approach of dealing with this problem is to fit an open population model. You can subdivide your study into several periods and fit a spatial version of Pollock's robust design capture-recapture model, which can estimate population size/density for each of these periods (in this context also called primary periods) using models of demographic closure. Alternatively, we may consider fully dynamic models, which contain explicit parameters of survival and recruitment (Chapter 16). These models can be quite computationally expensive, and if you wanted a faster, partial check you could alternatively fit a spatial Cormack-Jolly-Seber model, which only estimates survival. The magnitude of the survival estimates gives you some partial information about population closure in your study—if survival is close to 1 there is little evidence of losses of individuals, either through permanent emigration or death. These and other open population models are presented in Chapter 16. Finally, if your data are too sparse to fit a full-blown open population model, you can subdivide your study into $t = 1, 2, \ldots, T$ primary periods and estimate abundance separately for each period's data, possibly sharing the detection parameters across periods, if you can safely assume they remain constant. You can do that by either letting N_t be independent from each other, or by specifying an underlying distribution for all N_t in a multi-session framework as described in Chapter 14.

10.7 Summary and outlook

Design of capture-recapture studies in the context of *spatial* models is an important problem, but solutions to this problem are mostly *ad hoc* or incomplete at the present time. As a general rule, we always recommend scenario analysis by Monte Carlo simulation (Efford and Fewster, 2012; Sollmann et al., 2012; Sun, 2013). This takes a lot of time but it guarantees forward progress, or at least not choosing the dumbest from among several design options. We discussed some examples from the literature that assess trap spacing and evaluate trap clustering and rotating coverage strategies for sampling large areas. The nice thing about simulation studies is that we can generate data for any complex situation we desire, even if we can't fit the model effectively. Thus, we can always characterize worst-case situations under pathological model misspecifications and understand the implications such departures have for parameters estimation.

When designing a spatial capture-recapture study for a single species, trap spacing and the size of the array can (and should) be tailored to the spatial behavior of that species to ensure adequate data collection. However, some trapping devices like camera traps may collect data on more than one species and researchers may want to analyze these data, too. The flexibility in study design of SCR models makes such multi-species approaches more feasible. Independent of the trapping device used, study design will in most cases face a limit in terms of the number of traps available or logistically manageable. As a consequence, researchers need to find the best compromise between trap spacing and the overall grid area.

Particularly for large mammal research, SCR models have much more realistic requirements in terms of area coverage than non-spatial CR models. In the latter, density estimates can be largely inflated with small trapping grids relative to individual movement (Maffei and Noss, 2008)—covering at least four times the average home range is recommended. Further, we need consistent coverage of the entire study area, as all individuals in the population of interest must have some probability >0 to be captured. In contrast, SCR models work well in study areas that are similar in size to an individual's home range (as long as sufficient data is collected) (Sollmann et al., 2012; Marques et al., 2011), and they provide unbiased estimates for sampling designs that do not expose all individuals in the sampled population to detectors, i.e., that have "holes" (Efford and Fewster, 2012). These results, however, should not encourage researchers to design non-invasive trap arrays based on minimum area requirements and with a minimum number of detectors. Study design should still strive to expose as many individuals as possible to sampling and obtain adequate data on individual movement. Large amounts of data, both individuals and recaptures, do not only improve precision of parameter estimates (Sollmann et al., 2012; Efford et al., 2004), they also allow including potentially important covariates (such as sex or time effects in the black bear example—see also Chapter 7) into SCR models.

Beyond the traditional grid-based sampling design, the flexibility of SCR models allows for different spatial detector arrangements such as dispersed clusters of detectors. How well these different designs perform, comparatively, remains to be explored.

An alternative to simulation-based scenario analysis for constructing spatial designs is a formal model-based strategy in which we seek the configuration of design points (trap locations) $\mathbf{x}_1, \ldots, \mathbf{x}_J$ that is optimal for some formal information-based objective function. This is a standard approach in classical sampling and experimentation, yet it has not gained widespread use in ecology. In our view, model-based design under SCR models has great potential due to its coherent formulation and flexibility. On this topic, we have just barely scratched the surface here, showing how to formulate a criterion that is a function of the design, and then optimizing the criterion over the design space. Our cursory analysis of model-based design in a single situation did reveal an important aspect of design that has not been discussed in the literature. That is, the optimal spacing of traps in an array depends on the *density* of traps in the state-space. In our analysis, the spacing of 11 and 21 trap optimal designs was quite different. Therefore, this should be considered in practical SCR design exercises.

Conceptually, the information in SCR studies comes in two parts: Recaptures of individuals at different traps (spatial recaptures) and the total sample size of individuals. Maximizing both of these things as objectives induces an explicit trade-off in the construction of capture-recapture designs. We need designs that are good for estimating $\bar{p}$ and also designs that obtain a high sample size, n. Designs that are extremely good only for one or the other will are probably bad SCR designs—estimators of density with low precision—or designs in which N is not estimable due to a lack of spatial recaptures. One possible exception is when telemetry data (or other auxiliary data) are available. In Chapter 13 we discuss SCR models that integrate auxiliary information on resource selection obtained by telemetry. Telemetry data are directly informative about the coefficient of the distance term (σ or α_1) which, in fact, can be estimated from telemetry data alone. It stands to reason that, when telemetry data are available, this should affect considerations related to trap spacing. Conceivably even, one might be able to build SCR designs that don't yield any formal spatial recaptures because all of the information about σ is provided by the telemetry data. We have done limited evaluations of the trap spacing problem in the presence of telemetry data, and the results suggest that, while efficient designs that involve telemetry data have a larger trap spacing than without telemetry data, the realization of some spatial recaptures is important even when telemetry data are available. With the **R** code we provide in Chapter 13, you should be able to carry out your own custom evaluation of these types of design problems.

PART

Advanced SCR Models

CHAPTER

Modeling Spatial Variation in Density

11

Underlying every SCR model is a spatial point process that describes the number and distribution of animal activity centers. Spatial point processes are characterized by two key elements: a spatial domain (or state-space), $\mathcal{S}$, and an intensity function which returns the expected density of points at any location in $\mathcal{S}$. If the intensity is constant throughout $\mathcal{S}$, the point process is said to be homogeneous. Thus far we have focused our attention on homogeneous point processes whose realized values are the locations of the N activity centers. When a Poisson prior is placed on N, the model is known as a homogeneous Poisson point process, which is the classic model of "complete spatial randomness." A similar model, that we often use in conjunction with data augmentation and MCMC, places a binomial prior on N. This is also a model of spatial randomness, and in this chapter we will compare and contrast the two.

The spatial randomness assumption is often viewed as restrictive because ecological processes such as habitat selection can result in non-uniform distributions of organisms. We have argued, however, that this assumption is less restrictive than may be recognized because a homogeneous point process actually allows for infinite possible "point patterns," or realized configurations of activity centers. Furthermore, given enough data, the uniform prior will have very little influence on the estimated locations of activity centers. Nonetheless, a homogeneous point process does not allow one to model population density using covariates, which is an important objective in much ecological research. For example, even when assuming a homogeneous point process for the activity centers, an estimated density surface may strongly suggest that individuals are more abundant in one habitat than another; however, such results do not provide the basis for formally testing hypotheses about spatial variation in density, and they could not be used to make predictions about habitat-specific abundance in other regions. A more direct approach is to replace the homogeneous model with an inhomogeneous model in which the point process intensity is allowed to vary spatially.

In this chapter, we cover methods for fitting inhomogeneous Poisson and binomial point process models so that density can be modeled as a function of covariates in much the same way as is done in generalized linear models. The covariates we consider differ from those covered in previous chapters, which were typically attributes of the animal (e.g., sex or age) or the trap (e.g., baited or not) and were used to model

Spatial Capture-Recapture. http://dx.doi.org/10.1016/B978-0-12-405939-9.00011-6

movement or encounter rate. In contrast, here we wish to model covariates that are defined at all points in $\mathcal{S}$, and so we will refer to them as state-space covariates or density covariates. These may include continuous covariates such as elevation, or categorical covariates such as habitat type. Typically, these state-space covariates are formatted as raster images with a prescribed resolution and extent.

One thing to keep in mind when modeling density is that the SCR definition of density is different than what is perhaps a more common definition of density in ecology. In SCR models, density is defined as the number (or expected number) of *activity centers* in some region, whereas in other ecological studies, density is often defined as the number of *individuals* in some region at some instant in time. The latter definition is closer to the quantity being estimated in distance sampling studies. So which definition is better? Does it make more sense to contemplate activity centers or individuals at an instant in time? From our perspective, either definition may suffice for a given objective, but we note that there exists a formal relationship between the two since an activity center is the *average* of an individual's locations during some time period. As such, an activity center may be a better descriptor of an individual's preferences than is a location during a single instant in time. Moreover, with SCR models we can model both the distribution of activity centers (as we will do in this chapter) as well as the distribution of individuals during specific instances in time, as is demonstrated in Chapter 15.

Inhomogeneous Poisson point process models were discussed in the original formulation of SCR models by Efford (2004) and were described in more detail by Borchers and Efford (2008). We will show that an inhomogeneous point process with a binomial prior on N is quite similar to the Poisson model, but is more easily implemented in MCMC algorithms. To do so, we will define the data augmentation parameter ψ in terms of the point process intensity function, and we will replace the uniform prior on the activity centers with a prior that is also derived from the intensity function. Development of this prior, which does not have a standard form, is a central component of this chapter. First we begin with a review of homogeneous point process models.

11.1 Homogeneous point process revisited

The homogeneous Poisson point process is *the* model of complete spatial randomness and is often used in ecology as a null model to test for departures from randomness (Cressie, 1991; Diggle, 2003; Illian et al., 2008). The Poisson model asserts that the number of points in $\mathcal{S}$ is Poisson distributed: $N \sim \text{Poisson}(\mu||\mathcal{S}||)$ where $\mu > 0$ is the intensity parameter and $||\mathcal{S}||$ is the area of the state-space. The intensity parameter μ is the density of points, and thus multiplying the intensity by the area of some region yields the expected number of points in that region. As with all homogeneous point process models, the N points are distributed uniformly, which implies that they do not interact with each other in any way—for example, they neither attract nor repel one another.

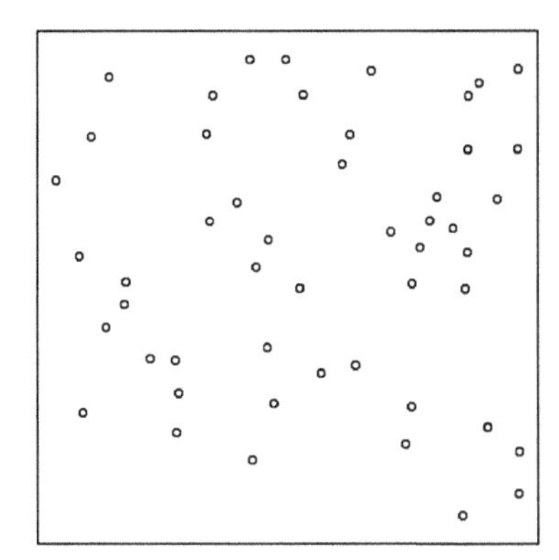

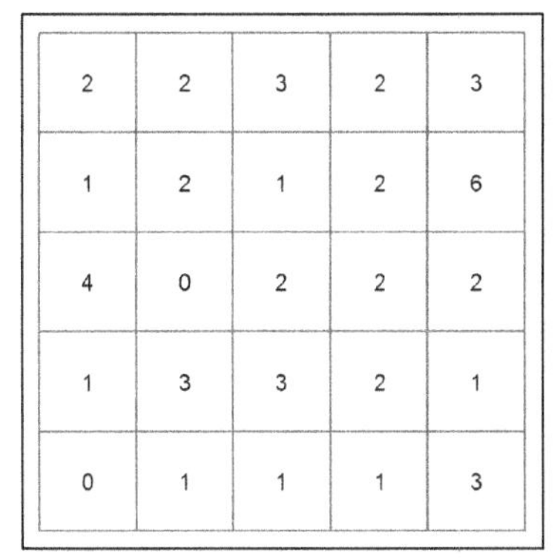

FIGURE 11.1

Homogeneous binomial point process with $N = 50$ points represented in continuous and discrete space.

Unlike the Poisson point process, the binomial point process assumes that N is fixed, not random. The distinction is illustrated by this simple **R** code that generates realizations from Poisson and binomial point processes in the unit square ($\mathcal{S} = [0, 1] \times [0, 1]$):

```
> Area <- 1                              # Area of unit square
> muP <- 4                               # intensity
> nP <- rpois(1, muP*Area)               # number of points: random
> PPP <- cbind(runif(nP), runif(nP)) # Poisson point pattern
> nB <- 4                                # number of points: fixed
> muB <- nB/Area                         # intensity
> BPP <- cbind(runif(nB), runif(nB)) # binomial point pattern
```

Both of these models are homogeneous because the intensity parameter is constant ($\mu = 4$ in both cases) and the locations of the N points are mutually independent and uniformly distributed. The key distinction is that N is random in the former and fixed in the latter.

Another difference between the Poisson and binomial models is that if the state-space is divided into K disjunct regions, the number of points in each region $n(\mathcal{B}_k)$: $k = 1, \ldots, K$; is independent and identically distributed (*iid*) under the Poisson model but not under the binomial model. In the Poisson case, the counts are $n(\mathcal{B}_k) \sim \text{Poisson}(\mu||\mathcal{B}_k||)$, where $||\mathcal{B}_k||$ is the area of the region $\mathcal{B}_k$. For the binomial model, $n(\mathcal{B}_k) \sim \text{Binomial}(N, \pi(\mathcal{B}_k))$ where $\pi(\mathcal{B}_k)$ is the proportion of the state-space in $\mathcal{B}_k$; however, these counts are not *iid* because the number of points in one region is informative about the number of points in another region. For example, if $N = 10$ and if there are seven points outside the region $\mathcal{B}_1$, then we can say with certainty that $\mathcal{B}_1 = 10 - 7 = 3$.

Figure 11.1 is meant to further illustrate the characteristics of the binomial model. The left panel shows a point pattern realized from a homogeneous binomial point process with $N = 50$. The right panel shows the same realization, except that the state-space has been discretized into 25 equally sized disjunct regions, or pixels, and the counts in each pixel are shown. Since the pixels are of the same size, we have that $\pi(\mathcal{B}_k) = 1/25$, and the expected number of points in each pixel is $\mathbb{E}(n(\mathcal{B}_k)) = N\pi(\mathcal{B}_k) = 50/25 = 2$, which happens to be the empirical mean in this instance. However, as previously stated, these counts are not independent realizations from a binomial distribution since $\sum_k n(\mathcal{B}_k) = N$. Rather, the model for the entire vector is multinomial: $\{n(\mathcal{B}_1), n(\mathcal{B}_2), \ldots, n(\mathcal{B}_k)\} \sim \text{Multinomial}(N, \{p(\mathcal{B}_1), p(\mathcal{B}_2), \ldots, p(\mathcal{B}_K)\})$ (Illian et al., 2008). If you need a refresher on the multinomial distribution, refer to Section 2.2.3, and consider the following **R** code, which generates counts similar to those seen in Figure 11.1:

```
> n.Bk <- rmultinom(1, size=50, prob=rep(1/25, 25))
> matrix(n.Bk, 5, 5)

     [,1] [,2] [,3] [,4] [,5]
[1,]    2    2    2    2    1
[2,]    2    4    0    5    0
[3,]    0    3    2    4    1
[4,]    1    2    1    4    1
[5,]    4    2    4    1    0
```

The dependence among counts has virtually no practical consequence when the number of pixels is large. For example, if there are 100 pixels, the number of points in one pixels carries very little information about the expected number of points in another pixel. However, if there are only two pixels, then clearly the number of points in one pixel allows one to determine how many points are in the other pixel.

The discrete representation of space shown in Figure 11.1 is not only helpful for understanding the properties of a point process, it is also of practical importance when fitting SCR models because spatial covariates are almost always represented as rasters, i.e., grids with predetermined extent and resolution. In such cases, the definition of the prior for the point locations can be changed from the probability that a point occurs at some location in space to the probability that it occurs in some pixel of the raster. As we will explain in Section 11.4.2, this typically involves changing the prior from a uniform distribution to a multinomial or categorical distribution.

Having sketched out the basic characteristics of homogeneous Poisson and binomial point process models, we will now review their relevance to SCR models before moving on to the inhomogeneous models. In an SCR model with a homogeneous point process, the intensity parameter μ is interpreted as population density, and N is interpreted as population size (i.e., the number of activity centers in $\mathcal{S}$). These interpretations are true regardless of whether we consider the Poisson model or the binomial model, but since N is always unknown, one might wonder why we are discussing the binomial model at all.

In our work, we typically adopt the binomial model simply because it is easy to implement using MCMC and data augmentation. And while N is truly unknown, we use an upper bound, M, which is fixed. Thus, the standard point process we use in

Bayesian analyses can be regarded in two ways. First, it is a binomial point process with M points. Second, in terms of N, it is a thinned binomial point process, where ψ is the thinning parameter. With this in mind, the only real difference between the Poisson and binomial models, as implemented in SCR contexts, is that in the former, we have $N \sim \text{Poisson}(\mu||\mathcal{S}||)$, and in the latter we have $N \sim \text{Binomial}(M, \psi)$. In other words, we just have a different prior on N, and when using MCMC, the binomial prior is much more convenient because it fixes the size of the parameter space and makes it easy to extend the model in each of the ways discussed in this book. It is also worth remembering that the Poisson distribution is the limit of the binomial distribution when M is very large and ψ is very small (Chapter 2), and thus the two models are much more similar than they may appear.

You might have noticed that the intensity parameter μ was not shown for the binomial prior $N \sim \text{Binomial}(M, \psi)$. Instead, we see the data augmentation parameter ψ, which has been used throughout this book, but without much mention of the point process intensity. What then is the relationship between ψ and μ? As first discussed in Chapter 4, under data augmentation, the expected value of N is $\mathbb{E}[N] = M\psi$. But, from this chapter, we also know that the expected value of N can be written in terms of μ as $\mathbb{E}[N] = \mu||\mathcal{S}||$. Therefore, $\psi = \mu||\mathcal{S}||/M$ and hence we can directly estimate μ rather than ψ if we so choose. This will be demonstrated in the next section, where the objective is to model μ as a function of spatially referenced covariates. First, consider the following **R** code, which illustrates some of the concepts we just covered:

```
> Area <- 1                    # Area of state-space
> M <- 100                     # Data augmentation size
> mu <- 10                     # Intensity(points per area)
> psi <- (mu*Area)/M           # Data augmentation parameter(thinning rate)
> N <- rbinom(1, M, psi)       # Realized value of N under binomial prior
> cbind(runif(N), runif(N)) # Point pattern from thinned binomial model

             [,1]        [,2]
[1,]   0.52779588  0.84306878
[2,]   0.11529168  0.80635046
[3,]   0.06777632  0.66072116
[4,]   0.18694649  0.56761245
[5,]   0.30176929  0.03159091
[6,]   0.84352724  0.89691452
[7,]   0.52766808  0.08871199
[8,]   0.73007529  0.63184825
[9,]   0.01119023  0.69807029
```

11.2 Inhomogeneous point processes

The principal difference between homogeneous and inhomogeneous point processes is that the intensity parameter μ is allowed to vary spatially in the inhomogeneous model. Thus, rather than μ being a fixed constant, it is now a function defined at each point $\mathbf{s} \in \mathcal{S}$. A vast number of options exist for modeling spatial variation in the intensity of a point process (Cox, 1955; Stoyan and Penttinen, 2000; Illian et al., 2008), but here we focus on modeling μ as a function of spatially referenced covariates and a vector of regression coefficients $\boldsymbol{\beta}$; a function we will denote $\mu(\mathbf{s}, \boldsymbol{\beta})$. To be clear, $\mu(\mathbf{s}, \boldsymbol{\beta})$, is a function that returns the expected density of activity centers at location $\mathbf{s}$, given the covariate values at $\mathbf{s}$.[1] Since the intensity must be positive, and because the natural logarithm is the canonical link function of the Poisson generalized linear model (McCullagh and Nelder, 1989), it is natural to consider the following model:

$$\log(\mu(\mathbf{s}, \boldsymbol{\beta})) = \beta_0 + \sum_{v=1}^{V} \beta_v C_v(\mathbf{s}), \tag{11.2.1}$$

which states that there are V covariates and β_v is the regression coefficient for covariate $C_v(\mathbf{s})$. This covariate, $C_v(\mathbf{s})$, could be any variable defined at all points in the state-space, such as habitat type or elevation. Equation (11.2.1) should look familiar because it is the standard linear predictor used in Poisson regression. As with other GLMs, one could consider alternative link functions.

Recall from the previous section that for a homogeneous point process, the expected number of points in the state-space was simply the intensity parameter multiplied by area: $\mathbb{E}(N) = \mu||\mathcal{S}||$. But now that we are regarding the intensity as a function, rather than a scalar, this equation is not very useful. So what is $\mathbb{E}(N)$ for an inhomogeneous point process? Contemplating a discrete state-space is useful for figuring this out. Imagine that the state-space is represented as a raster with many tiny pixels. In this case, we will associate $\mathbf{s}$ with pixel ID, i.e., $\mathbf{s}$ just references some pixel with V covariate values associated with it. The expected number of individuals in this pixel, say $\mathbb{E}(n(\mathbf{s}))$, can intuitively be found by evaluating the intensity function (Eq. (11.2.1)) at $\mathbf{x}$ and multiplying it by the area of the pixel. In other words, we compute the expected number of individuals in a pixel by multiplying the expected value of density for that pixel by the area of the pixel. If we do this for each pixel in the state-space, then summing up these values gives us what we are after, the expected value of N. Specifically, $\mathbb{E}(N) = \sum_{\mathbf{s} \in \mathcal{S}} \mathbb{E}(n(\mathbf{x}))$. As the area of the pixels approaches zero, such that we move from discrete space back to continuous space, the summation is replaced with an integration of the form:

$$\mathbb{E}(N) = \int_{\mathcal{S}} \mu(\mathbf{s}, \boldsymbol{\beta}) d\mathbf{s}. \tag{11.2.2}$$

Together, Eqs. (11.2.1) and (11.2.2) describe a model for spatial variation in density as well as population size. The key task in fitting such inhomogeneous point process models is to estimate the $\boldsymbol{\beta}$ parameters.

We have now described an approach for modeling the point process intensity, yet in order to define the likelihood or to develop an MCMC algorithm for the inhomogeneous model, we need to specify the prior distribution for the activity centers. Recall that under the homogeneous point process, the prior was $\mathbf{s}_i \sim \text{Uniform}(\mathcal{S})$, for $i = 1, \ldots, N$, or equivalently:

$$[\mathbf{s}_i] = 1/||\mathcal{S}||, \tag{11.2.3}$$

[1] The use of $\mathbf{x}$ to denote any point in the state-space could cause confusion because we use $\mathbf{x}_j$ as the location of a trap, but it is standard notation, and the distinction should be made evident by the context.

where, as before, $||\mathcal{S}||$ is the area of the state-space. This simply indicates that an activity center is just as likely to occur at any location as another. However, if animals exhibit habitat selection or simply occur in one region more often than another, it would be preferable to replace this prior with one describing the spatial variation in density. Clearly this prior should be determined in some way by the spatially varying intensity function $\mu(\mathbf{s}, \boldsymbol{\beta})$. Since the integral of a probability density function (pdf) must be unity, we can convert $\mu(\mathbf{s}, \boldsymbol{\beta})$ into a pdf by dividing it by a normalizing constant. In this case, the normalizing constant is found by integrating $\mu(\mathbf{s}, \boldsymbol{\beta})$ over the entire state-space. The probability density function of the new prior is therefore:

$$[\mathbf{s}_i|\boldsymbol{\beta}] = \frac{\mu(\mathbf{s}_i, \boldsymbol{\beta})}{\int_{\mathcal{S}} \mu(\mathbf{s}, \boldsymbol{\beta})d\mathbf{s}}. \tag{11.2.4}$$

Substituting the uniform prior with this new distribution allows us to fit inhomogeneous binomial point process models to spatial capture-recapture data.

As a practical matter, note that the integral in the denominator of Eq. (11.2.4) is evaluated over space, and since we always regard space as two-dimensional (the state-space is planar), this is a two-dimensional integral that can be approximated using the methods discussed in Chapter 9, which include Monte Carlo integration and Gaussian quadrature. Alternatively, if our state-space covariates are in raster format, i.e., they are in discrete space, the integral can be replaced with a summation over all the pixels in the raster,

$$[\mathbf{s}_i|\boldsymbol{\beta}] = \frac{\mu(\mathbf{s}_i, \boldsymbol{\beta})}{\sum_{\mathbf{s}\in\mathcal{S}} \mu(\mathbf{s}, \boldsymbol{\beta})}, \tag{11.2.5}$$

where $\mathbf{s}_i$ is now defined as "pixel ID" rather than a point in space.

Although the discrete space approach is standard practice, it is technically unjustified because covariate values must be known for all points in space, and a raster is simply a set of spatially referenced covariate values at an evenly spaced subset of points (the pixel centers). This same problem is present anytime that we have a sample of the spatial covariates, rather than a function defining their value for all points in space. In such cases, it may be necessary to interpolate the values of the covariates for points in space where they were not measured. One option would be to use a Kriging interpolator, as demonstrated by Rathbun (1996). Another option is to sample the spatial covariates using probabilistic sampling methods, which allow for design-based estimators of their values for the entire study area (Rathbun et al., 2007). Either option could be implemented within maximum likelihood or MCMC estimation methods; however, we do not demonstrate them here because it seems likely that they will be inconsequential in most cases where the raster data are of high resolution, such that the loss of information is negligible when going from continuous space to discrete space. Furthermore, the validity of this assertion, and the level of resolution required to adequately approximate continuous space can often be assessed by checking the consistency of the parameter estimates among varying levels of resolution, as was demonstrated in Chapter 5.

We now have all the tools needed to fit inhomogeneous point process models. Likelihood-based inference for inhomogeneous Poisson point process models was described by Borchers and Efford (2008). Another example is demonstrated in the next section, but first we focus on the binomial model, which we favor when conducting Bayesian inference. In the previous section we noted that the data augmentation parameter ψ can be expressed in terms of the intensity parameter μ. The same is true for inhomogeneous models. Specifically, rather than $\mathbb{E}(N) = \psi M$ as before, we use the expected value of N shown in Eq. (11.2.2), which results in

$$\psi = \frac{\int_{\mathcal{S}} \mu(\mathbf{s}, \boldsymbol{\beta}) d\mathbf{s}}{M}. \tag{11.2.6}$$

Note that the data augmentation limit M must be high enough so that it is greater than the numerator—i.e., the expected value of N must be less than M.

In the next sections we walk through a few examples, building up from the simplest case where we actually observe the activity centers as though they were data. In the second example, we fit the inhomogeneous model to simulated data in which density is a function of a single continuous covariate. The next example shows an analysis in discrete space using both `secr` (Efford, 2011a) and **JAGS** (Plummer, 2003), and in the final example, we model the intensity of activity centers for a real data set collected on jaguars (*Panthera onca*) in Argentina.

11.3 Observed point processes

In SCR models, the points (activity centers) are not directly observed, but in other contexts they are. Examples include the locations of disease outbreaks, the locations of trees in a forest, and the locations of radio-tracked animals. In such cases, it is straightforward to fit inhomogeneous point process models and estimate the parameters $\boldsymbol{\beta}$ from Eq. (11.2.1), as we will do in the following example.

Suppose we knew the locations of N animal activity centers, perhaps as the result of an extensive telemetry study. If we assume N is Poisson distributed and the points are mutually independent of one another, we can fit the inhomogeneous Poisson point process model. The likelihood of this model has two components: $[\{\mathbf{s}_1, \ldots, \mathbf{s}_N\}|N]$ and $[N]$. The pdf of the first part is given by Eq. (11.2.4), and with the Poisson assumption we have:

$$\begin{aligned}
\mathcal{L}(\boldsymbol{\beta}|\{\mathbf{s}_1, \ldots, \mathbf{s}_N\}) &= [\{\mathbf{s}_1, \ldots, \mathbf{s}_N\}|N][N] \\
&= \left\{\prod_{i=1}^{N} \frac{\mu(\mathbf{s}_i, \boldsymbol{\beta})}{\int_{\mathcal{S}} \mu(\mathbf{s}, \boldsymbol{\beta}) d\mathbf{s}}\right\} \frac{e^{-\int_{\mathcal{S}} \mu(\mathbf{s}, \boldsymbol{\beta}) d\mathbf{s}} \int_{\mathcal{S}} \mu(\mathbf{s}, \boldsymbol{\beta}) d\mathbf{s}^N}{N!}.
\end{aligned}$$

This can be simplified by noting that the denominator in the first component of the model cancels with the corresponding piece in the numerator of the second component. And, since N is observed and thus does not depend on the parameters, $N!$

can be omitted as well. After log-transforming the remaining pieces, we have the log-likelihood often seen in textbooks, such as Diggle (2003, p. 104):

$$\ell(\boldsymbol{\beta}|\{\mathbf{s}_i\}) = \sum_{i=1}^{N} \log(\mu(\mathbf{s}_i, \boldsymbol{\beta})) - \int_{\mathcal{S}} \mu(\mathbf{s}, \boldsymbol{\beta})d\mathbf{s}.$$

Having arrived at the likelihood we could choose a prior distribution for $\boldsymbol{\beta}$ and obtain the posterior distribution using Bayesian methods, or we can find the maximum likelihood estimates (MLEs) using standard numerical methods as is demonstrated below.

First, we simulate some data under the model $\mu(\mathbf{s}, \boldsymbol{\beta}) = \exp(\beta_0 + \beta_1 \text{ELEV}(\mathbf{s}))$, where ELEV($\mathbf{s}$) is a spatial covariate, say elevation, and $\beta_0 = -6$ and $\beta_1 = 1$. It is worth emphasizing that a spatial covariate must be defined at any location in the state-space, as is true of the following covariate `elev.fn`:

```
> elev.fn <- function(s) {          # spatial covriate
+    s <- matrix(s, ncol=2)         # Force s to be a matrix
+    (s[,1] + s[,2] - 100)/ 40.8 # Return(standardized) "elevation"
+ }
> # intensity function
> mu <- function(s, beta0, beta1) exp(beta0 + beta1*elev.fn(s=s))
> beta0 <- -6 # intercept of intensity function
> beta1 <- 1 # effect of elevation on intensity
> # Next line computes integral
> EN <- cuhre(2, 1, mu, beta0=beta0, beta1=beta1,
+            lower=c(0,0), upper=c(100,100))$value
```

The function `elev.fn` returns the value of elevation at any location `s`. The standardization bit is not necessary, but helps with the model fitting below. The next lines of the code define the intensity function $\mu(\mathbf{s}, \boldsymbol{\beta})$ in terms of elevation and the regression coefficients. The last line uses the `cuhre` function in the `R2Cuba` package (Hahn et al., 2010) to compute the expected value of N in a $[0, 100] \times [0, 100]$ square state-space, which is the two-dimensional integral of Eq. (11.2.4). This integral could also be computed using a fine grid of points as we have done in previous chapters, but it is useful to gain familiarity with more efficient integration functions in **R**.

The **R** code above demonstrates how to obtain the expected value of N given a spatial covariate and the coefficients defining the intensity function. Now we need to generate a realized value of N and distribute the N points in proportion to the intensity function. This is not as simple as it was to simulate data from a homogeneous point process because the points are no longer uniformly distributed within the state-space. Instead, one must resort to methods such as rejection sampling, which involves simulating data from a standard distribution and then accepting or rejecting each point using probabilities defined by the distribution of interest. For more information, on rejection sampling readers should consult an accessible text such as Robert and Casella (2010). In our example, we simulate from a uniform distribution and then accept or reject using the (scaled) probability density function $[\mathbf{s}_i|\boldsymbol{\beta}]$ (Eq. (11.2.4)). The following **R** commands demonstrate the use of rejection sampling to simulate an inhomogeneous point process for the elevation covariate depicted in Figure 11.2.

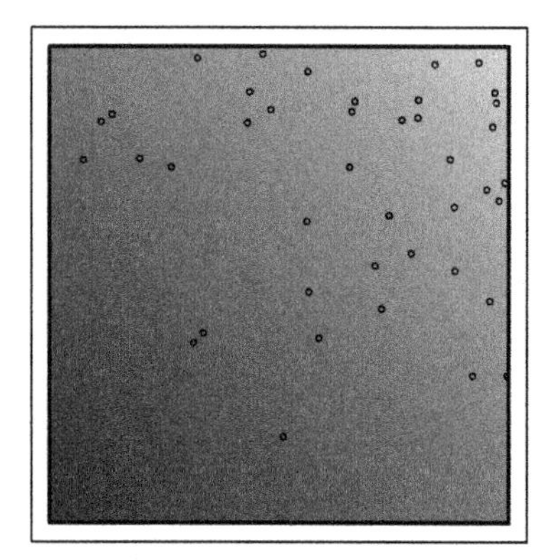

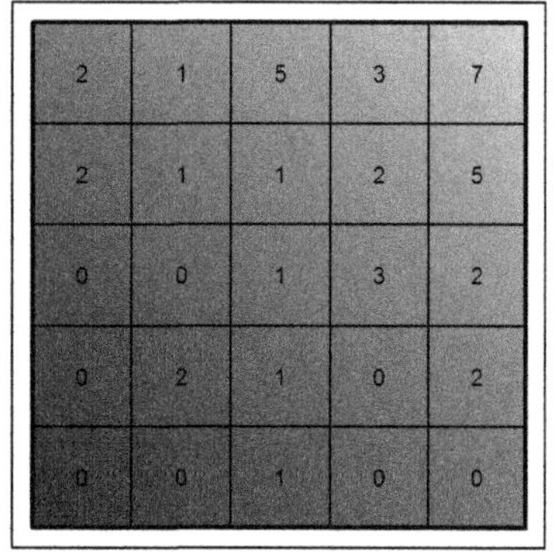

FIGURE 11.2

An example of a spatial covariate, say elevation, and a realization from an inhomogeneous Poisson point process with $\mu(\mathbf{s}, \boldsymbol{\beta}) = \exp(\beta_0 + \beta_1 \text{ELEV}(\mathbf{s}))$, where $\beta_0 = -6$ and $\beta_1 = 1$.

```
> set.seed(31025)
> beta0 <- -6 # intercept of intensity function
> beta1 <- 1 # effect of elevation on intensity
> # Next line computes integral, which is expected value of N
> EN <- cuhre(2, 1, mu, beta0=beta0, beta1=beta1,
+          lower=c(0,0), upper=c(100,100))$value
> EN
[1] 39.96634
> N <- rpois(1, EN) # Realized N
> s <- matrix(NA, N, 2) # This matrix will hold the coordinates
> elev.min <- elev.fn(c(0,0))
> elev.max <- elev.fn(c (100, 100))
> Q <- max(c(exp (beta0 + beta1*elev.min),
+         exp(beta0 + beta1*elev.max)))
> counter <- 1
> while(counter <= N) {
+    x.c <- runif(1, 0, 100); y.c <- runif(1, 0, 100)
+    s.cand <- c(x.c,y.c)
+    pr <- mu(s.cand, beta0, beta1) #/ EN
+    if(runif(1) < pr/Q) {
+      s[counter,] <- s.cand
+      counter <- counter+1
+      }
+    }
```

Similar methods are also implemented in the **R** package `spatstat` (Baddeley and Turner, 2005).

The 41 simulated points are shown in Figure 11.2. High elevations are represented by light gray and low elevations are darker. The density of points is apparently higher in lighter regions, suggesting that these simulated animals prefer high elevations. Given these points, we will now estimate β_0 and β_1 by minimizing the negative log-likelihood using **R**'s `optim` function.

```
> nll <- function(beta) {
+     beta0 <- beta[1]
+     beta1 <- beta[2]
+     EN <- cuhre(2, 1, mu, beta0=beta0, beta1=beta1,
+             lower=c(0,0), upper=c(100,100))$value
+     -(sum(beta0 + beta1*elev.fn(s)) - EN)
+ }
> starting.values <- c(-10, 0)
> fm <- optim(starting.values, nll, hessian=TRUE)
> cbind(Est=fm$par, SE=sqrt(diag(solve(fm$hessian)))) # estimates and SEs

            Est        SE
[1,] -5.9335547 0.2204693
[2,]  0.9545532 0.1771507
```

Maximizing the Poisson likelihood took a fraction of a second, and we obtained estimates of $\hat{\beta}_0 = -5.93$ and $\hat{\beta}_1 = 0.95$, which are very close to the data-generating values. The 95% confidence interval for $\hat{\beta}_1$ is [0.61, 1.3] and since it does not include zero, the null hypothesis that $\beta_1 = 0$, i.e., that there is no effect of elevation on density, can be rejected. In addition to testing hypotheses, these results can be used to predict population size in unsampled regions or create predicted density surface maps by plugging the parameter estimates into Eqs. (11.2.1) and (11.2.2).

You might wonder if the results would differ if we assumed a binomial rather than a Poisson distribution for N. This can be checked using the following code:

```
> nllBin <- function(beta, M=100) {
+    beta0 <- beta[1]
+    beta1 <- beta[2]
+    EN <- cuhre(2, 1, mu, beta0=beta0, beta1=beta1,
+            flags=list(verbose=0),
+            lower=c(0,0), upper=c(100,100))$value
+    N <- nrow(s)
+    psi <- EN/M
+    -(sum(beta0 +beta1*elev.fn(s) - log(EN)) +
+      dbinom(N, M, psi, log=TRUE))
+ }
> cbind(Est=fmBin$par, SE=sqrt(diag(solve(fmBin$hessian)))) # est and SE
            Est        SE
[1,] -5.9339490 0.1965479
[2,]  0.9545742 0.1771962
```

which indicates that the MLEs are almost identical, and supports the claim that the prior on N has little influence in SCR models. Notice, however, that the standard error for β_0 is smaller under the binomial model than it was under the Poisson model—a difference that will dissipate as M tends toward infinity.

This example demonstrates that if we had the data we wish we had, i.e., if we knew the coordinates of the activity centers, we could easily estimate the parameters governing the underlying point process and make inferences about spatial variation in density and abundance. Unfortunately, in many animal ecology studies, the locations of the N animals, or the N activity centers, cannot be directly observed. Thus, we need extra information to estimate the locations of these unobserved points, which in the case of SCR, comes from the locations where each animal is captured.

11.4 Fitting inhomogeneous point process SCR models

11.4.1 Continuous space

In this example, we will use the same set of points simulated in the previous section to generate spatial capture-recapture data. Specifically, we overlay a grid of 49 traps on the map shown in Figure 11.2 and simulate capture histories conditional on the activity centers. Then, we will attempt to estimate the activity center locations as though we did not know where they were, as is the case in real applications. We will also estimate β_0 and β_1 as before and see how the estimates compare when the points are not actually observed. The following **R** code simulates encounter histories under a Poisson observation model (see Chapter 9), which would be appropriate in camera trapping studies or when using other methods in which animals could be detected multiple times at a trap during a single occasion.

```
> xsp <- seq(20, 80, by=10); len <- length(xsp)
> X <- cbind(rep(xsp, each=len), rep(xsp, times=len)) # traps
> ntraps <- nrow(X); noccasions <- 5
> y <- array(NA, c(N, ntraps, noccasions)) # capture data
> sigma <- 5 # scale parameter
> lam0 <- 1 # basal encounter rate
> lam <- matrix(NA, N, ntraps)
> set.seed(5588)
> for (i in 1:N) {
+     for (j in 1:ntraps) {
+        # The object "s" was simulated in previous section
+        distSq <- (s[i,1]-X[j,1])^2 + (s[i,2] - X[j,2])^2
+        lam[i,j] <- exp(-distSq/(2*sigma^2)) * lam0
+        y[i,j,] <- rpois(noccasions, lam[i,j])
+     }
+ }
```

Now that we have a simulated capture-recapture data set `y`, we can simulate the posterior distributions of the model parameters using MCMC. A commented Gibbs sampler written in **R** is available in the accompanying **R** package `scrbook` (see `?scrIPP`). This function is not meant to be an all-purpose tool for fitting SCR models using MCMC. Instead, it is presented so that interested readers can better understand the computational aspects of the problem and can modify it for their purposes. The function can be used as so:

```
> fm1 <- scrIPP(y, X, M=150, 10,000, xlims=c(0,100), ylims=c(0,100),
+          space.cov=elev.fn, tune=c(0.4, 0.2, 0.3, 0.3, 7))
> plot(mcmc(fm1$out))
```

This code requests 10,000 posterior samples and estimates the effect of the spatial covariate, elevation, on density. The argument `space.cov` accepts any spatial covariate that returns a real value for any location in the rectangular state-space defined

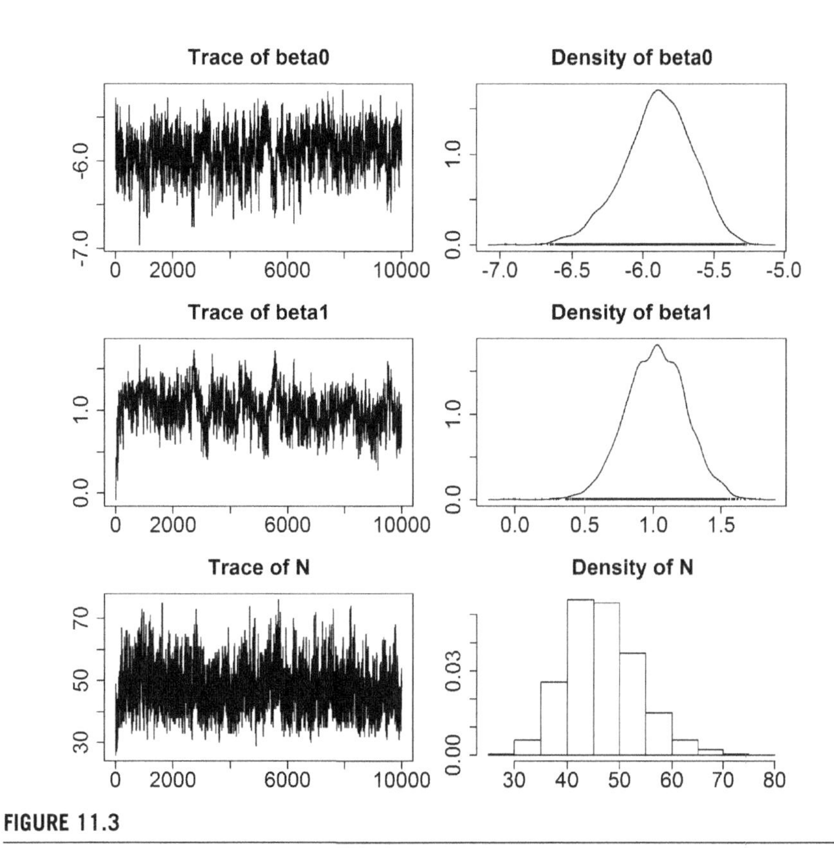

FIGURE 11.3

Trace plots and posterior distributions from MCMC analysis of SCR model with inhomogeneous point process. Analysis was conducted using the `scrIPP` function in the accompanying **R** package `scrbook`.

by the `xlims` and `ylims` arguments. Currently, the function places uniform priors on the parameters σ, λ_0, β_0, and β_1, although this could easily be modified. The `tune` argument specifies the tuning parameters used in the Metropolis-within-Gibbs steps of the algorithm. These should be chosen using trial and error to achieve an acceptance rate between 0.4 and 0.6, roughly. See Chapter 17 for more details about MCMC.

Results of the analysis are shown in Figure 11.3 and Table 11.1. Figure 11.3 displays the trace plots of the Markov chains as well as the posterior distributions for three parameters. The chains appear to converge rapidly but may need to be run longer to reduce Monte Carlo error. Summaries of the posterior distributions are presented in Table 11.1. The posterior means for β_0 and β_1 are quite similar to the MLEs from the analysis in the previous section in which we assumed no observation error. However, we see that the confidence intervals are wider. With respect to the other parameters in the model, we see that all of the data-generating parameters fall within

Table 11.1 Summary of posterior distributions from SCR model with inhomogeneous point process.

Parameter	Mean	SD	2.5%	97.5%
$\sigma = 5$	5.232	0.310	4.681	5.858
$\lambda_0 = 1$	0.802	0.119	0.595	1.049
$\beta_0 = -6$	−5.856	0.254	−6.376	−5.393
$\beta_1 = 1$	0.985	0.209	0.575	1.378
$N = 41$	47.615	8.041	35.000	66.000
$\mathbb{E}(N) = 39.9$	47.551	10.992	29.837	71.332

the 95% credible intervals. One thing to note is that, although the point estimates for the expected and realized values of N are quite similar, the posterior for the realized value of N is more precise. This is to be expected because the uncertainty associated with the realized value of N is entirely determined by the sampling error. That is, if we could perfectly detect all of the individuals in $\mathcal{S}$, there would be no uncertainty about N. In contrast, the variance for the expected value of N is composed of both process error and sampling error. See Chapter 5 and Efford and Fewster (2012) for additional discussion on the difference between realized and expected values of abundance.

Fitting continuous space inhomogeneous point process models is somewhat difficult in **BUGS** because the "IPP" prior $[\mathbf{s}_i|\boldsymbol{\beta}]$, unlike the uniform prior, is not one of the available distributions that comes with the software. It is possible to add new distributions in **BUGS**, but it is somewhat cumbersome. `secr` allows users to fit continuous space models using linear or polynomial functions of the easting and northing coordinates, but it does not accept truly continuous covariates that are functions of space. However, these are not really important limitations because discrete space versions of the model are straightforward, and virtually all spatial covariates are, or can be, defined as such.

11.4.2 Discrete space

To fit inhomogeneous point process models using covariates in discrete space, i.e., in raster format, we follow the same steps as outlined in Chapter 9—we define $\mathbf{s}_i$ as pixel ID, and we use the categorical distribution as a prior. This effectively changes the problem from estimating the coordinates of an activity center, to estimating the pixel in which an activity center is located. As pixel size approaches zero, these two become equivalent. A good example is found in (Mollet et al., In review). Here we present an analysis of the simulated data shown in the Figure 11.4. The spatial covariate, let's call it forest canopy height (CANHT), was simulated using the code shown on the help page `ch11` in `scrbook`. The points are the number of activity centers in each pixel, generated from a single realization of the inhomogeneous point process model with intensity $\mu(\mathbf{s}, \boldsymbol{\beta}) = \exp(\beta_0 + \beta_1 \text{CANHT}(\mathbf{s})) \times \text{pixelArea}$, where $\beta_0 = -6$ and $\beta_1 = 1$.

```
model{
sigma ~ dunif(0, 20)
lam0 ~ dunif(0, 5)
beta0 ~ dunif(-10, 10)
beta1 ~ dunif(-10, 10)
for(j in 1:nPix) {
  mu[j] <- exp(beta0 + beta1*CANHT[j])*pixArea
  probs[j] <- mu[j]/EN
}
EN <- sum(mu[]) # Expected value of N, E(N)
psi <- EN/M
for(i in 1:M) {
  z[i] ~ dbern(psi)
  s[i] ~ dcat(probs[])
  x0g[i] <- grid[s[i],1]
  y0g[i] <- grid[s[i],2]
  for(j in 1:ntraps) {
    dist[i,j] <- sqrt(pow(x0g[i]-traps[j,1],2) + pow(y0g[i]-traps[j,2],2))
    lambda[i,j] <- lam0*exp(-dist[i,j]*dist[i,j]/(2*sigma*sigma)) * z[i]
    y[i,j] ~ dpois(lambda[i,j])
    }
  }
N <- sum(z[]) # Realized value of N
}
```

PANEL 11.1

BUGS model specification for the inhomogeneous point process model in discrete space. A nearly equivalent formulation would involve omitting β_0 and modeling the expected number of activity centers as $\mathbb{E}(N) = M\psi$ with $\psi \sim \text{Uniform}(0, 1)$.

The **BUGS** description of the model is shown in Panel 11.1. The vector `probs[]` is the prior probability defined by Eq. (11.2.5), which is the probability that an individual's activity center is located at pixel $\mathbf{x}$. `grid` is the matrix of coordinates for each pixel.

This model can also be fit in `secr`, which refers to the raster data as a "habitat mask." The habitat mask is essentially a `data.frame` with attributes. The `data.frame` itself has two columns for the coordinates of each of the pixel centers. The attributes of the object include information such as the area of the pixels and the spacing between pixel centers. If there are covariates, these too are stored as an attribute of the habitat mask, and are formatted as a `data.frame` with one row per pixel and one column per covariate. Once the data have been formatted correctly, fitting the model in `secr` is as simple as:

```
> secr1 <- secr.fit(ch, model=D~canht, mask=msk)
```

where `D~canht` indicates that we want to model density as a function of canopy height, which is defined in the `msk` object. **R** code to format the data and fit the models

Table 11.2 Comparison of `secr` and **JAGS** results. Point estimates from the Bayesian analysis are posterior means. Intervals are lower and upper 95% CIs.

Parameter	Truth	Software	Mean	SD	2.5%	97.5%
λ_0	1.00	**JAGS**	1.04	0.087	0.88	1.22
	1.00	`secr`	1.08	0.089	0.92	1.27
σ	10.00	**JAGS**	10.16	0.373	9.46	10.94
	10.00	`secr`	9.84	0.350	9.18	10.55
β_1	1.00	**JAGS**	1.20	0.350	0.50	1.88
	1.00	`secr`	1.09	0.316	0.47	1.71
N	30.00	**JAGS**	26.63	2.585	23.00	33.00
	30.00	`secr`	28.19	3.037	24.49	37.39
$\mathbb{E}(N)$	32.30	**JAGS**	26.39	5.048	17.25	36.96
	32.30	`secr`	28.19	6.117	18.52	42.93

using `secr` and **JAGS** is available in `scrbook`, found by issuing the command: `help(ch11secr-jags)`.

Results of fitting the model in **JAGS** and `secr` are shown in Table 11.2 and are similar as expected. The differences that do exist are likely due to the differences in Bayesian and frequentist estimation methods, as discussed in Chapter 3.

11.5 Argentina jaguar study

Estimating density of large felines has been a priority for many conservation organizations, but few robust methodologies existed before the advent of SCR. Distance sampling is not feasible for such rare and cryptic species, and traditional capture-recapture methods yield estimates that are highly sensitive to the subjective choice of the effective survey area. SCR models provide a powerful alternative because density can be estimated directly and data can be collected using non-invasive methods such as camera traps or hair snares.

In this example, we show how inhomogeneous point process models can be used to test important hypotheses regarding the factors affecting density. The data come from an 8-year camera trapping study designed to assess the impacts of poaching on jaguar density in Argentina, near the borders of Brazil and Paraguay. Additional information about the study is presented in Paviolo et al. (2008, 2009). The expected effect of poaching is a decline in jaguar density due to the direct removal of individuals and the depletion of its main prey species. To conserve jaguars and related species, protected areas have been established and three levels of protection are recognized, as depicted in Figure 11.5. The dark gray area is the Iguazú National Park, which is patrolled regularly by law enforcement officials. The medium gray areas are not protected and rarely patrolled. Finally, the white areas are large soybean monocultures, cities, and dams, which do not provide habitat for jaguars.

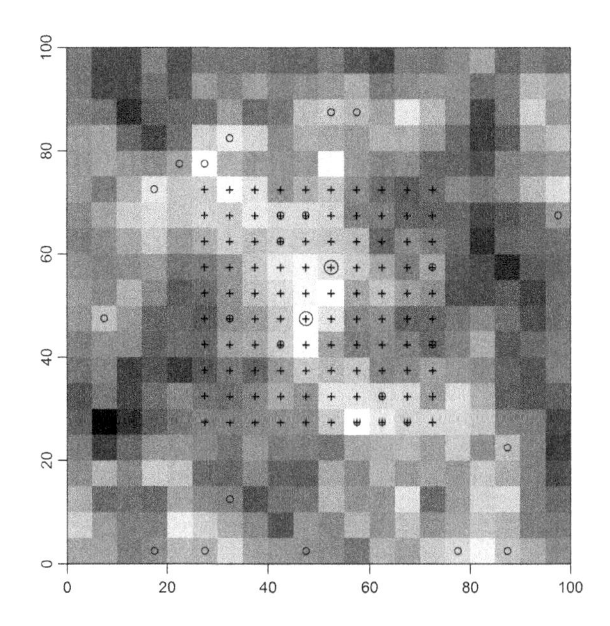

FIGURE 11.4

Simulated activity centers in discrete space. The spatial covariate, canopy height, is highest in the lighter areas and density increases with canopy height. A single activity center is shown as a small circle, and larger circles represent two activity centers in a pixel. Trap locations are shown as crosses.

To test for differences in density between the three regions, we modeled the point process intensity parameter as a function of protection status (PROTECT), which we treated as an ordinal variable:

$$\mu(\mathbf{s}, \boldsymbol{\beta}) = \exp(\beta_0 + \beta_1 \text{PROTECT}(\mathbf{s})) \times \text{pixelArea}.$$

We predicted that β_1 would be greater than zero, indicating that jaguar density increases with protection status. In addition to modeling spatial variation in density, we also modeled the scale parameter of the half-normal (or Gaussian) encounter model as sex-specific because male cats typically have larger home ranges than females (Sollmann et al., 2011). Since sex is an individual-specific covariate, and not observed for the individuals that were not captured, a prior distribution is required for the sex of uncaptured individuals. We used a Bernoulli prior with probability 0.5 to describe our uncertainty about sex ratio (see Chapter 7). Another equivalent option is to augment the data with an equal number of males and females and let the MCMC algorithm determine which of these individuals are actually members of the population.

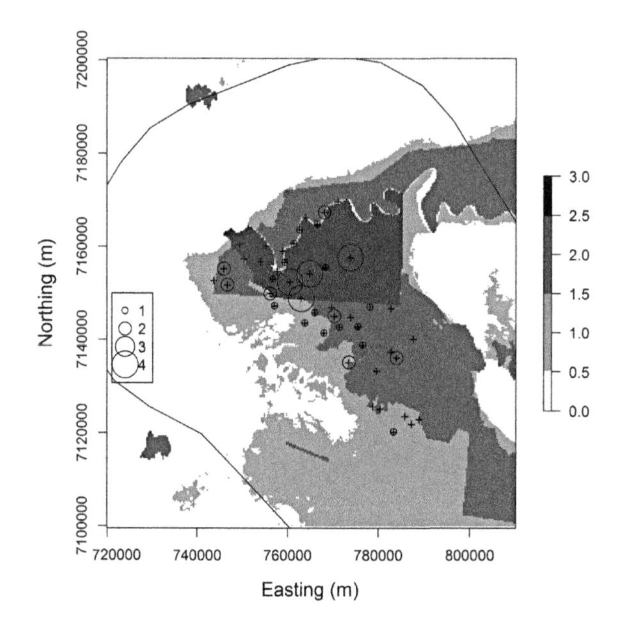

FIGURE 11.5

Jaguar detections at 46 camera trap stations. The three levels of protection status are no protection (light gray), some protection (gray), and Iguazú National Park (dark gray). Non-habitat (soybean monocultures) is shown in white.

An additional unique aspect of this study is the highly irregular state-space. Unlike in the examples of simulated data, the geometry of this state-space is not a simple rectangular region. Instead, it is the area south of the Iguazú River, which runs along the northern border of the park shown in dark gray in Figure 11.5, and it excludes the large soybean monocultures. Fitting models in highly convoluted spatial regions raises the question: How does one integrate Eq. (11.2.4) over this irregular space? Earlier we used the function `cuhre` in **R** for the two-dimensional integration, but its `lower` and `upper` arguments essentially assume that the state-space is rectangular. There are methods of transforming the state-space that might allow us to work around this problem, but once again we find that it is most convenient to work in discrete space and sum over all the pixels defining $\mathcal{S}$.

We fit the model to data from a single year in which 46 camera stations were operational, each consisting of a pair of cameras placed along roads or small trails. Forty-five detections of 16 jaguars (8 males and 8 females) were made over a 95-day sampling period. The mean number of sampling days at each camera station was 48.2. The raw capture data shown in Figure 11.5 suggest that the highest number of captures was in the national park, but there were also several traps in the park with no captures. Furthermore, few cameras were placed far from the protected area, making it somewhat difficult to detect differences in density. **R** code to fit the model

Table 11.3 Summaries of posterior distributions from the model of jaguar density. N is population size, σ is the scale parameter of the half-normal detection function, λ_0 is baseline encounter rate, β_1 is the effect of protection status on jaguar density, ρ is the sex-ratio. The last three parameters are the density estimates (jaguars/100 km^2) for the three levels of protection.

	Mean	SD	2.5%	97.5%
N	35.819	7.975	23.000	54.000
σ_{female}	5501.204	876.877	4142.276	7578.569
σ_{male}	6452.570	915.362	4970.321	8505.522
λ_0	0.006	0.002	0.003	0.010
ψ	0.355	0.094	0.200	0.564
β_0	−4.686	0.260	−5.235	−4.213
β_1	0.174	0.350	−0.510	0.864
ρ	0.489	0.055	0.382	0.600
D_{low}	0.906	0.326	0.381	1.668
D_{med}	0.770	0.284	0.270	1.439
D_{high}	1.370	0.307	0.831	1.995

is available in `scrbook` on the help page `jaguarDataCh11`. Parameter estimates are shown in Table 11.3.

The results indicate that there was little evidence that density differed among the 3 levels of protection (the posterior probability that $\beta_1 > 0$ was only). Figure 11.6 shows the estimated density surfaces. The first map is the expected density in each of the three protection categories, which was computed by plugging in the posterior mean values of β_0 and β_1 into the log-linear intensity function. The second map is the realized density surface—the conditional-on-N probability distribution of the number of activity centers in each pixel of the discretized state-space. The expected values would be used if we were interested in making inferences about other areas or time periods, whereas the realized map is the best description of the system during the study period.

We note that there is room for improvement in our analysis, and our results should be considered preliminary. The political boundaries used to demarcate protected areas are not as concrete as we might like. In reality, poaching pressure is likely higher near remote park boundaries than in well-guarded park interiors. One option for addressing this would be to use a continuous measure of poaching pressure such as distance from the nearest town, or some other accessibility metric. It would also be worthwhile to model density separately for each sex because many of the detections outside of the park were of males, and thus it is possible that the sexes use habitat differently (Conde et al., 2010). Other extensions warranting investigation include treating PROTECT as a categorical rather than an ordinal variable, and assessing the effects of roads and trails on jaguar movement using the methods described in Chapter 12. Developing

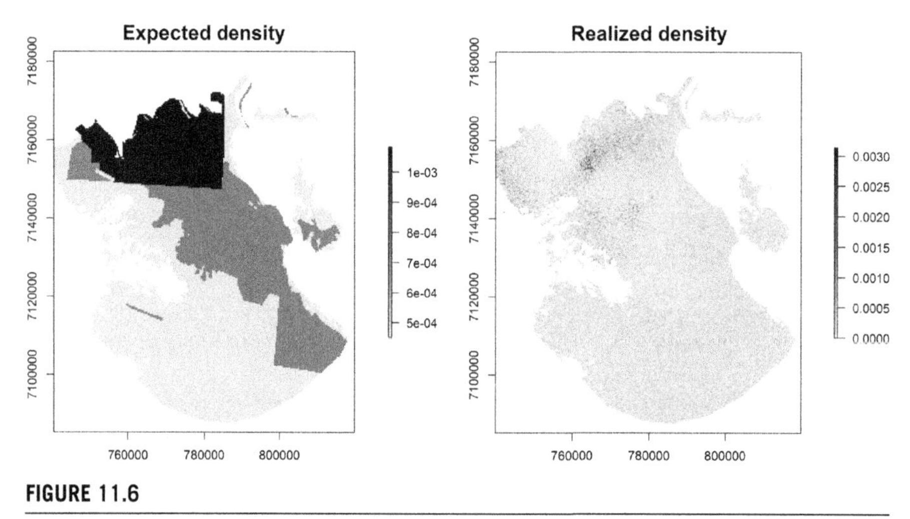

FIGURE 11.6

Estimated density (activity centers/pixel) surfaces from the analysis of the jaguar data.

models for these extensions could be readily accomplished by modifying the fitting functions found in the **R** package `scrbook`.

11.6 Summary and outlook

One of the distinguishing features of spatial capture-recapture models is that they allow for inference about spatial variation in density without relying on ad hoc approaches for determining the amount of area surveyed. The approach described in this chapter involves modeling the locations of activity centers as outcomes of an inhomogeneous point process with intensity determined by covariates defined at all locations in the state-space. Covariate effects can be evaluated in exactly the same way as is done in generalized linear models, making it easy to interpret the results.

All the examples in this section included a single state-space covariate, but this was for simplicity only. Including multiple covariates poses no additional challenges. Similarly, additional model structure such sex-specific encounter rate parameters or behavioral responses can be accommodated and fit using `secr`, or one of the **BUGS** dialects, or by extending the functions in `scrbook`. It is also possible to consider covariates that affect both density and ecological distance, as will be described in the next chapter. The ramifications of this are enormous for applied ecological research and conservation efforts because researchers can use capture-recapture data to identify areas where both density and landscape connectivity are high (Royle et al., 2013). Addressing such questions is simply not possible using standard, non-spatial capture-recapture methods.

Although we focused on modeling the point process intensity as a function of covariates, other options for fitting inhomogeneous models exist (Illian et al., 2008). Cox processes are models in which the point process intensity is a function of spatial random effects. Such methods are useful for accommodating overdispersion, but it seems unlikely that most SCR data sets could support such complexity. Gibbs processes are another important class of models that are distinguished by the interactions of points. Although little work has been done on such models in the context of SCR studies (Reich et al., In review), we expect they will receive more attention because they can be used to model processes such as territoriality (points repel one another) or aggregation (points attract one another). Neyman-Scott processes are another option for modeling aggregation or clustering, and could be useful for studying gregarious species.

CHAPTER

Modeling Landscape Connectivity

12

Every spatial capture-recapture model that we have considered so far has expressed encounter probability as a function of the Euclidean distance between individual activity centers **s** and trap locations **x**. As a practical matter, models based on Euclidean distance imply circular, symmetric, and stationary home ranges of individuals, which are not often biologically realistic. While these simple encounter probability models are often sufficient for many purposes, especially in small data sets, sometimes developing more complex models of the detection process that relates to space usage of individuals will be useful. Animals may not judge distance in terms of Euclidean distance but, rather, according to the configuration of habitat patches, quality of local habitat, perceived mortality risk, and other considerations. Together, the degree to which these factors facilitate or impede movement determines landscape connectivity (Tischendorf and Fahrig, 2000), which is widely recognized to be an important component of population viability (With and Crist, 1995; Compton et al., 2007). Moreover, because encounter probability and the distance metric upon which it is based represent outcomes of individual movements about their home range, ecologists might have explicit hypotheses about how environmental variables affect the distance metric. It is therefore desirable to incorporate these hypotheses directly into SCR models so that they may be formally evaluated statistically.

Although much theory has been developed to predict the effects of decreasing connectivity, few empirical studies have been conducted to test these predictions due to the paucity of formal methods for estimating connectivity parameters (Cushman et al., 2010; Hanks and Hooten, 2013). Instead, ecologists often rely on expert opinion or *ad hoc* methods of specifying connectivity values, even in important applied settings (Adriaensen et al., 2003; Beier et al., 2008; Zeller et al., 2012). In addition, no methods are available for simultaneously estimating population density and connectivity parameters, in spite of theory predicting interacting effects of density and connectivity on population viability (Tischendorf et al., 2005; Cushman et al., 2010). In this chapter, following Royle et al. (2013a), we provide a framework for modeling landscape connectivity using SCR models, by parameterizing models for encounter probability based on "ecological distance." A natural candidate framework for modeling ecological distance is the least-cost path which is used widely in landscape ecology for modeling connectivity, movement, and gene flow (Adriaensen et al., 2003; Manel et al., 2003; McRae et al., 2008). In practical applications, variables that

Spatial Capture-Recapture. http://dx.doi.org/10.1016/B978-0-12-405939-9.00012-8

influence landscape connectivity, or the effective cost of moving across the landscape, include things like highways (e.g., Epps et al., 2005), elevation (Cushman et al., 2006), ruggedness (Epps et al., 2007), snow cover (Schwartz et al., 2009), distance to escape terrain (Shirk et al., 2010), range limitations (McRae and Beier, 2007), or distance from urban areas, human disturbance, or other factors that animals might avoid.

Royle et al. (2013a) provided an SCR framework based on least-cost path methods for modeling landscape connectivity. They parameterized encounter probability based not on Euclidean distance but, rather, on the least-cost path between an individual's activity center and a trap location. This is parameterized in terms of one or more parameters that relate the *resistance* of the landscape to explicit covariates. In this way, SCR models can explicitly accommodate landscape structure and account for connectivity of the landscape. Using this methodological extension of SCR models, it is possible to make formal statistical inferences about movement and connectivity from capture-recapture studies that generate sparse individual encounter history data without subjective prescription of resistance or cost surfaces. While we believe there should be much ecological interest in developing SCR models that account for landscape connectivity, it is also important for obtaining more accurate estimates of density; under simple models of landscape connectivity, incorrectly fitting the basic model SCR0 produces substantial bias in estimates of N and hence density (Royle et al., 2013a).

12.1 Shortcomings of Euclidean distance models

In the standard SCR models, encounter probability is modeled as a function of Euclidean distance. For example, using the binomial observation model (Chapter 5), let y_{ij} be individual- and trap-specific binomial counts with sample size K and probabilities p_{ij}. The Gaussian encounter probability model is

$$p_{ij} = p_0 \exp(-d_{ij}^2/(2\sigma^2)), \tag{12.1.1}$$

where $d_{ij} = \|\mathbf{x}_j - \mathbf{s}_i\|$ is Euclidean distance. As usual, we will sometimes adopt the log-scale parameterization based on $\log(p_{ij}) = \alpha_0 + \alpha_1 d_{ij}^2$ where $\alpha_0 = \log(p_0)$ and $\alpha_1 = -1/(2\sigma^2)$.

The main problem with using the Euclidean distance metric in this encounter probability model is that it is unaffected by habitat or landscape structure, and it implies that the space used by individuals is stationary and symmetric, which may be unreasonable assumptions for some species. By stationary we mean in the formal sense of invariance to translation. That is, the properties of an individual home range centered at some point $\mathbf{s}$ are exactly the same as any other point say $\mathbf{s}'$. As an example, if the common encounter probability model based on a bivariate normal probability distribution function is used, then the implied space usage by *all* individuals, no matter their location in space or local habitat conditions, is symmetric with circular contours of usage intensity.

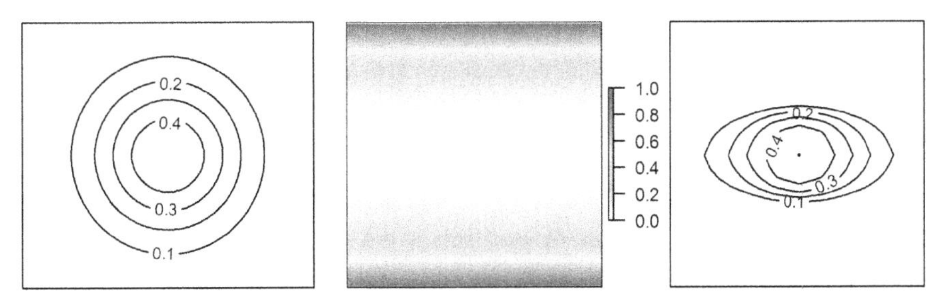

FIGURE 12.1

A symmetric home range (left), a habitat variable (center) such as representing an elevation gradient, and a non-symmetric home range (right) resulting from the cost imposed on movement by the habitat variable.

In the framework of Royle et al. (2013a), SCR models explicitly incorporate information about the landscape so that a unit of distance is variable depending on identified covariates, say $C(\mathbf{s})$. Thus, where an individual lives on the landscape, and the state of the surrounding landscape, will determine the nature of its use of space. In particular, they suggest distance metrics, based on least-cost path, that imply irregular, asymmetric, and non-stationary home ranges of individuals. As an example, Figure 12.1 shows a typical symmetric home range (left panel), and a compressed home range (right panel) resulting from the effect of an environmental variable (center panel) on an animal's movement behavior. We might think of the environmental variable as representing an elevation gradient of a valley and so, for a species that avoids high elevation (gray), space usage will be concentrated in flatter terrain at lower elevations (white) therefore producing the elliptical home range shape.

12.2 Least-cost path distance

We adopt a cost-weighted distance metric here, which defines the effective distance between points by accumulating pixel-specific costs determined using a cost function defined by the user. The idea of cost-weighted distance to characterize animal use of landscapes is widely used in landscape ecology for modeling connectivity, movement, and gene flow (Beier et al., 2008). For reasons of computational tractability we consider a discrete landscape defined by a raster of some prescribed resolution. The distance between any two points $\mathbf{s}$ and $\mathbf{s}'$ can be represented by a sequence of line segments connecting neighboring pixels, say $\mathbf{l}_1, \mathbf{l}_2, \ldots, \mathbf{l}_m$. Then the cost-weighted distance between $\mathbf{s}$ and $\mathbf{s}'$ is

$$d(\mathbf{s}, \mathbf{s}') = \sum_{i=1}^{m-1} \text{cost}(\mathbf{l}_i, \mathbf{l}_{i+1}) \|\mathbf{l}_i - \mathbf{l}_{i+1}\|, \tag{12.2.1}$$

where $\text{cost}(\mathbf{l}_i, \mathbf{l}_{i+1})$ is the user-defined cost to move from pixel $\mathbf{l}_i$ to neighboring pixel $\mathbf{l}_{i+1}$ in the sequence. Given the cost of each pixel, it is a simple matter to compute the cost-weighted distance between any two pixels, along *any* path, simply by accumulating the incremental costs weighted by distances. In the context of spatial capture-recapture models (and, more generally, landscape connectivity) we are concerned with the *minimum* cost-weighted distance, or the *least-cost path*, between any two points, which we will denote by d_{lcp} This is the sequence $\mathcal{P} = (\mathbf{l}_1, \mathbf{l}_2, \ldots, \mathbf{l}_m)$ that minimizes the objective function defined by Eq. (12.2.1). That is,

$$d_{lcp}(\mathbf{s}, \mathbf{s}') = \min_{\mathcal{P}} \sum_{i=1}^{m-1} \text{cost}(\mathbf{l}_i, \mathbf{l}_{i+1}) \|\mathbf{l}_i - \mathbf{l}_{i+1}\|. \tag{12.2.2}$$

The least-cost path distance can be calculated in many geographic information systems and other software packages, including the **R** package `gdistance` (van Etten, 2011) which we use below.

The key ecological aspect of least-cost path modeling is the development of models for pixel-specific cost. A natural approach is to model cost as a function of one or more covariates defined on every pixel of the raster. For example, using a single covariate $C(\mathbf{s})$ we define the cost of moving from some pixel $\mathbf{s}$ to neighboring pixel $\mathbf{s}'$ as

$$\log(\text{cost}(\mathbf{s}, \mathbf{s}')) = \alpha_2 \left(\frac{C(\mathbf{s}) + C(\mathbf{s}')}{2} \right). \tag{12.2.3}$$

Thus, if $\alpha_2 = 0$, then substituting $\text{cost}(\mathbf{s}, \mathbf{s}') = \exp(0) = 1$ into Eq. (12.2.2) will produce the ordinary Euclidean distance between points.

The use of least-cost path models to model landscape connectivity has been around for a long time. And, although α_2 is rarely known, conservation biologists design linkages that require this resistance value as input (see Beier et al., 2008, and articles cited therein). However, formal inference (e.g., estimation) of parameters is not often done. Instead, in many existing applications of least-cost path analysis, the parameter α_2 is fixed by the investigator, or based on expert opinion (Beier et al., 2008), although recently researchers have begun to define costs based on resource selection functions,[1] animal movement (Tracy, 2006; Fortin et al., 2005), or genetic distance data (e.g., Gerlach and Musolf, 2000; Epps et al., 2007; Schwartz et al., 2009). To formalize the use of cost-weighted distance in SCR models, we substitute Eq. (12.2.2) for Euclidean distance in the expression for encounter probability (Eq. (12.1.1)) and maximize the resulting likelihood (see below). In doing so, we can directly estimate parameters of the least-cost path model, evaluate how landscape covariates influence connectivity, and test explicit hypotheses about these things using only individual-level encounter history data from capture-recapture studies.

[1] We address the integration of resource selection models based on telemetry data with SCR models in Chapter 13.

12.2.1 Example of computing cost-weighted distance

As an example of the cost-weighted distance calculation, consider the following landscape comprised of 16 pixels with unit spacing identified as follows, along with the pixel-specific cost:

```
    pixel ID                 Cost
 4  8  12  16          100    1    1  1
 3  7  11  15          100  100    1  1
 2  6  10  14          100  100  100  1
 1  5   9  13          100  100    1  1
```

We assume the scale is such that the distance between neighboring pixels in any cardinal direction is 1 unit, and the distance between neighbors on a diagonal is $\sqrt{2}$ units. We assigned low cost of 1 to "good habitat" pixels (or pixels we think of as "highly connected" by virtue of being in good habitat) and, conversely, we assign high cost (100) to "bad habitat." This simple cost raster is shown in Figure 12.2. The **R** commands for creating this are as follows (which can be run using the **R** script SCRed):

```
> library(raster)
> library(gdistance)
> r <- raster(nrows=4,ncols=4)
> projection(r) <- "+proj=utm +zone=12 +datum=WGS84" # Sets the projection
> extent(r) <- c(.5,4.5,.5,4.5) #sets the extent of the raster
> costs1 <- c(100,100,100,100,1,100,100,100,1,1,100,1,1,1,1,1)
> values(r) <- matrix(costs1,4,4,byrow=FALSE) #assign the costs to the raster
> par(mfrow=c(1,1))
> plot(r)
```

This produces Figure 12.2.

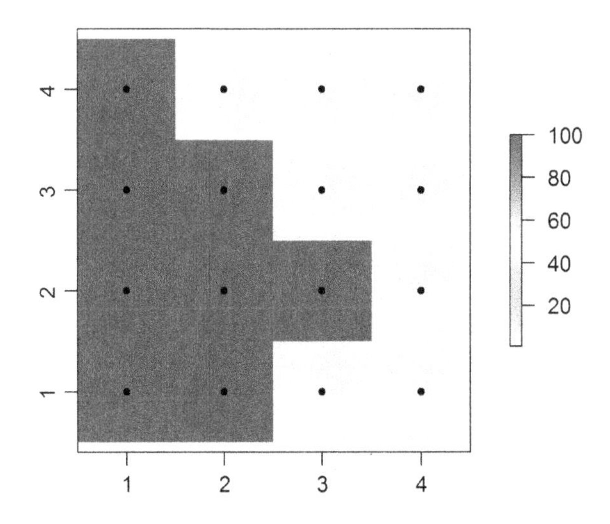

FIGURE 12.2

A 4 × 4 raster depicting a binary cost surface, with cost = 1 (white) or 100 (shaded) to represent ease of movement through a pixel.

For this simple case we can easily compute the shortest cost-weighted distance between any pixels "by eye." For example, the shortest cost-weighted distance between pixels 5 and 9 in this example is 50.5 units: $1 \times (100 + 1)/2 = 50.5$, the shortest distance between pixels 4 and 8 is also 50.5, while the shortest cost-distance between 4 and 12 is 51.5. What is the shortest distance between 7 and 16? Suppose an individual at pixel 7 can move diagonal (which has distance $\sqrt{2}$) and pay $\sqrt{2}(100 + 1)/2$, and then move once to the right to pay 1 additional unit cost, for a total of 72.4. However, if the individual instead moved one unit to the right, to pixel 11, and then diagonally, the total cost is 51.914 which is the minimum cost-weighted distance in getting from pixel 7 to 16. These two ways of moving from 7 to 16 have the same Euclidean distance, but different cost-weighted distances according to our cost function.

The least-cost path distances can be computed with just a few **R** commands, and these commands can be inserted directly into the likelihood construction for an ordinary spatial capture-recapture model. The **R** package `gdistance` calculates least-cost path using Dijkstra's algorithm (Dijkstra, 1959) from the `igraph` package (Csardi and Nepusz, 2006). To compute the least-cost path, or the minimum cost-weighted distances between every pixel and every other pixel, we make use of the helper function `transition`, which calculates the cost of moving between neighboring pixels. It operates on the inverse-scale ("conductance"), and so the `transitionFunction` argument is given as $1/mean(x)$. The function `geoCorrection` modifies this object depending on the projection of the coordinate system (e.g., it corrects for curvature of the earth's surface if longitude/latitude coordinates are used). The result is fed into the function `costDistance` to compute the pairwise distance matrix. For that, we define the center points of each raster, here these are just integers on $[1, 4] \times [1, 4]$. The commands altogether are as follows:

```
> tr1 <- transition(r,transitionFunction=function(x) 1/mean(x),directions=8)
> tr1CorrC <- geoCorrection(tr1,type="c",multpl=FALSE,scl=FALSE)
> pts <- cbind( sort(rep(1:4,4)),rep(4:1,4))
> costs1 <- costDistance(tr1CorrC,pts)
> outD <- as.matrix(costs1)
```

Now we can look at the results and see if it makes sense to us. Here we produce the first 5 columns of this distance matrix to illustrate a couple of examples of calculating the minimum cost-weighted distance between points:

```
>    outD[1:5,1:5]
          1          2          3          4          5
1    0.0000   100.0000   200.0000   205.2426   100.0000
2  100.0000     0.0000   100.0000   200.0000   141.4214
3  200.0000   100.0000     0.0000   100.0000   126.1604
4  205.2426   200.0000   100.0000     0.0000   105.2426
5  100.0000   141.4214   126.1604   105.2426     0.0000
```

An interesting case is that between point 1 and 4. Note that simply taking the shortest Euclidean distance, weighted by cost, produces a cost-weighted distance of 100×1 to move from pixel 1 to pixel 2, and similarly from 2 to 3 and 3 to 4, producing a total cost-weighted distance of 300. However, the actual *least-cost path* has cost-weighted distance 205.2426.

The key point here is that, once we can compute this distance matrix, we can use it as the distance matrix in computing the encounter probability between activity centers and traps, and we can use our existing MLE technology (Chapter 6) to fit models that are based on ecological distance.

12.3 Simulating SCR data using ecological distance

Royle et al. (2013a) simulated capture-recapture data such that landscape connectivity was governed by a cost function having a single covariate, and they considered two hypothetical covariate landscapes (Figure 12.3). The landscape is a 20×20 pixel raster, with extent = $[0.5, 4.5] \times [0.5, 4.5]$. For example, think of each pixel as representing, say, a 1×1 km grid cell with something like "percent developed" or "trail/road density" as the covariate. For sampling by capture-recapture, imagine that 16 camera traps are established at the integer coordinates $(1, 1), (1, 2), \ldots, (4, 4)$. The two covariates were constructed as follows (see `?make.EDcovariates` for the **R** commands): First is an increasing trend from the NW to the SE ("systematic covariate"), where $C(\mathbf{s})$ is defined as $C(\mathbf{s}) = row(\mathbf{s}) + col(\mathbf{s})$ and $row(\mathbf{s})$ and $col(\mathbf{s})$ are just the row and column, respectively, of the raster. This might mimic something related to distance from an urban area or a gradient in habitat quality due to land use, or environmental conditions such as temperature or precipitation gradients. In the second case, the covariate was generated using spatially correlated noise to

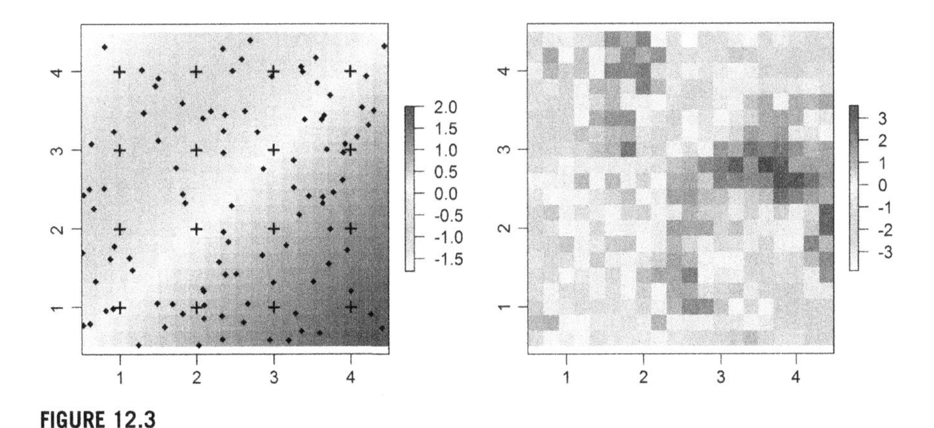

FIGURE 12.3

Two covariates (defined on a 20×20 grid) used in simulations. Left panel shows a covariate with systematic structure meant to mimic distance from some feature, and the right panel shows a "patchy" covariate. A hypothetical realization of $N = 100$ activity centers is superimposed on the left figure, along with 16 trap locations (indicated by "+").

emulate a typical patchy habitat covariate ("patchy covariate") such as tree or understory density.

For both covariates we use a cost function in which transition from pixel $\mathbf{s}$ to $\mathbf{s}'$ is given by:

$$\log(\text{cost}(\mathbf{s}, \mathbf{s}')) = \alpha_2 \left(\frac{C(\mathbf{s}) + C(\mathbf{s}')}{2} \right),$$

where $\alpha_2 = 1$ for simulating the observed data. Remember that with $\alpha_2 = 0$ the model reduces to one in which the cost of moving across each pixel is constant, and therefore Euclidean distance is operative. In the left panel of Figure 12.3, a sample realization of $N = 100$ activity centers is shown. While encounter probability is assumed to be related to landscape connectivity according to the single-variable cost function, individual activity centers are assumed to be uniformly distributed, although we can modify this assumption (See Section 12.8).

When distance is defined by the cost-weighted distance metric given by Eq. (12.2.2) then individual space usage varies spatially in response to the landscape covariate(s) used in the distance metric. As a consequence, home range contours are no longer circular, as in SCR models based on Euclidean distance. For example, using the patchy covariate (Figure 12.3, right panel) with a Gaussian encounter probability model but having distance metric defined by Eq. (12.2.2), produces home ranges such as those shown in Figure 12.4.

To simulate data, we have to load the `scrbook` package and call the function `make.EDcovariates` to generate our raster covariates (see the help file for how that is done). We process the covariate into a least-cost path distance matrix, and then simulate observed encounter data using standard methods which we have used many times previously in this book. The complete set of **R** commands is:

```
### Grab a covariate
> library(scrbook)
> set.seed(2013)
> out <- make.EDcovariates()
> covariate <- out$covariate.patchy

### prescribe some settings
> N <- 200
> alpha0 <- -2
> sigma <- .5
> alpha1 <- 1/(2*sigma*sigma)
> alpha2 <-1
> K <- 5
> S <- cbind(runif(N,.5,4.5),runif(N,.5,4.5))

# make up some trap locations
> xg <- seq(1,4,1); yg<-4:1
> traplocs <- cbind(sort(rep(xg,4)),rep(yg,4))
> points(traplocs,pch=20,col="red")
> ntraps <- nrow(traplocs)
```

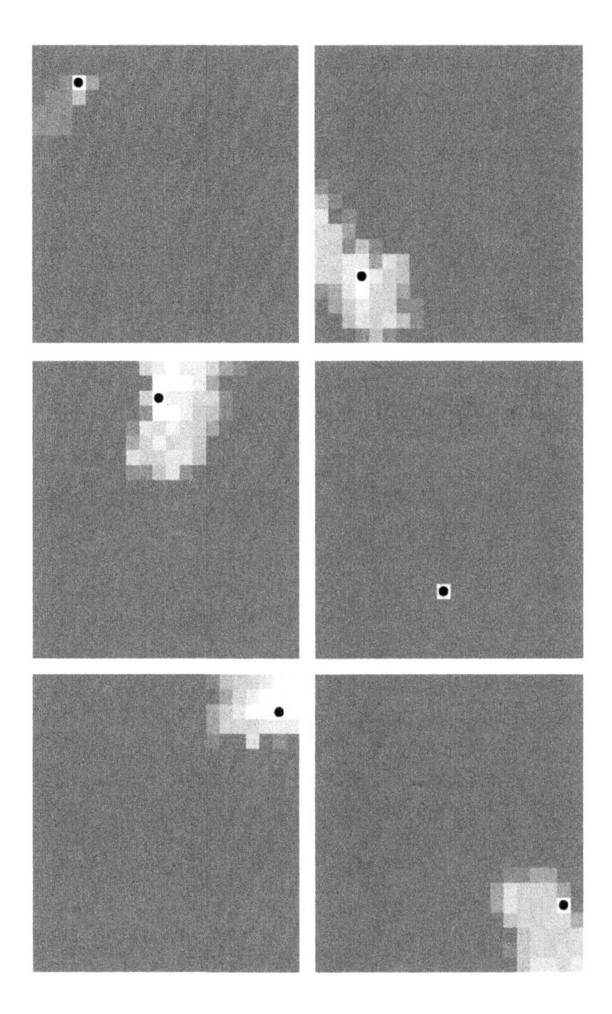

FIGURE 12.4

Typical home ranges for six individuals based on the cost surface shown in the right panel of Figure 12.3 with $\alpha_2 = 1$. The black dot indicates the home range center and the pixels around each home range center are shaded according to the probability of encounter, if a trap were located in that pixel. (For interpretation of the references to color in this figure legend, the reader is referred to the web version of this book.)

```
### make a raster and fill it up with the "cost"
> r <- raster(nrows=20,ncols=20)
> projection(r) <- "+proj=utm +zone=12 +datum=WGS84"
> extent(r) <- c(.5,4.5,.5,4.5)
> cost <- exp(alpha2*covariate)

### compute least-cost path distance
> tr1 <- transition(cost,transitionFunction=function(x) 1/mean(x),directions=8)
> tr1CorrC <- geoCorrection(tr1,type="c",multpl=FALSE,scl=FALSE)
```

```
> D <- costDistance(tr1CorrC,S,traplocs)
> probcap <- plogis(alpha0)*exp(-alpha1*D*D)

# now generate the encounters of every individual in every trap
# discard uncaptured individuals
> Y <- matrix(NA,nrow=N,ncol=ntraps)
> for(i in 1:nrow(Y)){
+   Y[i,] <- rbinom(ntraps,K,probcap[i,])
+ }
> Y <- Y[apply(Y,1,sum)>0,]
```

The matrix Y is the `nind` × `ntraps` matrix of observed encounter frequencies.

12.4 Likelihood analysis of ecological distance models

Throughout much of this book we rely on Bayesian analysis by MCMC mostly using the BUGS language, but sometimes (as in Chapter 17) we develop our own MCMC implementations. However, occasionally we prefer to use likelihood estimation, such as when we can compare a set of models directly by likelihood, either to do a direct hypothesis test of a parameter, or to tabulate a bunch of AIC values. For the class of models that use least-cost path, we also prefer likelihood methods not because they have any conceptual or methodological benefit, but simply because they are more computationally efficient to implement (Royle et al., 2013a).

There are no technical considerations in adapting our formulation of maximum likelihood estimation (Borchers and Efford, 2008) from Chapter 6 for the class of models based on least-cost path (see the appendix in Royle et al. (2013a) for complete details). The likelihood analysis is really just a straightforward adaptation in which we replace the Euclidean distance with least-cost distance. Consider the Bernoulli model in which the individual- and trap-specific observations have a binomial distribution conditional on the latent variable $\mathbf{s}_i$:

$$y_{ij}|\mathbf{s}_i \sim \text{Binomial}(K, p_{\boldsymbol{\alpha}}(d_{lcp}(\mathbf{s}_j, \mathbf{s}_i; \alpha_2); \alpha_0, \alpha_1), \qquad (12.2.4)$$

where we have indicated the dependence of p on the parameters $\boldsymbol{\alpha} = (\alpha_0, \alpha_1, \alpha_2)$, and d_{lcp} which itself depends on α_2, and the latent variable $\mathbf{s}_i$. We note that the only difference between likelihood analysis of this model and the standard Bernoulli model is the use of d_{lcp} here. For the random effect we have $\mathbf{s}_i \sim \text{Uniform}(\mathcal{S})$, for which we can easily compute the integrated (marginal) likelihood of an encounter history. The likelihood is given in the `scrbook` package as the function `intlik3ed`. The help file provides an example of its use and for simulating data. To use this function the cost covariate $C(\mathbf{s})$ has to be of class `RasterLayer`, which requires packages `sp` (Pebesma and Bivand, 2011) and `raster` (Hijmans and van Etten, 2012) to manipulate.

12.4.1 Example of SCR with least-cost path

Now we use the **R** function `nlm` along with our `intlik3ed` function to obtain the MLEs of the model parameters for the data simulated in Section 12.3. We'll do that

Table 12.1 Summary output from fitting models based on Euclidean and least-cost path distance to simulated data using the `intlik3ed` function (see `?intlik3ed`). Data were simulated based on the least-cost path model using the "patchy" covariate shown in Figure 12.3.

Distance Metric	−loglik	α_0	α_1	$\log(n_0)$	α_2
True value		−2.000	2.000	4.644	1.000
Euclidean	133.495	−1.885	1.247	3.549	–
Least-cost path (truth)	70.119	−1.780	2.471	4.459	0.046

for both the standard Euclidean distance and then for the ecological distance based on the "patchy" covariate using the following commands:

```
> frog1<-nlm(intlik3ed,c(alpha0,alpha1,3)),hessian=TRUE,y=Y,K=K,X=traplocs,
             distmet="euclid",covariate=covariate,alpha2=1)

> frog2<-nlm(intlik3ed,c(alpha0,alpha1,3,-.3),hessian=TRUE,y=Y,K=K,X=traplocs,
             distmet="ecol",covariate=covariate,alpha2=NA)
```

The summary output for the two model fits is shown in Table 12.1. The model based on least-cost path (the data-generating model) appears to be much preferred in terms of negative log-likelihood. The data-generating parameter values were $\alpha_0 = -2, \alpha_1 = 2$, and $\alpha_2 = 1$. The simulated sampling produced a sample of 96 individuals (N $= 200$) and so the number of individuals not captured is $n_0 = 104$, and $\log(n_0) = 4.64$. We see that the MLEs of the least-cost path model are pretty close, whereas they are not so close under the misspecified model based on Euclidean distance.

12.5 Bayesian analysis

While implementation of these ecological distance SCR models is reasonably straightforward, the model cannot be fitted in the **BUGS** engines because least-cost path distance cannot be computed. It would be possible to fit the models in **BUGS** if the parameter α_2 was fixed. In that case, one could compute the least-cost distance matrix ahead of time and reference the required elements for a given $\mathbf{s}$. Alternatively, it would be possible to write a custom MCMC routine using the methods we present in Chapter 17, although we have not yet developed our own MCMC implementation of SCR models with ecological distance metrics.

12.6 Simulation evaluation of the MLE

Royle et al. (2013a) carried out a limited simulation study to evaluate the general statistical performance of the density estimator under this new model, the effect of misspecifying the model with a normal Euclidean distance metric, and to assess the general bias and precision properties of the MLE using the systematic and patchy landscapes shown in Figure 12.3. We reproduce a subset of the results from Royle et al. (2013a) in Table 12.2 in order to highlight some key points. The results show extreme bias in estimates of N when the misspecified Euclidean distance is used, and only minor small-sample bias of 3–5% in the MLE of N using the least-cost distance which becomes negligible as the expected sample size increases (either due to increasing K, or larger population sizes). The performance of estimating the other parameters, including the cost parameter α_2, mirrors the results for estimating N.

Table 12.2 Simulation results for estimating population size N for a prescribed state-space with $N = 100$ or $N = 200$ and various levels of replication (K) using the "patchy" landscape shown in Figure 12.3. For each simulated data set, the SCR model was fitted by maximum likelihood with standard Euclidean distance ("euclid"), or least-cost path ("lcp"), which was the true data-generating model. The summary statistics of the sampling distribution reported are the mean, standard deviation ("SD"), and quantiles (0.025, 0.50, 0.975).

			$N = 100$		
	Mean	**SD**	**0.025**	**0.50**	**0.975**
$K = 3$					
euclid	78.68	18.12	49.40	76.34	125.47
lcp	110.96	28.65	69.55	106.98	181.84
$K = 5$					
euclid	77.85	11.55	59.17	77.44	101.14
lcp	104.44	15.79	78.38	101.47	139.55
$K = 10$					
euclid	78.01	5.26	68.00	77.96	87.81
lcp	100.42	7.56	86.72	100.34	115.47
			$N = 200$		
$K = 3$					
euclid	154.34	33.74	107.00	146.34	221.43
lcp	208.77	49.29	141.68	197.89	325.77
$K = 5$					
euclid	153.39	15.57	129.31	149.54	185.38
lcp	200.91	20.78	164.42	200.47	246.46
$K = 10$					
euclid	156.27	8.51	142.17	156.05	174.55
lcp	198.45	11.44	180.06	198.04	219.52

12.7 Distance in an irregular patch

We provide another illustration of how to employ ecological distance calculations in SCR models using an example meant to mimic a situation with a hard habitat boundary, such as a habitat corridor or park unit or some other block of relatively homogeneous good-quality habitat for some species. This particular system (shown in Figure 12.5) could be habitat surrounded by a suburban wasteland of DonutTownes and Beer-Marts, much less hospitable habitat for most species. For our purposes, we suppose that individuals live within the buffered "f-shaped" region, although we could also imagine the negative aspect of the situation in which individuals live outside of the region, so that the polygon represents a barrier (a lake) or bad habitat (an urban area) or similar. We describe the steps for creating this landscape shortly, so that you can use a similar process to generate more relevant landscapes for your own problems.

In this case we're not going to estimate any parameters of the cost function (though you could adapt the analyses of the previous sections to do that) but instead we're going to use ecological distance ideas only to constrain movement within (or to avoid) landscape features. Note that, normally, distance "as the crow flies" would not be suitable for irregular habitat patches such as that shown in Figure 12.5.

12.7.1 Basic geographic analysis in R

In practical applications our landscape will contain polygons which delineate good or bad habitat or other important characteristics of the landscape. These might exist as GIS shapefiles or merely as a text file with coordinates defining polygon boundaries.

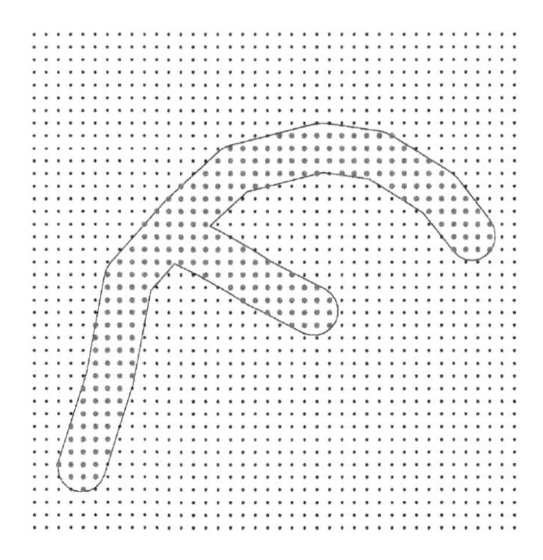

FIGURE 12.5

A fake wildlife corridor or reserve. The boundary outlines a polygon of suitable habitat surrounded by suburban development.

To work with polygons in the context of SCR models we need to create a raster, overlay the polygon, and assign values to each pixel depending on whether pixels are in the polygon or not, or how far they are from polygon boundaries. These operations are relatively easy to do within a GIS system but we need to be able to do them in **R** in order to compute the least-cost paths needed in the likelihood evaluation. Some additional geographic analyses are discussed in Section 17.7 where we talk about reading in the shapefile and using it in SCR analyses.

In practice we will have GIS shapefiles that define polygons but, here, we create a set of polygons by buffering and joining some line segments. In the **R** package `scrbook`, we provide a function `make.seg` which allows you to make such line segments given a specific trap region. To use `make.seg` we first create a plot region and then call `make.seg` which has a single argument being the number of points used to define the line. The user clicks on the visual display until the required number of points has been obtained by `make.seg`. In the following set of commands we generate two line segments, `l1` consisting of 9 points and `l2` consisting of 5 points, and these reside in a geographic region enclosed by $[0, 10] \times [0, 10]$.

```
> library(scrbook)
> library(sp)
> plot(NULL,xlim=c(0,10),ylim=c(0,10))
> l1 <- make.seg(9)
> plot(l1)
> l2 <- make.seg(5)
> plot(l1)
> lines(l2)
```

We used this function to create Fig. 12.5, a habitat corridor composed of line segments of class `SpatialLines` from the **R** package `sp`. The corridor can be loaded from `scrbook` by typing the command `data(fakecorridor)`. This data list has two line files in it (`l1` and `l2`) and a trap locations file (`traps`). We use some functions from the **R** packages `sp` and `rgeos` (Bivand and Rundel, 2011) to join and buffer (by 0.5 units) the two segments. The commands are as follows:

```
> data(fakecorridor)
> library(sp)
> library(rgeos)

> buffer <- 0.5
> par(mfrow=c(1,1))
> aa <- gUnion(l1,l2)
> plot(gBuffer(aa,width=buffer),xlim=c(0,10),ylim=c(0,10))
> pg <- gBuffer(aa,width=buffer)
> pg.coords <- pg@polygons[[1]]@Polygons[[1]]@coords

> xg <- seq(0,10,,40)
> yg <- seq(10,0,,40)
```

```
> delta <- mean(diff(xg))
> pts <- cbind(sort(rep(xg,40)),rep(yg,40))
> points(pts,pch=20,cex=.5)

> in.pts <- point.in.polygon(pts[,1],pts[,2],pg.coords[,1],pg.coords[,2])
> points(pts[in.pts==1,],pch=20,col="red")
```

The point of this example is to compute ordinary Euclidean distance but restricted by the boundaries of the corridor (or patch geometry in general) and thus not distance "as the crow flies." To do this, we imagine that animals will tend to avoid leaving the buffered habitat zone. Therefore, we assign `cost` $= 1$ if a pixel is within the buffer, and `cost` $= 10{,}000$ if a pixel is outside of a buffer. Therefore the cost to move to a neighboring pixel outside of the buffered area is 5,000.5 compared to the cost of 1 to move to a neighboring pixel inside the buffer. With this cost specification, we can compute the least-cost path distance matrix one time and modify our likelihood code to accept the distance matrix as input. We provide that likelihood in the package `scrbook` as the function `intlik3edv2`. We note also that this function accepts a habitat mask in the form of a vector of 0's and 1's that define any potential state-space restrictions. i.e., 1 if the pixel is an element of the state-space and 0 if it is not, and so additional modifications to the geometry of the region could be made. Here we simulate a population of $N = 200$ individuals in the corridor system and so we restrict our state-space accordingly for purposes of fitting the model. The code for doing all of this is in the help file for `intlik3edv2`, which contains the likelihood function and sample **R** script (`?intlik3edv2`).

```
### Define the cost structure
> cost <- rep(NA,nrow(pts))
> cost[in.pts==1]<-1     # low cost to move among pixels but not 0
> cost[in.pts!=1]<-10000 # high cost

### Stuff costs into a raster
> library("raster")
> r <- raster(nrows=40,ncols=40)
> projection(r) <- "+proj=utm +zone=12 +datum=WGS84"
> extent(r) <- c(0-delta/2,10+delta/2,0-delta/2,10+delta/2)
> values(r) <- matrix(cost,40,40,byrow=FALSE)

# check what it looks like
> plot(r)
> points(pts,pch=20,cex=.4)

# compute ecological distances:
> library("gdistance")
> tr1 <- transition(r,transitionFunction=function(x) 1/mean(x),directions=8)
> tr1CorrC <- geoCorrection(tr1,type="c",multpl=FALSE,scl=FALSE)
> costs1 <- costDistance(tr1CorrC,pts)
> outD <- as.matrix(costs1)
```

In the next block of code we simulate data and fit the model to the simulated data. Note that the object `traps` is loaded with `data(fakecorridor)` along with the data which define the f-shaped patch in Figure 12.5:

```
> library(scrbook)
> traplocs <- traps$loc
> trap.id <- traps$locid
> ntraps <- nrow(traplocs)

> set.seed(2013)
> N <- 200
> S.possible <- (1:nrow(pts))[in.pts==1]
> S.id <- sample(S.possible,N,replace=TRUE)
> S <- pts[S.id,]

> Dtraps <- outD[trap.id,]
> Deuclid <- e2dist(pts[trap.id,],pts)

> alpha0 <- -1.5
> sigma <- 1.5
> alpha1 <- 1/(2*sigma*sigma)
> K <-10

> probcap <- plogis(alpha0)*exp(-alpha1*D*D)
> Y <- matrix(NA,nrow=N,ncol=ntraps)
> for(i in 1:nrow(Y)){
+  Y[i,] <- rbinom(ntraps,K,probcap[i,])
> }
> Y <- Y[apply(Y,1,sum)>0,]

> frog1 <- nlm(intlik3edv2,c(-2.5,2,log(4)),hessian=TRUE,y=Y,K=K,X=traplocs,
+             S=pts,D=Dtraps,inpoly=in.pts)
> frog2 <- nlm(intlik3edv2,c(-2.5,2,log(4)),hessian=TRUE,y=Y,K=K,X=traplocs,
+             S=pts,D=Deuclid,inpoly=in.pts)
```

The output from fitting the two models; one with the correctly specified ecological distance constrained by the patch boundaries (constrained), and another using ordinary Euclidean distance (misspecified), are summarized in Table 12.3. We find little difference between the two models. In particular, 150 individuals were captured, leaving 50 individuals uncaptured. Therefore, $\log(n_0) = 3.9$. The correct model produces a slightly less accurate estimate, and it is favored by only 0.7 negative log-likelihood units. Therefore, for this single instance, the results are not too different. This is likely because the distance between individuals and traps that they are likely to be captured in is well approximated by Euclidean distance.

Table 12.3 Summary output from fitting models to simulated data in which movement is restricted by the habitat corridor shown in Figure 12.5. The two models fitted were those based on distance constrained by the corridor boundary ("constrained") and a misspecified model based on ordinary Euclidean distance which is "as the crow flies," and cuts through some boundaries. See `?fakecorridor` for the **R** commands to fit these models.

Distance	neg. LL	α_0	α_1	$\log(n_0)$
Constrained	−21.892	−1.338	0.332	4.353
Euclidean	−21.128	−1.307	0.382	4.212

12.8 Ecological distance and density covariates

Habitat characteristics that affect spatial variation in density can also affect home range size and movement behavior. For example, a species that occurs at high density in a forest may be reluctant to venture from a forest patch into an adjacent field. Thus, even if a trap placed in a field is located very close to an animal's activity center, the probability of capture may be very low. In this case, forest cover is a covariate of both density and encounter probability, and we could model it as such by combining the methods described in this chapter with those described in Chapter 11.

To demonstrate, we continue with our analysis of the data shown in Section 11.4.2. Once again, we suppose that density increases with canopy height, but this time, we also allow home range size to decrease as density increases. This commonly observed phenomenon can be explained by numerous factors such as intra-specific competition (Sillett et al., 2004) or optimal foraging behavior (Tufto et al., 1996; Saïd and Servanty, 2005).

A question that arises is: Is it possible to estimate the effect of the covariate on density (β_1) and α_2 using standard SCR data? In other words, can we model spatial variation in density and connectivity at the same time, using standard SCR data? Currently, it is not possible to model least-cost distance using **JAGS** or `secr`, so we wrote the function, `scrDED`, to fit the model using maximum likelihood. An example analysis is provided on the help page for the function in our **R** package `scrbook`. We briefly note here that the function requires the capture history data, the trap locations, and the raster data formatted using the `raster` package. The linear model for the intensity parameter $\mu(\mathbf{s}, \beta)$ and the least-cost distance function $\text{lcd}(\boldsymbol{\alpha})$ are specified using **R**'s formula interface. An example function call is

```
> fm <- scrDED(y, traplocs=X, den.formula=~elev, dist.formula=~elev,
+              rasters=elev.raster)
```

To assess the possibility of estimating both $\boldsymbol{\beta}$ and α_2, we conducted a small simulation study, generating 500 data sets from the model with both parameters set to 1, which corresponds to the conditions described above. The results indicate that it is possible to estimate both parameters (Figure 12.6), with MLEs appearing approximately unbiased.

12.9 Summary and outlook

Almost all published applications of SCR models to date have been based on models for the encounter probability that are functions of the Euclidean distance between individual activity centers and traps. The obvious limitations of such models are that Euclidean distance is unaffected by landscape or habitat structure and implies stationary, isotropic, and symmetrical home ranges. These are standard criticisms of the basic SCR model which we have seen many times in referee reports, or heard in discussions with colleagues. However, this should not be seen as criticism that

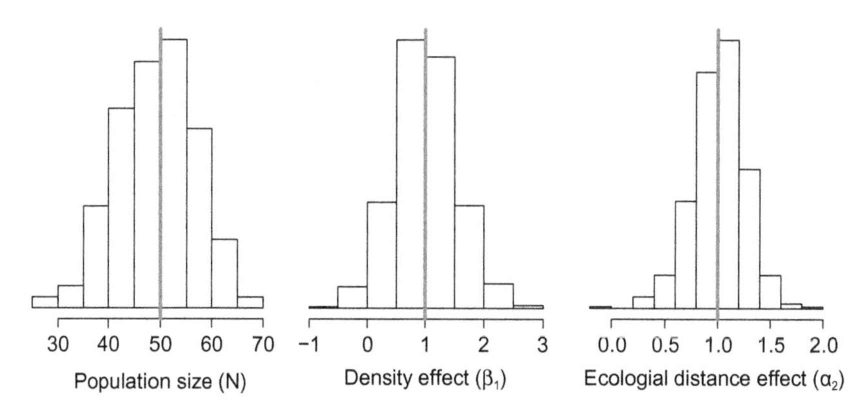

FIGURE 12.6

Histograms of parameter estimates (MLEs) from 500 simulations under the model in which both density and ecological distance are affected by the same covariate, canopy height. The vertical lines indicate the data-generating value.

is inherent to the basic conceptual formulation of SCR models because, as we have shown here, one can modify the Euclidean distance metric to accommodate more realistic formulations of distance that allow for inference to be made about landscape connectivity, and allow for the modeling of "distance" as a function of local habitat characteristics. As such, the effective distance between individual home range centers and traps varies depending on the local landscape.

How animals use space and therefore how distance to a trap is perceived by individuals is not something that can ever be known. We can only ever conjure up models to describe this phenomenon and fit those models to limited data from a sample of individuals during a limited amount of time. Here we have shown that there is hope to estimate connectivity parameters that describe how animals use space, from capture-recapture data alone, thereby allowing for irregular home range geometry that is influenced by landscape structure.

In the presence of functional landscape connectivity, misspecification of the model by an ordinary SCR model based on Euclidean distance produces biased estimates of model parameters (Royle et al., 2013a). This is expected because the effect is similar to failing to model heterogeneity, i.e., if we misspecify "model M_h" with "model M_0" (Otis et al., 1978) then we will expect to underestimate N. The effect of misspecifying the ecological distance metric with a standard homogeneous Euclidean distance has the same effect. In our view, this bias is not really the most important reason to consider models of ecological distance. Rather, inference about the structure of ecological distance is fundamental to many problems in applied and theoretical ecology related to modeling landscape connectivity, corridor and reserve design, population viability analysis, gene flow, and other phenomena. Models based on ecological distance allow investigators to evaluate landscape factors that influence movement of individuals over the landscape from capture-recapture data. Therefore SCR models based on ecological distance metrics might aid in understanding aspects of space usage and movement in animal populations and, ultimately, in addressing conservation-related problems such as corridor and reserve design.

CHAPTER

Integrating Resource Selection with Spatial Capture-Recapture Models

13

In Chapter 5 we briefly discussed the notion of how SCR encounter probability models relate to models of space usage. When using symmetric and stationary encounter probability models, SCR models imply that space usage is a decreasing function of distance from an individual's home range center. This is not a very realistic model in most applications. In this chapter, we extend SCR models to incorporate models of resource selection, such as when one or more explicit landscape covariates are available which the investigator believes might affect how individual animals use space within their home range. This is what Johnson (1980) called *third-order* selection—a term emphasizing the hierarchical nature of resource selection. An appealing feature of SCR models is that they provide a mechanism for modeling multiple levels of the resource selection hierarchy. For instance, Johnson (1980) defined *second-order* selection as the process determining the location of home ranges on a landscape, which is exactly the process being modeled using the methods presented in Chapter 11. Thus, SCR provides a way of studying the density and distribution of home range centers, while at the same time allowing for inferences about the use of resources within home ranges.

Our treatment follows Royle et al. (2013b) who integrated a standard family of resource selection models based on auxiliary telemetry data into the capture-recapture model for encounter probability. They argued that SCR models and resource selection models (Manly et al., 2002) are based on the same basic underlying model of space usage. The important distinction between SCR and RSF studies is that, in SCR studies, encounter of individuals is imperfect (i.e., "$p < 1$"); whereas, with RSF data obtained by telemetry, encounter is perfect. SCR and telemetry data can therefore be combined in the same likelihood by formally recognizing this distinction in the model.

There are two important motives for considering a formal integration of RSF models with capture-recapture. The first is to integrate models of resource use by individuals with models of population size or density. There is relatively little in the literature on this topic, although Boyce and McDonald (1999) describe a procedure in which (an estimate of) population size is used to scale resource selection functions to produce a population density surface. The second reason is to allow for the integration of auxiliary data from telemetry studies with capture-recapture data. Telemetry data have been widely used in conjunction with capture-recapture data in standard non-spatial models. For example, White and Shenk (2001) and Ivan (2012) suggested using telemetry data to estimate the probability that an individual is exposed to capture-recapture sampling. However, their estimator requires that individuals are

Spatial Capture-Recapture. http://dx.doi.org/10.1016/B978-0-12-405939-9.00013-X

telemetry-tagged in proportion to this unknown quantity (probability of exposure), which seems impossible to achieve in many studies. In addition, they do not directly integrate the telemetry data with the capture-recapture model so that common parameters may be jointly estimated.

Formal integration of capture-recapture with telemetry data for the purposes of modeling resource selection has a number of immediate benefits. For one, telemetry data provide direct information about σ (Sollmann et al., 2013a, c). As a result, model parameter estimates are improved which, as we see in Section 10.7, also has important implications for the design of SCR studies. In addition, active resource selection by animals induces a type of heterogeneity in encounter probability, which is misspecified by standard SCR encounter probability models. Animals that use more space due to the configuration of habitat or landscape features stand to be exposed to more traps than animals that use less space. Ignoring the heterogeneity in encounter probability resulting from resource selection can lead to bias in estimates of population size or density (Royle et al., 2013b). Finally, because the resource selection model translates directly to a model for encounter probability for spatial capture-recapture data, the implication of this is that it allows us to estimate resource selection model parameters directly from SCR data, i.e., *absent* telemetry data. This fact should broaden the practical relevance of spatial capture-recapture not just for estimating density, but also for directly studying movement and resource selection.

13.1 A model of space usage

Assume that the landscape is defined in terms of a discrete raster of one or more covariates, having the same dimensions and extent. Let $\mathbf{s}_1, \ldots, \mathbf{s}_G$ identify the center coordinates of G pixels that define a landscape, organized in the matrix $\mathbf{X}_{G\times 2}$. Let $C(\mathbf{s})$ denote a covariate defined for every pixel $\mathbf{s}$. We suppose that individual members of a population utilize space in some manner related to the covariate $C(\mathbf{s})$. We define one plausible model below.

As a biological matter, space use is the outcome of individuals moving around their home range (Hooten et al., 2010), i.e., where an individual is at any point in time is the result of some movement process. However, to understand space usage, it is not necessary to entertain explicit models of movement, just to observe the outcomes, and so we don't elaborate further on what could be sensible or useful models of movement, but we imagine existing methods of hierarchical or state-space models are suitable for this purpose (Ovaskainen, 2004; Jonsen et al., 2005; Forester et al., 2007; Ovaskainen et al., 2008; Hooten et al., 2010; McClintock et al., 2012). We consider explicit movement models in the context of SCR models in the later chapters of this book (Chapters 15 and 16). Here we adopt more of a phenomenological formulation of space usage as follows: If an individual appears in pixel $\mathbf{s}$ at some instant, this is defined as a decision to "use" pixel $\mathbf{s}$. Over any prescribed time interval, if we sample some number of times say R, then let m_{ij} be the use frequency of pixel j by individual i—i.e., the number of times individual i used pixel j. We assume the vector of use frequencies $\mathbf{m}_i = (m_{i1}, \ldots, m_{iG})$ has a multinomial distribution:

$$\mathbf{m}_i \sim \text{Multinomial}(R, \boldsymbol{\pi}_i),$$

where $R = \sum_j m_{ij}$ is the total number of "use decisions" made by individual i and

$$\pi_{ij} = \frac{\exp(\alpha_2 C(\mathbf{s}_j))}{\sum_{\mathbf{s}} \exp(\alpha_2 C(\mathbf{s}))},$$

for each $j = 1, 2, \ldots, G$ pixels. This is a standard RSF model used to model telemetry data. In particular, this is "protocol A" of Manly et al. (2002) where all available landscape pixels are censused (i.e., known without error), and used pixels are sampled randomly for each individual. The parameter α_2 is the effect of the landscape covariate $C(\mathbf{x})$ on the relative probability of use. Thus, if α_2 is positive, the relative probability of use increases as the covariate increases.

In practice, we don't observe m_{ij} for all individuals but, instead, only for a small subset which we capture and telemeter. For the telemetered individuals, we assume they use resources according to the same RSF model as the population as a whole.

To extend this model to make it more realistic, and consistent with the formulation of SCR models, let $\mathbf{s}$ denote the center of an individual's home range and let $d_{ij} = ||\mathbf{s}_j - \mathbf{s}_i||$ be the distance from the home range center of individual i, $\mathbf{s}_i$, to pixel j, $\mathbf{s}_j$. We modify the space usage model to accommodate that space use will be concentrated around an individual's home range center:

$$\pi_{ij} = \frac{\exp\left(-\alpha_1 d_{ij}^2 + \alpha_2 C(\mathbf{s}_j)\right)}{\sum_x \exp\left(-\alpha_1 d_{ij}^2 + \alpha_2 C(\mathbf{s})\right)}. \quad (13.1.1)$$

The parameters α_1, α_2 and the activity centers $\mathbf{s}$ can be estimated directly from telemetry data using standard likelihood methods based on the multinomial likelihood (Johnson et al., 2008b). Normally this model is expressed in terms of the scale parameter σ, where $\alpha_1 = 1/(2\sigma^2)$. The multinomial model Eq. (13.1.1) can be understood as a compound model of space usage, governed by distance-based "availability" according to a Gaussian kernel, and use conditional on availability (Johnson et al., 2008b; Forester et al., 2009). In other words, the model suggests a kind of distance-based availability in which a pixel is less available to an individual if it is located further away from $\mathbf{s}_i$.

Equation (13.1.1) resembles standard SCR encounter probability models that we have used previously, but here the model includes an additional covariate $C(\mathbf{s})$ (see Chapter 9). In particular, under this model for space usage or resource selection, if we have no covariates at all, or if $\alpha_2 = 0$, then the probabilities π_{ij} are directly proportional to the SCR model for encounter probability. In other words, setting $\alpha_2 = 0$, the probability of use for pixel j is:

$$p_{ij} \propto \exp\left(-\alpha_1 d_{ij}^2\right).$$

Clearly, whatever function of distance we use in the RSF model implies an equivalent SCR model for encounter probability and vice versa. In particular, for whatever encounter probability model we choose for p_{ij} in an ordinary SCR model, we can modify the distance component in the RSF function in Eq. (13.1.1) to be consistent with that model by setting:

$$\pi_{ij} \propto \exp(\log(p_{ij}) + \alpha_2 C(\mathbf{s}_j))$$

(see Forester et al., 2009). Therefore, to contemplate integrating RSFs based on telemetry data with SCR data, we think of SCR data as representing a thinning (or sampling) of the potentially observable ("perfect") use frequencies m_{ij} obtainable by telemetry. The above expression identifies the manner in which parameters of the SCR encounter probability model are shared by the parameters of the RSF model.

13.1.1 A simulated example

For a simulated landscape (shown in Figure 13.1), Royle et al. (2013b) depicted some typical space usage patterns under the model described above, which we reproduce here in Figure 13.2. The covariate in this case was simulated using a kriging model of correlated random noise with the following **R** commands:

```
> set.seed(1234)
> gr <- expand.grid(1:40,1:40)
> Dmat <- as.matrix(dist(gr))
> V <- exp(-Dmat/5)
> C <- t(chol(V))%*%rnorm(1600)
```

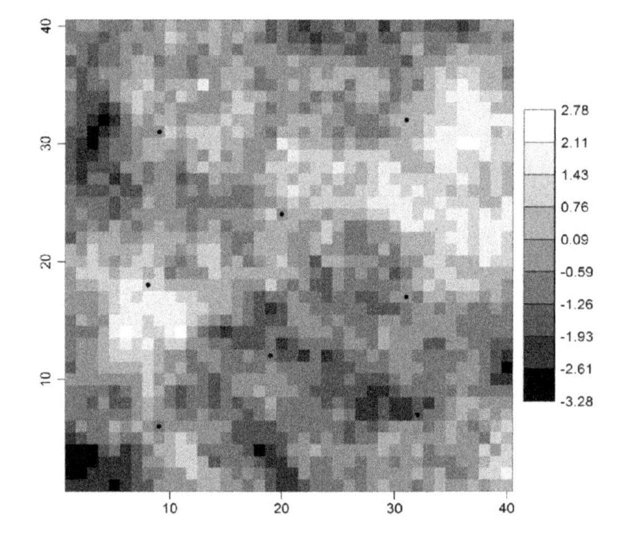

FIGURE 13.1

A typical habitat covariate reflecting habitat quality or hypothetical utility of the landscape to a species under study. Home range centers for 8 individuals are shown with black dots.

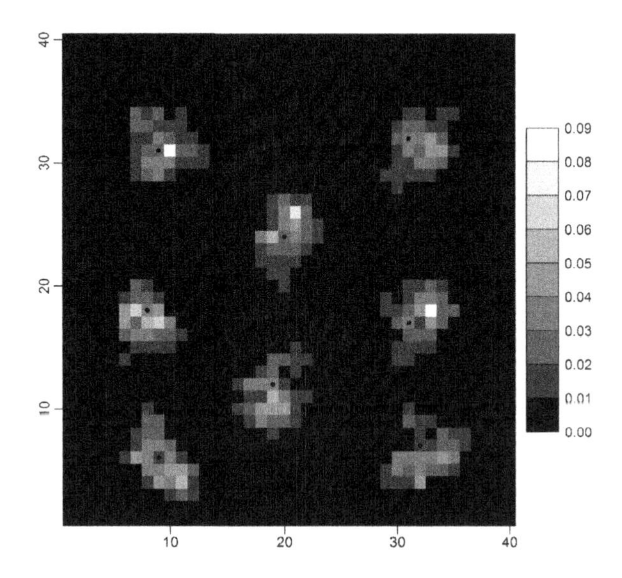

FIGURE 13.2

Space usage patterns of 8 individuals under a space usage model that contains a single covariate which is shown in Figure 13.1. The plotted value is the multinomial probability π_{ij} for pixel j under the model in Eq. (13.1.1).

The resulting covariate vector $\mathbf{C}$ is multivariate normal with mean 0 and variance-covariate matrix $\mathbf{V}$ which, here, has pairwise correlations that decay exponentially with distance. The use densities shown in Figure 13.2 were simulated with $\alpha_1 = 1/(2\sigma^2)$, $\sigma = 2$, and the coefficient on $C(\mathbf{s})$ set to $\alpha_2 = 1$. The resulting space usage densities—or "home ranges"—exhibit clear non-stationarity in response to the structure of the underlying covariate, and they are distinctly asymmetrical. We note that if α_2 were set to 0, the 8 home ranges shown here would be proportional to a bivariate normal kernel with $\sigma = 2$.[1] The commands for generating the covariate, and producing Figure 13.1 are in the package `scrbook` (see `?RSF_example`).

13.1.2 Poisson model of space use

A natural way to motivate the multinomial model of space usage is to assume that individuals make a sequence of resource selection decisions so that the outcomes m_{ij} are *independent* Poisson random variables:

$$m_{ij} \sim \text{Poisson}(\lambda_{ij}),$$

where

$$\log(\lambda_{ij}) = a_0 - \alpha_1 d_{ij}^2 + \alpha_2 C(\mathbf{s}_j).$$

[1] This is why we have always referred to the similar-looking model for encounter probability as the Gaussian or bivariate normal model, instead of half-normal.

In this case, the number of visits to any particular cell is affected by the covariate $C(\mathbf{s})$ but has a baseline rate, $\exp(a_0)$, related to the amount (in an expected value sense) of movement occurring over some time interval. This is an equivalent model to the multinomial model given previously in the sense that, if we condition on the total sample size $R = \sum_j m_{ij}$, then the vector $\mathbf{m}_i$ has a multinomial distribution with probabilities given by Eq. (13.1.1).

In practice, we never observe "truth," i.e., the actual use frequencies m_{ij}. Instead, we observe a sample of the actual use outcomes by an individual. As formulated in Section 5.4, we assume a binomial ("random") sampling model for encounter frequencies obtained from an SCR study:

$$y_{ij} \sim \text{Binomial}(m_{ij}, p_0).$$

We can think of these counts as arising by thinning the underlying point process (here, aggregated into pixels) where p_0 is the thinning rate of the point process. In this case, the marginal distribution of the observed counts y_{ij} is also Poisson but with mean

$$\log(\mathbb{E}(y_{ij})) = \log(p_0) + a_0 - \alpha_1 d_{ij}^2 + \alpha_2 C(\mathbf{s}_j).$$

Thus, the space-usage model (RSF) for the thinned counts y_{ij} is the same as the space usage model for the original variables m_{ij}. This is because if we remove m_{ij} from the conditional model by summing over its possible values, then the vector $\mathbf{y}_i$ is *also* multinomial with cell probabilities

$$\pi_{ij} = \frac{\lambda_{ij}}{\sum_j \lambda_{ij}},$$

where any constant (the intercept term a_0 and thinning rate p_0) cancels from the numerator and denominator. Thus, the underlying multinomial RSF model applies to the frequencies $\mathbf{m}_i$ and those produced from thinning or sampling, $\mathbf{y}_i$.

13.2 Integrating capture-recapture data

The key to combining RSF data with SCR data is to note that the Poisson model of space usage given above is exactly our Poisson encounter probability model from Chapter 9, but with a spatial covariate $C(\mathbf{s})$, and some arbitrary intercept or off-set related to the sampling rate of the telemetry device. We've used exactly this model for SCR data (Chapter 9), but with a different intercept, α_0, unrelated to the intercept of the Poisson use model for telemetry described above, but, rather, to the efficiency of the capture-recapture encounter device. In other words, we view camera traps (or other devices) located in some pixel $\mathbf{s}$ (or multiple pixels) as being equivalent to being able to turn on a type of (less perfect) telemetry device, and only in some pixels. Therefore, data from camera trapping are Poisson random variables for every pixel j where a trap is located:

$$y_{ij}|\mathbf{s}_i \sim \text{Poisson}(\lambda_{ij})$$

with

$$\log(\lambda_{ij}) = \alpha_0 - \alpha_1 d_{ij}^2 + \alpha_2 C(\mathbf{s}_j).$$

The parameters α_1 and α_2 are shared with the multinomial resource selection model for the telemetry data. A key point here is that if resource selection is happening, then it appears as a covariate on encounter rate (or encounter probability) in the same way as ordinary covariates which we discussed in Chapter 7.

Alternatively, the SCR study can produce binary encounters depending on the type of sampling being done, where $y_{ij} = 1$ if the individual i visited the pixel containing a trap and was detected. In this case, we imagine that y_{ij} is related to the latent variable m_{ij} being the event $m_{ij} > 0$, which occurs with probability

$$p_{ij} = 1 - \exp(-\lambda_{ij}) \tag{13.2.1}$$

and then the observed encounter frequencies for individual i and trap j, from sampling over K occasions, are binomial:

$$y_{ij}|\mathbf{s}_i \sim \text{Binomial}(K, p_{ij}).$$

To construct the likelihood for SCR data when we have direct information on space usage from telemetry data, we regard the two samples (SCR and RSF) as independent of one another, and we form the likelihood for each set of observations as a function of the same underlying parameters. The joint likelihood then is the product of the two components. In particular, let $\mathcal{L}_{scr}(\alpha_0, \alpha_1, \alpha_2, N; \mathbf{y})$ be the likelihood for the SCR data in terms of the basic encounter probability parameters and the total (unknown) population size N, and let $\mathcal{L}_{rsf}(\alpha_1, \alpha_2; \mathbf{m})$ be the likelihood for the RSF data based on telemetry which, because the sample size of telemetered individuals is fixed, does not depend on N. Assuming independence of the two data sets, the joint likelihood is the product of these two pieces:

$$\mathcal{L}_{rsf+scr}(\alpha_0, \alpha_1, \alpha_2, N; \mathbf{y}, \mathbf{m}) = \mathcal{L}_{scr}(\alpha_0, \alpha_1, \alpha_2, N; \mathbf{y}) \times \mathcal{L}_{rsf}(\alpha_1, \alpha_2; \mathbf{m}),$$

where $\mathcal{L}_{scr}$ is the standard integrated likelihood (Chapter 6), and the RSF likelihood contribution is the multinomial likelihood having cell probabilities given by Eq. (13.1.1). The **R** code for maximizing the joint likelihood was given in the supplement to Royle et al. (2013b), and we include a version of this in the `scrbook` package, see `?intlik3rsf`, which also shows how to simulate data and fit the combined SCR + RSF model.

13.3 SW New York black bear study

Royle et al. (2013b) applied the integrated SCR + RSF model to data from a study of black bears[2] (*Ursus americanus*) in a region of approximately 4, 600 km^2 in southwestern New York (Sun, in press). These data can be loaded from the `scrbook` package with the command `data(sunbears)`.

[2]This is a different data set from the Fort Drum bear study which we've analyzed in previous chapters.

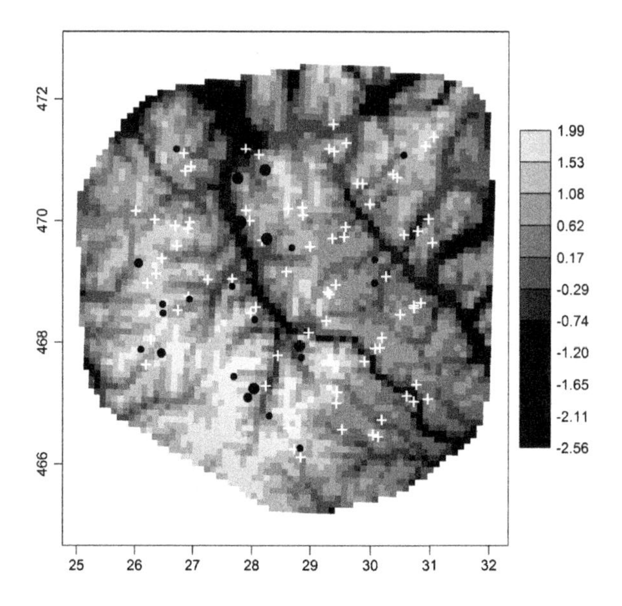

FIGURE 13.3

Elevation (standardized), and hair snare locations marked by the number of individuals captures at each site according to the size of the black circle. White crosses represent hair snare locations with no captures.

The data are based on a noninvasive genetic capture-recapture study using 103 hair snares in June and July, 2011. Hair snares were baited, scented and checked weekly (Sun, in press). The study yielded sparse encounter histories of 33 individuals with a total of 14 recaptures and 27 individuals captured 1 time only. Telemetry data were collected on three individuals, which produced locations for each bear approximately once per hour. Telemetry locations were thinned to once per 10 h to produce movement outcomes that might be more independent. This produced 195 telemetry locations used in the RSF component of the model. Elevation was used as a covariate for this model, a standardized version of which is shown in Figure 13.3 along with the number of individuals captured at each hair snare site.

There are a number of models that could be fitted to these data based on the combination of SCR and RSF data as well as the elevation covariate. The models fitted here are based on the Gaussian hazard encounter probability model, including an ordinary SCR model with no covariates or telemetry data, the SCR model with elevation affecting either baseline encounter probability or density $D(\mathbf{s})$ (Chapter 11), and models that use telemetry data. The six models and notation describing each are as follows:

Model 1, p(·): ordinary SCR model

Model 2, p(elev): ordinary SCR model with elevation as a covariate on baseline encounter probability.

Table 13.1 Summary of model-fitting results for the black bear study. Parameter estimates are for the intercept (α_0), logarithm of σ, the scale parameter of the Gaussian hazard encounter model, β is the coefficient of elevation on density, α_2 is the coefficient of elevation in the encounter probability/space-usage model, and the total population size of the state-space is N. Standard errors are in parentheses. The SCR data are based on $n = 33$ individuals, and the telemetry data are based on 3 individuals.

Model	α_0	$\log(\sigma)$	α_2	N	β	−loglik
p(elev)	−2.860	−1.117	0.175	95.80		122.738
	(0.390)	(0.139)	(0.248)	(22.99)		
p(·)	−2.729	−1.122		93.90		122.990
	(0.345)	(0.140)		(22.06)		
p(·) + D(elev)	−2.715	−1.133		94.20	1.247	118.007
	(0.353)	(0.139)		(21.90)	(0.408)	
p(elev) + D(elev)	−2.484	−1.157	−0.384	103.50	1.571	117.075
	(0.391)	(0.142)	(0.276)	(26.56)	(0.463)	
p(elev) + RSF	−3.068	−0.814	−0.281	81.60		1271.739
	(0.272)	(0.036)	(0.118)	(17.65)		
p(elev) + D(elev) + RSF	−3.070	−0.810	−0.371	89.10	1.273	1266.700
	(0.272)	(0.037)	(0.124)	(20.55)	(0.411)	

Model 3, p(·) + D(elev): ordinary SCR model with elevation as a covariate on density only.
Model 4, p(elev) + D(elev): ordinary SCR model with elevation as a covariate on both baseline encounter probability and density.
Model 5, p(elev) + RSF: SCR model including data from three telemetered individuals.
Model 6, p(elev) + D(elev) + RSF: SCR model including telemetered individuals and with elevation as a covariate on density.

Parameter estimates for the six models are given in Table 13.1 (reproduced from Royle et al. (2013b), see also the help file `?sunbears`). It is tempting to want to compare these different models by AIC but, because models 5 and 6 involve additional data, they cannot be compared with models 1–4.

By looking at Table 13.1, it is clear based on the negative log-likelihood of Models 1–4, that those containing an elevation effect on density (parameter β) are preferred (Models 3 and 4). The parameter estimates indicate a positive effect of elevation on density, which seems to be consistent with the raw capture data shown in Figure 13.3. Despite this strong effect of elevation, the estimates of N under each of these models only ranged from 93 to 103 bears for the 4,600 km^2 state-space, and so estimated density is somewhat consistent across models. If we consider not just density, but space usage (i.e., looking at the parameter α_2), the effect of elevation is negative. Thus, elevation appears to affect density and space usage differently. It was suggested that density operates at the second-order scale of resource selection and ".... is largely related to the spacing of individuals and their associated home ranges across the landscape. On the other hand, our RSF was defined based on selection of resources within the home range (third-order)" (Royle et al., 2013b). The positive effect of elevation on density is consistent with some other studies on black bears (e.g., Frary et al., 2011), and the negative effect of elevation on space usage can be attributed to seasonal variation in food availability, usage of corridors, or environmental conditions.

Models 5 and 6 include the additional telemetry data, thus the negative log-likelihoods are not directly comparable to the first four models, but we can still make a few important observations. First is that the parameter estimates under these two models are consistent with Model 4 in that elevation had a strong effect on both density and space usage. In comparing Models 5 and 6, the latter model which includes elevation as an effect on density reduces the negative log-likelihood by 5 units. Additionally, including the telemetry data reduces the standard errors (SE) of the density and space usage parameters and as we would expect, the incorporation of telemetry data also reduces the SE for σ. Model 6 (p(elev) + D(elev) + RSF) was used to produce maps of density (Figure 13.4) and space usage (Figure 13.5) showing the effect of elevation on both components of the model. The map of space usage shows the

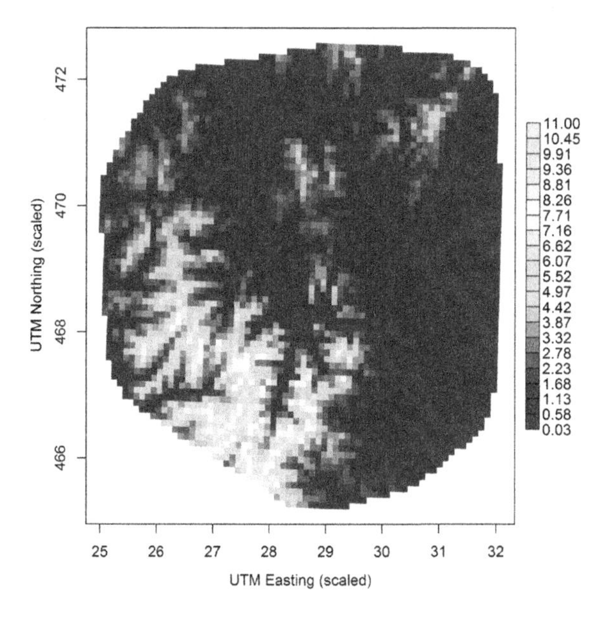

FIGURE 13.4

Predicted density of black bears (per 100 km^2) in southwestern New York study area.

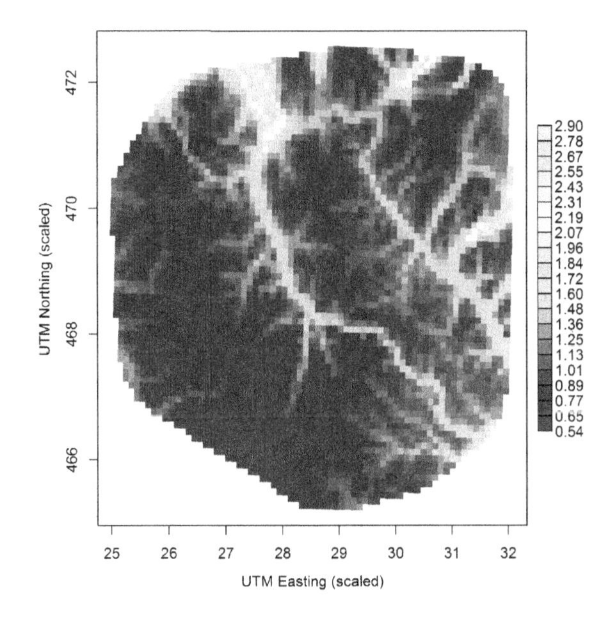

FIGURE 13.5

Relative probability of use of pixel **x** compared to a pixel of mean elevation, at a constant distance from the activity center.

relative probability of using a pixel **x** relative to one having the mean elevation, given a constant distance to the individual's activity center.

13.4 Simulation study

Using the simulated landscape shown in Figure 13.1, Royle et al. (2013b) presented results of a simulation study considering populations of $N = 100$ and $N = 200$ individuals exposed to encounter by a 7×7 array of trapping devices, with $K = 10$ sampling occasions, using the Gaussian hazard model (Eq. (13.2.1)) with

$$\log(\lambda_{ij}) = -2 - \frac{1}{2\sigma^2} d_{ij}^2 + 1 \times C(\mathbf{s}_j),$$

where $\sigma = 2, \alpha_0 = -2$ and $\alpha_2 = 1$. They investigated the effect of misspecification of the resource selection model with an ordinary model SCR0 (i.e., no habitat covariates affecting the trap encounter model), and the performance of the MLEs, under SCR + telemetry designs having 2, 4, 8, 12, and 16 telemetered individuals (with 20 independent telemetry fixes *per* individual). Three models were fitted: (i) the SCR only model, in which the telemetry data were not used; (ii) the integrated SCR/RSF model which combined all of the data for jointly estimating model parameters; and (iii) the RSF only model which just used the telemetry data alone (and therefore

Table 13.2 This table summaries the sampling distribution of the MLE of model parameters for models fitted to data generated under a resource selection model. The models fitted include the misspecified model, which is a basic model SCR0 (with no covariate), the SCR model with the covariate on encounter probability (SCR + C(x)), and the SCR model including the covariate and a sample of telemetered individuals (SCR/RSF; n is the number of individuals telemetered). Data were simulated with $N = 200$ individuals, $\alpha_2 = 1$ and $\sigma = 2$.

	$\hat{N}$	RMSE	$\hat{\alpha}_2$	RMSE	$\hat{\sigma}$	RMSE
$n = 2$						
SCR + C(x)	199.11	14.28	0.99	0.09	2.00	0.090
SCR/RSF	199.11	13.80	0.99	0.09	2.00	0.079
SCR0	161.48	39.98	–	–	1.84	0.180
$n = 4$						
SCR + C(x)	199.67	13.87	1.00	0.09	2.00	0.090
SCR/RSF	199.65	13.59	1.00	0.09	2.00	0.072
SCR0	161.32	40.00	–	–	1.83	0.191
$n = 8$						
SCR + C(x)	199.24	15.49	0.99	0.10	2.01	0.093
SCR/RSF	199.55	14.17	0.99	0.08	2.00	0.063
SCR0	161.46	40.06	–	–	1.84	0.184
$n = 12$						
SCR + C(x)	200.41	15.16	0.99	0.10	2.00	0.086
SCR/RSF	200.95	13.04	1.00	0.08	2.00	0.051
SCR0	162.40	38.95	–	–	1.84	0.185
$n = 16$						
SCR + C(x)	199.16	15.62	1.00	0.09	2.00	0.095
SCR/RSF	199.63	13.38	1.00	0.07	2.00	0.052
SCR0	160.93	40.44	–	–	1.84	0.190

the parameters α_0 and N are not estimable). An abbreviated version of the results from Royle et al. (2013b) are summarized in Table 13.2. We provide an **R** script (see `?RSFsim`) that can be modified for further analysis and exploration.

One thing we see is a dramatic negative bias in estimating N if the model SCR0 is fitted (interestingly, there is much less bias in estimating σ). Overall, though, when either the SCR model with covariate or the joint SCR + RSF model is fitted, the MLEs exhibit little bias for the parameter values simulated here. In terms of RMSE, there is only a slight $\approx$5–10% reduction in RMSE of the estimator of N when we have at least 2 telemetered individuals. Thus, estimating N benefits only slightly from the addition of telemetry data. This may be reasonable when we consider that information about the intercept, α_0, comes only from the capture-recapture data. However, there is a large improvement in precision (50–60%) for estimating the scale parameter σ. While this doesn't translate into improved estimation of N, it suggests that telemetry data should be relevant to the design of SCR studies for which trap spacing is one of the main considerations (Chapter 10). It is conceivable even that spatial recaptures are not needed if some telemetry data are available (in Chapter 19, in the context of mark-resight models, we show a case study of raccoons where additional telemetry data allows estimating model parameters in spite of a very low number of spatial recaptures (Sollmann et al., 2013a)). The resource selection parameter α_2 is well estimated even *without* telemetry data. The fact that parameters of resource selection can be estimated from spatial capture-recapture data alone should have considerable practical relevance in the study of animal populations and landscape ecology. For the highest sample size of telemetered individuals ($n = 16$), the RMSE for estimating this parameter only decreases from about 0.09 to 0.07 compared to SCR alone.

13.5 Relevance and relaxation of assumptions

In constructing the combined likelihood for RSF and SCR data, we assumed the data from capture-recapture and telemetry studies were independent of one another. This may be reasonably satisfied if the two samples are random samples from the population of individuals with activity centers on the state-space (see Chapter 19 for additional discussion of this). We cannot foresee situations in which violation of this assumption should be problematic or invalidate the estimator under the independence assumption.

Our model pretends that we do not know anything about the telemetered individuals in terms of their encounter history in traps. In principle it should not be difficult to admit a formal reconciliation of individuals between the two lists. In that case, we just combine the two conditional likelihoods before we integrate $\mathbf{s}$ from the product likelihood. This would be almost trivial to do if *all* individuals were reconcilable between the SCR and telemetry samples (or none, as in the case we have covered here). But, in general, we think you will often have an intermediate case, i.e., only a subset of telemetered animals will be known and there will be some individuals encountered by traps or devices whose telemetry marking status cannot be determined. That case, which is basically a type of marking uncertainty or misclassification, is clearly more difficult to deal with (see Chapter 19 for some additional context).

We presented the model in a discrete landscape which regarded potential trap locations and the covariate $C(\mathbf{s})$ as being defined on the same set of points. In practice, trap locations may be chosen independently of the definition of the raster and this does not pose any challenge or novelty to the model as it stands. In that case, the covariate(s) need to be defined at each trap location in order to fit the models. The model should be applicable also to covariates that are naturally continuous (e.g., distance-based covariates) although, in practice, it will usually be sufficient to work with a discrete representation of such covariates.

The multinomial RSF model for telemetry data assumes independent observations of resource selection. This would certainty be reasonable if telemetry fixes are made far apart in time (or thinned sufficiently). However, as noted by Royle et al. (2013b), the independence assumption is *not* an assumption of spatially independent movement outcomes in geographic space. Active resource selection should probably lead to the appearance of spatially dependent outcomes, regardless of how far apart in time the telemetry locations are. Even if resource selection observations are dependent, use of the independence model probably yields unbiased estimators while understating the variance. Continued development of integrated SCR + RSF models that accommodate more general models of movement is needed.

13.6 Summary and outlook

How animals use space is of fundamental interest to ecologists and is important in the conservation and management of many species. Investigating space use is normally done using telemetry and associated models referred to as resource selection functions (Manly et al., 2002). But, in all of human history, animal resource selection has never been studied using capture-recapture models. Instead, essentially all applications of SCR models have focused on density estimation. It is intuitive, however, that space usage or resource selection should affect encounter probability and thus it should be highly relevant to density estimation in SCR applications, and, vice versa, SCR applications should yield data relevant to resource selection questions. The development in this chapter shows clearly that these two ideas can be unified within the SCR methodological framework so that classical notions of resource selection modeling can be addressed simultaneous to modeling of animal density. What we find is that if animal resource selection is occurring, this can be modeled as covariates on encounter probability, with or without the availability of auxiliary telemetry data. If telemetry data do exist, we can estimate parameters jointly by combining the two likelihood components—that of the SCR data and that of the telemetry data.

Active resource selection by individuals induces a type of heterogeneous encounter probability, and this induces (possibly severe) bias in the estimated population size for a state-space when default symmetric encounter probability models are used. As such, it is important to account for resource selection when relevant covariates are known to influence resource selection patterns. Aside from properly modeling this selection-induced heterogeneity, integration of RSF data from telemetry with SCR models achieves a number of useful advances: First, it leads to an improvement in our ability to estimate density, and an improvement in our ability to estimate parameters of the RSF function. As many animal population studies have auxiliary telemetry information, the incorporation of such information into SCR studies has broad applicability. Secondly, the integrated model allows for the estimation of RSF model parameters directly from SCR data *alone*. Therefore, SCR models *are* explicit models of resource selection. In our view, this greatly broadens the utility and importance of capture-recapture studies beyond their primary historical use of estimating density or population size. Finally, we note that telemetry information provide direct information about the home range shape parameter, σ, in our analyzes above, and its estimation is greatly improved with even moderate amounts of telemetry data (see also Sollmann et al. (2013a) and Sollmann et al. (2013c)). This should have some consequences in terms of the design of capture-recapture studies (Chapter 10), especially as it relates to trap spacing.

Simultaneously conducting telemetry studies with capture-recapture is extremely common in field studies of animal populations. However, the simultaneous, integrated analysis of the two sources of data is uncommon. The new class of integrated SCR/RSF models allows researchers to model how the landscape and habitat influence the movement and space use of individuals around their home range, using capture-recapture data that can be augmented with telemetry data. This should improve our ability to understand, and study, aspects of space usage and it might, ultimately, aid in addressing conservation-related problems such as reserve or corridor design.

CHAPTER

Stratified Populations: Multi-Session and Multi-Site Data

14

In this chapter, we describe SCR models for situations when we have multiple distinct sample groups, strata, or "sessions" (the term used in `secr`) each with a population size parameter N_g, for group g. The modeling objectives in such cases are (1) to combine the data into a single, unified model for purposes of improved estimation of parameters that can be shared across groups and (2) to model variation in group-specific population sizes or, in the case of SCR models, density. Such "stratified" populations are commonplace in capture-recapture studies, especially in the context where the strata represent distinct spatial regions, yet most SCR applications have been based on models that are distinctly single-population models. This is done either by analyzing separate data sets one-at-a-time, producing many independent estimates of abundance, or by pooling data from multiple study areas. A standard example that arises frequently is that in which multiple habitat patches (often refuges, parks, or reserves) are sampled independently with the goal of estimating the population size of some focal species in each reserve. If there are parameters that can be shared across sessions or groups, it makes sense to combine the data together into a single model that permits the sharing of information about some parameters, but provides individual estimates of abundance for each land unit. A similar situation is that in which a number of replicate trap arrays are located within a landscape, sometimes for purposes of evaluating the effects of management actions or landscape structure on populations. This is a common situation in studies of small mammals (Converse et al., 2006a,b; Converse and Royle, 2012), or in mist-netting of birds (DeSante et al., 1995), but there are examples of large-scale monitoring of carnivores and other species too, e.g., tigers (Jhala et al., 2011).

In previous chapters, we've analyzed data for a number of examples that have a natural stratification or group structure. In Chapter 9, we analyzed the ovenbird data as an example of a multi-catch (independent multinomial) model, where we used year as the stratification variable, and the possum data (illustrating the single-catch situation) in which the group structure arose from the use of five distinct trap arrays. In Chapters 7 and 8 we fitted models with sex specificity of parameters using multi-session models, where the stratification variable in that case was sex. In this chapter, we focus on Bayesian analysis of stratified SCR models using data augmentation (Converse and Royle, 2012; Royle et al., 2012b). The technical modification of data augmentation to deal with such models is that it is based on a model for the joint distribution of

Spatial Capture-Recapture. http://dx.doi.org/10.1016/B978-0-12-405939-9.00014-1

the stratum-specific population sizes, N_g, conditional on their total. This results in a multinomial distribution for all N_g, which we can analyze in some generality using data augmentation. As a practical matter, specification of this multinomial distribution for the N_g parameters induces a distribution for an individual covariate, say g_i, which is "group membership." This is extremely handy to analyze by MCMC in the various **BUGS** engines that you are familiar with by now. The flexibility of model specification in **BUGS** is why we focus a whole chapter here on Bayesian analysis of stratified population models using data augmentation. However, we have noted previously that the **R** package `secr` fits a class of multi-session models which we applied to the ovenbird and possum data (Chapter 9), and models with sex-specific parameters in Chapters 7 and 8.

In the stratified population models considered here, an individual is assumed to be a member of a single stratum, so that the population sizes N_g for the g strata are independent of one another. However, stratified or multi-session SCR models are also directly relevant when the stratification index is time, either involving distinct periods within a biological season, or even across years. In this case, individuals might belong to multiple of the strata, but, the models discussed in this chapter do not acknowledge that explicitly. Unlike the case in which the strata represent spatial units, with temporally defined strata, a fully dynamic, or demographically open model for N might be appropriate—one that involves survival and recruitment. We deal with those models specifically in Chapter 16. However, the stratified models covered here can be thought of as a primitive type of model for open systems in which the population sizes are assumed to be independent across temporal strata, and we might still find them useful in cases where the strata are temporal periods or sessions.

14.1 Stratified data structure

We suppose that $g = 1, 2, \ldots, G$ strata (or groups), having population sizes N_g, and state-spaces $\mathcal{S}_g$, are sampled using some capture-recapture method producing sample sizes of n_g unique individuals and encounters y_{ijk} for individual $i = 1, 2, \ldots, \sum_{g=1}^{G} n_g$. Right now we won't be concerned with the details of every type of capture-recapture observation model so, for context, and to develop some technical notions, we consider a Bernoulli encounter model in which individual- and trap-specific encounter frequencies are binomial counts: $y_{ij} \sim \text{Binomial}(K, p_{ij})$. Let g_i be a covariate (integer-valued, $1, \ldots, G$) indicating the group membership of individual i. This covariate is observed for the sample of captured individuals but not for individuals that are never captured.

To illustrate the prototypical data structure for stratified SCR data, we suppose that a population comprised of four groups is sampled $K = 5$ times. Then, a plausible data set has the following structure:

```
      individual (i):  1  2  3  4  5  6  7  8  9  10
total encounters (y):  1  1  3  1  1  2  2  4  1   1
           group (g):  1  1  1  2  3  3  3  3  4   4
```

This data set indicates that three individuals were captured in group 1 (captured 1, 1, and 3 times), a single individual was captured in group 2, four individuals were captured in group 3, and two individuals were captured in group 4.

A key idea discussed shortly is that the assumption of certain models for the collection of abundance variables N_g implies a specific model for the group membership variable g_i. Then, the data from all groups can be pooled, and analyzed as data from a single population with the appropriate model on g_i, without having to deal with the N_g parameters in the model directly. In this way, we can easily build hierarchical models for stratified populations, using an *individual*-level parameterization of the model. Obviously, this is important for SCR models as they all possess at least one individual-level random effect in the form of the activity center **s**. In the context of stratified or multi-session type models, the group membership variable g_i is a *categorical* type of individual covariate (Huggins, 1989; Alho, 1990). Before considering SCR models specifically, in the next section we talk a little bit about the technical formulation of data augmentation for stratified populations in the context of ordinary closed population models.

14.2 Multinomial abundance models

One of the key ideas to Bayesian analysis of stratified population models is that we make use of multinomial models for allocating individuals into strata or sessions. We do this because it allows us to analyze the models by data augmentation (Converse and Royle, 2012; Royle and Converse, in review), and it has a natural linkage to the Poisson model, which is commonly used throughout ecology to model variation in abundance.

To motivate the technical framework, consider sampling $g = 1, 2, \ldots, G$ groups having unknown sizes N_g, and we wish to impose model structure on the group-specific population size variables using a Poisson distribution:

$$N_g \sim \text{Poisson}(\lambda_g) \tag{14.2.1}$$

with

$$\log(\lambda_g) = \beta_0 + \beta_1 C_g, \tag{14.2.2}$$

where C_g is some measured attribute for group g. For example, it might be a covariate describing landuse, habitat or landscape structure, or a variable defining some treatment regime applied to a sample of small-mammal trapping arrays. We could generalize this a bit by considering a random effect in Eq. (14.2.2), producing overdispersed population sizes N_g. For the special case of adding log-gamma noise, this results in a negative binomial model for N_g.

To develop a data augmentation scheme for this group-structured model, let's think about doing data augmentation separately for each population, by assuming that

$$N_g \sim \text{Binomial}(M_g, \psi),$$

where $\psi \sim \text{Uniform}(0, 1)$ as usual. We could do this multi-population data augmentation by just picking each M_g to be some large integer (as we always do when using data augmentation; see Section 5.7). However, we want to pick M_g in a way that induces the desired structure on N_g. If we want to enforce our Poisson model on N_g from above, we naturally choose M_g to be Poisson also, in which case the marginal distribution of N_g is Poisson with mean $\psi \exp(\beta_0 + \beta_1 C_g)$. For multiple groups that we want to model jointly, the key point is that we impose the structure that we desire for N_g on the super-population parameters M_g.

We cannot apply data augmentation directly to this stratified model, having Poisson augmented population sizes M_g, because the M_g values are not fixed (remember, the motivation for data augmentation is to fix the size of the data set). To resolve this problem, we remove the M_g variables from the model by conditioning on the "super-population" size $M_T = \sum_g M_g$. Then, the vector $\mathbf{M} = (M_1, \ldots, M_G)$ has a multinomial distribution:

$$\mathbf{M}|M_T \sim \text{Multinomial}(M_T, \boldsymbol{\pi}), \tag{14.2.3}$$

where $\pi_g = \lambda_g / \sum_g \lambda_g$. It will be easy to apply data augmentation to the larger super-population comprised of the populations from all G strata, having size M_T. Given this multinomial super-population structure, we relate our stratum-specific population sizes N_g to each M_g by a binomial sampling ("random thinning") model. Then the vector of N_g parameters has a multinomial distribution with cell probabilities $\psi \times \pi_g$.

When we apply data augmentation to the multinomial joint distribution, the ψ parameter of the binomial sampling model that relates each N_g to M_g takes the place of N_T, the total population size (across all groups or strata). In addition, by constructing the model conditional on the total, N_T, we lose information about the intercept β_0[1] but this is recovered in the data augmentation parameter ψ. Thus, one of these parameters (ψ or β_0) has to be fixed. We can set $\beta_0 = 0$ or else we can fix ψ (see Chapter 11). The constraint can be specified by noting that, under the binomial data augmentation model, $\mathbb{E}(N_T) = \psi M_T$ and, under the Poisson model, $\mathbb{E}(N_T) = \sum_g \exp(\beta_0 + \beta_1 C_g)$ and so we can set

$$\psi = \frac{1}{M_T} \sum_g \exp(\beta_0 + \beta_1 C_g).$$

The linkage of β_0 and ψ was also discussed in Chapter 11 in the context of building spatial models for density. In that case, β_0 was the intercept of the intensity function and one could choose to estimate either β_0 or the data augmentation parameter ψ.

14.2.1 Implementation in BUGS

The **BUGS** implementation of data augmentation for stratified populations is straightforward. For each individual in the super-population we introduce a latent variable g_i to indicate which group or stratum the individual belongs too, and we introduce a second variable z_i to indicate whether the individual is a "real" individual (uncaptured

[1] A technical argument is that the total N_T is the sufficient statistic for β_0 in the multinomial model and so, by conditioning on the total, β_0 is no longer a free parameter.

individuals are sampling zeros) or a structural zero. So, the multinomial structure for the M_g variables, and the binomial sampling of those super-population sizes, is equivalently represented by the latent variable pair (g_i, z_i) where g_i is categorical with prior probabilities π_g and $z_i \sim \text{Bernoulli}(\psi)$. In other words, the multinomial assumption for the latent variables M_g is formulated in terms of "group membership" for each individual in the super-population of size M_T according to:

$$g_i \sim \text{Categorical}(\boldsymbol{\pi})$$

with $\boldsymbol{\pi} = (\pi_1, \ldots, \pi_G)$ and $\pi_g = \lambda_g/(\sum_g \lambda_g)$ where λ_g is the model describing the variation in expected population size among the G strata. The data augmentation is described by the binary variables $z_1, \ldots, z_{M_T}$ such that

$$z_i \sim \text{Bernoulli}(\psi),$$

where ψ is constrained as noted in the previous section. Specifying the multinomial model with data augmentation in terms of the 2 individual variables (g_i, z_i) takes just a couple of lines of **BUGS** code shown here:

```
psi ~ dunif (0,1)
for (g in 1:G){
   pi[g] <- lambda[g]/sum(lambda[])
}
g[i] ~ dcat(pi[1:G])
z[i] ~ dbern(psi)
```

The complete **BUGS** specification for this individual-level formulation of the model is shown in Panel 14.1 for an ordinary closed population model (model M_0). This actually shows two equivalent formulations. In the left panel we have ψ and β_0 as free parameters. The right panel shows the equivalent model but recognizing the constraint between ψ and β_0. Running these models using the `multisession.sim` function, you can verify that the two parameters are not uniquely estimable. In particular, using the model (representation 1) in the left-hand side of Panel 14.1, you will see that draws of β_0 appear to be draws from the prior distribution, uninformed by the data, supporting the point we made previously that ψ and β_0 are not uniquely informed by the data when we apply data augmentation to the super-population size across all strata.

14.2.2 Groups with no individuals observed

In practical settings, when the groups represent small populations, it will sometimes happen that some strata have no encountered individuals or even that $N_g = 0$. This is dealt with implicitly in the development of the model shown in Panel 14.1 in the sense that the *prior* for N_g has the proper dimension (namely, G multinomial cells of non-zero probability) and thus some posterior mass may occur on non-zero values of N_g even if the *data* contain no representatives from stratum g. You can try this out to verify for yourself.

Implementation 1

```
model {
# This will show that psi and beta0
#   are confounded.
  p ~ dunif(0,1)
  beta0 ~ dnorm(0,.1)
  beta1 ~ dnorm(0,.1)
  psi ~ dunif(0,1)
  for(j in 1:G){
    log(lam[j]) <- beta0+beta1*C[j]
    gprobs[j]<-lam[j]/sum(lam[1:G])
  }
  for(i in 1:M){
    g[i] ~ dcat(gprobs[])
    z[i] ~ dbern(psi)
   mu[i] <- z[i]*p
   y[i] ~ dbin(mu[i],K)
  }
  N <- sum(z[1:M])
}
```

Implementation 2

```
model {
# This version constrains psi with
#   the intercept parameter
  p ~ dunif(0,1)
  beta0 ~ dnorm(0,.1)
  beta1 ~ dnorm(0,.1)
  psi <- sum(lam[])/M
  for(j in 1:G){
    log(lam[j]) <- beta0+beta1*C[j]
    gprobs[j]<-lam[j]/sum(lam[1:G])
  }
  for(i in 1:M){
    g[i] ~ dcat(gprobs[])
    z[i] ~ dbern(psi)
   mu[i] <- z[i]*p
   y[i] ~ dbin(mu[i],K)
  }
  N <- sum(z[1:M])
}
```

PANEL 14.1

BUGS model specification for a capture-recapture model with constant encounter probability and Poisson population sizes, N_g, with mean depending on a single covariate `C[j]`. Two versions of the model: The first one describes the model in terms of the intercept β_0 and DA parameter ψ, which are confounded. The required constraint is indicated in the specification under Implementation 2.

14.2.3 The group-means model

Under the Poisson model for group abundance N_g, even with a constant mean λ, each stratum or group may have a different realized population size, and this comes at the low price of a single parameter in the model (λ or, equivalently, the data augmentation parameter ψ). However, the implied mean/variance relationship of the Poisson model may be inadequate in some applications.

To accommodate more flexibility than afforded by the single-parameter Poisson model, there are a couple of choices: (1) We could allow the mean to be group specific such as: $N_g \sim \text{Poisson}(\lambda_g)$ where each λ_g is its own free parameter, independent of each other. This produces a model with G distinct "fixed" parameters, and effectively renders the Poisson assumption irrelevant as it doesn't induce any "Bayesian shrinkage" (Sauer and Link, 2002) or impose any group structure on the population sizes N_g. However, some information might be borrowed from the different groups for estimating the encounter probability parameters (Converse and Royle, 2012). Under this model, we constrain one of the λ_g parameters to be 0, and N_g for that group is taken up by the data augmentation parameter ψ; (2) Alternatively, we could identify specific fixed covariates which might explain variation across groups, such as habitat or landscape composition, land use, or an experimental treatment. Each additional covariate adds only one additional fixed parameter to the model; (3) A flexible formulation that provides something of an intermediate model, between that of a constant λ and independent group-specific λ_g parameters, is that in which we put a prior on λ_g. For example, if we assume

$$\lambda_g \sim \text{Gamma}(a, b)$$

this corresponds to imposing a Dirichlet compound-multinomial model on the population size vector, or, marginally, a negative binomial model on N_g. See Takemura (1999) for some discussion of such models relevant to data augmentation. For this model, we impose the constraint $b - 1$ to account for conditioning on the total population size N_T in order to use data augmentation.

14.2.4 Simulating stratified capture-recapture data

It is helpful, as always, to simulate some data in order to understand this group-structured model. Suppose we cracked the conservation lotto jackpot and obtained funding to carry out a camera trapping study of some flashy carnivore in 20 forest patches or reserves, using a 5×5 array of camera traps in each reserve. Here we will consider an ordinary closed population model, model M_0, and we suppose there is some forest level covariate, say Dist = disturbance regime, perhaps measured by an index of trail density. We imagine a model for patch-level population size such as the following:

$$N_g \sim \text{Poisson}(\lambda_g),$$
$$\log(\lambda_g) = \beta_0 + \beta_1 \text{Dist}_g.$$

We simulate some population sizes and encounter data under this model as follows:

```
> set.seed(2013)
> G <- 20                           # G = 20 groups or strata
> beta0 <- 3                        # Abundance model parameters
> beta1 <- .6
> p <- .3                           # Encounter probability
> K <- 5                            # Sample occasions for capture-recapture
> Dist <- rnorm(G)                  # Simulate covariate
> lambda <- exp(beta0+beta1*Dist)   # Simulate population sizes
> N <- rpois(G,lambda=lambda)

> y <- NULL                         # Simulate model M0 data
> for(g in 1:G){
+  if(N[g]>0)
+    y <- c(y,  rbinom(N[g],K,p))
+  }
> g <- rep(1:G,N)

> ##  Now keep the group id and encounter frequency only for
> ##          individuals that are captured
> g <- g[y>0]
> y <- y[y>0]
```

That's it! We just simulated a population sizes and capture-recapture data for population inhabiting $G = 20$ forest patches (the "groups" in this situation). To fit this model, we need to augment the `g` and `y` data objects, and then we can fit the model in **JAGS** or **WinBUGS** using the code given in Panel 14.1. See the help file `?multisession.sim` for analyzing the simulated data.

14.3 Other approaches to multi-session models

The multinomial super-population model allows for the joint modeling of a collection of population sizes using data augmentation. However, as we demonstrated in Section 9.2.4, we can analyze the models by putting independent binomial priors on each N_g and applying data augmentation independently for each population by itself. This is not any more or less difficult than the multinomial formulation described in this chapter but, we imagine, it could be slightly less efficient computationally. In this case we could build in among-group structure by modeling the DA parameter ψ as being variable for each individual in the augmented data set, as a function of group-specific variables (see Hendriks et al., 2013, for an example). For example, if C_g is the value of some covariate for group g, then we could have $z_i \sim \text{Bernoulli}(\psi_i)$ with

$$\text{logit}(\psi_i) = \beta_0 + \beta_1 C_{g_i}.$$

This implies a binomial model for the stratum population sizes:

$$N_g \sim \text{Binomial}(M, \psi_g).$$

If M is large then the N_g are approximately independent Poisson random variables with means $\psi_g M$.

As we noted in Chapter 6, the multi-session models in `secr` are based on a Poisson prior for N_g with mean Λ_g, and then among-group structure is modeled in the parameter Λ_g (e.g., using a log-linear model that depends on a group-specific covariate). In our view, either model (binomial based on data augmentation, or the explicit Poisson assumption of `secr`) is satisfactory for applications of capture-recapture to stratified populations. The main advantage of the formulation we provided here over that implemented in `secr` is we have quite a bit more flexibility in specifying models of all sorts, either in the population size model for N_g, or for the capture-recapture model. For example, Royle and Converse (in review) fitted a model having random group effects on encounter probability, and treatment and stratum effects on abundance.

14.4 Application to spatial capture-recapture

Although we developed the implementation of Bayesian models for stratified populations using ordinary closed population models, the underlying ideas are general and can be applied equally to spatial capture-recapture models without any novel considerations. We already discussed (Chapter 4) that SCR models are ordinary closed population models but with an individual covariate which is the activity center $\mathbf{s}_i$, and the observation model has to be defined for each trap. With this in mind, we can see how the model described in Section 14.2 can be modified to accommodate a group-structured SCR situation. Specifically, we include the prior distribution for $\mathbf{s}_i$ and the observation model that relates $\mathbf{s}_i$ to the probability of encounter for individual i and trap j.

One practical consideration in developing models for stratified populations is constructing the state-space for the super-population, and describing it in the BUGS language. If the groups represent replicate samples in time made on a single population (as in the ovenbird data set), so that there is a single state-space and a consistent set of trap locations, then implementation proceeds as in the ordinary closed population situation of Section 14.2, but extending the model to include the individual activity center, as we have done so many times in previous chapters. On the other hand, when the groups represent distinct spatial units such as small mammal trapping grids (or the possum data), each group has its own state-space, and trap locations, and this has to be accommodated in the model specification. This can be slightly challenging to describe in the **BUGS** language if the trap arrays have different sizes or geometry, but it is not so difficult if the trap arrays are similar in structure, e.g., 7×7 grids of unit spacing (Royle and Converse, in review). If the groups are close together, then the different arrays can share a single, enlarged state-space (as we did in Section 9.3.2).

14.4.1 Multinomial ("multi-catch") observations

We discuss Bayesian analysis of the multi-session model using data augmentation in the context of a multinomial observation model such as for a multi-catch sampling situation.[2] For context, we return to the ovenbird data set, from the **R** package `secr`, which we introduced in Chapter 9. Another example can be found in Royle and Converse (in review), who applied the model to a small-mammal trapping problem which involved replicate single-catch arrays of traps, in a study of the effects of forest management practices on small-mammal densities. In their analysis, they used the independent multinomial encounter model (multi-catch), recognizing that as a misspecification (see Section 9.3) of the true single-catch system. The ovenbird data is a type of multi-catch observation model where the group index variable is "year", and, in our earlier analyses, we analyzed the data set using independent binomial priors for N_g in **JAGS**, as well as with a Poisson prior in `secr` using the multi-session models. We mimic the `secr` analysis here, by using data augmentation and the multinomial distribution for N_g described above.

To refresh your memory about the multinomial observation model, let $\mathbf{y}_{ik} = (y_{i1k}, y_{i2k}, \ldots, y_{iJk}, y_{i,J+1,k})$ be the spatial encounter history for individual i, during

[2]It might be slightly confusing that we are considering multinomial observation models *and* multinomial models for group-specific abundance parameters N_g, but we will take care to be clear about this along the way.

sample occasion k where the last element $y_{i,J+1,k}$ corresponds to "not captured." For mist nets, an individual can be captured in at most one net during each occasion. Then, the vector $(y_{i1k}, y_{i2k}, \ldots, y_{iJk}, y_{i,J+1,k})$, contains a single 1 and the remaining values are 0. This $(J+1) \times 1$ vector, $\mathbf{y}_{ik}$, is a multinomial trial:

$$\mathbf{y}_{ik} \sim \text{Multinomial}(n = 1, \boldsymbol{\pi}_{ik}),$$

where $\boldsymbol{\pi}_{ik}$ is a $(J+1) \times 1$ vector where each element represents the probability of being encountered in a trap (for elements $1, \ldots, J$) or not captured at all (element $J+1$).

For the multi-catch observation model, the encounter probability vector is a function of distance between trap locations and individual activity centers, modeled on the multinomial logit scale. The Gaussian encounter probability model is:

$$\text{mlogit}(\pi_{ij}) = \eta_{ij} = \alpha_0 - \alpha_1 \text{d}(\mathbf{x}_j, \mathbf{s}_i)^2, \tag{14.4.1}$$

where $\alpha_1 = 1/(2\sigma^2)$ and σ^2 is the scale parameter of the Gaussian model. Then,

$$\boldsymbol{\pi}_{ij} = \exp(\eta_{ij})/[1 + \sum_j \exp(\eta_{ij})]$$

for each $j = 1, 2, \ldots, J$, and the last cell corresponding to the event "not captured" is:

$$\pi_{i,J+1} = 1 - \sum_{j=1}^{J} \pi_{ij}.$$

There are no novel technical considerations in order to model covariates of any kind. For example, in many studies we are concerned with a behavioral response to physical capture. This is typical in small-mammal trapping studies, where individuals often exhibit a trap-happy response due to baiting of traps, and in mist net studies of birds where individuals exhibit net avoidance after first capture. For this, let C_{ik} be a covariate of previous encounter (i.e., $C_{ik} = 0$ until the occasion of first capture, and $C_{ik} = 1$ thereafter), then we include this covariate in our multinomial observation model as follows:

$$\text{mlogit}(\pi_{ijk}) = \eta_{ijk} = \alpha_0 - \alpha_1 \text{d}(\mathbf{x}_j, \mathbf{s}_i)^2 + \alpha_2 C_{ik}.$$

In this case, the multinomial probabilities depend not only on individual and trap, but also on sample occasion. The additional array dimension of the multinomial cell probability can lead to extremely slow MCMC sampling using the various **BUGS** engines, as we saw with similar models in Chapter 7.

14.4.2 Reanalysis of the ovenbird data

Here we use Bayesian analysis by data augmentation to fit a model that approximates the Poisson model with expected value $\mathbb{E}(N_g) = \lambda_g$, and we model effects on the log-mean scale according to:

$$\log(\lambda_g) = \beta_0 + \beta_1 C_g.$$

```
model {
 alpha0 ~ dnorm(0,.01)                        # Prior distributions
 sigma ~ dunif(0,200)
 alpha1 <- 1/(2*sigma*sigma)
 psi <- sum(lambda[])/bigM
 for(t in 1:5){
    beta0[t] ~ dnorm(0,0.01)                  # Year-specific abundances
    log(lambda[t]) <- beta0[t]
    pi[t] <- lambda[t]/sum(lambda[])          # Calculate multinomial probs
 }
 for(i in 1:bigM){
    z[i] ~ dbern(psi)
    yrid[i] ~ dcat(pi[])
    S[i,1] ~ dunif(xlim[1],xlim[2])           # Activity centers
    S[i,2] ~ dunif(ylim[1],ylim[2])
    for(j in 1:ntraps){
        d2[i,j] <- pow(pow(S[i,1]-X[j,1],2) + pow(S[i,2]-X[j,2],2),1)
    }
   for(k in 1:K){
      Ycat[i,k] ~ dcat(cp[i,k,])
        for(j in 1:ntraps){                  # Construct trap enc. probs.
        lp[i,k,j] <- exp(alpha0 - alpha1*d2[i,j])*z[i]*(1-died[i,k])
        cp[i,k,j] <- lp[i,k,j]/(1+sum(lp[i,k,1:ntraps]))
       }
      cp[i,k,ntraps+1] <- 1-sum(cp[i,k,1:ntraps])  # last cell = not captured
   }
 }
}
```

PANEL 14.2

BUGS model specification for a stratified (multi-session) SCR model using data augmentation. This shows a multinomial ("multi-catch") type of observation model, used to analyze the ovenbird data. Some code to tally up the derived population sizes and density parameters is omitted. See `ovenbird.ms` script.

We considered two models: A model with year-specific abundance, and a model with a linear trend in density over time, so $C_g \equiv$ Year. However, using the Kuo and Mallick (1998) indicator variable selection idea (see Chapter 8), the linear trend term was found to have little or no posterior probability of inclusion, so we do not reproduce analyses of that here (but see the `ovenbird.ms` function for the **R** script). We show the **BUGS** model specification for the year-specific abundance model in Panel 14.2. Note the construction of the multinomial cell probabilities that distribute individuals among years, based on the year-specific mean λ_t. On the log scale, each of these parameters has a diffuse normal prior: `beta0[t] ~ dnorm(0,0.01)`. A few lines of model specification that compute the derived population size parameters and density are not

Table 14.1 Posterior summaries for the Bayesian stratified population (multi-session) model fitted to the ovenbird data. Results are based on three chains, each with 5,000 iterations (first 1,000 discarded), for a total of 12,000 iterations saved. Parameters are density (D_t), population size (N_t), parameters of the encounter probability model (α_0 and α_1) and year effect parameters of the log-linear model for population size ($\beta_{0,t}$). ψ is the data augmentation parameter and σ is the square root of the Gaussian encounter probability scale parameter.

	Mean	SD	2.5%	50%	97.5%	Rhat
D_1	0.883	0.191	0.562	0.868	1.308	1.002
D_2	0.972	0.200	0.624	0.954	1.418	1.001
D_3	1.146	0.224	0.758	1.125	1.638	1.001
D_4	0.836	0.183	0.538	0.819	1.247	1.001
D_5	0.705	0.167	0.428	0.685	1.088	1.001
N_1	72.208	15.596	46.000	71.000	107.000	1.002
N_2	79.478	16.367	51.000	78.000	116.000	1.001
N_3	93.725	18.327	62.000	92.000	134.000	1.001
N_4	68.399	14.952	44.000	67.000	102.000	1.001
N_5	57.665	13.659	35.000	56.000	89.000	1.001
α_0	−3.465	0.159	−3.779	−3.465	−3.155	1.004
α_1	0.000	0.000	0.000	0.000	0.000	1.009
$\beta_{0,1}$	4.250	0.244	3.754	4.257	4.710	1.001
$\beta_{0,2}$	4.349	0.233	3.872	4.356	4.786	1.001
$\beta_{0,3}$	4.516	0.220	4.059	4.522	4.930	1.001
$\beta_{0,4}$	4.194	0.248	3.697	4.202	4.664	1.001
$\beta_{0,5}$	4.013	0.275	3.456	4.022	4.524	1.001
σ	77.918	6.314	66.963	77.240	91.583	1.009
ψ	0.371	0.051	0.281	0.367	0.482	1.001

shown, but you can look at the **R** script `ovenbird.ms` in `scrbook` to run this analysis, and produce the posterior summaries shown in Table 14.1.

We previously analyzed these data in Section 9.2.4 using `secr`. To reproduce those results for a model that is similar to the one we just produced in **JAGS**, we execute this command:

```
> ovenbird.model.DT<-secr.fit(ovenCH,model=list(D~session),buffer=300)
```

Note, small values of `buffer` can produce a warning that it is too small relative to the indicated value of σ (which has posterior mass up to near $\sigma = 100$). The `secr` results are shown in Table 14.2. There are, as always, slight differences between the MLEs shown in Table 14.2 and the posterior summaries shown in Table 14.1. The absolute differences between the MLEs and the Bayesian posterior means were .037, −.011, −.006, −.004, and −.004 for years 1–5, respectively.

Table 14.2 Estimates for the multi-session model fitted to the ovenbird data using `secr`. The model had a year-specific density parameter (D), and constant encounter probability parameters shared across all 5 years.

	Link	Estimate	SE Estimate	LCL	UCL
2005					
D	log	0.920	0.228	0.571	1.484
g_0	logit	0.028	0.004	0.021	0.037
σ	log	78.566	6.379	67.025	92.095
2006					
D	log	0.963	0.238	0.598	1.553
g_0	logit	0.028	0.004	0.021	0.037
σ	log	78.566	6.379	67.025	92.095
2007					
D	log	1.139	0.282	0.706	1.836
g_0	logit	0.028	0.004	0.021	0.037
σ	log	78.566	6.379	67.025	92.095
2008					
D	log	0.832	0.206	0.516	1.341
g_0	logit	0.028	0.004	0.021	0.037
σ	log	78.566	6.379	67.025	92.095
2009					
D	log	0.701	0.173	0.435	1.130
g_0	logit	0.028	0.004	0.021	0.037
σ	log	78.566	6.379	67.025	92.095

14.5 Spatial or temporal dependence

The models described in previous sections of this chapter, and including the multi-session formulation used in `secr`, assume that the population sizes N_g are *independent* (in a limiting sense, under data augmentation). As a practical matter, this precludes the sharing of individuals among populations (i.e., the same individual cannot be captured in multiple groups), which can be violated in a number of situations. First, when the groups represent sampling in distinct time periods (seasons, years) but of the same functional population (a standard "robust design" situation), it is possible that some individuals remain in the population from one time period to the next. In this situation, by disregarding individual identity across groups, the models ignore a slight bit of dependence of N_g which may entail some incremental loss of efficiency. We imagine this should have little practical effect unless survival probability is extremely high between the periods. Estimators of parameters obtained by assuming independence should be unbiased (or, rather, ignoring the dependence should not affect the bias of the estimator much if at all).

A second distinct situation of non-independence is that in which the stratification variable is *spatial*, and the strata (e.g., trap arrays or other sampling mechanism) are in proximity to one another so that individuals can sometimes be encountered by more than one array (e.g., the possum data, see Figure 9.2). This case is somewhat easier to deal with in the analysis because we can build a model in which the state-space is the joint state-space enclosing all of the trapping arrays, so that we can preserve individual identity in an ordinary SCR model, just with a larger array of traps that is the union of the trap arrays of all sample groups. This may be impractical when the trap arrays are far apart creating only a slight bit of overlap of populations, because, in that case, the combined state-space may contain a huge population that one has to deal with in the MCMC (remember that increasing the size of the augmented population, M, increases computation time). Royle et al. (2011a) confronted this problem in an analysis of data from a sample of 1 km quadrats using a search-encounter type model (discussed in the following chapter). Even in the case of spatial dependency the independent N_g model is probably not too detrimental to inferences that apply to explaining marginal variation in N_g, such as habitat or landscape effects that are modeled on the expected value of N_g.

14.6 Summary and outlook

Capture-recapture data are not always collected in single isolated studies but, instead, are often grouped or stratified in some natural way, either because a number of distinct trap arrays are used, or sampling occurs in several habitat patches, or repeatedly over time. Often this is motivated by specific objectives, e.g., the trap arrays or units represent experimental replicates, but sometimes replicate trap arrays are used to derive more valid estimates of density by obtaining a representative (ideally, random) sample of space within some region. The fact that data are grouped in such a way raises the obvious technical problem of having to combine data from multiple arrays, sites, or otherwise defined groups in a single unified model that accommodates explicit sources of variation in density among these groups. This is naturally accomplished by developing an explicit model for variation in N, e.g., a Poisson GLM or similar (Converse and Royle, 2012; Royle et al., 2012b).

In this chapter, we outlined an approach to Bayesian analysis of multi-session models using data augmentation (Converse and Royle, 2012; Royle and Converse in review). This approach gives us one method for building explicit models for N_g and provides flexibility in specifying the encounter model using standard or novel capture-recapture modeling considerations. Certain types of multi-session models can be fitted easily in `secr` (see Chapter 9) and we suspect that platform will be satisfactory for many problems you encounter. However, as always, we believe the flexible model-building platform of the **BUGS** language can be beneficial in many situations.

A common applied context of these multi-session models is when replicate arrays are used to address explicit hypotheses about the effects of landscape variation or modification on abundance. For example, in studies of forestry practices and their effects on local fauna, small-mammal grids are used as experimental units, and the "dependent variable" is N (or density) of small mammals (or some small-mammal focal species) for each trap array, which is not observable. Thus, hierarchical models are needed to directly address the basic hypotheses of such studies. Another distinct context for the application of multi-session models is when the populations are temporally structured (e.g., the ovenbird data), such as when sampling occurs in distinct seasons or years. In these applications, we view multi-session models as a simplified type of open population model, an open model *without* explicit Markovian dynamics. They are analogous to what is usually referred to as models of random temporary emigration (Kendall et al., 1997; Chandler et al., 2011). The models are not incorrect, just simplified, reduced versions of more general Markovian models, and with fewer parameters to estimate, and they are likely more appropriate when there is complete or near complete population turnover between replicate sessions (e.g., survival is low, immigration/emigration is high, etc.). We cover general open population models in Chapter 16.

CHAPTER

Models for Search-Encounter Data

15

In this chapter we discuss models for what we call search-encounter data. These models are useful in situations where the locations of individuals, say $\mathbf{u}_{ik}$ for individuals i and sample occasions k, are observed directly by searching space (often delineated by a polygon) in some fashion, rather than restricted to fixed trap locations. In all the cases addressed in this chapter, both detection probability and parameters related to movement can be estimated using such models. To formalize this notion a little bit using some of the ideas we've introduced in previous chapters, most of the SCR models we've talked about in the book involve just two components of a hierarchical model, the observation component, which we denote by $[y|\mathbf{s}]$ (e.g., Bernoulli, Poisson, or multinomial), and the process component $[\mathbf{s}]$, describing the point process model for the activity centers. The search-encounter models described in this chapter involve an additional component for the locations conditional on the activity centers. We write this as follows: The observation model has the form $[y|\mathbf{u}]$, and the process model has two components, a movement model $[\mathbf{u}|\mathbf{s}]$, which describes the individual encounter locations conditional on $\mathbf{s}$, and the point process model $[\mathbf{s}]$. Because we can resolve parameters of the $[\mathbf{u}|\mathbf{s}]$ component using designs described in this chapter, search-encounter models are slightly more complicated, and more biologically realistic. Conversely, when we have an array of fixed trap locations, the movement process is completely confounded with the encounter process because the potential observation locations are prescribed, a priori, independent of any underlying movement process.

A few distinct types of situations exist where search-encounter models come in handy. The prototypical, maybe ideal, situation (Royle et al., 2011a) is where we have a single search path through a region of space from which observations are made (just as in the typical distance sampling situation, using a transect). As we walk along the search path, we note the location of each individual that is detected, *and their identity* (this is different from distance sampling in that sense), and sample on multiple occasions to obtain repeated encounters of some individuals. Alternatively, we could delineate a search area, and conduct a systematic search of that region. An example is that of Royle and Young (2008), which involved a plot search for lizards. They assumed the plot was uniformly searched which justified an assumption of constant encounter probability, p, for all individuals within the plot boundaries.

Spatial Capture-Recapture. http://dx.doi.org/10.1016/B978-0-12-405939-9.00015-3

The recent paper by Efford (2011a) discusses likelihood analysis of similar models. In the terminology of `secr` such models are referred to as *polygon detector* models.

15.1 Search-encounter designs

Before we discuss models for search-encounter data, we'll introduce some types of sampling situations that produce individual location data by searching space. We imagine there are a lot more sampling protocols (and variations) than identified here, but these are some of the standard situations that we have encountered over the last few years in developing applications of SCR models. For our purposes, we recognize two basic sampling designs, each of which might have variations due to modification of the basic sampling protocol.

15.1.1 Design 1: fixed search path

A useful class of models arises when we have a fixed search path or line, or multiple such lines, in some region (Figure 15.1) from which individual detections are made. We assume the survey path is laid out *a priori* in some manner that is done independent of the activity centers of individuals and the collection of data does not affect the survey path. The purpose of this assumption, in the models described subsequently,

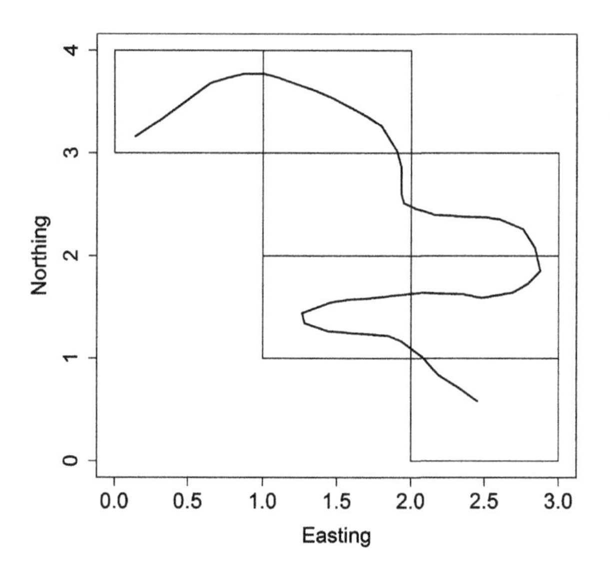

FIGURE 15.1

A survey line through parts of seven quadrats in a hypothetical landscape. An observer travels the transect and identifies individuals in the vicinity of the line, recording their identity and location.

is to allow us to assume that the activity centers are uniformly distributed on the prescribed state-space. Alternatively, explicit models could be entertained to mitigate a density gradient or covariate effects (see Chapter 11). The situation depicted in Figure 15.1 shows the search path traversing several delineated polygons, although the polygon boundaries may or may not affect the potential locations of individuals (see Section 15.2.4).

A number of variations of this fixed search path situation are possible, and these produce slightly different data structures and corresponding modifications to the model, although we do not address all of these from a technical standpoint here:

Protocol (1a). We record the locations of individuals, but not the location on the path where the encounter is made by the observer.
Protocol (1b). We record the location of individuals and the location on the search path where we first observed the individual.
Protocol (1c). We record the closest perpendicular distance from the search path to the individual location. This is a typical distance sampling situation, and this is a type of hybrid SCR/distance sampling model.

15.1.2 Design 2: uniform search intensity

In the uniform search intensity model (or just "uniform search"), we have one or more well-defined sample areas (polygons), such as a quadrat or a transect, and we imagine that the area is uniformly searched so that encounter probability is constant for all individuals within the search area. This type of sampling method is often called "area search" in the bird literature (Bibby et al., 1992). Sampling produces locations of individuals within the well-defined boundaries of the sample area. The polygon boundaries defining the sample unit are important because they tell us that $p = 0$, by design, outside of the plot boundary.

Using the example Figure 15.1, but ignoring the survey line through the plot (pretend it doesn't exist), we imagine that each of the identified quadrats is uniformly searched, which is to say, we assume that each individual within the boundaries of a quadrat has an equal and constant probability of being detected. In the context of replicate sampling occasions (e.g., on consecutive days), individuals may move on or off the plot, and so individuals may have different probabilities of being available to encounter, based on the closeness of their activity center to the quadrat boundaries. However, given that they're available, the uniform search model assumes they have constant encounter probability.

15.2 A model for fixed search path data

In contrast to most of the models described in this book, we develop models for encounter probability that depend explicitly on the instantaneous location $\mathbf{u}_{ik}$, for individual i at sample occasion k. We denote this encounter probability by

$p_{ik} \equiv p(\mathbf{u}_{ik}) = \Pr(y_{ik} = 1|\mathbf{u}_{ik})$. Note that we observe a biased sample of the location variables $\mathbf{u}$ because they are not observed for individuals that are not captured. That is, for the $y = 0$ observations. Therefore, we cannot analyze the conditional-on-$\mathbf{u}$ likelihood directly as if we had a complete data set of randomly sampled locations of individuals. Instead, we regard $\mathbf{u}$ as random effects and assume a model for it, which allows us to handle the problem of missing $\mathbf{u}_{ik}$ values (Section 15.4.1). We assume that individuals do not move *during* a sampling occasion or, if they do, the individual is not added to the data set twice.

To develop encounter probability models for this problem we cannot just use the SCR models discussed in previous chapters, because the "trap" is actually a line or collection of line segments (e.g., Figure 15.1), which we describe by the matrix of points $\mathbf{X}$. We should be able to describe any survey path accurately using a large number of discrete points, such as can be obtained from a GPS logger. Intuitively, $\Pr(y_{ik} = 1|\mathbf{u}_{ik})$ should increase as $\mathbf{u}_{ik}$ comes "close" to the line $\mathbf{X}$. It seems reasonable to express closeness by some distance metric $||\mathbf{u}_{ik} - \mathbf{X}||$ as the distance between locations $\mathbf{u}_{ik}$ and $\mathbf{X}$, and then relate encounter probability to distance to the line. For example, using a logit model of the form:

$$\text{logit}(p_{ik}) = \alpha_0 + \alpha_1||\mathbf{u}_{ik} - \mathbf{X}||.$$

For the case where $\mathbf{X}$ describes a meandering line, the average distance from $\mathbf{u}$ to the line might be reasonable; possible alternatives include the closest distance or the mean over specific segments of the line (within some distance), etc.

15.2.1 Modeling total hazard to encounter

Because the operative sampling "device", the line $\mathbf{X}$, is not a single point (like a camera trap) we have to somehow describe the total encounter probability to the line. A natural approach is to model the total hazard to capture (Borchers and Efford, 2008), which is standard in survival and reliability modeling, and distance sampling (Hayes and Buckland, 1983; Skaug and Schweder, 1999). Naturally, covariates are modeled as affecting the hazard rate and we think of distance to the line as a covariate acting on the hazard. Let $h(\mathbf{u}_{ik}, \mathbf{x})$ be the hazard of individual i being encountered by sampling at a point $\mathbf{x}$ on occasion k. For example, one possible model assumes, for all points $\mathbf{x} \in \mathbf{X}$,

$$\log(h(\mathbf{u}_{ik}, \mathbf{x})) = \alpha_0 + \alpha_1 \times ||\mathbf{u}_{ik} - \mathbf{x}||. \qquad (15.2.1)$$

Additional covariates could be included in the hazard function in the same way as for any model of encounter probability that we've discussed previously. The total hazard to encounter anywhere along the survey path, for an individual located at $\mathbf{u}_{ik}$, say $H(\mathbf{u}_{ik})$, is obtained by integrating over the surveyed line, which we can evaluate numerically by a discrete sum, evaluating the hazard at the set of points $\mathbf{x}_j$ along the surveyed path:

$$H(\mathbf{u}_{ik}) = \exp(\alpha_0)\left\{\sum_{j=1}^{J} \exp(\alpha_1 \, ||\mathbf{u}_{ik} - \mathbf{x}_j||)\right\}, \tag{15.2.2}$$

where $\mathbf{x}_j$ is the jth row of $\mathbf{X}$ defining the survey path as a collection of line segments which can be arbitrarily dense, but should be regularly spaced. Then the probability of encounter on a given sampling occasion is

$$p_{ik} \equiv p(\mathbf{u}_{ik}) = 1 - \exp(-H(\mathbf{u}_{ik})). \tag{15.2.3}$$

It is possible that the search path could vary by sampling occasion, say $\mathbf{X}_k$, which can easily be accommodated in the model simply by calculating the total hazard to encounter for each distinct search path.

This is a reasonably intuitive type of encounter probability model in that the probability of encounter is large when an individual's location $\mathbf{u}_{ik}$ is close to the line in the average sense defined by Eq. (15.2.2), and vice versa. Further, consider the case of a single survey point, i.e., $\mathbf{X} \equiv \mathbf{x}$, which we might think of as a camera trap location. In this case, note that Eq. (15.2.3) is equivalent to

$$\log(-\log(1 - p_{ik})) = \alpha_0 + \alpha_1 \, ||\mathbf{u}_{ik} - \mathbf{x}||$$

which is to say that distance is a covariate on detection that is linear on the complementary log-log scale, which is similar to the "trap-specific" encounter probability of our Bernoulli encounter probability model (see Chapter 5). The difference is that, here, the relevant distance is between the "trap" (i.e., the survey lines) and the individual's present location, $\mathbf{u}_{ik}$, which is observable. On the other hand, in the context of camera traps, the distance is that between the trap and a latent variable, $\mathbf{s}_i$, representing an individual's home range or activity center, which is not observed.

A key assumption of this formulation of the model is that encounter at each point along the line, $\mathbf{X}$, is independent of each other point. Then, the event that an individual is encountered at all is the complement of the event that it is not encountered anywhere along the line (Hayes and Buckland, 1983). In this case, the probability of not being encountered at point j is: $1 - p(\mathbf{u}_{ik}, \mathbf{x}_j) = \exp(-h(\mathbf{u}_{ik}, \mathbf{x}_j))$ and so the probability that an individual is not encountered at all is $\prod_j \exp(-h(\mathbf{u}_{ik}, \mathbf{x}_j))$. The encounter probability is therefore the complement of this, which is precisely the expression given by Eq. (15.2.3).

Any model for encounter probability can be converted to a hazard model so that encounter probability based on total hazard can be derived. We introduced this model above:

$$\log(h(\mathbf{u}_{ik}, \mathbf{x})) = \alpha_0 + \alpha_1 \, ||\mathbf{u}_{ik} - \mathbf{x}||.$$

which is usually called the Gompertz hazard function in survival analysis, and it is most often written as $h(t) = a\exp(b \times t)$, with parameters a and b, in which case $\log(h(t)) = \log(a) + b \times t$. In the context of survival analysis, t is "time" whereas, in SCR models, we model hazard as a function of distance. The Gaussian model has a squared-distance term:

$$\log(h(\mathbf{u}_{ik}, \mathbf{x})) = \alpha_0 + \alpha_1 \, ||\mathbf{u}_{ik} - \mathbf{x}||^2.$$

Borchers and Efford (2008) use this model:

$$h(\mathbf{u}_{ik}, \mathbf{x}) = -\log(1 - \text{expit}(\alpha_0) \exp(\alpha_1 \, ||\mathbf{u}_{ik} - \mathbf{x}||^2)),$$

which produces the Gaussian model for probability of detection at the point level, i.e., $\Pr(y = 1) = 1 - \exp(-h) = p_0 \exp(\alpha_1 \, ||\mathbf{u}_{ik} - \mathbf{x}||^2)$ where $p_0 = \text{logit}^{-1}(\alpha_0)$. Another model is:

$$\log(h(\mathbf{u}_{ik}, \mathbf{x})) = \alpha_0 + \alpha_1 \, \log(||\mathbf{u}_{ik} - \mathbf{x}||),$$

which is the Weibull hazard function.

15.2.2 Modeling movement outcomes

We have so far described the model for the encounter data in a manner that is conditional on the locations $\mathbf{u}_{ik}$, some of which are unobserved. Naturally, we should specify a model for these latent variables, i.e., a movement model, which is, naturally, specified conditional on the latent activity center. Such models can be analyzed using either Bayesian methods and MCMC (Royle and Young, 2008; Royle et al., 2011a) or by marginal likelihood (Efford, 2011a). To develop the movement model, we adopt what is now customary in SCR models—we assume that individuals are characterized by a latent variable, $\mathbf{s}_i$, which represents their activity center. This leads to some natural models for the movement outcomes $\mathbf{u}_{ik}$ conditional on the activity center $\mathbf{s}_i$. Royle and Young (2008) used a bivariate normal model:

$$\mathbf{u}_{ik}|\mathbf{s}_i \sim \text{BVN}(\mathbf{s}_i, \sigma^2_{move}\mathbf{I}),$$

where $\mathbf{I}$ is the 2×2 identity matrix. We consider alternatives below. This is a primitive model of individual movements about their home range, but we believe it will be adequate in many capture-recapture studies which are often limited by sparse data.

We adopt our default assumption for the activity centers $\mathbf{s}$:

$$\mathbf{s}_i \sim \text{Uniform}(\mathcal{S}); \quad i = 1, 2, \ldots, N.$$

The usual considerations apply in specifying the state-space $\mathcal{S}$—either choose a large rectangle, or prescribe a habitat mask to restrict the potential locations of $\mathbf{s}$.

15.2.3 Simulation and analysis in JAGS

Here we will simulate a sample data set under the situation depicted in Figure 15.1 and then analyze the data in **JAGS**. We begin by defining the state-space containing all of the grid cells in the rectangle $[-1, 4] \times [-1, 5]$, which contains 30 1×1 potential sample cells. The survey line in Figure 15.1 traverses 7 of those 1×1 grid cells. We define the total population to average 4 individuals per grid cell, which we can vary to simulated data sets of varying sizes. To set this up in **R**, we do this:

```
> xlim <- c(-1, 4)
> ylim <- c(-1, 5)
> perbox <- 4
> N <- 30*perbox   # Total of 30 1x1 quadrats
```

The line in Figure 15.1 is an irregular mesh of points obtained by a manual point-and-clicking operation, which mimics the way in which GPS points come to us. In order to apply our model we need a regular mesh of points. We can create a regular mesh of points from the unequally spaced points by using functions in the packages `rgeos` (Bivand and Rundel, 2011) and `sp` (Pebesma and Bivand, 2011), especially the function `sample.Line`, which produces a set of equally spaced points along a line. The **R** commands are as follows (the complete script is given in the function `snakeline` in the `scrbook` package):

```
> library(rgeos)
> library(sp)
> line1 <- source("line1.R")

> line1 <- as.matrix(cbind(line1$value$x,line1$value$y))
> points <- SpatialPoints(line1)

> sLine <- Line(points)
> regpoints <- sample.Line(sLine,250,type="regular") # Key step!
```

Next, we set a random number seed, simulate activity centers, and set some model parameters required to simulate encounter history data. In the following commands you can see where the regular mesh representation of the sample line is extracted from the `regpoints` object which we just created:

```
> set.seed(2014)
> sx <- runif(N,xlim[1],xlim[2])
> sy <- runif(N,ylim[1],ylim[2])

> sigma.move <- .35
> sigma <-.4
> alpha0 <- .8
> alpha1 <- 1/(2*(sigma^2))
> X <- regpoints@coords
> J <- nrow(X)
```

Next we're going to simulate data, which we do in two steps: For each individual in the population and for each of K sample occasions, we simulate the location of the individual as a bivariate normal random variable with mean $\mathbf{s}_i$ and $\sigma_{move} = 0.35$. Next, we compute the encounter probability model using Eq. (15.2.3), with the Gaussian hazard model, and then retain the data objects corresponding to individuals that get captured at least once. All of this is accomplished using the following commands:

```
> K <- 10  ## Sample occasions = 10
> U <- array(NA,dim=c(N,K,2)) ## Array to hold locations
> y <- pmat <- matrix(NA,nrow=N,ncol=K) ## Initialize
> for(i in 1:N){
```

```
+    for(k in 1:K){
+     U[i,k,] <- c(rnorm(1,sx[i],sigma.move),rnorm(1,sy[i],sigma.move))
+     dvec <- sqrt( (U[i,k,1] - X[,1])^2 + (U[i,k,2] - X[,2])^2  )
+     loghaz <- alpha0 - alpha1*dvec*dvec
+     H <- sum(exp(loghaz))
+     pmat[i,k] <- 1-exp(-H)
+     y[i,k] <- rbinom(1,1,pmat[i,k])
>    }
>  }
> Ux <- U[, ,1]
> Uy <- U[, ,2]
> Ux[y==0] <- NA
> Uy[y==0] <- NA
```

In the commands shown above, we define matrices, Ux and Uy, that hold the observed locations of individuals during each occasion. Note that, if an individual is *not* captured, we set the value to NA. We pass these partially observed objects to **JAGS** to fit the model.

Finally, we do the data augmentation and we make up some initial values for the location coordinates that are missing. In practice, we might think about using the average of the observed locations, which we do below.

```
> ncap <- apply(y,1,sum)
> y <- y[ncap>0,]
> Ux <- Ux[ncap>0,]
> Uy <- Uy[ncap>0,]

> M <- 200
> nind <- nrow(y)
> y <- rbind(y,matrix(0,nrow=(M-nrow(y)),ncol=ncol(y)))
> Namat <- matrix(NA,nrow=(M-nind),ncol=ncol(y))
> Ux <- rbind(Ux,Namat)
> Uy <- rbind(Uy,Namat)
> S <- cbind(runif(M,xlim[1],xlim[2]),runif(M,ylim[1],ylim[2]))
> for(i in 1:nind){
+     S[i,] <- c(mean(Ux[i,],na.rm=TRUE),mean(Uy[i,],na.rm=TRUE))
> }
> Ux.st <- Ux
> Uy.st <- Uy
> for(i in 1:M){
+     Ux.st[i,!is.na(Ux[i,])]<-NA
+     Uy.st[i,!is.na(Uy[i,])]<-NA
+     Ux.st[i,is.na(Ux[i,])]<-S[i,1]
+     Uy.st[i,is.na(Uy[i,])]<-S[i,2]
+ }
```

The **BUGS** model specification is shown in Panel 15.1, although we neglect the standard steps showing how to bundle the data, inits, and farm all of this stuff out to **JAGS** (see the help file for snakeline for the complete script). Simulating the data as described above, and fitting the model in Panel 15.1 produces the results summarized in Table 15.1.

```
model {

 alpha0~dunif(-25,25)              # Priors distributions
 alpha1~dunif(0,25)
 lsigma~dunif(-5,5)
 sigma.move<-exp(lsigma)
 tau<-1/(sigma.move*sigma.move)
 psi~dunif(0,1)

 for(i in 1:M){ # Loop over individuals
   z[i]~dbern(psi)
   s[i,1]~dunif(xlim[1],xlim[2])    # Activity center model
   s[i,2]~dunif(ylim[1],ylim[2])
   for(k in 1:K){                   # Loop over sample occasions
      ux[i,k] ~ dnorm(s[i,1],tau)   # Movement outcome model
      uy[i,k] ~ dnorm(s[i,2],tau)
      for(j in 1:J){ # Loop over each point defining line segments
        d[i,k,j]<-  pow(pow(ux[i,k]-X[j,1],2) + pow(uy[i,k]-X[j,2],2),0.5)
        h[i,k,j]<-exp(alpha0-alpha1*d[i,k,j]*d[i,k,j])
      }
     H[i,k]<-sum(h[i,k,1:J])         # Total hazard H
     p[i,k]<- z[i]*(1-exp(-H[i,k]))
     y[i,k] ~ dbern(p[i,k])
   }
}
 # Population size is a derived quantity
 N<-sum(z[])
}
```

PANEL 15.1

BUGS model specification for the fixed search path model, from Royle et al. (2011a). See the help file `?snakeline` for the **R** code to simulate data and fit this model.

15.2.4 Hard plot boundaries

The previous development assumed that locations of individuals can be observed anywhere in the state-space, determined only by the encounter probability model as a function of distance from the survey path. However, in many situations, we might delineate a plot with boundaries that restrict where individuals might be observed (as in the situation considered by Royle and Young (2008)). For such cases we truncate the encounter probability function at the plot boundary, according to:

$$p(\mathbf{u}_{ik}) = (1 - \exp(-H(\mathbf{u}_{ik})))\mathrm{I}(\mathbf{u}_{ik} \in \mathcal{X}), \qquad (15.2.4)$$

where $\mathcal{X}$ is the surveyed polygon, and the indicator function $\mathrm{I}(\mathbf{u}_{ik} \in \mathcal{X}) = 1$ if $\mathbf{u}_{ik} \in \mathcal{X}$ and 0 otherwise. That is, the probability of encounter is identically 0 if an individual

Table 15.1 Posterior summary statistics for the simulated fixed search path data. These are based on three chains, and a total of 9,000 posterior samples. The data-generating parameter values were $N = 100; \sigma_{move} = 0.35; \sigma = 0.4$, and $\alpha_0 = 0.8$. The parameter $\alpha_1 = 1/(2\sigma^2)$, and ψ is the data augmentation parameter.

Parameter	Mean	SD	2.5%	50%	97.5%	Rhat
N	117.626	5.675	107.000	117.000	129.000	1.015
α_0	1.305	0.494	0.425	1.280	2.387	1.009
α_1	3.806	0.423	3.050	3.777	4.733	1.008
σ_{move}	0.347	0.008	0.332	0.347	0.364	1.023
σ	0.364	0.020	0.325	0.364	0.405	1.008
ψ	0.587	0.044	0.501	0.588	0.673	1.006

is located *outside* the plot at sample period k. We demonstrated how to do this in the **BUGS** language below for a model of uniform search intensity (i.e., area-search model).

15.2.5 Analysis of other protocols

In the situation elaborated on above (what we called "Protocol 1a"), the survey path is used to locate individuals and whether or not an individual is encountered is a function of the total hazard to encounter along the survey path. We think there are a number of variations of this basic design that might arise in practice. A slight variation (what we called "Protocol 1b") is based on recording location of individuals and the location on the transect where we observed the individual. The probability of encounter is computed from the cumulative hazard up to the point on the line where the detection occurred (Skaug and Schweder, 1999). This is exactly a distance sampling observation model, but with an additional hierarchical structure that describes the individual locations about their activity centers. There are no additional novel considerations in analysis of this situation compared to Protocol 1a, and so we have not given it explicit consideration here. Similarly, "Protocol 1c" is a slight variation of this—instead of recording the point on the line where the individual was first detected, we use the point on the line that has the shortest perpendicular distance. This is a classical distance sampling observation model, and it represents an intentional misspecification of the model (in the sense that the closest perpendicular point is not actually where the detection will usually occur) but it seems that the effect of this is relatively minor.

15.3 Unstructured spatial surveys

A common situation in practice is that in which sampling produces a survey path, but the path was not laid out *a priori* but, rather, evolves opportunistically during the course of sampling, a situation we'll call an unstructured spatial survey (Thompson et al., 2012; Russell et al., 2012). We imagine that the survey path evolves in response to information about animal presence, which could be both the number of unique individuals or the quantity of signs in the local search area. The motivating problem has to do with area searches using dog teams, in which the dogs usually wander around hunting scat, and their search path is based on how they perceive the environment and what they're smelling. This violates the main assumptions that the line is placed a priori, independent of density and unrelated to detectability.

The analysis framework implemented by Thompson et al. (2012) and Russell et al. (2012) is based on a heuristic justification wherein the sampling of space is imagined to have been grid-structured, with grid cells that are large enough so that dogs are not influenced by scat or sign beyond the specific cell being searched. Then, we assume the dog applies a consistent search strategy to each cell so that resulting cell-level detections can be regarded as independent Bernoulli trials with probability p_{ij} depending on the distance $||\mathbf{x}_j - \mathbf{s}_i||$ between the grid cell with center $\mathbf{x}_j$, and individual with activity center $\mathbf{s}_i$ and the amount of search effort (or length of the search route) within a cell. In other words, we use an ordinary SCR type of model but treat the center point of each cell as an effective "trap." The deficiency with this approach is that some of the "subgrid" resolution information about movement is lost, so we probably lose precision about any parameters of the movement model when the cells are large relative to a typical home range size. We discuss a couple of examples below.

15.3.1 Mountain lions in Montana

Russell et al. (2012) analyzed mountain lion (*Puma concolor*) encounter history data to assess the status of mountain lions in the Blackfoot Mountains of Montana. The data collection was based on opportunistic searching by hunters with dogs, who tree the lion (Figure 15.2). Tissue is extracted with a biopsy dart and analyzed in the laboratory for individual identity. The authors used 5 km × 5 km grid cells for binning the encounters, and the length of the search path in each grid cell as a covariate of effort (C_j) for each grid cell. They used a Gaussian hazard model with baseline encounter probability that depends on sex and effort in each grid cell, on the log scale:

$$\log(\lambda_{0,ij}) = \alpha_0 + \alpha_2 \log(C_j) + \alpha_3 \text{SEX}_i.$$

Note for grid cells that were not searched, $C_j = 0$ and, for those, the constraint $\lambda_{0,ij} = 0$ was imposed so that the probability of encounter was identically 0.

One problem encountered by Russell et al. (2012) in their analysis is the possibility of dependence in encounters because of group structure in the data (usually, juveniles in association with their mother). In this situation, in addition to dependence of encounter, multiple individuals have effectively the same activity center, thus violating a number of assumptions related to the ordinary SCR model. To resolve this problem, the authors made some assumptions about group association and fitted models where group served as the functional individual.

FIGURE 15.2

Mountain lion (*Puma concolor*). Run!
Photo credit: Bob Wiesner.

15.3.2 Sierra national forest fisher study

Here we consider data from a study of fishers (*Martes pennanti*) by Thompson et al. (2012), which took place in Sierra National Forest, California. In this study, the survey area was divided into 15 approximately 1,400-ha hexagons (Figure 15.3), which is roughly the size of a female fisher's home range, and each hexagon was surveyed three times by sniffer dog teams searching for scat. The dogs were given considerable latitude to determine their route. Thus, the search path was not laid out a priori but rather evolved opportunistically, based on what the dog sensed at a local scale. The authors divided the region into 1 km grid cells (also shown in Figure 15.3), and built an SCR model that regarded those grid cells as effective traps.

We provide the data from this study in the `scrbook` package, and it can be loaded with the command `data(fisher)`. The **R** script `SCRfisher` produces the posterior summary statistics shown in Table 15.2. One thing to note here is the relatively poor mixing of the Markov chains here due to sparse data and a fairly long run is probably necessary.

15.4 Design 2: Uniform search intensity

A special case of a search-encounter model arises when it is possible to subject a quadrat (or quadrats) to a uniform search intensity. This could be interpreted as an exhaustive search, or perhaps just a thorough systematic search of the available habitat. The example considered by Royle and Young (2008) involved searching a 9 ha plot for horned lizards (Figure 15.4) by a crew of several people. It was believed, in that case, that complete and systematic (i.e., uniform) coverage of the plot was achieved.

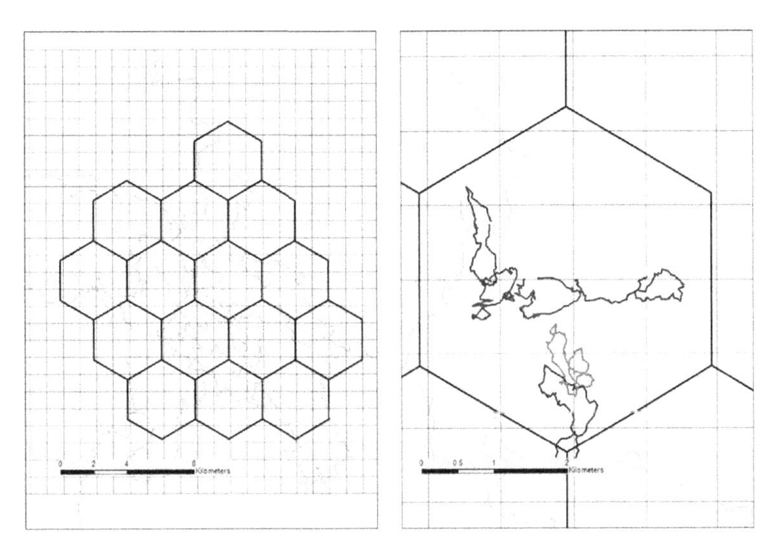

FIGURE 15.3

Fisher study area showing the gridding system (left panel). The larger hexagons are approximately the size of a typical female home range. The 1-km grid cells define the SCR model grid, where the center point of each one served as a "trap". The right panel shows the GPS trackline of the dog team through one of the polygons. The total length of the trackline was used as a covariate on encounter probability.
Credit: Craig Thompson, US Forest Service.

Table 15.2 Posterior summary statistics for the Fisher study data, based on 30,000 posterior samples. Here $\lambda_0 = \exp(\alpha_0)$. This example exhibits relatively poor mixing due to sparse data, and the Rhat statistic should be reduced by obtaining a larger posterior sample.

Parameter	Mean	SD	2.5%	50%	97.5%	Rhat
N	315.889	230.041	12.000	280.000	738.775	1.133
λ_0	0.003	0.033	0.000	0.000	0.016	1.097
α_1	0.188	0.170	0.005	0.138	0.641	1.002
σ	4.745	2.909	0.163	4.650	9.704	1.020
ψ	0.413	0.300	0.016	0.366	0.964	1.131

In general, however, we think you could have a random sample of area within the plot and approximate that as a uniform coverage—this is a design-based argument justifying the uniform search intensity model (we haven't simulated this situation, but it would be worth investigating).

It is clear that this uniform search intensity model is a special case of the fixed search path model in the sense that the probability of encounter of an individual is

FIGURE 15.4

A flat-tailed horned lizard showing its typical cryptic appearance in its native environment in the desert of southwestern Arizona, USA. Detection of flat-tailed horned lizards is difficult because they do not run when approached. Instead they shuffle under the sand or press down and remain motionless as shown in the picture. The horns are employed only as a last resort if the camouflage fails.
Photo credit: Kevin and April Young.

a constant p_0 *if* the individual is located in the polygon $\mathcal{X}$ during sample occasion k, i.e.,

$$p(\mathbf{u}_{ik}) = p_0 \mathrm{I}(\mathbf{u}_{ik} \in \mathcal{X}),$$

which resembles Eq. (15.2.4) except replacing the encounter probability function with constant p_0.

Subsequently, we give a simple analysis using simulated data, and simple movement models for $\mathbf{u}$, including a bivariate normal model and a random walk. For further examples and analyses, we refer you to Royle and Dorazio (2008), who reanalyzed the lizard data from Royle and Young (2008), Efford (2011b) and Marques et al. (2011).

15.4.1 Alternative movement models

As with the general fixed search path model ("Design 1"), we require a model to describe the movement outcomes $\mathbf{u}_{ik}$. In the analysis of Royle and Young (2008), a simple bivariate Gaussian movement model was used, in which

$$\mathbf{u}_{ik} | \mathbf{s}_i \sim \mathrm{BVN}(\mathbf{s}_i, \sigma^2_{move}\mathbf{I}),$$

However, clearly more general versions of the model can be developed. For example, imagine a situation where the successive surveys of a bounded sample polygon are relatively close together in time so that successive locations of individuals are not well approximated by the Gaussian movement model, which implies independence of locations. Naturally we might consider using an auto-regressive or random walk type of model in which the successive coordinate locations of individual i evolve as follows:

$$u_{1,i,k}|u_{1,i,k-1} \sim \text{Normal}(u_{1,i,k-1}, \sigma^2_{move})$$
$$u_{2,i,k}|u_{2,i,k-1} \sim \text{Normal}(u_{2,i,k-1}, \sigma^2_{move})$$

Here we use the notation u_1 and u_2 for the easting and northing coordinates, respectively (and, for clarity, we are using commas in the subscripting when we have to refer to time lags). In addition, we require that the initial locations have a distribution and, for that, we might begin with a simple model such as the uniformity model:

$$\mathbf{u}_{i,1} \sim \text{Uniform}(\mathcal{S})$$

which effectively takes the place of the model for $\mathbf{s}_i$ that we typically use. Under this model of Markovian movement, individuals don't have an activity center but, rather, they drift through space more or less randomly based just on their previous location. See Ovaskainen (2004) and Ovaskainen et al. (2008) for development and applications of similar movement models in the context of capture-recapture data. We could allow for dependent movements about a central location $\mathbf{s}_i$ using a bivariate auto-regression or similar type of model with parameter ρ, e.g.,

$$\mathbf{u}_{i,k}|\mathbf{s}_i \sim \text{BVN}(\rho\ (\mathbf{u}_{i,k-1} - \mathbf{s}_i), \sigma^2_{move}\mathbf{I}).$$

We don't have any direct experience fitting these movement models to real capture-recapture data, but we imagine they should prove effective in applications that yield large sample sizes of individuals and recaptures.

15.4.2 Simulating and fitting uniform search models

The **R** script `uniform_search`, in the `scrbook` package, provides a script for simulating and fitting search-encounter data using the *iid* Gaussian model and the random walk model. The **BUGS** model specification is shown in Panel 15.2 for the random walk situation. We encourage you to adapt this model and the simulation code for the auto-regression movement model. To fit this model to data, we set up the run with **JAGS** using the standard commands. We did not specify starting values for the missing coordinate locations, although **JAGS** should perform better if we provide decent ones, e.g., the last observed location or the average location. We imagine that resource selection could be parameterized in this movement model as well, perhaps using similar ideas to those described in Chapter 13.

The following script simulates a population of N individuals and their locations at each of four occasions to see if they are in a square [3,13] or not. Their locations follow a random walk model, perhaps resulting from sampling occasions that are close together in time. The initial locations are assumed to be uniformly distributed

```
model{
psi ~ dunif(0,1)                        # Prior distributions
tau ~ dgamma(.1,.1)
p0 ~ dunif(0,1)
sigma.move <- sqrt(1/tau)

for (i in 1:M){
  z[i] ~ dbern(psi)
  U[i,1,1] ~ dunif(0,16)                # Initial location
  U[i,1,2] ~ dunif(0,16)

   for (k in 2:n.occasions){
      U[i,k,1] ~ dnorm(U[i,k-1,1], tau)
      U[i,k,2] ~ dnorm(U[i,k-1,2], tau)
    }
   for(k in 1:n.occasions){
            # Test whether the actual location is in- or outside the
            #     survey area. Needs to be done for each grid cell
      inside[i,k] <- step(U[i,k,1]-3) * step(13-U[i,k,1]) *
                     step(U[i,k,2]-3) * step(13-U[i,k,2])
      Y[i,k] ~ dbern(mu[i,k])
      mu[i,k] <- p0 * inside[i,k] * z[i]
      }
   }
N <- sum(z[])                           # Population size, derived
}
```

PANEL 15.2

BUGS model specification for the uniform search intensity model similar to Royle and Young (2008), but with a random walk movement model. Also see the help file ?uniform_search in the **R** package scrbook.

on the state-space, which, in this case, is the square $[0, 16] \times [0, 16]$. We store the movement outcomes here in a three-dimensional array U, instead of in two separate two-dimensional arrays (one for each coordinate), as we did above. The **R** commands are as follows:

```
> N <- 100
> nocc <- 4
> Sx <- Sy <- matrix(NA,nrow=N,ncol=nocc)
> sigma.move <- .25

# Simulate initial coordinates on the square:
> Sx[,1] <- runif(N,0,16)
> Sy[,1] <- runif(N,0,16)

> for(t in 2:nyear){
```

```
+    Sx[,t] <- rnorm(N,Sx[,t-1],sigma.move)
+    Sy[,t] <- rnorm(N,Sy[,t-1],sigma.move)
+ }

# Now we generate encounter histories on a search rectangle
#     with sides [3,13]:
> Y <- matrix(0,nrow=N,ncol=nyear)
> for(i in 1:N){
+   for(t in 1:nyear){
+     # IF individual is in the sample unit we can capture it:
+     if(  Sx[i,t] > 3 & Sx[i,t]< 13 & Sy[i,t]>3 & Sy[i,t]<13)
+     Y[i,t] <- rbinom(1,1,.5)
+   }
+ }

# Subset data. If an individual is never captured,
> cap<- apply(Y,1,sum) > 0
> Y <- Y[cap,]
> Sx <- Sx[cap,]
> Sy <- Sy[cap,]

> Sx[Y==0] <- NA
> Sy[Y==0] <- NA

## Data augmentation:
> M <- 200
> Y <- rbind(Y,matrix(0,nrow=(M-nrow(Y)),ncol=nyear))
> Sx <- rbind(Sx,matrix(NA,nrow=(M-nrow(Sx)),ncol=nyear))
> Sy <- rbind(Sy,matrix(NA,nrow=(M-nrow(Sy)),ncol=nyear))

# Make 3-d array of coordinates "U"
> U <- array(NA,dim=c(M,nyear,2))
> U[, ,1] <- Sx
> U[, ,2] <- Sy
```

15.4.3 Movement and dispersal in open populations

In Chapter 16 we discuss many aspects of modeling open populations, including some aspects of modeling movement and dispersal, and the relevance of SCR models to these problems. However, the uniform search model above, is clearly relevant to modeling movement and dispersal in open populations. In particular, the model described in Panel 15.2 could easily be adapted to an open population by introducing a latent "alive state" with survival parameter ϕ_t. This would be a spatial version of the standard Cormack-Jolly-Seber model (Chapter 16.3).[1] In this case, i.e., of open populations, the bivariate normal model, or the random walk model, might serve as a model for the dynamics of activity centers over longer periods of time than apply to ordinary movement dynamics. In that sense, the same basic models apply to movement and dispersal (perhaps even to migration and other processes), but the operative time scales of these various processes are different.

[1] Some work related to this is currently being carried out by our colleagues Torbjørn Ergon and Michael Schaub.

15.5 Partial information designs

The prototype search-encounter (Design 1) and uniform search (Design 2) cases are ideal in the sense that they produce both precise locations of individuals and a precise characterization of the manner in which individuals are encountered by sampling space. We have seen a number of studies that, in an ideal world, would have generated data consistent with one of these situations but, for some practical reason, partial or no spatial information about the search area or the locations of individuals was collected (or retained), and so the models described above could not be used. We imagine (indeed, have encountered) at least three distinct situations:

a. The search path is not recorded, but locations of individuals are recorded.
b. The search path is recorded, but locations of individuals are not.
c. The search path is not recorded, and the locations are not recorded, we have only raw summary counts of individuals for prescribed areas or polygons.

For analysis of these search-encounter designs with partial information, we see a number of options of varying levels of formality, depending on the situation (and these are largely untested). For (a) we could assume uniform search intensity, which might be reasonable if the plots were randomly searched. Otherwise, the validity of this assumption would depend on the precise manner in which the search activity occurred. For (b) or (c), we could adopt the approach taken in the Fisher analysis above, and map the locations to the center of each plot, thinking of the plot as an effective trap, and using the search path length as a covariate (or not). A fourth case with even less information is that in which we don't record individual identity at all. Instead, we just have total count frequencies in each plot. This model is that considered by Chandler and Royle (2013), which is the focus of Chapter 18.

15.6 Summary and outlook

The generation of spatial encounter history data in ecological studies is widespread. While such data have historically been obtained mostly by the use of arrays of fixed traps (catch traps, camera traps, etc.), in this chapter we showed that SCR models are equally relevant to a large class of "search-encounter" problems, which are based on organized or opportunistic searches of spatial areas. Standard examples include "area searches" in bird population studies, use of detector dogs to obtain scat samples, from which DNA can be obtained to determine individual identity, or sampling along a fixed search path (or transects) by observers noting the locations of detected individuals (this is common in sampling for reptiles and amphibians). The latter situation closely resembles distance sampling but, with repeated observations of the same individual (on multiple occasions), it has a distinct capture-recapture element to it. In a sense, the fixed search path models are hybrid SCR-DS models.

Many models for search-encounter data have three elements in common. They contain: (1) a model for encounter conditional on locations of individuals; (2) a model that describes how these observable animal locations are distributed in space; and (3) a model for the distribution of activity centers. We interpret the second model component as an explicit movement model, and the existence of this component is distinct from most of the other models considered in this book. One of the key conceptual points is that, with these search-encounter types of designs, the locations of observations are *not* biased by the locations of traps but, rather, locations of individuals can occur anywhere within search plots or quadrats, or in the vicinity of a transect or search path. Because we can obtain direct observations of location—outcomes of movement—for individuals, it is possible to resolve explicit models of movement from search-encounter data. We considered the simple cases of the independent bivariate normal movement model, and a random walk type model, both of which can easily be fitted in the **BUGS** engines. We imagine that much more general movement models can be fitted, although we have had limited opportunities to pursue this and in most practical capture-recapture studies, sparse data may limit the complexity of the movement models that could be considered.

Search-encounter sampling is somewhat common, although we think that many people don't realize that it can produce encounter history data that are amenable to the development of formal SCR models for density, movement, and space usage. We believe that these protocols will become more appealing as methods for formal analysis of the resulting encounter history data become more widely known. At the same time, search-encounter models will increase in relevance in studies of animal populations because encounter history data based on DNA extracted from animal tissue or scat is easy to obtain by searching space opportunistically. As the cost of obtaining individual identity from scat or tissue decreases, the use of such information for developing spatial capture-recapture models can only increase.

CHAPTER

Open Population Models

16

In the previous chapters we focused mostly on closed population models for estimating density and for inference about spatial variation in density and space usage. However, a thorough understanding of population dynamics requires information about both spatial *and* temporal variation in population density and demographic parameters. In this chapter, we discuss modeling the processes governing spatial and temporal population dynamics, namely survival, recruitment, and movement over larger temporal scales (e.g., migration, dispersal, etc.). The ability to estimate these parameters is critical to both basic and applied ecological research (Knape et al., 2012). For example, testing hypotheses about life history trade-offs requires accurate estimates of both survival and fecundity (Caswell, 1989; Nichols et al., 1994). Inference about density-dependent population regulation, which has fascinated theoretical ecologists for well over a century, is likewise best accomplished by directly studying the factors affecting survival and fecundity, rather than the more common approach of modeling time series data (Nichols et al., 2000b). A mechanistic understanding of population changes, which is essential to address ecological and conservation related questions, requires useful models of vital rates. Furthermore, if we know how environmental variables affect demographic parameters, we can make predictions about population changes under different future scenarios. We can also assess the sensitivity of parameters such as population growth rate to variation in survival or fecundity. Although matrix population models are often used for these purposes (Caswell, 1989; Sæther and Bakke, 2000), the same objectives can be accomplished by computing posterior predictive distributions of projected population sizes as part of an MCMC algorithm.

The modeling framework we will develop in this chapter is based on a formulation of the classical Cormack-Jolly-Seber (CJS) and Jolly-Seber (JS) type models (Cormack, 1964; Jolly, 1965; Seber, 1965) which is amenable to modeling individual effects, including individual covariates. There is a long history of use of these models in fisheries, wildlife, and ecology studies (Pollock et al., 1990; Lebreton et al., 1992; Pradel, 1996; Williams et al., 2002; Schwarz and Arnason, 2005; Gimenez et al., 2007). Additionally, there have been many modifications and developments of the CJS and JS models including dealing with individuals that do not have a well-defined home range but instead are moving through the sampled area (transients), dealing with more than one site or state (multi-state models, where states may be geographic

Spatial Capture-Recapture. http://dx.doi.org/10.1016/B978-0-12-405939-9.00016-5

units, reproductive stage, etc.), and addressing individual movement through spatially implicit models.

For the first time, these models can fully integrate the movement of individuals in the vicinity of the trap array with their encounter histories to simultaneously estimate density, survival, and recruitment in a spatial model. For many species, such as those that are rare or not often observed by researchers, this allows inferences to be made about survival and recruitment without having to physically capture individuals. Additionally, another reason for extending SCR models to open populations arises purely from a sampling perspective. Longer time periods are often needed to sample rare or elusive species to ensure that enough captures and recaptures are produced. This prolonged sampling can quickly lead to violations in the assumption of population closure (see also Chapter 10). For example, the European wildcat study that was mentioned in Chapter 7 (see Kéry et al. (2011) for details) was conducted over a year-long period. While the researchers in that study used a closed population model, they did model variation in detection as a function of time to account for seasonal variation in behavior. Another approach would have been to use an open population model to account for possible changes in the population over time (however, open population SCR models had not been developed at the time of the wildcat study, so we'll forgive the authors for not having used this more appropriate model).

In this chapter, we present the traditional JS model and the spatial version, demonstrating both with an example of mist-netting ovenbirds, which was also analyzed in Chapter 9. Then we review the traditional CJS, multi-state CJS, and then describe the spatial model. To demonstrate the CJS models, we will use an example of American shad. Finally, we end by discussing some of the new approaches to modeling temporal dynamics including correlated movement and dispersal.

16.1 Background

16.1.1 Brief overview of population dynamics

The most basic formulation of models for population growth stems from an idea originally used in accounting, the balance sheet (see Conroy and Carroll (2009, Chapter 3) for a more complete description). To gain a mechanistic understanding of population dynamics, it is important to understand four fundamental processes that drive population size: births and immigrants (i.e., population "credits") and deaths and emigrants (i.e., population "debits"). The population at time $t+1$ is a function of these four components:

$$N(t+1) = N(t) + B(t) + I(t) - D(t) - E(t),$$

where $N(t)$ is the population size at time t, $B(t)$ and $I(t)$ are the credits (additions) from births and immigrants at time t, and $D(t)$ and $E(t)$ are the debits (losses) due to deaths and emigration. This balance equation model is known as the "BIDE model." A simple population growth model under density independence, assuming no immigration or emigration, can be derived as:

$$N(t+1) = N(t) + N(t)r(t),$$

where $r(t) = b(t) - d(t)$. Here, $b(t)$ and $d(t)$ are the per capita birth and death rates and thus $r(t)$ is the per capita growth rate. Models which are based only on the intrinsic population growth rate, "r," however, do not retain much information about the underlying drivers of the population dynamics. Density-dependent, age structured, stochastic effects on growth, spatially structured, and competition models (e.g., Lotka-Volterra) all are derivations of the basic BIDE model.

In closed population models, we focus on estimating population size, N, but in open population models we are interested in the dynamics that arise between years or seasons and thus we focus not only on $N(t)$ but on the processes that drive population changes. The parameters governing these processes – the parameters of the BIDE model – can be estimated using the JS and CJS models, described in more detail throughout this chapter. In the absence of movement, survival can be estimated in the CJS model and both survival and recruitment can be estimated in the JS model. However, in considering movement, it becomes difficult to distinguish births from immigrants in recruitment and deaths from emigrants in survival rate, because data are usually only collected in one area and when an animal leaves that area we cannot determine its fate.

For example, survival ($\phi(t)$) is defined as the probability of an individual surviving from time t to $t+1$, and often this is called "*apparent* survival" because deaths and emigration cannot be separated. Mortality, the probability of dying from time t to $t+1$ is $1 - \phi(t)$. Recruitment (γ), here defined as the probability of a new individual entering the population between t and $t+1$, includes both those born into the population and immigrants. This inability to distinguish between the different forms of losses and gains does not allow researchers to test specific hypotheses about population dynamics. To address this, Nichols and Pollock (1990) applied the robust design to a two age class situation in order to separate estimates of recruitment into immigration and *in situ* reproduction. While models that focus on the population growth rate tend to lose important information on population dynamics, more recent work has been done to estimate the contributions of survival and recruitment to the per capita growth rate, "r," using capture-recapture data and a reverse-time modeling approach (Pradel, 1996; Nichols et al., 2000a). All of these model improvements have provided invaluable information in the study of population dynamics, but none explicitly incorporate animal movement.

16.1.2 Animal movement related to population demography

One issue that arises frequently in traditional open population models is that movement can make it difficult to distinguish survival from emigration. For example, we know that movement of transients and temporary emigration will affect the estimates of survival, causing us to refer to estimates as "apparent survival" (Lebreton et al., 1992). This is because an animal that appears in the population for a short period of time and then leaves is going to appear as though it has died. Due to this problem, there has been a significant amount of work developing models to deal with temporary emigration and transients in both closed and open capture-recapture models (Kendall et al., 1997; Pradel et al., 1997; Hines et al., 2003; Clavel et al., 2008; Chandler et al., 2011; Gilroy et al., 2012). Because movement is modeled directly within the SCR framework, we can better understand the impact of animals moving onto and off of the trap array and hence we can improve our estimates of survival by combining the traditional CJS and JS models with the SCR model.

While demographic parameters such as survival rates, population growth, etc., are influenced by density (Fowler, 1981; Murdoch, 1994; Saether et al., 2002), it is also likely that movement of individuals can influence these parameters. It is generally accepted that population structure (i.e., age, stage, or size distribution) can affect both population size and growth over time (Caswell and Werner, 1978). We also know that how animals distribute themselves in space can directly influence the age or stage structure of a population—this can be behavioral, habitat related, or some combination of factors. For example, if habitat is limited, some younger members of the population might have trouble finding and/or defending a territory. Ultimately, this may lower survival for a certain age class in the population directly impacting the population structure.

Dispersal can also affect population structure. Dispersal can be related to access to reproduction, population regulation, habitat quality, as well as the linking of local populations in metapopulation ecology (Clobert et al., 2001; Ovaskainen, 2004; Ovaskainen et al., 2008). It is known that dispersal can be influenced by density dependence (Matthysen, 2005); for example, competition may cause individuals to be more likely to emigrate from an area, or individuals may leave an area in search of a mate. We discuss modeling dispersal with capture-recapture data a bit further in Section 16.4 at the end of the chapter.

16.2 Jolly-Seber models

16.2.1 Traditional Jolly-Seber models

The JS model was developed as a way to estimate not only detection and abundance, but survival and recruitment (new individuals coming into the population) based on capture-recapture data (Jolly, 1965; Seber, 1965). There are a number of ways that researchers have formulated the JS model (Cooch and White, 2006) and while all are slightly different, the resulting estimates of abundance and the driving parameters such as survival and some form of recruitment are consistent. Commonly used formulations are the Link-Barker (Link and Barker, 2005), Pradel-recruitment (Pradel, 1996), Burnham JS (Burnham, 1997), and the super-population formulation of Schwarz and Arnason (1996). In all of these models, the parameter of interest is recruitment, or how new individuals arrive into the population. Therefore one of the main differences between the various models is how new entrants into the population are parameterized.

Traditionally, sampling for the JS model included only one data collection event per primary occasion and this allowed for the estimation of survival and recruitment. However, without repeated visits within a primary occasion, there is not enough data to allow for variation in detection and this leads to potentially inaccurate estimates of population size. This led Pollock (1982) to devise the robust design in order to allow for heterogeneity in capture probability (by sex, age, social status, etc.) and trap response under the JS model. We present the robust design approach as it is more flexible, and generalizing to the spatial version of the JS model will be simpler. The basic idea is that there are primary occasions (e.g., years, seasons) and we allow the population to be "open" between the primary occasions. This means that individuals can enter and leave the population (i.e., births, deaths, immigration, emigration can occur) between the primary occasions; however, within a primary occasion, the population is assumed to be closed to these processes. The standard JS model does not allow for variation in detection probability between individuals or within a primary occasion because only one sample is collected per primary period. However, when multiple samples are taken within a primary occasion (we call these "secondary occasions"), then variation in detection probability can be modeled and thus our estimates of N can be improved. To that extent, we can envision the data as arising from repeated sampling over seasons or years (or *primary* periods) within which one or more samples (e.g., trap nights) might be taken (*secondary* periods). Figure 16.1 demonstrates the sampling process graphically. Comparing this with all of our previous work, the sample occasions (e.g., trap nights, weeks, etc.) described in the closed population chapters are called *secondary* sampling occasions in the context of open population models.

We can easily formulate a non-spatial JS model using the robust design. We define y_{ikt} as the encounter history for individual i at secondary occasion k during primary occasion t. If we have a Bernoulli encounter process then we can describe the observation model, specified conditional on the "alive state," $z(i, t)$, for individual i

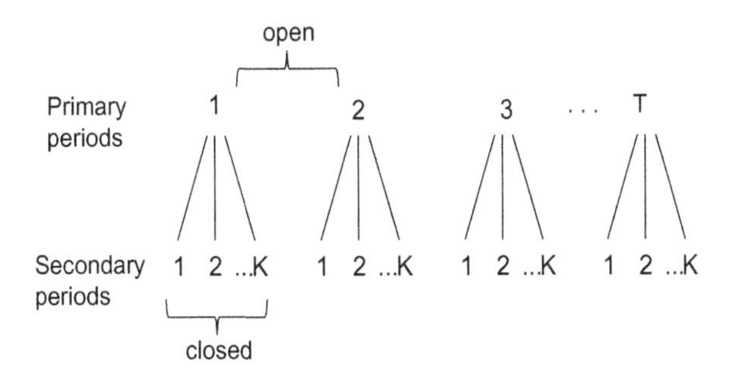

FIGURE 16.1

Schematic of the robust design with T primary sampling periods and K secondary periods. The populations are considered open between primary periods and closed within each.

at primary time t, as:

$$y_{ikt}|z(i,t) \sim \text{Bernoulli}(p_t z(i,t)).$$

(Note: throughout this chapter we will focus on changes in the alive state, so we will index z using parentheses in order to make the notation easier to read, where $z(i,t)$ is equivalent to z_{it}.) Thus, if individual i is alive at time t (i.e., $z(i,t) = 1$), then the observations are Bernoulli with detection probability p as before. Conversely, if the individual is not alive ($z(i,t) = 0$), then the observations must be fixed zeros with probability 1. Note our distinct use of the variable z here as representing the state of individuals (alive/dead) instead of our previous use of z as the data augmentation variable.

Survival and recruitment in the open population are manifest in a model for the latent state variables $z(i,t)$ describing individual mortality and recruitment events. An important aspect of the hierarchical formulation of the model that we adopt here is that the model for the state variables is described conditional on the total number of individuals ever alive during the study (a parameter which we label N) based on T periods, as in Schwarz and Arnason (1996). Data augmentation induces a special interpretation on the latent state variables $z(i,t)$. In particular, "not alive" includes individuals that have died, or individuals that have not yet been recruited. Royle and Dorazio (2008) showed that using this formulation simplifies the state model and allows it to be implemented directly in the **BUGS** language. For example, considering the case $T = 2$, the state model is composed of the following two components: First, the initial state is described by:

$$z(i,1) \sim \text{Bernoulli}(\psi),$$

and then a model describing the transition of individual states from $t = 1$ to $t = 2$:

$$z(i,2) \sim \text{Bernoulli}(\phi z(i,1) + \gamma(1 - z(i,1))).$$

If $z(i,1) = 1$, then the individual may survive to time $t = 2$ with probability ϕ whereas, if $z(i,1) = 0$, then the "pseudo-individual" may be recruited with probability γ.

We can then generalize this model for $T > 2$ time periods and allow survival and recruitment to be time dependent. Initialize the model for time $T = 1$ as we have done above and then the model describing the transition of individual states from t to $t+1$ is:

$$z(i,t+1) \sim \text{Bernoulli}(\phi_t z(i,t) + \gamma_t(1 - z(i,t))).$$

This parameterization results in $T - 1$ survival and recruitment parameters. The main difference here from the CJS model, described below, is that we include recruitment and are interested in estimating N for each t. Since this state model described above is conditional-on-N, we must deal with the fact that N is unknown, which is done through data augmentation similar to how we used it in the closed population models.

16.2.2 Data augmentation for the Jolly-Seber model

The fundamental challenge in carrying out a Bayesian analysis of this model is that the parameter N (the total number of individuals alive during the study) is not known. We have discussed and demonstrated data augmentation in many previous chapters; however, with the open population model, we have to take care that two issues are addressed: (1) the data augmentation is large enough to accommodate all potential individuals alive in the population during the entire study and (2) that individuals cannot die and then re-enter the population. Royle and Dorazio (2008) describe this formulation for open population models, including the non-spatial JS and robust design models (see also Kéry and Schaub 2012; Link and Barker 2010).

To begin, let's consider the role of recruitment, γ, in the model when we use data augmentation to estimate N. Data augmentation formally reparameterizes the model, replacing $N(1)$, the number of individuals alive in the first primary period with M, where we assume $N(1) \sim \text{Binomial}(M, \psi)$. That is, the expected value of $N(1)$ under the model is equal to ψM. As a result of this reparameterization, the recruitment parameters, γ_t, are also relative to the number of "available recruits" on the data augmented list of size M, and not directly related to the population size. We can derive N_t and R_t, the population size and number of recruits in year t, as a function of the latent state variables $z(i, t)$. For example, the total number of individuals alive at time t is

$$N_t = \sum_{i=1}^{M} z(i, t)$$

and the number of recruits is

$$R_t = \sum_{i=1}^{M} (1 - z(i, t - 1))z(i, t),$$

which is the number of individuals *not* alive at time $t - 1$ but alive at time t.

In the case of just two primary periods, this process is straightforward. When the number of primary sample occasions is greater than 2, we must formulate the model for recruitment by introducing another latent variable, in order to ensure that an individual can only be recruited once into the population. Here, this formulation of the model uses a set of indicator variables, labeled $A(i, t)$. Let $A(i, t) = 1$ if individual i is available to be recruited at time t, otherwise $A(i, t) = 0$. We define $A(i, t)$ for $i = 1, 2 \ldots, M$ and $t > 1$ as

$$A(i, t) = \left(1 - \mathrm{I}\left(\sum_{t=1}^{t=t-1} (z(i, t)) > 0\right)\right).$$

Each recruitment indicator is conditional on whether the individual was ever previously alive. This ensures that an individual cannot be recruited again after initial recruitment. Then, we can describe the state variables $z(i, t)$ by a first-order

Markov process. For $t = 1$, the initial states are fixed:

$$A(i, 1) \equiv z(i, 1)$$

and, for subsequent states, we have

$$z(i, t)|z(i, t - 1) \sim \text{Bernoulli}(\phi_t z(i, t - 1) + \gamma_t A(i, t)).$$

Thus, if an individual is in the population at time t (i.e., $z(i, t) = 1$), then that individual's status at time $t + 1$ is the outcome of a Bernoulli random variable with parameter (survival probability), ϕ_t. If the individual, however, has never been in the population previous to time t (i.e., $A(i, t) = 1$), then it is recruited into the population with probability γ_t. Note that γ_t is related to per-capita recruitment rate (say η) as $\gamma_t = N_{t-1}\eta / \sum_i^M A_{i,t-1}$. We define this model in **JAGS** by using the `sum()` and `step()` functions together to ascertain if a particular individual i was ever previously alive. The `step()` function is a logical test in **JAGS** for $x \geq 0$ such that `step(`$x \geq 0$`)` returns a 1, otherwise 0. Individuals that were ever previously alive are no longer eligible to be "recruited" into the population. The implementation of this model in **JAGS** is shown in Panel 16.1.

16.2.2.1 Ovenbird mist-netting study

We now return to the ovenbird data collected using mistnets at Patuxent Wildlife Research Center. We introduced these data in Chapters 9 and 14, and they are provided with the `secr` package (see, Efford et al. 2004, Borchers and Efford 2008). To refresh your memory: 44 mist nets spaced 30 m apart on the perimeter of a 600 m × 100 m rectangle (see Figure 16.2) were operated on 9 or 10 non-consecutive days in late May and June for 5 years from 2005 to 2009.

In Chapters 9 and 14, we dealt with this dataset as a type of spatial "multi-session" model where abundances in each year, N_t, were regarded as independent random variables either with a Poisson prior (as implemented in `secr`) or a binomial prior if analyzed using **BUGS** with data augmentation. This is the simplest approach for modeling data collected over multiple years, but it does not allow for inference about demographic processes, as does the JS model.

In the spatial multi-session model (S-MS) we did not use individual identity across years; however, we need to maintain the order of individuals across years to estimate the survival and recruitment of individuals into the population. We organize the data set so that each row in the array represents just one individual across all primary periods. For the ovenbird data set, we can organize the data by creating a master list of all individuals captured during the entire study. From this list, we can assign each individual a unique row in our data set (in the **R** commands, we do this by using the `unique()` function on the row names for each year of our three-dimensional array and use `pmatch()` to associate the data to the correct column). The resulting array is individual by secondary occasion by primary occasion, $M \times K \times T$. The **R** commands to organize the data in a way suitable for fitting a Jolly-Seber type model are included in the `scrbook` package using the function `ovenbirds.js()` and are not shown

```
model{

psi ~ dunif(0,1)
phi ~ dunif(0,1)
p.mean ~ dunif(0,1)

for(t in 1:T){
   N[t] <- sum(z[1:M,t])
   gamma[t] ~ dunif(0,1)
}

for(i in 1:M){
  z[i,1] ~ dbern(psi)            # Alive state for the first year
 cp[i,1] <- z[i,1]*p.mean
  Y[i,1] ~ dbinom(cp[i,1], K)     # Y are the number of encounters
  A[i,1] <- (1-z[i,1])

  for(t in 2:T){                    # For loop for years 2 to T
      a1[i,t] <- sum(z[i, 1:t])     # Sum over the alive states from 1 to t
       A[i,t] <- 1-step(a1[i,t] - 1)
  # A is the indicator if an individual is available to be recruited
       mu[i,t] <- (phi*z[i,t-1]) + (gamma[t]*A[i,t-1])
  # Alive state at t is dependent on phi and gamma
        z[i,t] ~ dbern(mu[i,t])
       cp[i,t] <- z[i,t]*p.mean
        Y[i,t] ~ dbinom(cp[i,t], K)
           }
       }
}
```

PANEL 16.1

JAGS model specification for the non-spatial Jolly-Seber model using data augmentation.

here. The key difference between this model and organization of the data and that in Chapter 9 is that, here, we have to preserve individual identity across years (in both the model and data structure).

The data augmentation must be large enough to include individuals alive during any of the time periods and to account for that, we set $M = 200$. There were 70 unique individuals captured over the 5-year period. For this example, we hold survival constant but allow recruitment to be time dependent (since γ is essentially a function of the data augmentation process as described above, it does not make sense to hold γ constant, even if per-capita recruitment is constant.).

To implement the model in Panel 16.1, the following commands are used:

```
 # Set initial values for the alive state, z
> zst <- c(rep(1,M/2),rep(0,M/2))
> zst <- cbind(zst,zst,zst,zst,zst)
```

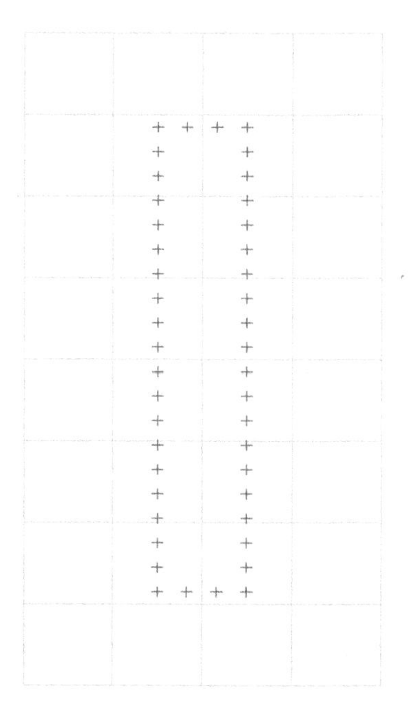

FIGURE 16.2

Arrangement of the mist nets in the ovenbird study. The nets are arranged in a 600 m by 100 m rectangle, spaced 30 m apart.

```
> inits <-function()list(z=zst,sigma=runif(1,25,100),gamma=runif(5,0,1))
> parameters <- c("psi","N","phi", "p.mean", "gamma")
> data <- list (K=10,Y=Ybin,M=M)

> library("rjags")
> out1 <- jags.model("modelNSJS.txt",data,inits,n.chains=3,n.adapt=500)
> out2NSJS <- coda.samples(out1,parameters,n.iter=20000)
```

In this non-spatial JS model, N_t is estimated to be between about 22 and 33 for each of the 5 years (see Table 16.1 for results). The posterior mean for detection (`p.mean` in the model) was 0.14. We did not include `p.mean` in the table because the SCR models do not have a parameter that directly corresponds to it. Instead, SCR models have a detection function that is related to distance.

16.2.2.2 *Shortcomings of the traditional JS models*

One of the biggest shortcomings of the non-spatial JS model is that we estimate N but have no explicit spatial area associated with it (so, in Table 16.1, the density estimate from the non-spatial JS model is listed as NA). Ignoring the spatial information in the data makes the estimation of density an informal process. As we saw in the closed models, the incorporation of spatial information in the model will allow us to make an explicit estimate of density. This improvement should also carry through to our

Table 16.1 Posterior means of model parameters for the non-spatial Jolly-Seber model (NS-JS), the spatial Jolly-Seber model (S-JS), and the spatial multi-session model (S-MS) fitted to the ovenbird data set. Density shown in individuals per hectare.

	NS-JS	S-JS	S-MS
D[1]	–	0.96	0.93
D[2]	–	1.00	1.00
D[3]	–	1.10	1.20
D[4]	–	1.10	0.89
D[5]	–	0.79	0.76
N[1]	26.46	33.43	32.42
N[2]	30.21	35.53	35.63
N[3]	33.05	39.23	42.10
N[4]	29.48	37.07	30.88
N[5]	21.69	27.57	26.15
alpha0	–	−2.90	−2.88
alpha1	–	1.2e-04	1.22e-04
sigma	–	6.40	6.44
gamma[1]	0.50	0.50	–
gamma[2]	0.09	0.09	–
gamma[3]	0.11	0.13	–
gamma[4]	0.13	0.16	–
gamma[5]	0.07	0.08	–
phi	0.48	0.53	–
psi	0.14	0.17	–
R2	–	14.78	–
R3	–	18.61	–
R4	–	19.77	–
R5	–	8.45	–

estimation of other demographic parameters such as survival and recruitment as the movement of individuals is directly accounted for in the model.

16.2.3 Spatial Jolly-Seber models

To parameterize the spatial JS models, we follow all of the same steps as the non-spatial model but also include the trap location information into our detection function. Basically, we are using the closed population SCR model to estimate the detection parameters and initial population size, and the open component is carried out in the process of how we model the transition of $z(i,t)$ to $z(i,t+1)$ which is the same as in the non-spatial JS model. To do so, we describe the Bernoulli observation model, specified conditional on $z(i,t)$, as has been done throughout the book:

$$y_{ijkt}|z(i,t) \sim \text{Bernoulli}(p_{ijk}z(i,t))$$

with

$$p_{ijk} = p_0 \exp(-\alpha_1 d_{ij}^2), \tag{16.2.1}$$

where $d_{ij} = ||\mathbf{x}_j - \mathbf{s}_i||$, the distance between activity center $\mathbf{s}_i$ and trap $\mathbf{x}_j$. As before, p_0 is the baseline encounter probability, for an individual with home range center located precisely at a trap, and $\alpha_1 = 1/(2\sigma^2)$ where σ is the scale parameter in this Gaussian encounter probability model.

If individual i is alive at time t ($z(i,t) = 1$), then the observations are Bernoulli. Conversely, if the individual is not alive ($z(i,t) = 0$), then the observations must be fixed zeros with probability 1. As always, other observation models can be considered in the context of a fully open JS-type model, such as the Poisson or multinomial models described in Chapter 9, and we can consider many alternative models of encounter probability.

We initialize the model for time $T = 1$ and then model the transition of individual states from t to $t + 1$ as:

$$z(i, t+1) \sim \text{Bernoulli}(\phi_t z(i,t) + \gamma_t(1 - z(i,t))).$$

Previously, in Section 16.2.2, we described how this formulation of the model uses a set of latent indicator variables $A(i,t)$ which describes if an individual is available to be recruited into the population during time interval $(t-1, t)$. We apply the same approach here, so that, as before, $A(i,t) = 1$ if individual i is recruited in time interval $(t-1, t)$; otherwise $A(i,t) = 0$.

The number of recruits into the population is calculated based on the alive state of the previous time steps $(1, 2, \ldots, t-1)$ and the current time step (t). For example, to estimate the number of recruits from time period 1 to 2, we count those individuals not in the population at time 1 ($z(i,1) = 0$) but alive at time 2 ($z(i,2) = 1$). We can determine if individual i has entered the population at time $t = 2$ by using the formula: $R_{i,2} = (1 - z(i,1))z(i,2)$ and then sum $R_{i,2}$ over M to get the total number of recruits. We can do this for all the primary periods in our study, as shown in the **JAGS** code in Panel 16.2. For this example, we model the activity centers as independent for each primary period, such that $s(i,t) \sim \text{Uniform}(S)$. However, we discuss alternative models for the activity centers in Section 16.4.

16.2.3.1 *Ovenbird mist-netting study*

In the previous analysis of the ovenbird data, we did not make use of the spatial location for each net the ovenbirds were captured in. However, there were 44 mist nets operational during each of the sampling occasions. We already organized the data so that the 3-D encounter histories are set up (see Section 16.2.2.1). The data set is then $M = 200$ individuals by $K = 10$ secondary occasions by $T = 5$ primary occasions. In the non-spatial version, we reduced the data to captured or not-captured; however, the encounter history array, `Yarr`, contains the number of the net that each individual was captured in and contains a "45" if the individual was not captured. The code above describes how the encounter history array is created, so we do not reproduce this piece of code here. To fit the model, use the following **R** code which

```
model {
psi ~ dunif(0,1)         # Prior distributions
phi ~ dunif(0,1)
alpha0 ~ dnorm(0,10)
sigma ~ dunif(0,200)
alpha1 <- 1/(2*sigma*sigma)
A <- ((xlim[2]-xlim[1]))*((ylim[2]-ylim[1]))  # Area of state-space

for(t in 1:T){
  N[t] <- sum(z[1:M,t])  # Calculate abundance for each year
  D[t] <- N[t]/A        # Calculate density for each year
  R[t] <- sum(R[1:M,t])    # Calculate the recruits for each year
  gamma[t] ~ dunif(0,1)  # Prior for time specific recruitment parameter
}

for(i in 1:M){
  z[i,1] ~ dbern(psi)
  R[i,1] <- z[i,1]       # To estimate the number of recruits
  R[i,2] <- (1-z[i,1])*z[i,2]
  R[i,3] <- (1-z[i,1])*(1-z[i,2])*z[i,3]
  R[i,4] <- (1-z[i,1])*(1-z[i,2])*(1-z[i,3])*z[i,4]
  R[i,5] <- (1-z[i,1])*(1-z[i,2])*(1-z[i,3])*(1-z[i,4])*z[i,5]

  for(t in 1:T){
     # Independent activity centers for each year
     S[i,1,t] ~ dunif(xlim[1],xlim[2])
     S[i,2,t] ~ dunif(ylim[1],ylim[2])
  for(j in 1:ntraps){
      d[i,j,t] <- pow(pow(S[i,1,t]-X[j,1],2) + pow(S[i,2,t]-X[j,2],2),1)
     }
  for(k in 1:K){
     for(j in 1:ntraps){
        lp[i,k,j,t] <- exp(alpha0 - alpha1*d[i,j,t])*z[i,t]
        cp[i,k,j,t] <- lp[i,k,j,t]/(1+sum(lp[i,k,,t]))
     }
    cp[i,k,ntraps+1,t] <- 1-sum(cp[i,k,1:ntraps,t])
    # Here, the last cell indicates not captured
    Ycat[i,k,t] ~ dcat(cp[i,k,,t])
  }
}
  A[i,1]<-(1-z[i,1])
for(t in 2:T){                   # For loop for years 2 to T
      a1[i,t] <- sum(z[i, 1:t])  # Sum over alive states from 1 to t
       A[i,t] <- 1-step(a1[i,t] - 1)
   # A indicates if individual is available to be recruited at time t
       mu[i,t] <- (phi*z[i,t-1]) + (gamma[t]*A[i,t-1])
   # Alive state at t is dependent on phi and gamma
        z[i,t] ~ dbern(mu[i,t])
          }
     }
}
```

PANEL 16.2

JAGS model specification for the fully spatial Jolly-Seber model. This extends the ordinary Jolly-Seber model by the inclusion of a spatial component to the encounter probability model.

sets the initial values for `z[i,t]`, the parameters to monitor, and calls **JAGS**. The code is also available in the `ovenbirds.js()` function:

```
> zst <- c(rep(1,n),rep(0,M-n))
> zst <- cbind(zst,zst,zst,zst,zst)

> inits <- function(){list(z=zst,sigma=runif(1,25,100),
        gamma=runif(5,0,1), S=Sst,alpha0=runif(1,-2,-1))}
> parameters <- c("psi", "alpha0", "alpha1", "sigma", "N",
                    "D", "phi", "gamma", "R")
> data <- list(X=as.matrix(X[[1]]), K=10, Ycat=Yarr,
              M=M, ntraps=ntraps, ylim=ylim, xlim=xlim)
> library("rjags")
> out1 <- jags.model("modelJS.txt", data, inits, n.chains=3,
             n.adapt=500)
> out2JS <- coda.samples(out1,parameters,n.iter=10000)
```

Our results for density, α_0, and α_1 are similar to those found in the multi-season analysis from Chapter 9. Since all of the parameters including α_0 and α_1 are shared between seasons, we would expect these results to be similar between the multi-season model and the JS model (see Table 16.1). There are some slight differences in the parameter estimates, for example, the density is lower in year 4 in the multi-season model than in the JS model. This may be due to a smaller sample size in that year; due to the Markovian relationship between abundances, the JS model is able to make use of the data more efficiently. This may also be due to the implied prior on N_t (see Royle and Dorazio 2008, Chapt. 10). Because we have defined the same state-space for the spatial JS model and multi-season, our estimates of N_t are directly comparable. However, the estimates of N_t under the non-spatial JS model are not directly comparable as we do not have a well-defined effective trapping area. We see from Table 16.1 that N_t is smallest for the non-spatial JS model across all years. This suggests that the actual effective trapping area is smaller than our state-space, but we cannot estimate the effective trapping area in a formal manner and thus we cannot make useful comparisons between the abundance estimates.

In the JS formulation of the model, we also estimate the recruitment for each year, and we can look at our derived values for recruitment (R_2, R_3, R_4, and R_5). R_2 is the number of new recruits from primary period 1 to 2; R_3 is the number of new recruits from primary period 2 to 3; and so forth. R_2 and R_3 are almost double that of R_4 and R_5, suggesting that fewer animals were recruited into the population in the latter years of the study. The density in the last year of the study was lower than previous years. It is good to check your results when you see a pattern like this—the number of recruits declining each year—because this could be an indication that the data augmentation was not large enough. In this example, we checked to make sure that $M = 200$ was sufficiently large by examining the recruitment parameter, γ. If γ is close to 1 during any of the time periods, then there are not enough augmented individuals in the overall data set. In this case, the 97.5% quantile of γ_5, the recruitment probability in the final

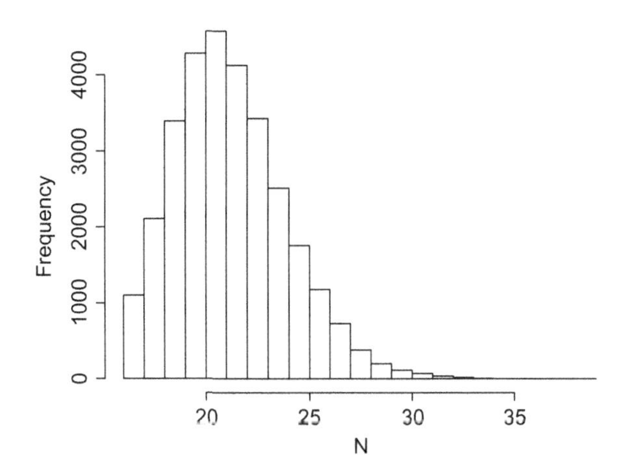

FIGURE 16.3

Posterior distribution of N_5 from the spatial JS model for the ovenbird dataset. This figure suggests that there is no truncation of the posterior of N_5 by M.

year of the study, was 0.14 and none of the other γs were close to 1 either. You can also look at the posterior distributions of the total N or each N_t to make sure they are not truncated, Figure 16.3 shows that the posterior distribution of N_5 is not truncated. The posterior mean for survival, ϕ, was 0.53. Although we did not do that in this example, it should be easy to see that we could allow survival to vary by time, as we did with recruitment. Our estimates of survival seem reasonable when compared with the ovenbird literature. Some studies have found annual male ovenbird survival to be around 0.62 (Porneluzi and Faaborg, 1999; Bayne and Hobson, 2002), whereas female ovenbird survival is much lower (0.21, Bayne and Hobson, 2002). With more individuals, we could run this model with survival estimated for each sex separately. However, researchers should be careful not to overparameterize models. The results indicate that the posterior mean estimate of ϕ was greater in the SCR model (0.53) than the non-spatial model (0.48) which suggests that the SCR model is partitioning some movement from survival this produces an estimate of true rather than apparent survival.

16.3 Cormack-Jolly-Seber models

16.3.1 Traditional CJS models

Cormack-Jolly-Seber (CJS) models are used extensively to model animal survival. There are two common ways to fit these models, using either a multinomial likelihood approach (Lebreton et al., 1992) or a state-space formulation of the model (Gimenez et al., 2007; Royle, 2008). The multinomial likelihood approach is based on summarizing the data to counts of unique encounter histories, which have a multinomial distribution. The data are summarized over individuals and so it is not feasible to build general models that contain individual covariates. On the other hand, the state-space formulation of the model preserves individual identity and, therefore, parameters can be modeled at the individual level, and individual effects (covariates or heterogeneity) can be included. In the present context of spatial capture-recapture models, we naturally think about including individual locations, or activity centers, as individual covariates.

We can adopt a state-space parameterization of the single-state, non-spatial CJS model in which the observation model is described conditional on the latent state variables $z(i,t)$ – the "alive state," which indicates whether individual i is alive ($z(i,t) = 1$) or not ($z(i,t) = 0$) during each of $t = 1, 2, \ldots, T$ *primary* periods. Let y_{it} indicate the observed encounter data of individual i in primary period t. The model, specified conditional on $z(i,t)$, is:

$$y_{it}|z(i,t) \sim \text{Bernoulli}(p_t z(i,t)).$$

Analogous to the JS model, if individual i is alive at time t ($z(i,t) = 1$), then the observations are Bernoulli with probability of detection p_t.

If the individual is not alive ($z(i,t) = 0$), then the observations are fixed zeros with probability 1. Contrary to the JS model, in the CJS model we condition on first capture, which means that $z(i,t)$ will be 1, with probability 1, when t is the primary period individual i is first captured in. We denote this by $z(i, f_i)$, where f_i indicates the primary occasion in which individual i is first captured, which can vary from $1 \ldots T$. This ensures that each individual is alive upon entering the data set.

The "alive state" at time t for each individual is a function of the state at the previous time step $t-1$. Because we condition on the first capture, the initial state is set to one:

$$z(i, f_i) = 1.$$

The model for the transition of individual states from t to $t+1$ for all $t > f_i$ is

$$z(i,t) \sim \text{Bernoulli}(\phi z(i,t-1)).$$

The individual survives with probability ϕ to time $f_i + 1$ and so forth. Once an individual leaves the population (i.e., $z(i,t) = 0$), there is no mechanism for the individual to return. In the CJS model we are not estimating N_t, so we do not need to make use of data augmentation here in order to account for uncaptured individuals (remember: we explicitly condition on first capture). This version of the model is easy to construct in **BUGS** (or **JAGS**), which is shown in Panel 16.3. Variations on this basic model and associated code for fitting the model in **BUGS** are described in detail in Kéry and Schaub (2012, Chapters 7–9).

16.3.1.1 *Movement and survival of American shad in the Little River*

As an example for the CJS model, we use data collected on American shad (*Alosa sapidissima*) in the Little River in North Carolina, USA (see photo in Figure 16.4).

```
model{
  phi ~ dunif(0,1)  # Survival (constant over time)

  for(t in 1:T){
    p[t] ~ dunif(0, 1)  # Detection (varies with time)
  }

 for(i in 1:M){
    z[i,first[i]] ~ dbern(1)
    for (t in (first[i]+1):T){
        tmp[i,t] <- z[i,t]*p[t]
          y[i,t] ~ dbern(tmp[i,t])
      phiUP[i,t] <- z[i,t-1]*phi
          z[i,t] ~ dbern(phiUP[i,t])
    }
  }
}
```

PANEL 16.3

JAGS model specification for the non-spatial basic Cormack-Jolly-Seber (CJS) model. Note that the first alive state of each individual, `z[i,first[i]]`, is not stochastic. It is equal to 1 with probability 1.

The Little River is a tributary to the Neuse River and the confluence is near Goldsboro, North Carolina, about 212 river kilometers from the Pamlico Sound. The motivation for this example stems from an interest in better understanding survival and movement of migratory fish. American shad are an anadromous fish that use rivers for spawning. The data were collected and analyzed as described in Raabe (2012). Using a resistance board weir near the river mouth, 315 fish were tagged with passive integrated transponders (PIT) in the spring of 2010. An array of seven upstream PIT antennas passively recaptured individuals during upstream and downstream migrations. Each time a fish passed over the antenna, it was recorded and the data were summarized weekly for 12 weeks. The fish do not necessarily move past all antennas and may remain in the river between antennas for more than a week, thus they are not all detected at each time period. The antennas do not always operate perfectly either and fish that pass may not be recorded at some times.

To apply the CJS model, we create the encounter history for each individual for the 12 weeks and we also create a vector to indicate the period (week) of first capture. The code is not shown here but is availabe in the `scrbook` package within the function `shad.cjs()`. This function contains all of the code to fit the non-spatial, multi-state, and spatial CJS models to the American shad data set.

FIGURE 16.4

American shad caught in North Carolina, USA. *Credit*: Joshua Raabe, North Carolina State University.

Table 16.2 shows the estimated detection probabilities for each of the 12 primary periods in the study. The posterior mean for detection probabilities ranges from 0.126 to 0.880, which could potentially be due to variation in water flow, stream depth, storms, timing of migration, etc. The weekly survival probability, ϕ, had a posterior mean estimate of 0.824. This estimate could be considered low for a weekly probability, but is likely due to the fact that the migration upstream can be quite energetically taxing and the fish are likely to only feed minimally in rivers (Leggett and Carscadden, 1978; Leonard and McCormick, 1999). Additionally, the CJS model is only estimating apparent survival and some fish may have left the stream temporarily or permanently heading back to the ocean or possibly to other tributaries that are not monitored. We demonstrate in Panel 16.3 how to allow p to vary by time, but we could also allow survival, ϕ, to vary by time by implementing it exactly as we do for p. As we move into the multi-state model, we can test for movement and survival by state, which allows more specific biological questions to be addressed.

16.3.2 Multi-state models

The standard CJS model only allows for estimation of survival and detection probabilities. However, researchers are often interested in addressing other ecological questions such as age-dependent survival rates, habitat-specific movements, etc. Multi-state models allow researchers to directly address such questions by incorporating more than one state that an individual may be in (Arnason (1972, 1973) and Brownie et al. (1993)). These possible states can be geographic location, age class, or reproductive status among many others. Instead of just having an encounter history for an individual, we will also have auxiliary information on the state of that individual at capture (e.g., breeder or non-breeder, or geographic location). Since our interest, in

Table 16.2 Posterior summaries of model parameters for the non-spatial CJS model fitted to the American shad data set.

	Mean	SD	2.5%	50%	97.5%
p[1]	0.499	0.289	0.026	0.499	0.975
p[2]	0.627	0.058	0.511	0.628	0.738
p[3]	0.762	0.036	0.689	0.763	0.829
p[4]	0.880	0.025	0.828	0.882	0.925
p[5]	0.548	0.043	0.465	0.548	0.633
p[6]	0.259	0.038	0.190	0.258	0.337
p[7]	0.126	0.031	0.072	0.124	0.194
p[8]	0.236	0.045	0.155	0.234	0.332
p[9]	0.237	0.049	0.148	0.234	0.341
p[10]	0.589	0.072	0.447	0.590	0.728
p[11]	0.834	0.063	0.700	0.839	0.942
p[12]	0.468	0.072	0.330	0.466	0.614
ϕ	0.824	0.011	0.802	0.825	0.846

the context of spatial modeling, is in movement of individuals, we will consider states that represent spatial units or geographic locations. For example, we might think that the movement rates between locations (or, in the present context, state transition rates) could be due to habitat features (or quality) and we can use multi-state models to help us address such a question. In addressing movement through a multi-state modeling approach, the movement is often parameterized as random or Markovian between patches (Arnason, 1972, 1973; Schwarz et al., 1993).

In the simplest version of the multi-state model we have just two states. Thus, individuals can be marked and recaptured in one of two states (we'll call them A and B here). We will assume that the two "states" are different geographic regions. In the single-state model above, an individual i was either alive ($z(i, t) = 1$) at time t or dead ($z(i, t) = 0$). Now, we must consider that the individual could be alive in a given state or dead, and that individuals can transition between states. An easy way to think about this is to look at the state-transition matrix in Table 16.3. Here, ϕ^{A} is the probability of surviving in state A from time t to $t + 1$ and ϕ^{B} is the analogous parameter for state B. The movement (i.e., state-transition) parameters are ψ^{AB} and ψ^{BA}, where ψ^{AB} is the probability that an individual, which survived from t to $t + 1$, moves to state B just before $t + 1$, and vice versa for ψ^{BA}. The movement could also be defined as occurring before the survival; i.e., ψ^{AB} is the probability that an individual moves from state A to state B shortly after time t.

Because individuals are not necessarily observed and the observation probability may depend on the state, detection probability should be estimated separately for each of the states. Hence, we also have p^{A} and p^{B}, the probability of detecting an individual in state A and state B, respectively. Also, at this point, we assume that there is no error in observed state (i.e., if the animal is observed, then the state is recorded correctly).

Table 16.3 Transition matrix for a multi-state model with two states where the rows are the departure state at time $t-1$ and the columns are the arrival state at time t.

	State A	State B	Dead
State A	$\phi^A(1-\psi^{AB})$	$\phi^A\psi^{AB}$	$1 - \phi^A$
State B	$\phi^B\psi^{BA}$	$\phi^B(1-\psi^{BA})$	$1 - \phi^B$
Dead	0	0	1

Classical multi-state models are closely related to closed population SCR models. In particular, keeping with the notion of having a discrete state variable and going back to the shad example from Section 16.3.1.1, instead of having a Markovian state-transition model, we imagine that each fish has a "home area", and the observation state is conditioned on that (constant) home area, **s**. We can express that model as

$$u_{it} \sim \text{Categorical}(\psi(s_i))$$

where u_{it} (a discrete state) is the current location of fish i, and **s** (also a discrete state) is its home area. We could imagine that the state-transition probability vector, $\psi(\mathbf{s})$, is related in some fashion to distance between possible observation states and home area. In this model, the current state, u_{it}, is not Markovian as it is in classical multi-state models but, rather, an independent sample from a distribution indexed by s. We see that, by letting the number of possible states increase to infinity, the model morphs into a continuous space SCR model, except u_{it} can take on any value in the state-space (this type of "search-encounter" model was discussed in Chapter 15). If we restrict the potential observation locations to some prescribed subset of the state-variable **s**, e.g., trap locations, then the model is precisely an SCR model for a fixed trap array. Therefore, SCR models can be viewed as multi-state models, but with a continuous state-variable (instead of discrete) - "space" - and with independent transitions between states in successive times (instead of Markovian).

To describe the two state model for **JAGS**, we define the observation as $y_{it} = 1, 2$, or 3 where 3 indicates "not observed." Additionally, we use $z(i, t)$ to indicate the true state of individual i such that $z(i, t) = 1, 2$, or 3 where 1 indicates alive and in state 1, 2 indicates alive and in state 2, 3 indicates "not alive." Using these definitions, we just need to set up the transition matrix based on Table 16.3 and define each item within the model specification, shown in Panel 16.4. Note that this can become quite cumbersome when dealing with models that have many states.

16.3.2.1 *Movement and survival of American shad in the Little River*

Previously, we analyzed the American shad data using a basic (i.e., non-spatial) CJS model. To demonstrate the two-state version of the multi-state CJS model, we will reanalyze the American shad data set, now defining two regions of the river ("downstream" and "upstream"). Any number of states could be selected for this

```
model{

# r is an index for state (excluding the 'not alive' state)
for(r in 1:2){
   phi[r] ~ dunif(0,1)
   psi[r] ~ dunif(0,1)
   p[r] ~ dunif(0,1)
}

for (i in 1:M){
    z[i,first[i]] <- y[i, first[i]]
for (t in (first[i]+1):T){
    z[i,t] ~ dcat(ps[z[i,t 1], i, ])
    y[i,t] ~ dcat(po[z[i,t], i, ])
   }
    ps[1, i, 1] <- phi[1] * (1-psi[1])
    ps[1, i, 2] <- phi[1] * psi[1]
    ps[1, i, 3] <- 1-phi[1]
    ps[2, i, 1] <- phi[2] * (1-psi[2])
    ps[2, i, 2] <- phi[2] * psi[2]
    ps[2, i, 3] <- 1-phi[2]
    ps[3, i, 1] <- 0
    ps[3, i, 2] <- 0
    ps[3, i, 3] <- 1

    po[1, i, 1] <- p[1]
    po[1, i, 2] <- 0
    po[1, i, 3] <- 1-p[1]
    po[2, i, 1] <- 0
    po[2, i, 2] <- p[2]
    po[2, i, 3] <- 1-p[2]
    po[3, i, 1] <- 0
    po[3, i, 2] <- 0
    po[3, i, 3] <- 1
 }
}
```

PANEL 16.4

JAGS model specification for a two-state version of the multi-state CJS model. Code modified from Kéry and Schaub (2012, Chapter 9).

example, and a logical choice might be 7, the number of antenna used in the study. However, here we will simplify problem by using just two states in order to demonstrate how to set up and run the model described in the previous section. Defining two states, downstream and upstream, allows us to explore potential differences in

Table 16.4 Results of the multi-state model for the American shad example. p^A is the detection probability in the first state (A), which in this case is the downstream area. ϕ^A is the weekly survival probability in state A and ψ^{AB} is the probability that an individual, which survived from t to $t+1$ in Site A, moves to State B just before $t+1$.

	Mean	SD	2.5%	50%	97.5%
p^A	0.777	0.045	0.689	0.777	0.866
p^B	0.434	0.027	0.382	0.434	0.489
ϕ^A	0.850	0.022	0.807	0.851	0.893
ϕ^B	0.782	0.019	0.743	0.782	0.820
ψ^{AB}	0.421	0.034	0.356	0.421	0.489
ψ^{BA}	0.927	0.014	0.897	0.937	0.952

survival based on the location of the fish in the stream. To set up the model, we first assigned each antenna to a state based on its location: those below 20 river kilometers were considered in the downstream state, all others were considered upstream. Each fish has an encounter history including whether or not the fish was detected during each week of the 12-week study, but also the "state" of capture ("downstream" or "upstream"). A vector to indicate the period of first capture was also created. Fish captured in more than one state during the week were assigned the state in which they were captured most during that week. And the model assumes that individuals observed in a state at consecutive primary periods did not move from that state within the primary period. The data manipulation and model specification for the multi-state CJS model are provided in the `scrbook` package under the function `shad.cjs()`.

Survival between the two areas was quite different (see Table 16.4). This might suggest that fish moving further upstream are expending more energy and are more likely to die. While survival in the two states was different, it is intuitive that the average of the survival probabilities for A and B is essentially the same as that from the basic non-spatial CJS ($\phi = 0.82$, see Table 16.2). Also, it should be noted that ψ^{BA} was very high, indicating that fish in this study are returning downstream after spawning in the upstream area. These results highlight the utility in using a multi-state model to understand movement between states; here, we used spatial states, but age, class, breeding status, etc., are all possibilities.

16.3.3 Spatial CJS models

In the above example on American shad, we lost a lot of information (about movement) by using a two-state model. As already mentioned, we could have used a seven-state model that would have allowed us to use the encounters at each antenna. However, as the number of states increases, so does the number of parameters, particularly the number of transition parameters. It stands to reason that highly parameterized transition probability matrices require huge amounts of data, which are often not available. Information is also lost in that we must reduce the encounter histories to be binary ("captured" or not during each sample occasion). In our shad example, fish can pass an antenna multiple times within a sample occasion but this information is typically not used in multi-state models. And finally, one other issue that multi-state models have not rectified is being in more than one state at a time. Again, in our shad example, we must decide what to do with a fish that is detected at more than one antenna. By reducing our example to two states of upstream and downstream, we reduced this problem to just a few cases. However, within the dataset, many fish are detected at two or more antenna during a week. This can be addressed sometimes by creating additional "states", but again, the number of states can grow quickly.

These issues are directly resolved by using a fully spatial CJS model in continuous space. We've established many times that various observation models allow for multiple detections in a given occasion, analogous to closed SCR models, so that information is not lost by reducing the data to binary encounter histories. Additionally, by not defining a distinct state, spatial CJS models directly address the issue of individuals only being able to be in one state at a time. The formulation as an SCR model also resolves the problem of estimating large transition probability matrices, by allowing us to essentially parameterize the whole matrix by "distance" and therefore reduce the dimensionality of the problem to just 1 or a few parameters.

To achieve a fully spatial CJS model, we build on the state-space and multi-state CJS models, but explicitly incorporate individual movement as an individual covariate (Royle, 2009). With this in mind, we need only make a few changes to the model. We will not have discrete states and thus the biggest difference is that individuals do not "transition" between a finite set of states, but instead are allowed to move in continuous space.

We may consider the same basic encounter models as described previously (i.e., Poisson, Bernoulli, or multinomial). In particular, let y_{ijkt} indicate the observed encounter data of individual i in trap j, during interval (secondary period or sub-sample) $k = 1, 2, \ldots, K$ and primary period t. We note that in some cases we may have only one interval ($K = 1$), which correspond to the design underlying a standard CJS or JS models, whereas the case $K > 1$ corresponds to the "robust design" (Pollock, 1982). The Poisson observation model, specified conditional on $z(i, t)$, is:

$$y_{ijkt} | z(i, t) \sim \text{Poisson}(\lambda_0 g_{ij} z(i, t)),$$

where λ_0 is the baseline encounter rate and g_{ij} is the detection model as a function of distance. If the individual is not alive ($z(i, t) = 0$), then the observations must be fixed zeros with probability 1. Remember that in the CJS formulation, we condition on first capture which means that $z(i, t)$ will be 1 when t is the first primary period of capture. As before in the non-spatial CJS model, we can denote this as $z(i, f_i)$ where f_i indicates the primary occasion in which individual i is first captured. Modeling time-effects either within or across primary periods is straightforward. For that, we define $\lambda_0 \equiv \lambda_0(k, t)$ and then develop models for $\lambda_0(k, t)$, as in our closed SCR models (we note that trap-specific effects could be modeled analogously).

To model survival, we follow the same description as that of the non-spatial version of the CJS (Section 16.3.1). In that version, we did not allow for survival to be time specific. However, it is easy to do so by allowing ϕ to vary with each time step:

$$z(i,t) \sim \text{Bernoulli}(\phi_t z(i,t-1)).$$

Under this model, and the one in Section 16.3.1, recruitment is not modeled.

16.3.3.1 *Movement and survival of American shad in the Little River*

Going back to our American shad example, we can consider that this is exactly a spatial capture-recapture problem. In stream networks, the placement of PIT antennas along the stream mimics the type of spatial data collected in terrestrial passive detector arrays such as camera traps, hair snares, acoustic recording devices, etc. The difference is that for fish and aquatic species, the stream constrains the movement of individuals to a linear network. Using the data from the array of seven PIT antennas and the number of times each fish passed over the antenna, we can apply the spatial CJS model to evaluate movement up and downstream of these fish. When we look at the individuals encountered at each antenna for each of the primary periods, the dimensions of the data are 315 individuals by 7 antennas by 12 sample occasions. Individuals can encounter any antenna any number of times during the week, which means we just sum the encounters over the week and eliminate any need for explicit secondary occasions in the model. The result is a three-dimensional array instead of a four-dimensional array. Given the structure of the encounters, we use a Poisson encounter model in this example shown in 16.5. The code to carry out this model is provided in the `scrbook` package using the function `shad.cjs()`.

The baseline encounter rate, λ_0, was allowed to vary by week and ranged from 0.188 to 5.555. We use the Poisson encounter model in this spatial CJS example rendering λ_0 not directly comparable to p_0 from the non-spatial and multi-state versions, which arises as the detection probability under the binomial encounter model. The posterior mean for ϕ was 0.784 (see Table 16.5), again showing that the weekly survival probability is rather low, just as we saw in the two previous example analyses of these data. Here, we are modeling survival probability as constant, but there is reason to believe that it might vary by time (similar to detection) and we might consider this additional parameterization in a more complete analysis of the data set. The other parameter of interest is σ, the movement parameter, which had a posterior mean of 13.954. Stream locations are recorded in river kilometers, so σ is in units of km. Our system here is linear, so we do not think of fish as having a home range radius, which would imply a circular home range. This example demonstrates how to carry out a spatial CJS model making use of all the data collected. However, in this version of the model, we have specified the activity centers as being independent across each primary period ($s(i,t) \sim \text{Uniform}(S)$). While this may be a useful model in many applications and a good starting place for demonstrating the model, here, we were hoping to gain more information about the movement of shad up and downstream. When an individual is not captured during a primary period, there

```
model {
# Priors
sigma ~ dunif(0,80)
sigma2 <- sigma*sigma
lam0 ~ dgamma(0.1, 0.1)
phi ~ dunif(0, 1)   # Survival (constant across time)
tauv~dunif(0, 30)
tau<-1/(tauv*tauv)

for (i in 1:M){
  z[i,first[i]] <- 1
  S[i,first[i]] ~ dunif(0,50)   #Fish enter the stream at 0, thus the
                  #first AC is set to the lower stream end

for(j in 1:nantenna) {
  D2[i,j,first[i]] <- pow(S[i,first[i]]-antenna.loc[j], 2)
       lam[i,j,first[i]] <- lam0*exp(- D2[i,j,first[i]]/(2*sigma2))
       tmp[i,j,first[i]] <- lam[i,j,first[i]]
         y[i,j,first[i]] ~ dpois(tmp[i,j,first[i]])
      }

   for (t in first[i]+1:T) {
         S[i,t] ~ dunif(xl, xu)#xl and xu are the upper and lower
                    limits of the stream
        for(j in 1:nantenna) {
            D2[i,j,t] <- pow(S[i,t]-antenna.loc[j], 2)
            lam[i,j,t] <- lam0 * exp(-D2[i,j,t]/(2*sigma2))
            tmp[i,j,t] <- z[i,t]*lam[i,j,t]
            y[i,j,t] ~ dpois(tmp[i,j,t])
 }
     phiUP[i,t] <-  z[i,t-1]*phi
       z[i,t] ~ dbern(phiUP[i,t])
}
}
}
```

PANEL 16.5

JAGS model specification for the spatial Cormack-Jolly-Seber (CJS) model for the American shad data set. Note that the first alive state of each individual, `z[i,first[i]]`, is not stochastic. It is equal to 1 with probability 1.

is no data to inform the activity center and thus the posterior is sampled directly from the prior. This is unsatisfying and we recognize that one could use the information from the preceding time period (or other variations, such as the following time period or combinations of the two) to inform the location information when the fish is not

Table 16.5 Posterior summary statistics of model parameters from the spatial Cormack-Jolly-Seber model fitted to the American shad data set.

	Mean	SD	2.5%	50%	97.5%
lam0[1]	5.555	0.224	5.125	5.553	6.003
lam0[2]	4.442	0.155	4.143	4.437	4.752
lam0[3]	1.892	0.068	1.763	1.891	2.031
lam0[4]	1.126	0.055	1.021	1.125	1.238
lam0[5]	0.949	0.058	0.838	0.948	1.067
lam0[6]	0.359	0.040	0.284	0.357	0.443
lam0[7]	0.188	0.031	0.133	0.186	0.254
lam0[8]	0.309	0.044	0.230	0.307	0.402
lam0[9]	0.363	0.052	0.269	0.361	0.471
lam0[10]	0.627	0.072	0.493	0.625	0.777
lam0[11]	1.611	0.109	1.408	1.607	1.835
lam0[12]	0.939	0.139	0.697	0.929	1.241
ϕ	0.784	0.012	0.760	0.785	0.807
σ	13.954	0.197	13.573	13.950	14.350

observed. In addition to providing information on the fish activity centers, this also holds promise for improving estimates of "true" ϕ as we can better separate emigration from mortality (see Section 20.1.7 for a short discussion). In the following section, we discuss further models for the movement dynamics that can be implemented and postulate a few models specifically related to the shad example.

16.4 Modeling movement and dispersal dynamics

Animal movement is both a nuisance and a fascinating process of interest in its own right. It can be a nuisance in the sense that ignoring it can bias estimators of density and survival. Numerous models (including SCR!) have been developed to deal with movement simply as a way of eliminating bias. However, movement is a key determinant of population viability, age structure, and distribution (Clobert et al., 2001), and rather than simply accounting for movement, we often want to develop explicit models for inference about processes such as dispersal and migration.

At this point, very little work has been done to model movement using SCR models; however, we expect that in the near future this will be one of the most exciting areas of research. While many sophisticated movement models exist, few of them are embedded within a framework that allows for estimation of density and other demographic parameters. This is one of the most important unrealized promises of SCR models, and in this section we will outline some possible avenues for moving forward. To begin, we will focus on the movement of activity centers among primary periods, which might result from dispersal or other large scale movements.

A primitive movement model might assume that activity centers change over time but are independent from year to year for a given individual such $\mathbf{s}(i, t) \sim \text{Uniform}(\mathcal{S})$. This is also how the spatial versions of the JS and CJS models were formulated above and this might be a reasonable model when there are large time lags between surveys, or if the individuals redistribute themselves frequently in the study population.

A more realistic model includes a movement (or dispersal) kernel of the form $[\mathbf{s}_t|\mathbf{s}_{t-1}]$. The bivariate normal distribution is one possibility (see Section 15.4), which we represent as $\mathbf{s}_t \sim \text{BVN}(\mathbf{s}_{t-1}, \boldsymbol{\Sigma})$. A symmetric kernel could be specified as $\boldsymbol{\Sigma} = \left(\begin{smallmatrix} \tau & 0 \\ 0 & \tau \end{smallmatrix}\right)$ where the scale parameter τ determines how far an individual is likely to disperse. For example, many adult passerines exhibit high site fidelity, and thus τ is likely to be very small for these individuals. However, juvenile passerines can disperse hundreds of kilometers and so τ might be much larger for these individuals.

Other approaches to analyzing movement in a mark-recapture framework focus on modeling individual locations rather than activity centers (Ovaskainen, 2004; Grimm et al., 2005; Ovaskainen et al., 2008; Hooten et al., 2010). The same could be done in SCR models by including a model component $[\mathbf{u}|\mathbf{s}]$ where $\mathbf{u}$ is location of an individual at some point in time. We could further allow the observed locations to follow an auto-regressive model such that $\mathbf{u}_{ikt} \sim \text{Normal}(\rho(\mathbf{u}_{i,k,t-1} - \mathbf{s}(i, t-1)), \Sigma_t^*)$. These are just a few simple examples; as more information becomes available and data are collected over longer time periods, we will be able to use more complex movement models in open SCR models.

16.4.1 Cautionary note

Under the Markovian movement models described above, activity centers can leave the state-space. This is a problem because it will result in an unjustified decrease in density over time. Several solutions exist, but the easiest one to implement is to truncate the dispersal kernel such that activity centers are not allowed to leave $\mathcal{S}$. In **JAGS**, this can be accomplished using the truncation function, `T()`. For example: `s[i,1,t] ~ dnorm(s[i,1,t-1], tau)T(xl, xu)` constrains the variable to lie within the range specified by `xl` and `xu`.

Another issue related to the state-space is that it should be large enough such that an individual whose activity center is located near a boundary during the initial time period should have a negligible probability of being captured. Thus, for individuals that move long distances, the state-space should be very large. Alternatively, the state-space may be designated based on knowledge of suitable habitat, thereby constraining the region within which individuals can move.

16.4.2 Thoughts on movement of American shad

In our American shad example above, we had reason to believe that individual movement is directly related to stream flow. When the stream flow is low, we might expect that the fish move very little, and when the stream flow is high, they might move more within the stream. In this case, we could model the effect of stream flow in a number of ways. First, we might allow σ to be a function of flow and to vary for each primary occasion, according to:

$$\log(\sigma_t) = \mu_\sigma + \alpha_2 \text{Flow}_t.$$

This model would only relate the movement of fish about their activity center with the flow, essentially only affecting the encounter probability. But if we think that the change in activity centers between primary periods might be related to the general pattern of fish migrating upstream more during high flow or staying closer to the same location in low flow, then we could allow the activity center locations to be a function of flow. For example, we might assume the activity centers move from the previous value depending on flow (i.e., larger movements in response to flow),

$$\mathbf{s}(i,t) \sim \text{Normal}(\mathbf{s}(i,t-1) + \beta \text{Flow}_t, \tau^2 \mathbf{I}),$$

We could extend this model too so that individuals might move at different rates upstream versus downstream based on their migration patterns. These are just a few thoughts on simple ways to model movement as a function of habitat variables which we have only started exploring on these data.

16.4.3 Modeling dispersal

Dispersal is widely studied in population ecology and is often of interest because it relates directly to population regulation, habitat quality, and connectivity of local populations (Clobert et al., 2001). However, studying dispersal with capture-recapture data can be difficult for a few reasons. One common issue with using capture-recapture data for dispersal estimation is that short distances are sampled more frequently than long distances. This is particularly true if we consider that most trap arrays are not large relative the potential dispersal distances of animals. In some cases, such as with small mammals, we may be able to capture both short and long distance dispersals in one trap array; in other cases, we may have discrete study sites set up across a larger area which capture individuals within and between sites. Either way, data are likely to be sparse for long distance dispersal events and this is particularly true if there are different habitat types which are sampled with different levels of effort (Ovaskainen et al., 2008), thus causing more difficulty in fitting models to data where much information is missing. In addition, determining if an individual has left an area or died can be difficult if the sampling does not cover the area an individual has moved to or if the sampling method has failed (e.g., a band or tag falls off or a mark is lost). As a result of this, dispersal biases estimates of survival probability obtained using standard JS or CJS models, and it is therefore important to model dispersal and survival explicitly (Schaub and Royle, 2013).

Regardless of these common sampling limitations, let's look at an ideal situation where we have the trap array large enough to observe some dispersal events (or possibly multiple trap arrays on the landscape where an individual is observed in different arrays). We sketch out a possible dispersal model but note that this is a simple example. In this case, each individual could have some probability of dispersing, say

η where $pd_{i,t} \sim \text{Bernoulli}(\eta)$ indicates if an individual disperses at time t and then

$$s_{i,t+1,1} = s_{i,t,1} + pd_{i,t}(ds_{i,t}\cos(\theta_{i,t})),$$
$$s_{i,t+1,2} = s_{i,t,2} + pd_{i,t}(ds_{i,t}\sin(\theta_{i,t})),$$

where ds_i is the dispersal distance for individual i and θ is the dispersal direction. Thus when $pd_i = 0$, then the activity centers remain the same as the previous time step and if $pd_i = 1$ then the individual disperses to a new activity center. For this specification, we have to provide a model for dispersal distance. One option is to let $ds_{i,t} \sim \text{exponential}(L)$ where L is the mean dispersal distance for individuals dispersing and let $\theta_{i,t} \sim \text{Uniform}(-\pi, \pi)$ where π is not a parameter in this case, but the mathematical constant. If all individuals are expected to move some distance between periods, then the pd indicator could be removed. More complex models involving non-Euclidean distance (Graves et al., 2013), weighted directional movement and different movement states could be fit (see Jonsen et al., 2005; Johnson et al., 2008a; McClintock et al., 2012).

16.5 Summary and outlook

In this chapter we have described a framework for making inference not only about spatial and temporal variation in population density, but also demographic parameters including survival, recruitment, and movement. The ability to model population vital rates is essential for ecology, management, and conservation; and the models described here allow researchers to examine the spatial and temporal dynamics governing those population parameters. While we have covered a lot of ground in this chapter, there are many variations of the basic JS and CJS models, such as dead recovery models or models that address transiency that we have not explicitly "converted" to a spatial framework. These areas provide a broad field of further model development. As open models are further developed, mechanisms for dealing directly with dispersal and transients will provide improved inference frameworks for understanding movement as well as the potential to estimate *true* survival instead of only *apparent survival*. This is a function of explicitly modeling movement, which means we can separate movement from mortality, as we sketched out in the model above for dispersal, providing a huge advantage over traditional models. Also, models of individual dispersal can be used to examine variation in population dynamics relative to habitat, density dependence, or climatic events.

Birth and death processes, as well as movement, all have the potential to be related to the space usage of animals in the landscape. Understanding the impact of spatially varying density on survival and recruitment will provide insights into the basic ecology of species. With the advent of non-invasive techniques, like camera trapping and genetic analysis of tissue, we can start to understand the population dynamics of species that are rarely observed in the wild. As more and more data are collected, we can use the models to explore the spatio-temporal patterns of survival, recruitment, density, and movement of species, providing incredibly useful biological and ecological information as we face broad changes in climate, land use, habitat fragmentation, etc. Rathbun and Cressie (1994) articulate a model for marked point processes where they separate out the spatial birth, growth, and survival processes for longleaf pine trees. Because of the application, these demographic parameters are slightly different than how they are often considered in wildlife and ecology, but still, there are analogies. Allowing birth, growth, and survival as well as density to arise from different spatially varying processes is the next stage in development of the open SCR models.

PART

Super-Advanced SCR Models IV

CHAPTER

Developing Markov Chain Monte Carlo Samplers 17

In this chapter we will dive a little deeper into Markov chain Monte Carlo (MCMC) sampling. We will construct custom MCMC samplers in **R**, starting with easy-to-code Generalized Linear (Mixed) Models (GL(M)Ms) and moving on to basic CR and SCR models. This material might seem slightly out of place here, as it does not deal with specific aspects or modifications of SCR models, but rather, with a particular way of implementing them (and other models, too). Knowing how to build an MCMC sampler is not essential for any of the SCR models we have covered so far, but we will need these skills to implement some models that come up in the last few chapters of this book. The aim of this chapter is to provide you with some working knowledge of building MCMC samplers. To this end, we will *not* provide exhaustive background information on the theory and justification of MCMC sampling—there are entire books dedicated to that subject and we refer you to Robert and Casella (2004, 2010). Rather, we aim to provide you with enough background and technical know-how to start building your own MCMC samplers for SCR models in **R**. You will find that quite a few topics that come up in this chapter have already been covered in previous chapters, particularly the introduction into Bayesian analysis in Chapter 3. To keep you from having to leaf back and forth we will in some places briefly review aspects of Bayesian analysis, but we try to focus on the more technical issues of building MCMC samplers relevant to SCR models.

17.1 Why build your own MCMC algorithm?

The standard programs we have used so far to do MCMC analyses are **WinBUGS** (Gilks et al., 1994) and **JAGS** (Plummer, 2003). The wonderful thing about these **BUGS** engines is that they automatically use appropriate and, most of the time, reasonably efficient forms of MCMC sampling for the model specified by the user.

The fact that we have such a Swiss Army knife type of MCMC machine begs the question: Why would anyone want to build their own MCMC algorithm? For one, there are a limited number of distributions and functions implemented in **BUGS**. While **JAGS** provides more options, some more complex models may be impossible to build within these programs. A very simple example from spatial capture-recapture that can give you a headache in **WinBUGS** is when your state-space is an

Spatial Capture-Recapture. http://dx.doi.org/10.1016/B978-0-12-405939-9.00017-7

irregular-shaped polygon, rather than an ideal rectangle that can be characterized by four pairs of coordinates. It is easy to restrict activity centers to any arbitrary polygon in **R** using an ESRI shapefile (and we will show you an example in Section 17.7), but you cannot use a shapefile in a **BUGS** model. Similarly, models of space usage that take into account ecological distance (Chapter 12) cannot be implemented in the **BUGS** engines.

Sometimes, implementing an MCMC algorithm in **R** may be faster and more memory-efficient than in **WinBUGS**—especially if you want to run simulation studies where you have hundreds or more simulated data sets, several years' worth of data or other large models, this can be a big advantage. Further, writing your own sampler gives you more control over which kind of updater is used (see following sections). Finally, building your own MCMC algorithm is a great exercise to understand how MCMC sampling works. So while using the **BUGS** language requires you to understand the structure of your model, building an MCMC algorithm requires you to think about the relationship between your data, priors, and posteriors, and how these can be efficiently analyzed and characterized. However, if you don't think you will ever sit down and write your own MCMC sampler, consider skipping this chapter—apart from coding, it will not cover anything SCR-related that is not covered by other, more model-oriented chapters as well.

17.2 MCMC and posterior distributions

MCMC is a class of simulation methods for drawing (correlated) random numbers from a target distribution, which in Bayesian inference is the posterior distribution. As a reminder, the posterior distribution is a probability distribution for an unknown parameter, say θ, given observed data and its prior probability distribution (the probability distribution we assign to a parameter before we observe data). The great benefit of having the posterior distribution of θ is that it can be used to make probability statements about θ, such as the probability that θ is equal to some value, or the probability that θ falls within some range of values. The posterior distribution summarizes all we know about a parameter and thus, is the central object of interest in Bayesian analysis. Unfortunately, in many if not most practical applications, it is nearly impossible to compute the posterior directly. Recall Bayes' theorem:

$$[\theta|y] = \frac{[y|\theta][\theta]}{[y]}, \tag{17.2.1}$$

where θ is the parameter of interest, y is the observed data, $[\theta|y]$ is the posterior, $[y|\theta]$ is the likelihood of the data conditional on θ, $[\theta]$ is the prior probability distribution of θ, and, finally, $[y]$ is the marginal probability distribution of the data, defined as

$$[y] = \int [y|\theta][\theta]d\theta.$$

This marginal probability is a normalizing constant that ensures that the posterior integrates to 1. Often, the integral is difficult or impossible to evaluate, unless you are dealing with a really simple model. For example, consider a normal model, with a set of n observations, $y_i; i = 1, 2, \ldots, n$:

$$y_i \sim \text{Normal}(\mu, \sigma^2),$$

where σ is known and our objective is to estimate μ. To fully specify the model in a Bayesian framework, we first have to define a prior distribution for μ. Recall from Chapter 3 that for certain data models, certain priors lead to conjugacy, i.e., if you choose a certain prior for your parameter, the posterior distribution will be of a known parametric form. More specifically, under conjugacy, the prior and posterior distributions are from the same parametric family. The conjugate prior for the mean of a normal model is also a normal distribution:

$$\mu \sim \text{Normal}(\mu_0, \sigma_0^2).$$

If μ_0 and σ_0^2 are fixed, the posterior for μ has the following form (for some of the algebra behind this, see Chapter 2 in Gelman et al. (2004)):

$$\mu|y \sim \text{Normal}(\mu_n, \sigma_n^2), \tag{17.2.2}$$

where

$$\mu_n = \left(\frac{\sigma^2}{\sigma^2 + n\sigma_0^2}\right) \times \left(\mu_0 + \frac{n\sigma_0^2}{\sigma^2 + n\sigma_0^2}\right) \times \bar{y}$$

and

$$\sigma_n^2 = \frac{\sigma^2\sigma_0^2}{\sigma^2 + n\sigma_0^2}.$$

We can directly obtain estimates of interest from this normal posterior distribution, such as its mean $\hat{\mu}$ (which is equivalent to an estimate of μ_n) and variance; we do not need to apply MCMC, since we can recognize the posterior as a parametric distribution, including the normalizing constant $[y]$. But generally we will be interested in more complex models with several, say m, parameters. In this case, computing $[y]$ from Eq. (17.2.1) requires m-dimensional integration, which can be difficult or impossible. Thus, the posterior distribution is generally only known up to a constant of proportionality:

$$[\theta|y] \propto [y|\theta][\theta].$$

The power of MCMC is that it allows us to approximate the posterior using simulation without evaluating the highdimensional integrals, and to directly sample from the posterior, even when the posterior distribution is unknown! The price is that MCMC is computationally expensive. Although MCMC first appeared in the scientific literature in 1949 (Metropolis and Ulam, 1949), widespread use did not occur until the 1980s when computational power and speed increased (Gelfand and Smith, 1990). It is safe to say that the advent of practical MCMC methods is the primary reason why Bayesian inference has become so popular during the past three decades.

In a nutshell, MCMC lets us generate sequential draws of θ (the parameter(s) of interest) from distributions approximating the unknown posterior over T iterations. The distribution of the draw at t depends on the value drawn at $t-1$; hence, the draws form a Markov chain.[1] As T goes to infinity, the Markov chain converges to the desired distribution, in our case the posterior distribution for θ. Thus, once the Markov chain has reached its stationary distribution, the generated samples can be used to characterize the posterior distribution, $[\theta|y]$, and point estimates of θ, its standard error and confidence bounds, can be obtained directly from this approximation of the posterior.

17.3 Types of MCMC sampling

There are several general MCMC algorithms in widespread use, the most popular being Gibbs sampling and Metropolis-Hastings sampling, both of which were briefly introduced in Chapter 3. We will be dealing with these two classes in more detail and use them to construct MCMC algorithms for SCR models. Also, we will briefly review alternative techniques that are applicable in some situations.

17.3.1 Gibbs sampling

Gibbs sampling was named after the physicist J.W. Gibbs by Geman and Geman (1984), who applied the algorithm to a Gibbs distribution.[2] The roots of Gibbs sampling can be traced back to the work of Metropolis et al. (1953), and it is actually closely related to Metropolis sampling (see Section 11.5 in Gelman et al, 2004, for the link between the two samplers). We will focus on the technical aspects of this algorithm, but if you find yourself hungry for more background, Casella and George (1992) provide a more in-depth introduction to the Gibbs sampler.

Let's go back to our example from above to understand the motivation and functioning of Gibbs sampling. Recall that for a normal model with known variance and a normal prior for μ, the posterior distribution of μ is also normal. Conversely, with a fixed (known) μ, but unknown variance, the conjugate prior for σ^2 is an inverse-gamma distribution with shape and scale parameters a and b:

$$\sigma^2 \sim \text{Inverse-Gamma}(a, b).$$

With fixed a and b, algebra reveals that the posterior $[\sigma^2|\mu, y]$ is also an inverse-gamma distribution, namely:

$$\sigma^2|\mu, y \sim \text{Inverse-Gamma}(a_n, b_n), \tag{17.3.1}$$

[1]Remember that for T random samples $\theta^{(1)}, \ldots, \theta^{(T)}$ from a Markov chain the distribution of $\theta^{(t)}$ depends only on the immediately preceding value, $\theta^{(t-1)}$.

[2]A distribution from physics we are not going to worry about, since it has no immediate connection with Gibbs sampling other than giving it its name.

where $a_n = n/2 + a$ and $b_n = (1/2)\sum_{i=1}^{n}(y_i - \mu)^2 + b$. However, what if we know neither μ nor σ^2, which is probably the more common case? The joint posterior distribution of μ and σ^2 now has the general structure

$$[\mu, \sigma^2|y] = \frac{[y|\mu, \sigma^2][\mu][\sigma^2]}{\int [y|\mu][\mu][\sigma^2] d\mu \, d\sigma^2}$$

or

$$[\mu, \sigma^2|y] \propto [y|\mu, \sigma^2][\mu][\sigma^2].$$

This cannot easily be reduced to a distribution we recognize. However, we can condition μ on σ^2 (i.e., we treat σ^2 as fixed) and remove all terms from the joint posterior distribution that do not involve μ to construct the full conditional distribution,

$$[\mu|\sigma^2, y] \propto [y|\mu][\mu].$$

The full conditional of μ again takes the form of the normal distribution shown in Eq. (17.2.2); similarly, $[\sigma^2|\mu, y]$ takes the form of the inverse-gamma distribution shown in Eq. (17.3.1), both distributions we can easily sample from. And this is precisely what we do when using Gibbs sampling: we break down high-dimensional problems into convenient one-dimensional problems by constructing the full conditional distributions for each model parameter separately; and we sample from these full conditionals, which, if we choose conjugate priors, are known parametric distributions. Let's put the concept of Gibbs sampling into the MCMC framework of generating successive samples, using our simple normal model with unknown μ and σ^2 and conjugate priors as an example. These are the steps you need in order to build a Gibbs sampler:

Step 0: Begin with some initial values for θ, say $\theta^{(0)}$. In our example, $\theta = (\mu, \sigma)$, so we have to specify initial values for μ and σ, for example by drawing a random number from some uniform distribution, or by setting them close to what we think they might be. (Note: This step is required in any MCMC sampling; chains have to start from somewhere. We will get back to these technical details a little later.)
Step 1: For iteration t, draw $\theta^{(t)}$ from the conditional distribution $[\theta_1^{(t)}|\theta_2^{(t-1)}, \ldots, \theta_d^{(t-1)}]$. Here, θ_1 is μ, which we draw from the normal distribution in Eq. (17.2.2) using the $\sigma^{(t-1)}$ value for σ.
Step 2: Draw $\theta_2^{(t)}$ from the conditional distribution $[\theta_2^{(t)}|\theta_1^{(t)}, \theta_3^{(t-1)}, \ldots, \theta_d^{(t-1)}]$. Here, θ_2 is σ, which we draw from the inverse-gamma distribution in Eq. (17.3.1), using the newly generated $\mu^{(t)}$ value for μ.
Step 3,…, d: Draw $\theta_3^{(t)}, \theta_4^{(t)}, \ldots, \theta_d^{(t)}$ from their conditional distribution $[\theta_3^{(t)}|\theta_1^{(t)}, \theta_2^{(t)}, \theta_4^{(t-1)}, \ldots, \theta_d^{(t-1)}], \ldots, [\theta_d^{(t)}|\theta_1^{(t)}, \ldots, \theta_{d-1}^{(t)}]$. In our example we have no additional parameters, so we only need Step 0 through to 2.
Repeat Steps 1 to d for T = a large number of samples.

```
Norm.Gibbs <- function(y=y,mu_0=mu_0,sigma2_0=sigma2_0,a=a,b=b,niter=niter){

ybar <- mean(y)
n <- length(y)
mu <- 1          #mean initial value
sigma2 <- 1      #sigma2 initial value
an <- n/2 + a    #shape parameter of IvGamma of sigma2
out <- matrix(nrow=niter, ncol=2)
colnames(out) <- c('mu', 'sig')

for (i in 1:niter) {

   #update mu
   mu_n <- ((sigma2/(sigma2+n*sigma2_0))*mu_0
   + (n*sigma2_0/(sigma2 + n*sigma2_0))*ybar)
   sigma2_n <- (sigma2*sigma2_0)/ (sigma2 + n*sigma2_0)
   mu <- rnorm(1,mu_n, sqrt(sigma2_n))

   #update sigma2
   bn <- 0.5 * (sum((y-mu)^2)) + b
   sigma2 <- 1/rgamma(1,shape=an, rate=bn)
   out[i,] <- c(mu,sqrt(sigma2))
}
return(out)
}
```

PANEL 17.1

R code for a Gibbs sampler for a normal model with unknown μ and σ and conjugate priors (normal and inverse-gamma, respectively) for both parameters.

Note that the order in which we update the parameters within the Gibbs algorithm does not matter. In terms of **R** coding, this means we have to write Gibbs updaters for μ and σ^2 and embed them into a loop over T iterations. The final code in the form of an **R** function is shown in Panel 17.1.

This is it! You can go ahead and simulate some data, e.g., $y \sim \text{Normal}(5, 0.5)$, and then use the function `NormGibbs` in the **R** package `scrbook` to run your first Gibbs sampler (note that the **R** function `rnorm` requires you to supply the standard deviation σ and we have written `NormGibbs` so that it returns σ instead of σ^2 so you can easily compare your input value and parameter estimate):

```
> set.seed(13)
#true mean and sd are 5 and 0.5
> y <- rnorm(1000, 5,0.5) #data

> mu_0 <- 0 #prior mean
> sigma2_0 <- 100 #prior variance
```

```
#inverse-gamma hyperparameters
> a <- 0.1
> b <- 0.1
```

```
> mod = Norm.Gibbs(y, mu_0, sigma2_0, a,b,niter=10000)
```

Your output, mod, will be a table with two columns, one per parameter, and T rows, one per iteration. For this 2-parameter example you can visualize the joint posterior by plotting samples of μ against samples of σ (Figure 17.1):

```
> plot(out[,1], out[,2])
```

The marginal distribution of each parameter is approximated by examining the samples of this particular parameter. You can visualize it by plotting a histogram of the samples (Figure 17.2 upper left and right):

```
> par(mfrow=c(1,2))
> hist(out[,1]); hist(out[,2])
```

Finally, recall an important characteristic of MCMC, namely, that the chain has to have converged (reached its stationary distribution) in order to regard samples as being from the posterior distribution. In practice, that means you have to throw out some of the initial samples called the burn-in. We will talk about this in more detail when we talk about convergence diagnostics. For now, you can use the plot(out[,1]) or plot(out[,2]) command to make a time series plot of the samples of each parameter and visually assess how many of the initial samples you should discard.

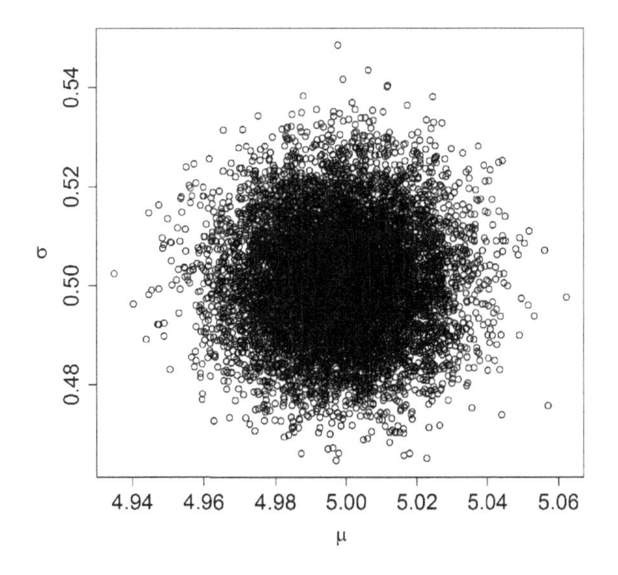

FIGURE 17.1

Joint posterior distribution of μ and σ from a normal model.

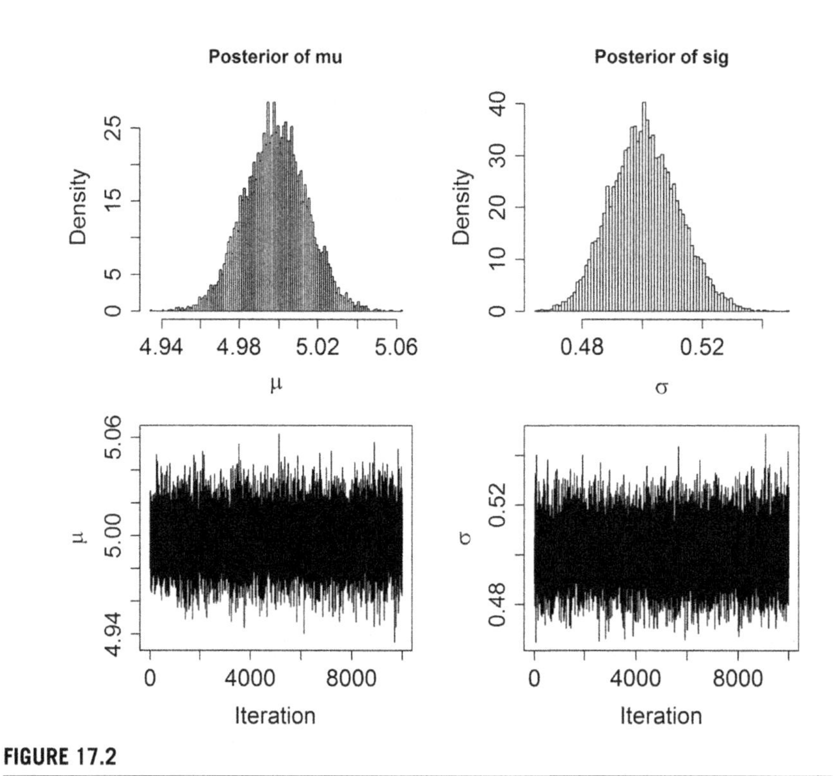

FIGURE 17.2

Plots of the posterior distributions of μ (upper left) and σ (upper right) from a normal model and time series plots of μ (lower left) and σ (lower right).

Figure 17.2 bottom left and right shows plots for the samples of μ and σ from our simulated data set; you see that in this simple example the Markov chain apparently reaches its stationary distribution very quickly—the chains look "grassy" seemingly from the start. It is hard to discern a burn-in phase visually (but we will see examples further on where the burn-in is clearer), and you may just discard the first 500 draws to be sure you only use samples from the posterior distribution. The mean of the remaining samples are your estimates of μ and σ, which, as we see below, are almost identical to the input values:

```
> summary(mod[501:10000,])
       mu                sig
 Min.   :4.935   Min.   :0.4652
 1st Qu.:4.988   1st Qu.:0.4930
 Median :4.998   Median :0.5006
 Mean   :4.998   Mean   :0.5008
 3rd Qu.:5.009   3rd Qu.:0.5084
 Max.   :5.062   Max.   :0.5486
```

17.3.2 Metropolis-Hastings sampling

Although it is applicable to a wide range of problems, the limitations of Gibbs sampling are obvious: what if we do not want to use conjugate priors or what if we cannot recognize the full conditional distribution as a parametric distribution, or simply do not want to worry about these issues? The most general solution is to use the Metropolis-Hastings (MH) algorithm, which also goes back to the work by Metropolis et al. (1953). You saw the basics of this algorithm in Chapter 3. In a nutshell, because we do not recognize the posterior $[\theta|y]$ as a parametric distribution, the MH algorithm generates samples from a known proposal distribution, say $h(\theta)$, that depends on the value of θ at the previous time step, $\theta^{(t-1)}$. The candidate value θ^* is accepted with probability

$$r = \min\left(1, \frac{[\theta^*|y]h(\theta^{(t-1)}|\theta^*)}{[\theta^{(t-1)}|y]h(\theta^*|\theta^{(t-1)})}\right).$$

Proposal distributions must be chosen so that reversibility is ensured. That means, it must be possible to go from any one value to any other. But within that criterion the proposal distribution can be absolutely anything! You can generate candidate values from a Normal(0, 1) distribution, from a Uniform(−3455, 3455) distribution, or anything of proper support. Note, however, that good choices of $h(\theta)$ are those that approximate the posterior distribution. Obviously, if $h(\theta) = [\theta|y]$ (i.e., the posterior) then you always accept the candidate value and it stands to reason that proposal distributions that are more similar to $[\theta|y]$ will lead to higher acceptance probabilities. Actually, when $h(\theta) = [\theta|y]$ we can draw samples of θ directly from $h(\theta)$, which brings us back to Gibbs sampling. Thus, Gibbs sampling is a special case of Metropolis-Hastings sampling.

The original Metropolis algorithm required $h(\theta)$ to be symmetric so that

$$h(\theta^*|\theta^{(t-1)}) = h(\theta^{(t-1)}|\theta^*).$$

In that case these two terms just cancel out from the MH acceptance probability and r is then just the ratio of the target density evaluated at the candidate value to that evaluated at the current value. A later development of the algorithm by Hastings (1970) lifted this condition. Since using a symmetric proposal distribution makes life a little easier, we are going to focus on this specific case. A type of symmetric proposal useful in many situations is the so-called *random walk* proposal distribution where candidate values are drawn from a normal distribution with the mean equal to the current value and some standard deviation, say δ, which is prescribed by the user (see below for further explanation).

Parameters with bounded support: Many models contain parameters that have bounded support, e.g., variance parameters live on $[0, \infty]$, parameters that represent probabilities live on $[0, 1]$, etc. For such cases, it is still sometimes convenient to use a random walk proposal distribution that can generate any real number (e.g., a normal random walk proposal). Under these circumstances you should not constrain the proposal distribution itself, but you can just reject parameters that are outside of the parameter space (Section 6.4.1 in Robert and Casella, 2010). You will see plenty of examples of updating parameters with bounded support in this chapter.

It is worth knowing that there are alternatives to the random walk MH algorithm. For example, in the independent MH, the proposal distribution h does not depend on $\theta^{(t-1)}$, while the Langevin algorithm (Roberts and Rosenthal, 1998) aims at avoiding the random walk by favoring moves toward regions of higher posterior probability density. The interested reader should look up these algorithms in Robert and Casella (2004, 2010).

Building a MH sampler can be broken down into several steps. We are going to demonstrate these steps using a different but still simple and common model: the logit-normal or logistic regression model. For simplicity, assume that

$$y|\theta \sim \text{Bernoulli}\left(\frac{\exp(\theta)}{1+\exp(\theta)}\right)$$

and

$$\theta \sim \text{Normal}(\mu, \sigma^2).$$

The following steps are required to set up a random walk MH algorithm:

Step 0: Choose initial values, $\theta^{(0)}$.
Step 1: Generate a proposed value of θ from $h(\theta|\theta^{(t-1)})$. We will use the random walk MH algorithm, so we draw θ^* from Normal$(\theta^{(t-1)}, \delta)$, where δ is the standard deviation of the normal proposal distribution, the tuning parameter that we have to set.
Step 2: Calculate the ratio of posterior densities for the proposed and the original value for θ:

$$r = \frac{[\theta^*|y]}{[\theta^{(t-1)}|y]}.$$

In our example,

$$r = \frac{\text{Bernoulli}(y|\theta^*) \times \text{Normal}(\theta^*|\mu, \sigma)}{\text{Bernoulli}(y|\theta^{(t-1)}) \times \text{Normal}(\theta^{(t-1)}|\mu, \sigma)}.$$

Step 3: Set

$$\begin{aligned}\theta^t &= \theta^* \text{ with probability } \min(r, 1)\\ &= \theta^{(t-1)} \text{ otherwise.}\end{aligned}$$

We can do this last step by drawing a random number u from a Uniform(0, 1) and accept θ^* if $u < r$. This is repeated for $t = 1, 2, \ldots, T$ a large number of samples. As in Gibbs sampling, the order in which we update parameters does not matter. The **R** code for this MH sampler is provided in Panel 17.2.

The reason why, in the **R** code, we sum the logs of the likelihood and the prior, rather than multiplying the original values, is simply computational. The product of small probabilities can be numbers very close to 0, which computers do not handle well. Thus, we exponentiate the sum of the logarithms to achieve the desired result. Similarly, in case you have forgotten, $x/y = \exp(\log(x) - \log(y))$, with the latter being favored for computational reasons.

```
Logreg.MH <- function(y=y, mu0=mu0, sig0=sig0, delta=delta, niter=niter) {

out <- c()

theta <- runif(1, -3,3) #initial value

for (iter in 1:niter){
   theta.cand <- rnorm(1, theta, delta)

   loglike <- sum(dbinom(y, 1, exp(theta)/(1+exp(theta)), log=TRUE))
   logprior <- dnorm(theta,mu0 ,sig0, log=TRUE)
   loglike.cand <- sum(dbinom(y, 1, exp(theta.cand)/(1+exp(theta.cand)),
   log=TRUE))
   logprior.cand <- dnorm(theta.cand, mu0, sig0, log=TRUE)

   if (runif(1)<exp((loglike.cand+logprior.cand)-(loglike+logprior))){
   theta <- theta.cand
}
out[iter] <- theta
}

return(out)
}
```

PANEL 17.2

R code to run a Metropolis sampler on a simple logit-normal model.

Unlike Gibbs sampling, where all draws from the conditional distribution are used, the MH algorithm discards a portion of the candidate values, which inherently makes it less efficient than Gibbs sampling—the price you pay for its increased generality. In Step 1 of the MH sampler we had to choose a standard deviation, δ, for the normal proposal distribution. Choice of the parameters that define our candidate distribution is also referred to as "tuning," and it is important since adequate tuning will make your algorithm more efficient. δ should be chosen (a) large enough so that each step of drawing a new proposal value for θ can cover a reasonable distance in the parameter space; otherwise, mixing of the Markov chain is inefficient and chains will tend to have strong autocorrelation; and (b) small enough so that proposal values are not rejected too often; otherwise the random walk will "get stuck" at specific values for too long. As a rule of thumb, the candidate value should be accepted in about 40% of all cases. Acceptance rates of 20–80% are probably ok, but anything below or above may well render your algorithm inefficient (this does not mean that it will give you wrong results, only that you will need more iterations to converge to the posterior

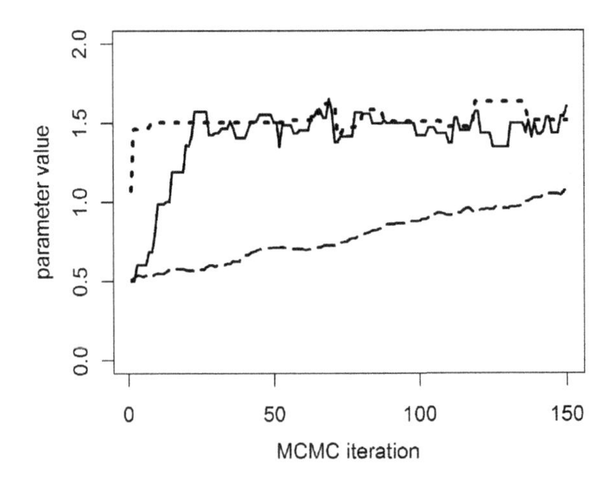

FIGURE 17.3

Time series plots of θ from a MH algorithm with tuning parameter $\delta = 0.01$ (dashed line), 0.2 (solid line), and 1 (dotted line).

distribution). In practice, tuning will require some "trial-and-error," some common sense and, with enough experience, some intuition. Or, one can use an adaptive phase, where the tuning parameter is automatically adjusted until it reaches a user-defined acceptance rate, at which point the adaptive phase ends and the actual Markov chain begins. This is computationally a little more advanced. Link and Barker (2010) discuss this in more detail. It is important that the samples drawn during the adaptive phase are discarded.

To illustrate the effects of tuning, we ran the Metropolis-Hastings algorithm in Panel 17.2 with $\delta = 0.01$, $\delta = 0.2$, and $\delta = 1$. The first 150 iterations for θ are shown in Figure 17.3. We see that for a very small δ (the dashed line) the burn-in is extremely slow—after 150 iterations the chain isn't even halfway there, while for the other two values of δ (solid and dotted lines) the burn-in phase seems to be over after only about 10 iterations. While $\delta = 0.2$ leads to reasonably good mixing, the chain clearly gets stuck on certain values with $\delta = 1$.

Other than graphically, you can easily check acceptance rates for the parameters you monitor (i.e., that are part of your output) using the `rejectionRate` function of the package `coda` (we will talk more about this package a little later on). Do not let the term "rejection rate" confuse you; it is simply 1—acceptance rate. There may be parameters—for example, individual values of a random effect or latent variables—that you do not want to save, though, and in our next example we will show you a way to monitor their acceptance rates with a few extra lines of code.

17.3.3 Metropolis-within-Gibbs

One weakness of the MH sampler is that formulating the joint posterior when evaluating whether to accept or reject the candidate values for θ becomes increasingly complex or inefficient as the number of parameters in a model increases. As you already saw in Chapter 3, in these cases you can combine MH sampling and Gibbs sampling. You can use the principles of Gibbs sampling to break down your high-dimensional parameter space into easy-to-handle one-dimensional conditional distributions and use MH sampling for these conditional distributions. Better yet, if you have some conjugacy in your model, you can use the more efficient Gibbs sampling for these parameters and one-dimensional MH for all the others. You have already seen the basics of how to build both types of algorithms, so we can jump straight into an example here and build a Metropolis-within-Gibbs algorithm.

GLMMs: Poisson regression with a random effect. Let's a model that gets us closer to the problem we ultimately want to deal with—a GLMM. Here, we assume we have Poisson counts, y_{ij}, from $j = 1, 2, \ldots, n$ plots in i different study sites, and we believe that the counts are influenced by some plot-specific covariate, $\mathbf{x}$, but that there is also a random site effect. So our model is:

$$\begin{aligned} y_{ij} &\sim \text{Poisson}(\lambda_{ij}), \\ \lambda_{ij} &= \exp(\alpha_i + \beta x_{ij}). \end{aligned}$$

Let's place normal priors on α and β,

$$\alpha_i \sim \text{Normal}(\mu_\alpha, \sigma_\alpha^2)$$

and

$$\beta \sim \text{Normal}(\mu_\beta, \sigma_\beta^2).$$

In this model, we do not specify μ_α and σ_α, but instead, estimate them as well, so we have to specify hyperpriors for these parameters:

$$\begin{aligned} \mu_\alpha &\sim \text{Normal}(\mu_0, \sigma_0^2), \\ \sigma_\alpha^2 &\sim \text{Inverse-Gamma}(a_0, b_0). \end{aligned}$$

Note that for simplicity we assume that β is constant across the i study sites, and for analysis we set μ_β and σ_β (i.e., we don't estimate these parameters from the data). With the model completely specified, we can compile the full conditionals, breaking the multi-dimensional parameter space into one-dimensional components:

$$\begin{aligned} [\alpha_1|\alpha_2, \alpha_3, \ldots, \alpha_i, \beta, \mathbf{y}_1] &\propto [\mathbf{y}_1|\alpha_1, \beta][\alpha_1] \\ &\propto \text{Poisson}(\mathbf{y}_1|\exp(\alpha_1 + \beta\mathbf{x}_1)) \times \text{Normal}(\alpha_1|\mu_\alpha, \sigma_\alpha), \end{aligned}$$

where $\mathbf{y}_1 = (y_{11}, y_{12}, \ldots, y_{1n})$ is the vector of observed counts for site $i = 1$ and, in general, $\mathbf{y}_i$ is the vector of all counts for site i; analogously, $\mathbf{x}_i$ is the vector of all observations of the covariate for site i. The other full conditionals for each α_i are constructed similarly:

$$[\alpha_2|\alpha_1, \alpha_3, \ldots, \alpha_i, \beta, \mathbf{y}_2] \propto [\mathbf{y}_2|\alpha_2, \beta][\alpha_2]$$
$$\propto \text{Poisson}(\mathbf{y}_2|\exp(\alpha_2 + \beta\mathbf{x}_2)) \times \text{Normal}(\alpha_2|\mu_\alpha, \sigma_\alpha),$$

and so on for all elements of $\boldsymbol{\alpha}$. The full conditional for β is:

$$[\beta|\boldsymbol{\alpha}, \mathbf{y}] \propto [\mathbf{y}|\boldsymbol{\alpha}, \beta][\beta]$$
$$\propto \text{Poisson}(\mathbf{y}|\exp(\boldsymbol{\alpha} + \beta\mathbf{x})) \times \text{Normal}(\beta|\mu_\beta, \sigma_\beta).$$

Finally, we need to update the hyperparameters for the random effects vector $\boldsymbol{\alpha}$:

$$[\mu_\alpha|\boldsymbol{\alpha}] \propto [\boldsymbol{\alpha}|\mu_\alpha, \sigma_\alpha][\mu_\alpha],$$
$$[\sigma_\alpha|\boldsymbol{\alpha}] \propto [\boldsymbol{\alpha}|\mu_\alpha, \sigma_\alpha][\sigma_\alpha].$$

Note that the likelihood contributions of the counts $\mathbf{y}$ at each site, when conditioned on $\boldsymbol{\alpha}$, do not depend on the hyperparameters μ_α and σ_α. As such, the full conditionals for these hyperparameters only depend on the collection of all $\boldsymbol{\alpha}$, not the data. Since we assumed $\boldsymbol{\alpha}$ to come from a normal distribution, the choice of priors for μ_α (normal) and σ_α^2 (inverse-gamma) leads to the same conjugacy we observed in our initial normal model, so that both hyperparameters can be updated using Gibbs sampling.

Now let's build the updating steps for these full conditionals. Again, for the MH steps that update $\boldsymbol{\alpha}$ and β we use normal proposal distributions with standard deviations δ_α and δ_β.

First, we set the initial values $\alpha^{(0)}$ and $\beta^{(0)}$. Then, starting with α_1, we draw $\alpha_1^{(1)}$ from Normal$(\alpha_1^{(0)}, \delta_\alpha)$, calculate the conditional posterior density of $\alpha_1^{(0)}$ and $\alpha_1^{(1)}$ and compare their ratios,

$$r = \frac{\text{Poisson}(\mathbf{y}_1|\exp(\alpha_1^{(1)} + \beta\mathbf{x}_1)) \times \text{Normal}(\alpha_1^{(1)}|\mu_\alpha, \sigma_\alpha)}{\text{Poisson}(\mathbf{y}_1|\exp(\alpha_1^{(0)} + \beta\mathbf{x}_1)) \times \text{Normal}(\alpha_1^{(0)}|\mu_\alpha, \sigma_\alpha)},$$

and accept $\alpha_1^{(1)}$ with probability min$(r, 1)$. We repeat this for all $\boldsymbol{\alpha}$.

For β, we draw $\beta^{(1)}$ from Normal$(\beta^{(0)}, \delta_\beta)$, compare the posterior densities of $\beta^{(0)}$ and $\beta^{(1)}$,

$$r = \frac{\text{Poisson}(\mathbf{y}|\exp(\boldsymbol{\alpha} + \beta^{(1)}\mathbf{x})) \times \text{Normal}(\beta^{(1)}|\mu_\beta, \sigma_\beta)}{\text{Poisson}(\mathbf{y}|\exp(\boldsymbol{\alpha} + \beta^{(0)}\mathbf{x})) \times \text{Normal}(\beta^{(0)}|\mu_\beta, \sigma_\beta)},$$

and accept $\beta^{(1)}$ with probability min$(r, 1)$.

For μ_α and σ_α^2, we sample directly from the full conditional distributions (Eq. (17.2.2) and Eq. (17.3.1)):

$$\mu_\alpha^{(1)} \sim \text{Normal}(\mu_n, \sigma_n^2),$$

where

$$\mu_n = \frac{\sigma_\alpha^{2(0)}}{\sigma_\alpha^{2(0)} + n_\alpha \sigma_0^2} \times \mu_0 + \frac{n_\alpha \sigma_0^2}{\sigma_\alpha^{2(0)} + n_\alpha \sigma_0^2} \times \bar{\alpha}^{(1)}$$

and

$$\sigma_n^2 = \frac{\sigma_\alpha^{2(0)} \sigma_0}{\sigma_\alpha^{2(0)} + n\sigma_0^2}.$$

Here, $\bar{\alpha}$ is the current mean of the vector $\boldsymbol{\alpha}$, which we updated before, and n_α is the length of $\boldsymbol{\alpha}$. For σ_α^2 we use

$$\sigma_\alpha^{2(1)} \sim \text{Inverse-Gamma}(a_n, b_n),$$

where

$$a_n = n_a/2 + a_0,$$

and

$$b_n = 0.5 \sum_{i=1}^{n_\alpha} (\alpha_i^{(1)} - \mu_\alpha^{(1)})^2 + b_0.$$

We repeat these steps over T iterations of the MCMC algorithm. Call the function `PoisGLMM` in `scrbook` to check out what this algorithm looks like in **R**.

In this example we may not want to save each individual α_i, but are only interested in their mean and standard deviation. Since these two parameters will change as soon as the value for one element in $\boldsymbol{\alpha}$ changes, their acceptance rates will always be close to 1 and are not representative of how well your algorithm performs. To monitor the acceptance rates of parameters you do not want to save, you simply need to add a few lines of code into your updater to see how often the individual parameters are accepted. The code for updating $\boldsymbol{\alpha}$ from our Poisson GLMM below shows one way to monitor acceptance of individual α_i.

```
#initiate counter for acceptance rate of alpha
alphaUps <- 0

#loop over sites, update intercepts alpha one at a time;
#only data at site i contributes information
#lev is the number of sites i
for (i in 1:lev) {
      alpha.cand <- rnorm(1, alpha[i], delta_alpha)
      loglike <- sum(dpois (y[site==i], exp(alpha[i] + beta*x[site==i]),
        log=TRUE))
      logprior <- dnorm(alpha[i], mu_alpha,sig_alpha, log=TRUE)
      loglike.cand <- sum(dpois (y[site==i], exp(alpha.cand + beta *x[site==i]),
        log=TRUE))
      logprior.cand <- dnorm(alpha.cand, mu_alpha,sig_alpha, log=TRUE)
      if (runif(1)< exp((loglike.cand+logprior.cand) -(loglike+logprior))) {
      alpha[i] <- alpha.cand
      alphaUps <- alphaUps+1
```

```
    }
}

#lets you check the acceptance rate of alpha at every 100th iteration
if(iter %% 100 == 0) {
     cat("    Acceptance rates\n")
     cat("      alpha =", alphaUps/lev, "\n")
}
```

17.3.4 Rejection sampling and slice sampling

While MH and Gibbs sampling are probably the most widely applied algorithms for posterior approximation, there are other options that work under certain circumstances and may be more efficient when applicable. **WinBUGS** applies these algorithms and we want you to be aware that there are more algorithms out there for approximating posterior distributions than just Gibbs and MH. One alternative algorithm is rejection sampling. Rejection sampling is not an MCMC method, since each draw is independent of the others. The method can be used when the posterior $[\theta|y]$ is not a known parametric distribution but can be expressed in closed form. Then, we can use a so-called envelope function, say, $g(\theta)$, that we can easily sample from, with the restriction that $[\theta|y] < Mg(\theta)$. We then sample a candidate value for θ from $g(\theta)$, calculate $r = [\theta|y]/Mg(\theta)$, and keep the sample with the probability r. M is a constant that has to be picked so that r lies between 0 and 1, for example by evaluating both $[\theta|y]$ and $g(\theta)$ at n points and looking at their ratios. Rejection sampling only works well if $g(\theta)$ is similar to $[\theta|y]$, and packages like **WinBUGS** use adaptive rejection sampling (Gilks and Wild, 1992), where a complex algorithm is used to fit an adequate and efficient $g(\theta)$ based on the first few draws. Though efficient in some situations, rejection sampling does not work well with high-dimensional problems, since it becomes increasingly hard to define a reasonable envelope function. For an example of rejection sampling in the context of SCR models, see Chapter 11, where we use it to simulate inhomogeneous point processes.

Another alternative is slice sampling (Neal, 2003). In slice sampling, we sample uniformly from the area under the plot of $[\theta|y]$. Consider a single univariate parameter. Let's define an auxiliary variable, $U \sim \text{Uniform}(0, [\theta|y])$. Then, θ can be sampled from the vertical slice of $[\theta|y]$ at U (Figure 17.4):

$$\theta|U \sim \text{Uniform}(B),$$

where $B = \{\theta : [\theta|y] \geq U\}$

Slice sampling can be applied in many situations; however, implementing an efficient slice sampling procedure can be complicated. We refer the interested reader to Robert and Casella (2010, Chapter 7) for a simple example. Both rejection sampling and slice sampling can be applied on one-dimensional conditional distributions within a Gibbs sampling setup.

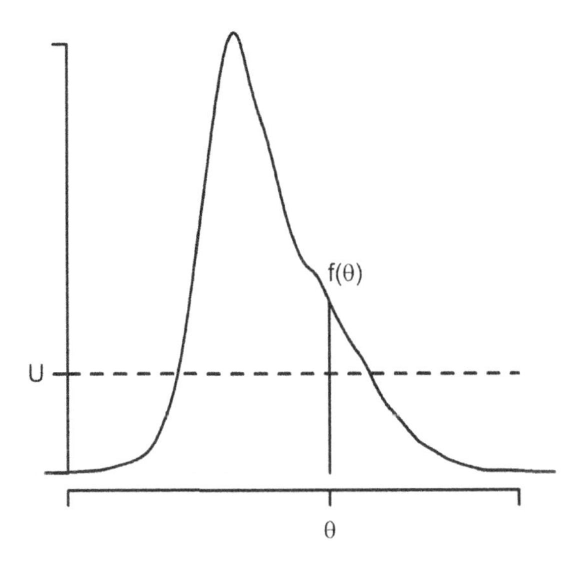

FIGURE 17.4

Slice sampling. For $U \sim$ Uniform(0,$[\theta|y]$), we can sample θ from the vertical slice of $[\theta|y]$ at U; $\theta|U \sim$ Uniform(B), where $B = \{\theta : [\theta|y] \geq U\}$.

17.4 MCMC for closed capture-recapture model M_h

By now you have seen MCMC samplers for some simple generalized (mixed) linear models. Now, to ease you into more complex models, we construct our own MCMC algorithm using a Metropolis-within-Gibbs sampler for the non-spatial model with individual heterogeneity in capture probability, model M_h, developed in Chapter 4.

To recapitulate: Under the non-spatial model, each of the n observed individuals is either detected ($y = 1$) or not ($y = 0$) during each of K sampling occasions. We estimate N using data augmentation and have a Bernoulli model for the data augmentation variables z_i

$$z_i \sim \text{Bernoulli}(\psi).$$

The binomial observation model is expressed conditional on the latent variables z_i

$$y_i \sim \text{Binomial}(K, z_i p_i).$$

Further, we prescribe a distribution for the capture probability p_i. Here we assume

$$\text{logit}(p_i) \sim \text{Normal}(\mu_p, \sigma_p^2).$$

As usual, we have to go through two general steps before we write the MCMC algorithm:

1. Identify the model with all its components (including priors).

2. Recognize and express the full conditional distributions for all parameters.

Our model components are as follows: $[y_i|p_i, z_i]$, $[p_i|\mu_p, \sigma_p]$, and $[z_i|\psi]$ for *each* $i = 1, 2, \ldots, M$, and then prior distributions $[\mu_p]$, $[\sigma_p]$, and $[\psi]$. The joint posterior distribution of all unknown quantities in the model is proportional to the joint distribution of all elements y_i, p_i, z_i, and also the prior distributions of the prior parameters:

$$\left\{\prod_{i=1}^{M}[y_i|p_i, z_i][p_i|\mu_p, \sigma_p][z_i|\psi]\right\}[\mu_p, \sigma_p, \psi].$$

For prior distributions, we assume that μ_p, σ_p, and ψ are mutually independent, and for μ_p and σ_p we use improper uniform priors; finally, $\psi \sim \text{Uniform}(0, 1)$. This is equivalent to Beta(1, 1), which will come in handy, as we will see in a moment. Note that the likelihood contribution for each individual, when conditioned on p_i and z_i, does not depend on ψ, μ_p, or σ_p. As such, the full conditional for the structural parameter ψ only depends on the collection of data augmentation variables z_i, and that for μ_p and σ_p will only depend on the collection of latent variables $p_i; i = 1, 2, \ldots, M$ (this is equivalent to what we saw in the Poisson regression with random intercept $\boldsymbol{\alpha}$, where hyperparameters for the distribution of $\boldsymbol{\alpha}$ did not depend on the observed data). The full conditionals for all the unknowns are as follows:

1. For p_i:

$$\begin{aligned}[p_i|y_i, \mu_p, \sigma_p, z_i] \propto [y_i|p_i][p_i|\mu_p, \sigma_p] &\text{ if } z_i = 1\\ [p_i|\mu_p, \sigma_p] &\text{ if } z_i = 0.\end{aligned}$$

2. for z_i:

$$[z_i|y_i, p_i, \psi] \propto [y_i|z_i\, p_i]\text{Bernoulli}(z_i|\psi).$$

3. For μ_p:

$$[\mu_p|p_i, \sigma_p] \sim \left\{\prod_i [p_i|\mu_p, \sigma_p]\right\} \times \text{const.}$$

4. For σ_p:

$$[\sigma_p|p_i, \mu_p] \sim \left\{\prod_i [p_i|\mu_p, \sigma_p]\right\} \times \text{const.}$$

5. For ψ:

$$[\psi|z_i] \propto \left\{\prod_i [z_i|\psi]\right\} \text{Beta}(1, 1).$$

Remember that Beta(1,1) is equivalent to Uniform(0,1). The beta distribution is the conjugate prior to the binomial and Bernoulli distributions, and the general form of a full conditional of a beta-binomial model with sample size n, $x_i \sim \text{Bernoulli}(\theta)$ and

$\theta \sim \text{Beta}(a, b)$ is

$$[\theta|\mathbf{x}] \propto \text{Beta}(a + \sum_i x_i, b + n - \sum_i x_i).$$

In our case that means

$$[\psi|z_i] \propto \text{Beta}(1 + \sum z_i, 1 + M - \sum z_i).$$

What we've done here is identify each of the full conditional distributions in sufficient detail to toss them into our Metropolis-Hastings algorithm. The constant terms in the full conditionals for μ_p and σ_p reflect the improper prior we chose for both parameters. Because of the choice of an improper prior, prior probability densities for both parameters $\propto 1$, i.e., constant, and these constants cancel out of the MH acceptance ratio (see updating step below and following example). Below, you see the updating step for the detection parameter $\mathbf{p}$. Note that (1) we draw candidate values on the logit scale and (2) instead of looping through $1 - M$ individuals to update all p_i, we update all elements of the vector of $\mathbf{p}$ in parallel, for computational efficiency.

```
### update the logit(p) parameters
lp.cand <- rnorm(M,lp,1) # 1 is a tuning parameter
p.cand <- plogis(lp.cand)
ll <- dbinom(ytot,K,z*p, log=T)
prior <- dnorm(lp,mu,sigma, log=T)
llcand <- dbinom(ytot,K,z*p.cand, log=T)
prior.cand <- dnorm(lp.cand,mu,sigma, log=T)

kp <- runif(M) < exp((llcand+prior.cand)-(ll+prior))
p[kp] <- p.cand[kp]
lp[kp] <- lp.cand[kp]
```

The parameters μ_p and σ_p are also updated using MH steps (see the code for μ_p below). In truth, we could also sample μ_p and σ_p^2 directly with certain choices of prior distributions. For example, if $\mu_p \sim \text{Normal}(0, 1000)$ then the full conditional for μ_p is also normal (see Section 17.3.1), etc.:

```
p0.cand<- rnorm(1,p0,.05)
if(p0.cand>0 & p0.cand<1){
    mu.cand<-log(p0.cand/(1-p0.cand))
    ll<-sum(dnorm(lp,mu,sigma,log=TRUE))
    llcand<-sum(dnorm(lp,mu.cand,sigma,log=TRUE))
    if(runif(1)<exp(llcand-ll)) {
     mu<-mu.cand
     p0<-p0.cand
    }
}
```

For ψ we can easily sample directly from the beta distribution:

```
psi <- rbeta(1, sum(z) + 1, M-sum(z) + 1)
```

To update the z_i we have opted for a MH updater (although they could be updated directly from their full conditional). Since z_i can only take the values of 0 or 1, we generate candidate values using `z.cand<-ifelse(z==1,0,1)`. The updating step for z_i is detailed in the next example. You can check out the full code by invoking `modelMh` from the **R** package `scrbook`.

17.5 MCMC algorithm for model SCR0

Conceptually, but also in terms of MCMC coding, it is only a small step from the non-spatial model M_h to a fully spatial capture-recapture model. Next, we will walk you through the steps of building your own MCMC sampler for the basic SCR model (i.e., without any individual-, site-, or time-specific covariates) with both a Poisson and a binomial encounter process. As usual, we will have to go through two general steps before we write the MCMC algorithm:

1. Identify the model with all its components (including priors) .
2. Recognize and express the full conditional distributions for all parameters.

It is worthwhile to go through all of Step 1 for an SCR model, but you have probably seen enough of Step 2 in our previous examples to get the essence of how to express a full conditional distribution. Therefore, we will exemplify Step 2 for some parameters and tie these examples directly to the respective **R** code snippets.

Step 1—Identify your model: Recall the components of the basic SCR model with a Poisson encounter process from Chapter 9: We assume that individuals i, or rather, their activity centers $\mathbf{s}_i$, are uniformly distributed across the state-space $\mathcal{S}$,

$$\mathbf{s}_i \sim \text{Uniform}(\mathcal{S}),$$

and that the number of times individual i encounters trap j, y_{ij}, is a Poisson variable with mean λ_{ij},

$$y_{ij} \sim \text{Poisson}(\lambda_{ij}).$$

The link between individual location, movement, and trap encounter rates is made by the assumption that λ_{ij} is a decreasing function of the distance between $\mathbf{s}_i$ and the location of j, $\mathbf{x}_j$, say

$$d_{ij} = ||\mathbf{x}_j - \mathbf{s}_i||,$$

of the Gaussian (or half-normal) form

$$\lambda_{ij} = \lambda_0 \exp(-d_{ij}^2/2\sigma^2),$$

where λ_0 is the baseline trap encounter rate at $d_{ij} = 0$ and σ is the scale parameter of the half-normal function.

As in the non-spatial example for model M_h, we estimate N, here the number of $\mathbf{s}_i$ in $\mathcal{S}$, using data augmentation (Section 4.2). We create $M - n$ all-zero encounter histories and estimate N by summing over the auxiliary data augmentation variables, z_i, which we assume are Bernoulli random variables,

$$z_i \sim \text{Bernoulli}(\psi).$$

To link the two model components, we modify our trap encounter model to

$$\lambda_{ij} = \lambda_0 \exp(-d_{ij}^2/2\sigma^2) z_i.$$

The model has the following structural parameters, for which we need to specify priors:

- ψ the Uniform(0, 1) is required as part of the data augmentation procedure and in general is a natural choice of an uninformative prior for a probability. It will also lead to conjugacy as we saw in the example of model M_h, so that we can update ψ directly from its full conditional distribution using Gibbs sampling;
- $\mathbf{s}_i$ since $\mathbf{s}_i$ is a pair of coordinates it is two-dimensional and we use a uniform prior limited by the extent of our state-space over both dimensions;
- σ we can conceive several priors for σ but let's assume an improper prior, one that is Uniform over $(0, \infty)$. As we already saw, this choice is convenient when updating the parameter, because the constant prior probability cancels out of the MH acceptance ratio;
- λ_0 analogously, we will use a Uniform$(0, \infty)$ improper prior for λ_0.

Step 2—Construct the full conditionals: Having completed Step 1, let's look at the full conditional distributions for some of these parameters. We saw that with improper priors, full conditionals are proportional only to the likelihood of the observations; for example, consider σ:

$$[\sigma|\mathbf{s}, \lambda_0, \mathbf{z}, \mathbf{y}] \propto \left\{ \prod_i [y_i|\mathbf{s}_i, \lambda_0, z_i, \sigma] \right\}.$$

The **R** code to update σ is shown below. Notice that we automatically reject negative candidate values, since σ cannot be <0:

```
sig.cand <- rnorm(1, sigma, 0.1) #draw candidate value
 if(sig.cand>0){ #automatically reject sig.cand that are <0
```

```
    lam.cand <- lam0*exp(-(d*d)/(2*sig.cand*sig.cand))
    ll <- sum(dpois(y, lam*z, log=TRUE))
    llcand <- sum(dpois(y, lam.cand*z, log=TRUE))
    if(runif(1) < exp(llcand - ll)){
     ll <- llcand
     lam <- lam.cand
     sigma <- sig.cand
    }
}
```

These steps are analogous for λ_0 and $\mathbf{s}_i$, and we will use MH steps for all of these parameters. Similar to the random intercepts in our Poisson GLMM, we update each $\mathbf{s}_i$ individually. Note that to be fully correct, the full conditional for $\mathbf{s}_i$ contains both the likelihood and prior component, since we did not specify an improper, but a proper uniform prior on $\mathbf{s}_i$. However, with a uniform distribution the probability density of any value is 1/(upper limit – lower limit) = constant. Thus, the prior components are identical for both the current and the candidate value so that when you calculate the ratio of posterior densities, r, the identical prior components appear both in the numerator and denominator and cancel each other out.

We still have to update z_i. The full conditional for z_i is

$$[z_i|y_i, \sigma, \lambda_0, \mathbf{s}_i] \propto [y_i|z_i, \sigma, \lambda_0, \mathbf{s}_i][z_i],$$

and since $z_i \sim \text{Bernoulli}(\psi)$, the term has to be taken into account when updating z_i:

```
zUps <- 0 #set counter to monitor acceptance rate
for(i in 1:M) {
#no need to update seen individuals, since their z =1
 if(seen[i])
   next
 zcand <- ifelse(z[i]==0, 1, 0)
 llz <- sum(dpois(y[i,],lam[i,]*z[i], log=TRUE))
 llcand <- sum(dpois(y[i,], lam[i,]*zcand, log=TRUE))

 prior <- dbinom(z[i], 1, psi, log=TRUE)
 prior.cand <- dbinom(zcand, 1, psi, log=TRUE)
 if(runif(1) < exp((llcand+prior.cand)-(llz+prior))){
  z[i] <- zcand
  zUps <- zUps+1
 }
}
```

The parameter ψ is a hyperparameter of the model, with an uninformative prior distribution of Uniform(0, 1) or Beta(1, 1), so that

$$[\psi|\mathbf{z}] \propto \text{Beta}(1 + \sum_i z_i, 1 + M - \sum_i z_i).$$

These are all the building blocks you need to write the MCMC algorithm for the spatial null model with a Poisson encounter process. You can find the full **R** code by calling the function `SCR0pois` in the **R** package `scrbook`:

17.5.1 SCR model with binomial encounter process

The equivalent SCR model with a binomial encounter process is very similar. Here, each individual i can only be detected once at any given trap j during a sampling occasion k. Thus,

$$y_{ij} \sim \text{Binomial}(K, p_{ij}),$$

where p_{ij} is some function of distance between $\mathbf{s}_i$ and trap location $\mathbf{x}_j$. Here we use:

$$p_{ij} = 1 - \exp(-\lambda_{ij}).$$

Recall from Chapter 3 that this is the complementary log-log (cloglog) link function, which constrains p_{ij} to fall between 0 and 1. For our MCMC algorithm that means that, instead of using a Poisson likelihood, $\text{Poisson}(y|\sigma, \lambda_0, \mathbf{s}, z)$, we use a binomial likelihood, $\text{Binomial}(y|\sigma, \lambda_0, \mathbf{s}, z; K)$, in all the conditional distributions. An exemplary updating step for λ_0 under a binomial encounter model is shown below. The full MCMC code for the binomial SCR with a cloglog link (`SCR0binom.cl`) can be found in the **R** package `scrbook`:

```
  lam0.cand <- rnorm(1, lam0, 0.1)
  #automatically reject lam0.cand that are <0
  if(lam0.cand >0){
  lam.cand <- p0.cand*exp(-(d*d)/(2*sigma*sigma))
  p.cand <- 1-exp(-lam.cand)
  ll<- sum(dbinom(y, K, pmat *z, log=TRUE))
  llcand <- sum(dbinom(y, K, p.cand *z, log=TRUE))
  if(runif(1) < exp(llcand - ll)){
      ll<-llcand
      pmat <- p.cand
      lam0 <- lam0.cand
  }
}
```

Another possibility is to model variation in the individual- and site-specific detection probability, p_{ij}, directly, without any transformation, such that

$$p_{ij} = p_0 \exp(-d_{ij}^2/(2\sigma^2))$$

and $p_0 \in [0, 1]$. This formulation is analogous to how detection probability is modeled in distance sampling under a half-normal detection function; however, in distance sampling p_0—detection of an individual on the transect line—is assumed to be 1 (Buckland et al., 2001). Under this formulation the updater for p_0 becomes:

```
p0.cand <- rnorm(1, p0, 0.1)
if(p0.cand > 0 & p0.cand < 1){
    #automatically rejects p0.cand that are not {0,1}
     p.cand <- p0.cand*exp(-(d*d)/(2*sigma*sigma))
     ll <- sum(dbinom(y, K, pmat *z, log=TRUE))
     llcand <- sum(dbinom(y, K, p.cand *z, log=TRUE))
     if(runif(1) < exp(llcand - ll)){
        ll <- llcand
          pmat <- p.cand
          p0 <- p0.cand
        }
   }
```

17.6 Looking at model output

Now that you have an MCMC algorithm to analyze spatial capture-recapture data with, let's run an actual analysis so we can look at the output. As an example, we will use the Fort Drum bear data set we first introduced in Chapter 1 and already analyzed in several preceding chapters. You can load the Fort Drum data (`data(beardata)`), extract the trap locations (`trapmat`) and detection data (`bearArray`), and build the augmented $M \times J$ array of individual encounter histories:

```
> M = 700
> trapmat <- beardata$trapmat
#summarizes captures across occasions
> bearmat <- apply(beardata$bearArray, 1:2, sum)
> Xaug <- matrix(0, nrow=M, ncol=dim(trapmat)[1])
> Xaug[1:dim(bearmat)[1],] <- bearmat   #create augmented data set
```

In addition to these data, we need to specify the outermost coordinates of the state-space. Since bears are wide-ranging animals we add a 20-km buffer to the maximum and minimum coordinates of the trap array:

```
> xl <- min(trapmat[,1])- 20
> yl <- min(trapmat[,2])- 20
> xu <- max(trapmat[,1])+ 20
> yu <- max(trapmat[,2])+ 20
```

Finally, use the MCMC code for the binomial encounter model with the cloglog link (`SCR0binom.cl`) and run 5000 iterations. This should take approximately 25 min (in real life we would, of course, run the algorithm a lot longer but for demonstration purposes let's stick to a number of iterations that can be run in a manageable amount of time):

```
> set.seed(13)
> mod0 <- SCR0binom.cl(y=Xaug, X=trapmat, M=M, xl=xl, xu=xu, yl=yl,
+                      yu=yu, K=8, delta=c(0.1, 0.05, 2), niter=5000)
```

Before, we used simple **R** commands to look at model results. However, there is a specific **R** package to summarize MCMC simulation output and perform some convergence diagnostics—package `coda` (Plummer et al., 2006). Download and install `coda`, then convert your model output to an mcmc object:

```
> chain <- mcmc(mod0)
```

which can be used by `coda` to produce MCMC-specific output.

17.6.1 Markov chain time series plots

Start by looking at time series plots of your Markov chains using `plot(chain)`. This command produces a time series plot and marginal posterior density plots for each monitored parameter, similar to what we did before using the `hist` and `plot` commands. Figure 17.5 shows an example of these plots for σ and λ_0. Time series plots will tell you several things: First, recall from Section 17.3.2 that the way the chains move through the parameter space gives you an idea of whether your MH steps are well tuned. If chains were constant over many iterations you would need to decrease the tuning parameter of the (normal) proposal distribution. If a chain moves along some gradient to a stationary state very slowly, you may want to increase the tuning parameter so that the parameter space is explored more efficiently.

Second, you will be able to see if your chains converged and how many initial simulations you have to discard as burn-in. In the case of the chains shown in Figure 17.5, we would probably consider the first 750–1,000 iterations as burn-in, as afterwards the chains seem to be fairly stationary.

17.6.2 Posterior density plots

The `plot` command also produces posterior density plots and it is worthwhile to look at those carefully. For parameters with priors that have bounds (e.g., uniform over some interval), you will be able to see if your choice of the prior is truncating the posterior distribution. In the context of SCR models, this will mostly involve our choice of M, the size of the augmented data set. If the posterior of N has a lot of mass concentrated close to M (or equivalently, the posterior of ψ has a lot of mass concentrated close to 1), as in the example in Figure 17.6, we have to re-run the analysis with a larger M. A diffuse posterior plot suggests that the parameter may not be well identified. There may not be enough information in your data to estimate model parameters and you may have to consider a simpler model, or your model may contain parameters that are confounded and therefore not identifiable. Finally, posterior density plots will show you if the posterior distribution is symmetrical or skewed—if the distribution has a heavy tail, using the mean as a point estimate of your parameter of interest may be biased and you may want to opt for the median or mode instead.

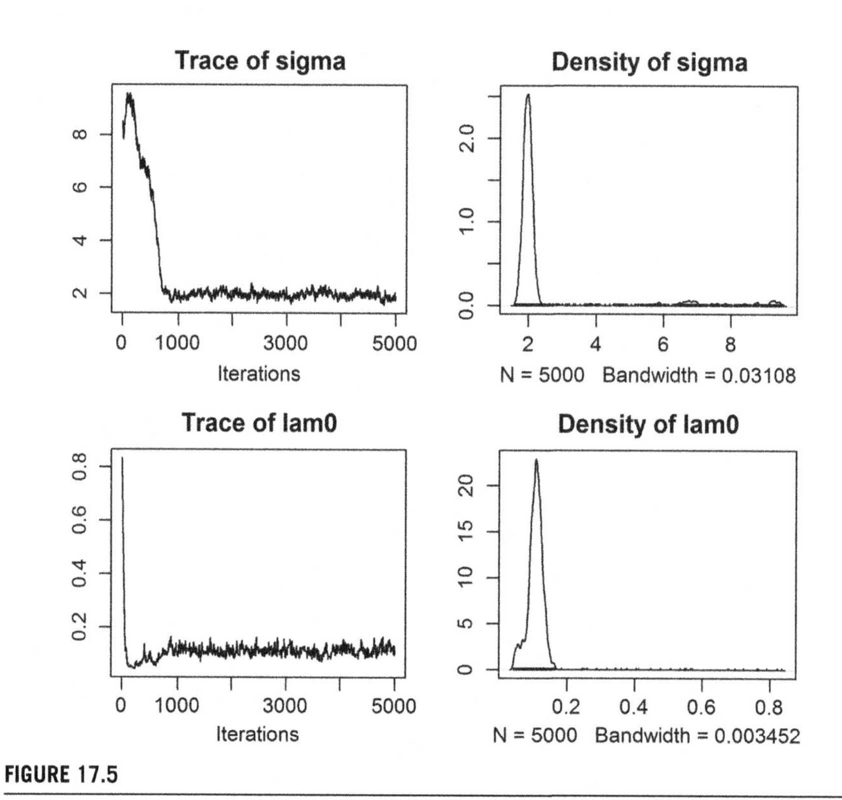

FIGURE 17.5

Time series and posterior density plots for σ and λ_0 for the Fort Drum black bear data.

17.6.3 Serial autocorrelation and effective sample size

Checking the degree of autocorrelation in the Markov chains and estimating the effective sample size the chain has generated should be part of evaluating your model output. If you use **WinBUGS** through the `R2WinBUGS` package, the `print` command will automatically return the effective sample size for all monitored parameters. In the `coda` package there are several functions you can use to do so. The function `effectiveSize` will directly give you an estimate of the effective sample size for the parameters:

```
> effectiveSize(window(chain, start=1001))
    sigma        lam0        psi          N
 93.89807   163.72311   51.96443   46.45394
```

Alternatively, you can use the `autocorr.diag` function, which will show you the degree of autocorrelation for different lag values (which you can specify within the function call, we use the defaults below):

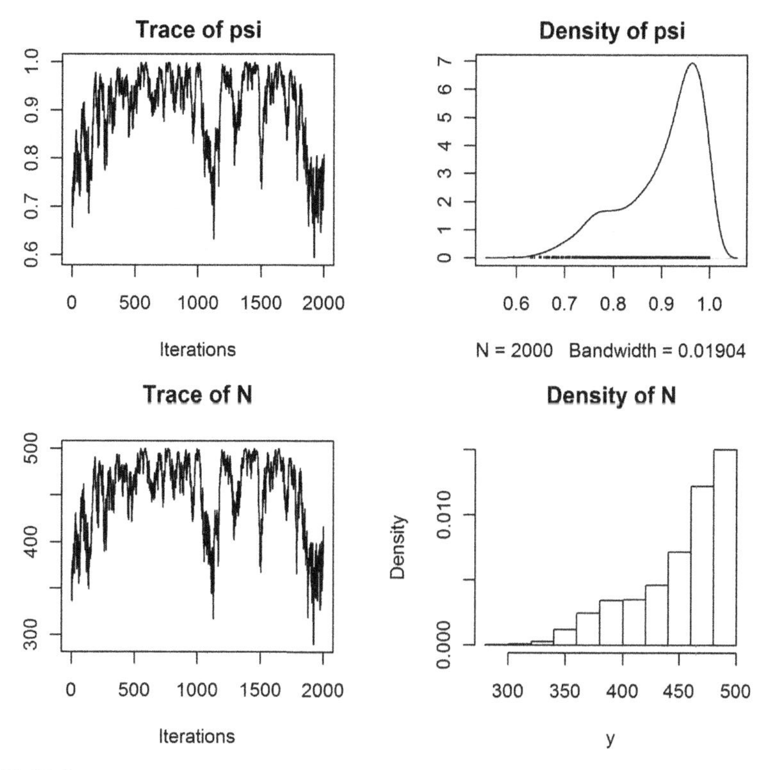

FIGURE 17.6

Time series and posterior density plots of ψ and N for the Fort Drum black bear data truncated by the upper limit of M (500).

```
> autocorr.diag(window(chain, start=1001))
            sigma        lam0         psi          N
 Lag 0    1.0000000  1.00000000  1.0000000  1.0000000
 Lag 1    0.9316928  0.91464875  0.9745833  0.9663320
 Lag 5    0.7603332  0.67445407  0.8525272  0.8500215
 Lag 10   0.6065374  0.48724122  0.7514657  0.7530124
 Lag 50   0.1122331  0.06564406  0.3811939  0.3823236
```

In the present case we see that autocorrelation is especially high for the parameter ψ and effective sample size for this parameter is only 52! This means we would have to run the model for much longer to obtain a reasonable effective sample size. For now, let's continue using this small number of samples to look at the output.

17.6.4 Summary results

Now that we checked that our chains apparently have converged and pretending that we have generated enough samples from the posterior distribution, we can look at the actual parameter estimates. The `summary` function will return two sets of results: the mean parameter estimates, with their standard deviation, the naive standard error—i.e., your regular standard error calculated for T (= number of iterations) samples without accounting for serial autocorrelation—and the time-series SE (in **WinBUGS** and earlier in this book referred to as MC error), which accounts for autocorrelation. Remember that this error decreases with increasing chain length and should be 1% or less of the parameter estimate. In **WinBUGS** the MC error can be extracted from a model object created with the `bugs` call, say `mod`, by using `mod$summary`. You should adjust the `summary` call by removing the burn-in from calculating parameter summary statistics. To do so, use the `window` command, which lets you specify at which iteration to start "counting." In contrast to **WinBUGS**, which requires you to set the burn-in length before you run the model, this command gives us full flexibility to make decisions about the burn-in after we have seen the trajectories of our Markov chains. For our example, `summary(window(chain, start=1001))` returns the following output:

```
Iterations = 1001:5000
Thinning interval = 1
Number of chains = 1
Sample size per chain = 4000

1. Empirical mean and standard deviation for each variable,
   plus standard error of the mean:
              Mean        SD    Naive SE   Time-series SE
   sigma    1.9697   0.12534   0.0019818         0.012792
   lam0     0.1124   0.01521   0.0002405         0.001311
   psi      0.7295   0.11794   0.0018648         0.015278
   N      510.9190  81.99868   1.2965130        10.580567
2. Quantiles for each variable:
              2.5%       25%        50%        75%      97.5%
   sigma    1.7288    1.8831     1.9666     2.0517     2.2240
   lam0     0.0863    0.1008     0.1112     0.1217     0.1449
   psi      0.5100    0.6423     0.7261     0.8170     0.9549
   N      359.0000  451.0000   508.0000   572.0000   668.0000
```

Looking at the MC errors (the column labeled `Time-series SE`), we see that in spite of the high autocorrelation, the MC error for σ is below the 1% threshold, whereas for all other parameters, MC errors are still above, another indication that for a thorough analysis we should run a longer chain.

Our algorithm gives us a posterior distribution of N, but we are usually interested in the density, D. Density itself is not a parameter of our model, but we can derive a posterior distribution for D by dividing each value of N (N at each iteration) by the area of the state-space (here 3032.719 km^2) and we can use summary statistics of the resulting distribution to characterize D:

```
> summary(window(chain[,4]/ 3032.719, start=1001))
```

```
Iterations = 1001:5000
Thinning interval = 1
Number of chains = 1
Sample size per chain = 4000

1. Empirical mean and standard deviation for each variable,
   plus standard error of the mean:

          Mean             SD       Naive SE  Time-series SE
     0.1684690      0.0270380      0.0004275       0.0034888

2. Quantiles for each variable:

      2.5%     25%      50%      75%     97.5%
    0.1184  0.1487   0.1675   0.1886    0.2203
```

We see that the mean density of $0.17/\text{km}^2$ is very similar to the estimate of $0.18/\text{km}^2$ obtained under the non-spatial model M_0 in Chapter 4.

17.6.5 Other useful commands

While inspecting the time series plot gives you a first idea of how well you tuned your MH algorithm, use `rejectionRate` to obtain the rejection rates (1—acceptance rates) of the parameters that are written to your output:

```
> rejectionRate(chain)
      sigma        lam0         psi           N
 0.42988598  0.78775755  0.00000000  0.03160632
```

Rejection rates should lie between 0.2 and 0.8 (Section 17.3.2), so our tuning seems to have been appropriate here. Draws of the parameter ψ are never rejected since we update it with Gibbs sampling, where all candidate values are kept. And since N is the sum of all z_i, all it takes for N to change from one iteration to the next are small changes in the z-vector, so the rejection rate of N is always low. If you have run several parallel chains, you can combine them into a single mcmc object using the `mcmc.list` command on the individual chains (note that each chain has to be converted to an mcmc object before combining them with `mcmc.list`). You can then easily obtain the Gelman-Rubin diagnostic (Gelman et al., 2004), in **WinBUGS** called Rhat, using `gelman.diag`, which will indicate if all chains have converged to the same stationary distribution (see Section 17.8.1 for an example). For details on these and other functions, see the `coda` manual, which can be found (together with the package) on CRAN.

17.7 Manipulating the state-space

So far, we have constrained the location of the activity centers to fall within the outermost coordinates of our rectangular state-space by posing upper and lower bounds for X and Y. But what if $\mathcal{S}$ has an irregular shape—maybe there is a large water body we would like to remove from $\mathcal{S}$, because we know our terrestrial study species does not occur there. Or the study takes place in a clearly defined area such as an island.

As mentioned before, this situation is difficult to handle in **BUGS** engines. In some simple cases we can adjust the state-space by setting one of the coordinates of $\mathbf{s}_i$ to be some function of the other and reject candidate $\mathbf{s}_i$ that do not fall within this modified state-space. In this manner, we can cut off corners of the rectangle to approximate the actual state-space.[3] To illustrate this approach, plot the following rectangle, representing your state-space polygon, and line, representing, for example, the approximation of a shore line:

```
> xlim <- c(-5,5)
> ylim <- c(-7,7)
> plot(xlim, ylim, type='n')
> abline(a=4, b=0.4)
```

The Y coordinates limiting your state-space to the habitat that is suitable to the species you study can now be expressed as a linear function of the X coordinates, in this case, $Y = 4 + 0.4 \times X$. To include this new limit in a **BUGS** model, we need to change the following:

```
#draw SX and SY as before
SX[i] ~ dunif(xlim[1],xlim[2])
SY[i] ~ dunif(ylim[1],ylim[2])
#calculate upper limit for Y given X
ymax[i] <- 4+0.4*SX[i]
# use step function to see if location [SX, SY]
# is below the Y limit (Pin = 1) or not (Pin = 0)
Pin[i] <- step(ymax[i] - SY[i])
In[i] ~ dbern(Pin[i])
```

The object `In` is a vector of M 1s, passed as data to the model. If Pin = 0, the likelihood will be 0 and the candidate [SX, SY] pair will be rejected. If Pin = 1, this bit of the likelihood is equal to 1, and whether or not the candidate pair of coordinates is accepted depends only on capture history of i. This approach can be very useful in some situations but is clearly restricted by the functional form of the relationship between SX and SY that it requires.

In **R**, we are much more flexible, as we can use the actual state-space polygon to constrain $\mathbf{s}_i$. To illustrate that, let's look at a camera trapping study of raccoons (*Procyon lotor*) conducted on South Core Banks, a barrier island within Cape Lookout National Seashore, North Carolina (details of the study can be found in Sollmann et al. (2013a) and in Chapter 19 where we present the analysis of this data set with spatial

[3]This idea was pitched to us by Mike Meredith, Biodiversity Conservation Society Sarawak and WCS Malaysia.

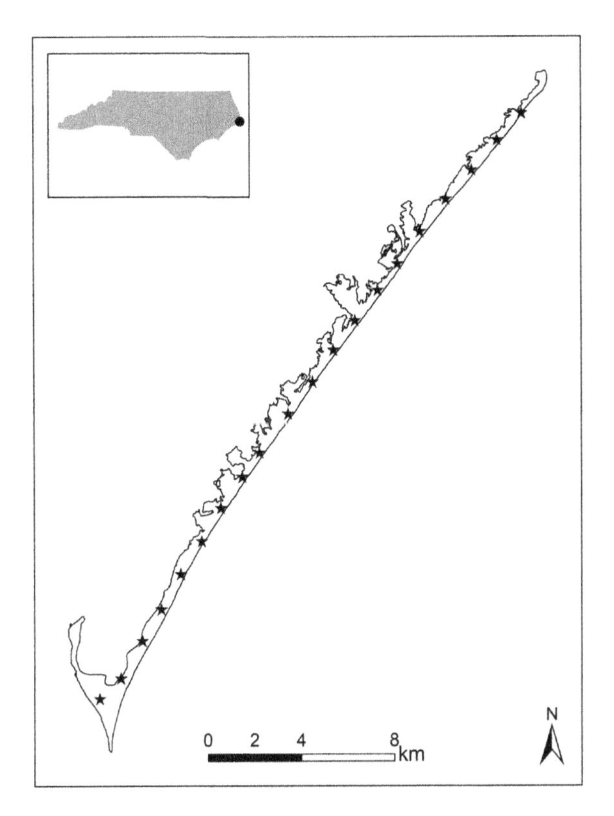

FIGURE 17.7

Camera traps (stars) set up on South Core Banks, a barrier island within Cape Lookout National Seashore, North Carolina (inset map) to estimate the raccoon population (see Chapter 19 for details).

mark-resight models). Since camera traps were spread across the entire length of the island, we set the state-space to be delineated by the shore line of the island (Figure 17.7), which clearly cannot easily be approximated as a rectangle. Instead, within **R** we can use an actual shapefile of the island.

In other circumstances you may still want to create the state-space as before, by adding some buffer to your trapping grid, but you may find that the resulting rectangle includes water bodies, paved parking lots, or any other kind of habitat you know is never used by the species you study. In order to precisely describe the state-space, these features need to be removed. You can create a precise state-space polygon in **ArcGIS** and read it into **R**, or create the polygon directly within **R**, by intersecting two shapefiles—one of the rectangle defining the outer limits of your state-space and one of the landscape features you want to remove. While you will most likely have to obtain the shapefile describing the landscape of and around your trapping grid (coastlines, water bodies, etc.) from some external source, the polygon shapefile buffering your outermost trapping grid coordinates can easily be written in **R**.

If xmin, xmax, ymin, and ymax mark the most extreme X and Y coordinates of your trapping grid and b is the distance you want to buffer with, load the package shapefiles (Stabler, 2006) and issue the following **R** commands:

```
> xl = xmin-b
> xu = xmax+b
> yl = ymin-b
> yu = ymax+b
              #create data frame with coordinate pairs
> dd <- data.frame(Id=c(1,1,1,1,1),X=c(xl,xu,xu,xl,xl),
+  Y=c(yl,yl,yu,yu,yl))
> ddTable <- data.frame(Id=c(1),Name=c("Item1"))
              #convert to shapefile, type polygon
> ddShapefile <- convert.to.shapefile(dd, ddTable, "Id", 5)
              # name and save to location of choice
> write.shapefile(ddShapefile, 'c:/Test', arcgis=T)
```

You can read shapefiles into by **R** loading the package maptools (Lewin-Koh et al., 2011) and using the function readShapeSpatial. Make sure you read in shapefiles in UTM format, so that units of the trap array, the movement parameter σ, and the state-space are all identical. Intersection of polygons can be done in **R** also, using the package rgeos (Bivand and Rundel, 2011) and the function gIntersect. The area of your (single) polygon can be extracted directly from the state-space object SSp:

```
> area <- SSp@polygons[[1]]@Polygons[[1]]@area /1000000
```

Note that dividing by 1,000,000 will return the area in km^2 if your coordinates describing the polygon are in UTM. If your state-space consists of several disjoint polygons, you will have to sum the areas of all polygons to obtain the size of the state-space. To include this polygon into our MCMC sampler we need one last spatial **R** package, sp (Pebesma and Bivand, 2011), which has a function, over, which allows us to check if a pair of coordinates falls within a polygon or not.[4] All we have to do is embed this new check into the updating steps for the $\mathbf{s}_i$:

```
    #draw candidate value
Scand <- as.matrix(cbind(rnorm(M, S[,1], 2), rnorm(M, S[,2], 2)))
     #convert to spatial points on UTM (m) scale
Scoord <- SpatialPoints(Scand*1000)
```

[4]Remember from Section 6.4.2 that the over function takes as its second argument (among others) an object of the class "SpatialPolygons" or "SpatialPolygonsDataFrame." The former produces a vector while the latter produces a data frame (e.g., in the example above), which is important for how you index the output.

```
      # check if scand is within the polygon
SinPoly <- over(Scoord,SSp)

for(i in 1:M) {
     #if scand falls within polygon, continue update
   if(is.na(SinPoly[i]) == FALSE) {
... [rest of the updating step remains the same]
```

Note that it is much more time efficient to draw all M candidate values for **s** and check once if they fall within the state-space, rather than running the `over` command for every individual pair of coordinates. To make sure that our initial values for **s** also fall within the polygon of $\mathcal{S}$, we use the function `runifpoint` from the package `spatstat` (Baddeley and Turner, 2005), which generates random uniform points within a specified polygon. You'll find this modified MCMC algorithm (`SCR0poisSSp`) in the **R** package `scrbook`.

Finally, observe that we are converting candidate coordinates of $\mathcal{S}$ back to meters to match the UTM polygon. In all previous examples, for both the trap locations and the activity centers we have used UTM coordinates divided by 1,000 to estimate σ on a km scale. This is adequate for wide-ranging species like bears. In other cases you may center all coordinates on 0. No matter what kind of transformation you use on your coordinates, make sure to always convert candidate values for $\mathcal{S}$ back to the original scale (UTM) before running the `over` command.

17.8 Increasing computational speed

Using custom written MCMC algorithms in **R** is not only more flexible but can also be faster than using programs such as **JAGS** and especially **WinBUGS**. Also, **R** tends to use much less memory than **JAGS**, which can be crucial if you are running a large model but only have limited memory available. **WinBUGS** is limited in the amount of memory it can access and thus will likely not max out your memory, but as a trade-off, it will take a long time to run such models. In this chapter we have provided you with the guidelines to write your own MCMC sampler. But beyond the material that we have covered there are a number of ways you can make your sampler more efficient, through parallel computing or by accessing an alternative computer language such as **C++**. Exploring these options exhaustively is beyond the scope of this book; instead, in this section we will give you some pointers to get started with these more advanced computational issues.

17.8.1 Parallel computing

If you are using a computer with several cores, you can make use of parallel computing to speed up overall computation. In parallel computing we execute commands simultaneously on different cores of the computer, instead of running them serially on one single core. For example, imagine you have 4 cores available and you want to implement a for-loop in **R**; instead of going through the loop iteration by iteration, you can prompt **R** to execute iterations 1–4 simultaneously on the four different cores. The core that finishes first will then continue with iteration 5, and so on. There are several packages in **R** that allow you to induce parallel computing, such as `snow` (Tierney et al., 2011) and `snowfall` (Knaus, 2010), and the more current versions of **R** (from 2.14.0 upwards) come with a pre-installed set of functions grouped under the name `parallel`.

The MCMC algorithms developed here and in other parts of this book come with plenty of opportunities to parallelize computation. In various instances within the algorithm, we have for-loops across our augmented data set of size M, or we may have for-loops across sampling occasions. We also have for-loops across iterations of the algorithm, but since one iteration of the Markov chain depends on the preceding iteration these should always be run serially, not in parallel. There is another dimension we can think of, and that is running multiple chains of an algorithm to assess convergence. This is a comparatively easy implementation of parallel computing and thus provides a good starting point to understand how it works in **R**.

Let's go back to the Fort Drum black bear data we analyzed above with the cloglog version of the binomial SCR model (Section 17.6) and run three parallel chains using `snowfall`. All we need to do is wrap our function `SCR0binom.cl` within another function that can then be executed in parallel, returning a list with one output matrix for each chain (install `snowfall` before executing the code below; we assume the data objects are already in your workspace from the previous analysis):

```
> library(snowfall)
## create wrapper function
> wrapper <- function(a){
+ out <- SCR0binom.cl(y=Xaug, X=trapmat, M=M, xl=xl, xu=xu, yl=yl,
+                     yu=yu, K=8, delta=c(0.1, 0.05, 2), niter=5000)
+ return(out)
+ }
```

After creating the wrapper function we need to initialize the cluster of cores, defining that we want computation to be implemented in parallel and how many cores we want it to be run on. Here, we assume we have (at least) 3 cores, but if your computer only has 2, make sure to adjust the code accordingly (i.e., set `cpus=2`). In that case, 2 of the 3 chains will be run in parallel and whichever core finishes first will then pick up the third chain. Further, we have to export all **R** libraries and data to all the cores, and set up a random number generator, so that we do not get identical results from the different cores:

```
> sfInit( parallel=TRUE, cpus=3) #initialize cluster
> sfLibrary(scrbook) #export library scrbook
> sfExportAll() #export all data in current workspace
> sfClusterSetupRNG() #set up random number generator
> outL = sfLapply(1:3,wrapper) # execute 'wrapper' 3 times
```

The object `outL` is a list of length 3, with one `out` matrix from the function `SCR0binom.cl` for each chain. After computation is complete, terminate the cluster using the command `sfStop`. Note that the intermediate output of current values and acceptance rates in the **R** console is suppressed when using parallel computing. We can now look at the output as described previously using the package `coda`, by first defining `outL` to be a list of `mcmc` objects:

```
> library(coda)
#turn output into MCMC list
> res <- mcmc.list(as.mcmc(outL[[1]]),as.mcmc(outL[[2]]),as.mcmc(outL[[3]]))
> summary(window(res, start=1001)) #remove first 1000 iterations as burn-in


[... some output removed ...]

              Mean        SD    Naive SE   Time-series SE
sigma       1.9723   0.13093   0.0011952        0.0087055
lam0        0.1115   0.01535   0.0001401        0.0009003
psi         0.7130   0.10787   0.0009847        0.0077910
N         499.6166  74.74934   0.6823650        5.4232653

2. Quantiles for each variable:

              2.5%       25%        50%        75%      97.5%
sigma      1.74339    1.8811     1.9637     2.0530     2.2618
lam0       0.08443    0.1007     0.1105     0.1211     0.1438
psi        0.52046    0.6350     0.7093     0.7814     0.9627
N        366.00000  446.0000   497.0000   547.0000   674.0000
```

Now that we have parallel chains we can also use the function `gelman.diag` to evaluate if chains have converged:

```
> gelman.diag(window(res, start=1001)) #assess chain convergence

Potential scale reduction factors:

       Point est.  Upper C.I.
sigma        1.01        1.04
lam0         1.01        1.02
psi          1.07        1.21
N            1.07        1.21

Multivariate psrf

1.05
```

We can see that estimates are similar to what we observed when running a single chain (see Section 17.6) and that all 3 chains appear to have converged, based on their point estimates of the $\hat{R}$ statistic, but, as already noted before, for a real analysis we might want to run this model for quite a bit longer, to bring down the upper confidence interval limits on $\hat{R}$ for ψ and N. If you have 3 cores then running these 3 parallel chains should not have taken longer than running a single chain. Yet if you look at the effective sample size now using `effectiveSize`, you can see that it has roughly tripled, as we would expect:

```
> effectiveSize(window(res, start=1001))
   sigma      lam0       psi         N
272.6935  411.8384  167.4192  168.3355
```

17.8.2 Using C++

Parallel computing is a great tool to speed up computations, but its usefulness is limited by how many cores you have available. Even with a decent number of cores, large models may still take a long time to run. A major reason for this is that for-loops in **R** are time consuming, whereas they are handled much more time-efficiently in other computer languages such as **C++**. As we saw above, MCMC algorithms consist of for-loops within for-loops, so that it stands to reason that implementing them in a language like **C++** should make those algorithms run much faster. Being avid **R** users, we cannot claim to be fluent in **C++** or to be aware of all the opportunities this language brings for faster computing. It is also beyond the scope of this book to go into the nuts and bolts of how **C++** works or provide a tutorial, and we refer you to the vast amounts of online and print material designed to give the interested user an introduction to **C++**. Just google "introduction C++" and you are sure to come across sites such as `http://www.cplusplus.com` that provide step-by-step instructions to get you started. Here, we only want to point out one approach to linking **R** with **C++**: the packages `inline` (Sklyar et al., 2010) and `RcppArmadillo` (François et al., 2011). These two packages provide a very convenient interface between the two languages, but there are other ways of calling **C++** functions from within **R**, such as the `.Call` command. If you are interested, we suggest you refer to the package manuals and vignettes, as well as the online document "Writing R extensions" (at `http://cran.r-project.org/doc/manuals/R-exts.html`) for a much more thorough treatment of this topic.

In order to use **C++** you need a compiler such as `g++` that (together with other compilers, for example for **C** and **FORTRAN**) comes with **Rtools**, which you can easily download from the web (at `http://cran.r-project.org/bin/windows/Rtools/`). All of these compilers are part of the GNU compiler collection (`http://gcc.gnu.org/`). Make sure the version of **Rtools** matches your version of **R** or you may run into compilation errors later on. To give you a taste of **C++** we will show you how to write a function that calculates the squared distances of individual activity centers to all traps, as is implemented in the `scrbook` package in the function `e2dist` (to be exact, `e2dist` calculates the distance, not the squared distance), and compare performance between **R** and **C++**. We will refer to these functions as "distance functions." First, let us set up dummy data—a matrix holding the coordinates of the trap array, outer limits of the state-space, and uniformly distributed activity centers for $M = 700$ individuals:

```
> gx <- seq(1,10,1)
> gy <- seq(1,10,1)
> X <- as.matrix(expand.grid(gx, gy))
```

```
> M <- 700
> J <- dim(X)[1]
> b <- 3
> xl <- min(gx)-b
> xu <- max(gx)+b
> yl <- min(gy)-b
> yu <- max(gy)+b
> S <- cbind(runif(M, xl, xu), runif(M, yl,yu))
```

Next, we can write a "pedestrian" version of `e2dist` and check how long it takes to calculate the squared distance matrix:

```
> Dfun <- function(M, J, S, X){
+ D2 <- matrix(0, nrow=M, ncol=J)
+ for(i in 1:M){
     + for(j in 1:J){
     + D2[i,j] <- (S[i,1]-X[j,1])^2 + (S[i,2]-X[j,2])^2
     + }}
+ return(D2)
+}

> system.time(
+ (D2R <- Dfun(M, J, S, X))
+)

user  system  elapsed
0.81    0.01     0.82
```

The code to implement the same function in **C++** using the `inline` and `RcppArmadillo` packages is shown in Panel 17.3. These packages allow you to use a range of data formats such as lists and matrices, and they take care of compiling the code in **C++** and loading the resulting function into **R**. This is also referred to compiling **C++** code "on the fly." You will see that the way the code is set up is reasonably similar to **R**. One difference that is worthy to point out is that in **C++** indices for vectors range from 0 to $n - 1$, *not* from 1 to n, as in **R**. Note that with `inline` we only need to write the core of the code and define the type of the variables we want to pass to the function, while the `cxxfunction` call takes care of the rest. Once your function is compiled and loaded you should check out the full **C++** code by calling `DfunArma@code`.

Executing this code shows that it is faster than the **R** version of the distance function or `e2dist`; in fact, it is too fast for the time resolution of the `system.time` function to even give us a time estimate:

```
### calculate squared distances using RcppArmadillo
library(inline)
library(RcppArmadillo)

#write core of function code
code<-'
/*define input, assign correct class (matrix, vector etc)*/
arma::mat Sn=Rcpp::as<arma::mat>(S);
arma::mat Xn=Rcpp::as<arma::mat>(X);
int Ntot=Rcpp::as<int>(M);
int ntraps=Rcpp::as<int>(J);
/*create matrix to hold squared distances*/
arma::mat D2(Ntot, ntraps);

/*loop over M and J to calculate distances*/
for (int i=0; i<Ntot; i++){
for(int j=0; j<ntraps; j++){
D2(i,j)= pow(Sn(i,0)-Xn(j,0), 2) + pow(Sn(i,1)-Xn(j,1), 2);
}
}
/*return D2 in R format*/
return Rcpp::wrap(D2);
'

# compile and load
DfunArma<-cxxfunction(signature(M="integer", J="integer", S="numeric",
X="numeric"), plugin="RcppArmadillo", body=code)
```

PANEL 17.3

Code to compute squared distance between individual activity centers and traps in **C++** from within **R** using `inline` and `RcppArmadillo`.

```
> system.time(
+ (out <- DfunArma(M,J,S,X)))

user  system  elapsed
   0       0        0
```

While speed differences of less than 1 s may seem negligible, remember that each command has to be executed at each iteration of the Markov chain. Especially with time-consuming models such as those for open populations (Chapter 16) or multi-session models (Chapter 14) we believe that **C++** holds large potential to make implementation of such models more feasible.

17.9 Summary and outlook

Programs like **JAGS** and **WinBUGS** do all the MCMC-related things that we went through in this chapter (and quite a bit more). Looking through your model, they determine which parameters they can use standard Gibbs sampling for (i.e., for conjugate full conditional distributions). Then, they determine whether to use adaptive rejection sampling, slice sampling, or—in the "worst" case—Metropolis-Hastings sampling for the other full conditionals (how the sampler is chosen differs among softwares). For MH sampling, they will automatically tune the updater so that it works efficiently.

Although these programs are flexible and extremely useful for performing MCMC simulations, it sometimes is more efficient to develop your own MCMC algorithm. Building an MCMC code follows three basic steps: Identify your model including priors and express full conditional distributions for each model parameter. If full conditionals are parametric distributions, use Gibbs sampling to draw candidate parameter values from those distributions; otherwise use Metropolis-Hastings sampling to draw candidate values from a proposal distribution and accept or reject them based on their posterior probability densities.

These custom-made MCMC algorithms give you more modeling flexibility than existing software packages, especially when it comes to handling the state-space: In **WinBUGS** and **JAGS** we define a continuous rectangular state-space using the corner coordinates to constrain the uniform priors on the activity centers **s**. But what if a continuous rectangle is an inadequate description of the state-space? In this chapter we saw that in **R** it only takes a few lines of code to use any arbitrary polygon shapefile as the state-space, which is especially useful when you are dealing with coastlines or large bodies of water that need removing from the state-space. Another example is the SCR **R** package `SPACECAP` (Gopalaswamy et al., 2012a) that was developed because implementation of an SCR model with a discrete state-space was inefficient in **WinBUGS**.

Another situation in which using a **BUGS** engine becomes increasingly complicated or inefficient is when using point processes other than the homogeneous binomial point process ("uniformity of density") which underlies the basic SCR model (see Section 5.10). In Chapter 11 you already saw an example of an inhomogeneous point process model and we briefly introduce a different point process, implemented using a custom-made MCMC algorithm, in Chapter 20. Finally, Chapter 19 deals with partially marked populations using custom-made MCMC algorithms to handle the (partially) latent individual encounter histories. While some of these models can be written in the **BUGS** language, they are painstakingly slow; others (for example, the classes of models considered in Chapter 12) cannot be implemented in **WinBUGS/JAGS** at all and we have to either use likelihood-based inference or develop our own MCMC algorithms. In conclusion, while you can certainly get by using **BUGS/JAGS** for standard SCR models, knowing how to write your own MCMC sampler gives you more flexibility to tailor these models to your specific needs.

CHAPTER

Unmarked Populations

18

Traditional capture-recapture models share the fundamental assumption that each individual in a population can be uniquely identified when captured. Often, this can be accomplished by marking individuals with color bands, ear tags, or some other artificial mark that subsequently can be read in the field. For other species, such as tigers (*Panthera tigris*) or marbled salamanders (*Ambystoma opacum*), individuals can be identified using only their natural markings. However, many species do not possess adequate natural markings and are difficult to capture, making it impractical to use standard capture-recapture techniques.

Estimating density when individuals are unmarked can be accomplished using a variety of alternatives to capture-recapture, such as distance sampling (Buckland et al., 2001) and N-mixture models (Royle, 2004b). These methods can be very effective when their assumptions are met, but when it is not possible to obtain accurate distance data, or when movement complicates the use of fixed-area plots, these methods may yield biased estimates of density (Chandler et al., 2011). Furthermore, some species are so rare and cryptic that it is nearly impossible to collect enough data using traditional survey methods.

In this chapter, we investigate spatially explicit alternatives for estimating density of unmarked populations, and we highlight the work of Chandler and Royle (2013) who demonstrated that the "individual recognition" assumption of traditional capture-recapture models is not a requirement of spatial capture-recapture models. They showed that, under certain conditions, spatially correlated count data are sufficient for making inferences about animal distribution and density even when no individuals are marked. The Chandler and Royle (2013) "spatial count model" (hereafter the SC model) requires neither distance data nor fixed-area plots. Instead, the observed data are trap- and occasion-specific counts, which are modeled as a reduced-information summary of the *latent* encounter histories. Because the model is formulated in terms of the data we wish we had, i.e., the typical encounter history data observed in standard capture-recapture studies of marked animals, the SC model is just a SCR model with a single extension to account for the fact that the encounter histories are unobserved. However, this results in a drastically different model than the models typically used for count data in ecology because the SC model is parameterized in terms of individuals, and specifically, their locations relative to the sampling device.

Spatial Capture-Recapture. http://dx.doi.org/10.1016/B978-0-12-405939-9.00018-9

The ability to fit SCR models to data from unmarked populations has important implications. For one, it means that SCR models can be applied to data collected using methods like point counts in which observers record simple counts of animals at an array of survey locations. The model can also be fitted to camera trapping data collected on unmarked animals, representing one of the first formal methods for estimating density from such data (but see Rowcliffe et al., 2008). So, is the SC model a free lunch? At face value, it sounds as though it allows for estimation of all the quantities of interest in standard capture-recapture studies, but with very little data. But of course the answer is no—lunch is still not free because with this model come new assumptions, and as was demonstrated by Chandler and Royle (2013), even with "perfect" data, parameter estimates will typically not be very precise. This should not be surprising given that we are asking so much from simple count data.

The real value of the SC model is twofold. First, it demonstrates an important theoretical result, namely that spatial correlation in count data carries information about density and distribution; a result that stands in stark contrast to a prevailing view of spatial correlation as a nuisance to be avoided or modeled out of unsightly residual plots. The second reason why this model is important is that it provides the basis for numerous model extensions that *can* yield precise density estimates. We will discuss some of these possibilities in this chapter, but perhaps the most useful extension—accommodating data from both marked and unmarked individuals—is treated separately in the next chapter. Here, we focus on situations in which all individuals are unmarked, and we begin by presenting the most basic formulation of the model. Then, we proceed by way of a few examples to consider extensions of the model in which ancillary information can be used to increase precision.

18.1 Existing models for inference about density in unmarked populations

When capture-recapture methods are not a viable option, ecologists often collect simple count data or even binary detection/non-detection data. These data are often treated as an index of abundance or occurrence and are analyzed using generalized linear models such as Poisson regression or logistic regression, perhaps with random effects (Zuur et al., 2009). However, index methods cannot be used to make unbiased inferences about abundance or occurrence unless strong assumptions about constant detection probability are valid (Williams et al., 2002; Sollmann et al., 2013b). In particular, index methods can be highly misleading when covariates affect both the ecological process of interest and the observation process. A classic example is given by Bibby and Buckland (1987) who found that songbird detection probability was negatively related to vegetation height, whereas density was positively associated with vegetation height in restocked conifer plantations. This intuitive phenomenon has been demonstrated repeatedly (Kéry, 2008; Sillett et al., 2012) and has led to the development of a vast number of models to estimate population size and occurrence probability when individuals are unmarked and detected imperfectly

(Buckland et al., 2001; Williams et al., 2002; MacKenzie et al., 2006; Royle and Dorazio, 2008). A review of these models is beyond the scope of this chapter, but we mention a few deficiencies of existing methods that warrant the exploration of alternatives for robust inference when standard capture-recapture methods do not apply.

Distance sampling (Buckland et al., 2001; Buckland, 2004), which we briefly introduced in Chapter 4, is perhaps the most widely used method for estimating population density when individuals are unmarked and detection probability is less than 1. This class of methods is known to work impeccably when estimating the number of beer cans in a field or the number of duck nests in a wetland. Distance sampling can also work very well in more interesting situations, and it is an extremely powerful method when the assumptions can be met. However, the assumptions that distance data can be recorded without error and that animals are distributed randomly with respect to the transect can be easily violated by common processes such as animal movement and measurement error. Although numerous methods have been proposed to relax some of these assumptions (Borchers et al., 1998; Royle et al., 2004; Johnson, 2010; Marques et al., 2010; Chandler et al., 2011), a more important issue is that distance sampling is simply not practical in many settings. For example, many species are so rare or elusive that they can only be reliably surveyed using "indirect" methods such as camera traps or hair snares.

In response to the increasing use of camera traps in studies of threatened species, and the problems associated with commonly used indices of abundance (Jennelle et al., 2002; O'Brien, 2011; Sollmann et al., 2013b), several density estimators have been developed for situations in which the population being studied is unmarked (Rowcliffe et al., 2008, 2011). These estimators assume that (1) cameras are randomly placed with respect to animal density, (2) animals neither avoid nor are attracted to the cameras, and (3) detection probability can be either modeled as a function of distance between the animal and the camera or as a function of movement velocity (which must be known or estimated using auxiliary data). Although these methods might represent an important improvement over index-based methods, the assumptions may not hold in many situations, especially when applied to data from standard designs in which camera stations are either baited or placed along trails—issues that can be dealt with directly using SCR models (see Chapters 12 and 13). Nonetheless, empirical studies have found that the assumptions do hold in some cases (Rowcliffe et al., 2008).

Other common approaches to estimating density when individuals are unmarked include double observer sampling, removal sampling, and repeated counts, for which custom models have been developed (Nichols et al., 2000b; Farnsworth et al., 2002; Royle, 2004b,a; Nichols et al., 2009; Fiske and Chandler, 2011). To obtain reliable density estimates using these methods, the area surveyed must be well defined and closed with respect to movement and demographic processes. Given a sufficiently short sampling interval, such as a 5 min point-count, the closure assumption may be reasonable. However, short sampling intervals limit the number of detections, so observers generally visit each survey location multiple times during a season. But then, animal movement may invalidate the closure assumption, and a model of temporary emigration is required (Kendall et al., 1997; Chandler et al., 2011).

Furthermore, distance-related heterogeneity in detection probability can introduce bias in these models, although this bias is negligible when the ratio of plot size to the scale parameter of the detection function is low (Efford and Dawson, 2009).

We mention these issues not to suggest that existing models do not have value—indeed we believe that they can be used to obtain reliable density estimates in many situations—rather, our aim is to highlight the need for alternative methods when the assumptions of existing models cannot be met and when spatially explicit inference is the objective.

18.2 Spatial correlation in count data

18.2.1 Spatial correlation as information

All of the previous methods require some sort of auxiliary information to model both abundance and detection. For instance, multiple observers, distance data, or repeated visits may be required to ensure that model parameters are identifiable (but see Lele et al., 2012; Sólymos et al., 2012). The same is true for the SC model, but the auxiliary information comes in the form of spatial correlation, which requires no extra effort to collect.

It is natural to be suspicious of the claim that spatial correlation is a good thing. In fact, elaborate methods have been devised to deal with spatial correlation as a nuisance (Lichstein et al., 2002; Dormann et al., 2007), and ecologists have been admonished for failing to obtain "real" replicates uncontaminated by spatial correlation (Hurlbert, 1984). The following heuristic may be helpful for seeing the value of spatial correlation in the context of density estimation.

Imagine a 10×10 grid of camera traps and a single unmarked individual exposed to "capture" whose home range center lies in the center of the trapping grid. If the individual has a small home range size relative to the extent of the trapping grid, we can envision what the spatial correlation structure of the encounters might look like. If the animal's home range is symmetric around the activity center then the number of times the individual is detected at each trap (the trap count) should decrease with the distance between the home range center and the trap; i.e., traps with the same distance from the activity center will yield counts that are more highly correlated with one another than traps located at different distances from the activity center. Thus, the correlation among the counts tells us something about the location of the activity center. It is relatively intuitive that spatial correlation carries information about distribution, but what about density?

Imagine now that there are two activity centers located in the trapping grid. Using trap counts alone, is it possible to determine the number and location of these activity centers? The answer is yes, at least under certain circumstances. Figure 18.1 shows the locations of the two hypothetical activity centers, and the total counts obtained at each trap after 10 survey occasions. Assuming that animals have bivariate normal home ranges, the fact that there are two areas in the map with high counts that dissipate with distance suggests that the most likely number of individuals given these data is 2.

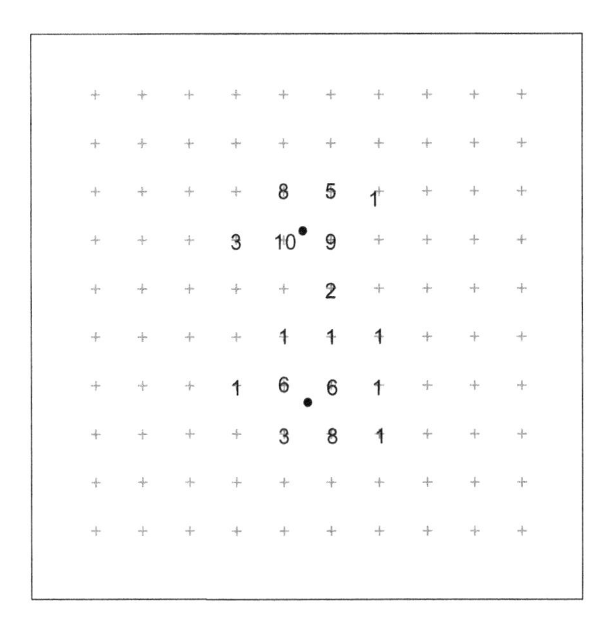

FIGURE 18.1

Simulated count data at each of 100 camera traps (crosses) after $K = 10$ sampling occasions. The black dots are the locations of two animal activity centers. The spatial count model estimates both the location and number of activity centers exposed to sampling using such spatially referenced count data.

Furthermore, the degree to which the counts dissipate from the two areas of highest intensity is information about the parameter governing home range size. These two pieces of information are enough to estimate the number of individuals exposed to sampling, given that a bivariate normal home range is a valid assumption. Of course, the data could just as well have been generated by a single individual whose home range is bimodal, and thus *as always* the assumptions of our model need to be carefully examined using our biological knowledge of the system. If the assumptions do not hold, it is almost always possible to relax them, for instance by allowing for non-stationary home ranges as we demonstrated in Chapters 12 and 13.

18.2.2 Types of spatial correlation

The spatial correlation allowed by the SC model is assumed to arise from animal movement; however, this is just one type of spatial correlation that may exist in ecological count data. Another common type of spatial correlation results from the spatial correlation of environmental covariates. Habitat variables, such as, the percent cover of deciduous forest in North America, will often be patchy rather than randomly distributed, and this can result in spatial correlation in abundance, and hence in count data. Often, this type of spatial correlation can be dealt with by simply including the habitat covariate in the model. For example, a simple, non-spatial species distribution model with only a few habitat variables can result in a distribution map that reflects the spatial correlation in abundance (Sillett et al., 2012; Royle et al., 2012). The point is that the relevant assumption of non-spatial models (e.g., GLMs) is that no spatial correlation exists in the *residuals*, and often, any spatial correlation apparent in the counts can be accounted for using covariates. This may be obvious, but it is a point that seems to be frequently misunderstood.

Of course, sometimes spatial correlation exists in residuals even after including covariate effects. This may be due to unobserved covariates or unmodeled processes such as dispersal. When mechanistic models cannot be developed to describe these processes, several options exist for handling spatial correlation as a nuisance (Besag and Kooperberg, 1995; Zuur et al., 2009; Wikle, 2010). In the context of SCR models, including the SC model dealt with in this chapter, movement-induced spatial correlation is always explicitly modeled, and other sources of spatial correlation can be accounted for as well. For instance, environmentally induced spatial correlation can be modeled by adopting an inhomogeneous point process model for the activity centers. That is, the point process intensity can be modeled as a function of observed covariates, and theoretically, it should be possible to allow for spatially correlated random effects to deal with unobserved covariates. See Chapter 11 for details.

18.3 Spatial count model

18.3.1 Data

Whereas traditional SCR models require spatially referenced individual encounter histories, the SC model requires simple, spatially reference count data. Let n_{jk} be the count data at sampling location j on occasion k. The entire $J \times K$ matrix of counts will be denoted $\mathbf{n}$. A sampling location in this context could be any device capable of recording count data, such as a human observer or a camera trap, and one of the benefits of the SC model is that it can be applied to data collected using many different survey methods. For ease of presentation, we will refer to sampling devices as traps, but remember that a trap is just something capable of recording count data. As in all SCR models, we also require the coordinates of the J traps, and we denote the location of trap j by $\mathbf{x}_j$. In some instances, additional data might be available such as trap-specific covariates, state-space covariates, information on the identities of a subset of individuals, or perhaps even distance data. We consider some of these model extensions in Section 18.6, but for the time being we ignore these possibilities so that we can focus on the basic model.

18.3.2 Model

The state model is exactly the same as the one we have dealt with throughout this book. It is a point process describing the number and distribution of activity centers in the state-space $\mathcal{S}$. Although it might be possible to fit inhomogeneous point process models using the methods described in Chapter 11, given the simplicity of the data, we concentrate on a homogeneous point process $\{\mathbf{s}_i, \ldots, \mathbf{s}_N\} \sim \text{Uniform}(\mathcal{S})$ where $\mathbf{s}_i$ is the activity center of individual i in the population of size N. For the moment, we will assume that N is known.

The observation model is the same as in other SCR models in the sense that it describes the probability of encountering individual i at trap j, conditional on the location of the individual's activity center. The specific encounter process will depend on the sampling method, and here we consider the standard camera trapping situation in which an individual can be encountered at multiple traps during a single occasion—e.g., one night during a camera-trapping study—and it can be detected multiple times at a single trap during an occasion. This is the Poisson encounter model (a.k.a. the count detector case) described in Chapter 9. Our experience with alternative observation models such as the Bernoulli and multinomial models suggests that the parameters of the model may not be identifiable in these cases, at least when no additional information is available. This is a subject of ongoing research.

As before, we define y_{ijk} as the encounter data for individual i at trap j on occasion k, which we model as:

$$y_{ijk} \sim \text{Poisson}(\lambda_{ij}), \tag{18.3.1}$$

where λ_{ij} is the encounter rate. A common encounter rate model is the Gaussian, or half-normal, model:

$$\lambda_{ij} = \lambda_0 \exp(-\|\mathbf{x}_j - \mathbf{s}_i\|^2/2\sigma^2)$$

in which λ_0 is the baseline encounter rate, $\|\mathbf{x}_j - \mathbf{s}_i\|$ is the Euclidean distance between the trap and activity center, and σ is the scale parameter determining the rate at which encounter rate decreases with distance. In this context, σ also determines the amount of correlation among the counts because if σ is low relative to the trap spacing then it is unlikely that an individual will be detected at multiple traps.

When individuals cannot be uniquely identified, the encounter histories cannot be directly observed, which seems like a massively insurmountable problem of epic proportions. The solution of Chandler and Royle (2013) is the same one we routinely apply when we cannot directly observe the process of interest—we regard the encounter histories as latent variables. This leaves the remaining task of specifying the relationship between the count data and the encounter histories, i.e., we need a model of $[\mathbf{n}|\mathbf{y}]$ where $\mathbf{y}$ represents the entire collection of encounter histories. In this case, there is only one possibility because, by definition, the count data are simply a reduced-information summary of the latent encounter histories. That is, they are the sample- and trap-specific totals, aggregated over all individuals:

$$n_{jk} = \sum_{i=1}^{N} y_{ijk}. \tag{18.3.2}$$

So, unlike most model development problems faced in this book, we don't have to consider competing probability models for $[\mathbf{n}|\mathbf{y}]$, but instead, we recognize the fact that the relationship between the counts and the latent encounter histories is deterministic. This deterministic constraint poses some computational challenges, which we discuss below. But first we present some alternative formulations of the model.

Recall from Chapter 2 that the sum of two or more Poisson random variables is also a Poisson random variable. Specifically, if $x_1 \sim \text{Poisson}(\lambda_1)$ and $x_2 \sim \text{Poisson}(\lambda_2)$, then $(x_1 + x_2) \sim \text{Poisson}(\lambda_1 + \lambda_2)$. Thus, under this Poisson model for the latent encounter histories, the count data can be modeled as Poisson:

$$n_{jk} \sim \text{Poisson}(\Lambda_j), \tag{18.3.3}$$

where

$$\Lambda_j = \lambda_0 \sum_i \exp(\|\mathbf{x}_j - \mathbf{s}_i\|^2/2\sigma^2),$$

and because Λ_j does not depend on k, we can aggregate the replicated counts, defining $n_{j.} = \sum_k n_{jk}$ and then

$$n_{j.} \sim \text{Poisson}(K\Lambda_j).$$

As such, K and λ_0 serve equivalent roles as affecting baseline encounter rate. Formulating the model in terms of the aggregated count data demonstrates that the model can be applied to data from a single sampling occasion ($J \equiv 1$) as has been noted elsewhere for standard SCR models (Efford et al., 2009b). In the context of studying marked populations, the model parameters will only be identifiable in the $J \equiv 1$ case if an animal can be captured at multiple traps during a single occasion. The SC model essentially requires the same thing, which is to say that it requires correlation in the count data resulting from an individual being captured in multiple, closely spaced traps.

This formulation of the model in terms of the aggregate count also simplifies computations as the latent encounter histories do not need to be updated in the MCMC estimation scheme; however, retaining them in the formulation of the model is important if some individuals are uniquely marked. This is because uniquely identifiable individuals produce observations of some of the y_{ijk} variables, which we elaborate on in the next chapter.

18.3.3 On *N* being unknown

Even though there are no observed encounter histories in the situation we consider here, we can still use data augmentation (Tanner and Wong, 1987; Liu and Wu, 1999) to resolve the problem that N is unknown. In fact, we are actually using two different types of data augmentation since we first augment the observed data with latent encounter histories, and then we augment this latent data array with a set of all-zero encounter histories. This approach turns out to be very similar to other data augmentation schemes used to model spatial dependence in other contexts (Wolpert and Ickstadt, 1998; Best et al., 2000).

Although the process of data augmentation should be familiar by now, we briefly review the basics. For homogeneous point process models, N is typically modeled as $N \sim \text{Binomial}(M, \psi)$, which is equivalent (marginally) to a discrete uniform prior on N if $\psi \sim \text{Uniform}(0, 1)$. Since a binomial model is equivalent to a series of M independent Bernoulli trials, we can rewrite $N \sim \text{Binomial}(M, \psi)$ as $z_i \sim \text{Bernoulli}(\psi)$ where z_i is an auxiliary variable indicating if individual i is a member of the population, such that $N = \sum_{i=1}^{M} z_i$. Having expanded the model to include a prior on N, we can summarize the SC model, with a Gaussian observation model, as follows:

$$
\begin{aligned}
z_i &\sim \text{Bernoulli}(\psi),\\
y_{ijk} &\sim \text{Poisson}(\lambda_{ijk} z_i),\\
\lambda_{ijk} &= \lambda_0 \exp(-\|\mathbf{x}_j - \mathbf{s}_i\|^2)/(2\sigma^2),\\
n_{jk} &= \textstyle\sum_{i=1}^{M} y_{ijk}.
\end{aligned}
$$

18.3.4 Inference

Bayesian analysis can proceed once suitable priors have been put on the hyperparameters ψ, σ, and λ_0. Chandler and Royle (2013) provided **R** code for fitting the model using MCMC, and they evaluated the model's performance with uniform priors on the three hyperparameters. They also discussed the possibilities and effects of including prior knowledge about σ into the model. In Section 18.5, we explain how the model can be implemented using **JAGS**, but first we briefly contemplate the viability of classical analysis of this model.

The obvious challenge faced when conducting a classical analysis of this model is that the number of latent variables is huge. In all SCR models, the activity centers are latent, but now, even the encounter histories are latent. Maximizing likelihoods with latent variables (random effects) involves integrating (or summing) over all possible values of the latent variables. For the activity centers, this is typically accomplished by integrating the conditional-on-s likelihood $[\mathbf{y}_i|\mathbf{s}_i]$ over the two-dimensional state-space $\mathcal{S}$ (Chapter 6). However, with the SC model, we have to sum over all possible encounter histories meeting the constraint of Eq. (18.3.2). The number of possible encounter histories will, in general, be too high to make the likelihood tractable, and thus we do not think that maximum-likelihood is a viable option for analyzing this model. However, one might be able to obtain approximate maximum-likelihood estimates using simulation-based methods (Lele et al., 2010), which will typically be more computationally challenging than the Bayesian analysis.

18.4 How much correlation is enough?

In Chapter 10, we noted that if trap spacing is too wide relative to the encounter rate parameter σ, then few spatial recaptures will be realized and the model parameters will be estimated poorly. The same principle applies here—σ shouldn't be too small or too large relative to trap spacing or else the counts will be *iid* Poisson random variables. So how much correlation is enough? Phrased differently, what is the ideal ratio of σ to trap spacing to ensure correlation and minimize the variance of the posterior distributions? We see two options for answering this questions, both of which are topics in need of additional research. The first approach is to use the methods described in Chapter 10, i.e., by either conducting simulation studies with various trap spacing to σ ratios, or to analytically minimize a variance criterion for a given set of sampling conditions and effort. The former approach was used by Chandler and Royle (2013) whose limited simulation study indicated that an ideal ratio is approximately 2. This agrees with findings from previous research on the optimal design of SCR studies (Chapter 10), as it should.

A second approach that may be of use if a data set has already been collected is to use standard techniques from spatial statistics to determine if adequate correlation exists in the counts. For example, one might compute Ripley's K-statistic or generate (semi-) variograms (Illian et al., 2008). We have not studied the utility of such approaches, but they seem worthy of investigation.

18.5 Applications

18.5.1 Simulation example

Simulating data under the SC model proceeds by first simulating standard SCR encounter history data and then collapsing it into count data. The following blocks of **R** code generate data from the model, with parameters $\sigma = 5$, $\lambda_0 = 0.4$, and $N = 50$. The state-space is a $[0, 100] \times [0, 100]$ square, and a grid of 100 traps is centered in the middle. The first block of code generates the trap coordinates X and the $N = 50$ activity centers:

```
> tr <- seq(15, 85, length=10)
> X <- cbind(rep(tr, each=length(tr)),
+            rep(tr, times=length(tr)))    # 100 trap coords
> set.seed(10)
> xlim <- c(0, 100); ylim <- c(0, 100)    # S is [0,100]x[0,100] square
> A <- (xlim[2]-xlim[1])*(ylim[2]-ylim[1])/1e4 # Area of S
> mu <- 50                                # Density(animals/unit area)
> N <- rpois(1, mu*A)                     # Generate N=50 as Poisson deviate
[1] 50
> s <- cbind(runif(N, xlim[1], xlim[2]), runif(N, ylim[1], ylim[2]))
```

We could have set $N = 50$ directly, but instead we treated density as a fixed parameter ($\mu = 50$) and generated N as a random variable—it just so happens that with the specified random seed, N equals 50.

Now we can generate the encounter histories under the Poisson observation model. Let's suppose that sampling is conducted over $K = 5$ nights:

```
> sigma <- 5
> lam0 <- 0.4
> J <- nrow(X)
> K <- 5
> y <- array(NA, c(N, J, K))
> for(j in 1:J) {
```

```
+        dist <- sqrt((X[j,1]-s[,1])^2 + (X[j,2] - s[,2])^2)
+        lambda <- lam0*exp(-dist^2/(2*sigma^2))
+        for(k in 1:K) {
+            y[,j,k] <- rpois(N, lambda)
+        }
+ }
```

The object y is the $N \times J \times K$ array of encounter data, which cannot be directly observed if the animals are unmarked. Converting the encounter data to count data can be accomplished using a single apply command:

```
> n <- apply(y, c(2,3), sum)
> dimnames(n) <- list(paste("trap", 1:J, sep=""),
+                     paste("night", 1:K, sep=""))
> n[1:4,]
       night1  night2  night3  night4  night5
trap1       1       0       0       0       0
trap2       1       2       2       0       1
trap3       1       0       0       1       0
trap4       0       0       0       0       0
```

This displays the first four rows of n, the $J \times K$ matrix of counts.

The question now is: Is it possible to estimate the parameters? In our simulated dataset we have $J \times K = 500$ data points, but how many parameters do we need to estimate? A frequentist might say that there are only three parameters: λ_0, σ, and N (or density μ) because inference about the latent parameters is carried out using prediction methods after the three hyperparameters have been estimated. However, a Bayesian would probably say that each **s** and each element of the latent encounter array **y** is a parameter in need of a posterior. From this perspective there are far more parameters than data points, and thus it would appear as though the situation is dire. Whether or not the parameters are actually estimable is a rather difficult question to answer. One simplistic, but not definitive, approach for addressing the question is to conduct a simulation study and evaluate the frequentist performance of the model by asking how often the data-generating values are included in confidence/credible intervals, and how biased are point estimates. Chandler and Royle (2013) conducted such a simulation study and found that, while the variance of the posterior distribution was high by most standards, the bias of the posterior mode of N was small and the coverage of the credible intervals was close to nominal. Moreover, they found no evidence that the posterior distributions were dominated by the priors, further supporting the conclusion that spatial correlation in the count data is sufficient for estimating density and encounter probability parameters.

At this point in time the SC model can only be fit using one of the **BUGS** engines, or using custom software like the **R** code accompanying Chandler and Royle (2013). Although **BUGS** might provide the most flexible option for fitting the model, it is not straightforward because of the constraints in the model. **JAGS** has a distribution called dsum that was designed for this type of situation in which the observed data are a sum of random variables. Panel 18.1 shows the **JAGS** code, but we abbreviated the

```
model{
sigma ~ dunif(0, 200) # Tailor this to your state-space
lam0 ~ dunif(0, 5)    # consider dgamma() as an alternative
psi ~ dbeta(1,1)
for(i in 1:M) {
   z[i] ~ dbern(psi)
   s[i,1] ~ dunif(xlim[1], xlim[2])
   s[i,2] ~ dunif(ylim[1], ylim[2])
   for(j in 1:J) { # Number of traps
       distsq[i,j] <- (s[i,1] - X[j,1])^2 + (s[i,2] - X[j,2])^2
       lam[i,j] <- lam0 * exp(-distsq[i,j] / (2*sigma^2))
       for(k in 1:K) { # Number of occasions
           y[i,j,k] ~ dpois(lam[i,j]*z[i])
           }
       }
   }
for(j in 1:J) {
   for(k in 1:K) {
       n[j,k] ~ dsum(y[1,j,k], y[2,j,k], ..., y[200,j,k])
       }
   }
N <- sum(z[])   # Realized population size
A <- (xlim[2]-xlim[1])*(ylim[2]-ylim[1]) # Area of state-space
D <- N / A      # Realized density
ED <- (M*psi)/A # Expected density
}
```

PANEL 18.1

JAGS code defining the spatial count model. This version includes the latent encounter histories. Note the abbreviated arguments to dsum().

arguments to `dsum` because in practice you need to provide all M of them. The code looks slightly unwieldy if M is large, but you can easily create it using the `paste` function in **R**. Here is an example, with an unrealistically small value of $M = 10$:

```
> paste("y[", 1:10, ",j,k]", sep="", collapse=", ")
[1] "y[1,j,k], y[2,j,k], y[3,j,k], y[4,j,k], y[5,j,k], y[6,j,k],
y[7,j,k], y[8,j,k], y[9,j,k], y[10,j,k]"
```

The **JAGS** model in Panel 18.1 can be used to fit the version of the model in which the latent encounters are updated at each Monte Carlo iteration. One challenge faced when using this version of the model is that **JAGS** cannot autogenerate initial values that honor the constraints in the model, so it is necessary to provide them. The following code presents one fairly general way of creating acceptable starting values and formatting the data for analysis using the `rjags` package:

```
> library(rjags)
> dat1 <- list (n=n, X=X, J=J, K=K, M=200, xlim=xlim, ylim=ylim)
> init1 <- function() {
+    yi <- array(0, c(dat1$M, dat1$J, dat1$K))
+    for(j in 1:dat1$J) {
+         for(k in 1:dat1$K) {
+            yi[sample(1:dat1$M, dat1$n[j,k]),j,k] <- 1
+         }
+    }
+    list(sigma=runif(1, 1, 2), lam0=runif(1),
+         y=yi, z=rep(1, dat1$M))
+ }
> pars1 <- c("lam0", "sigma", "N", "mu")
```

The code in Panel 18.1 is useful because it shows how closely this model is related to standard SCR models, and it provides the basis for including data on both marked and unmarked individuals, as will be discussed in the next chapter. However, this model runs very slowly, even when using a fast 64 bit machine with chains run

```
model{
sigma ~ dunif(0, 200)
lam0 ~ dunif(0, 5)
psi ~ dbeta(1,1)
for(i in 1:M) {
   z[i] ~ dbern(psi)
   s[i,1] ~ dunif(xlim[1], xlim[2])
   s[i,2] ~ dunif(ylim[1], ylim[2])
   for(j in 1:J) { # Number of traps
       distsq[i,j] <- (s[i,1] - X[j,1])^2 + (s[i,2] - X[j,2])^2
       lam[i,j] <- lam0 * exp(-distsq[i,j] / (2*sigma^2)) * z[i]
       }
   }
for(j in 1:J) {
   bigLambda[j] <- sum(lam[,j])
   for(k in 1:K) {
       n[j,k] ~ dpois(bigLambda[j])
       }
   }
N <- sum(z[])
}
```

PANEL 18.2

JAGS code defining the spatial count model. This version does not include the latent encounter histories, and thus runs much faster than the code in Panel 18.1.

in parallel. The code in Panel 18.2 runs much faster because it does not include the latent encounter histories.

An even faster (but perhaps less efficient) alternative is to use the scrUN function in scrbook. The usage is as follows:

```
> out1 <- scrUN(n=n, X=X, M=300, niter=25000, xlims=xlim, ylims=ylim,
              inits=list(lam0=0.3, sigma=rnorm(1, 5, 0.1)), updateY=TRUE,
              tune=c(0.004, 0.09, 0.35))
```

where n is the matrix of counts, X is the trap coordinate matrix, M sets the size of the data-augmented latent data, xlims and ylims define the rectangular state-space, inits is a list of starting values, and updateY determines if the latent encounter histories are updated as part of the MCMC algorithm. In general, we recommend using the option updateY=FALSE because the Markov chains tend to mix better. Even so, it can be important to fiddle with the tuning parameters until the acceptance rates are between 40% and 60%. Otherwise, the Markov chains will exhibit extremely high autocorrelation. This is one reason to favor **JAGS** over our implementation in scrbook since **JAGS** finds suitable tuning parameters automatically during the adaptive phase (when using Metropolis updates).

We fit the model to the simulated data using both formulations—with and without the latent encounter histories—using both **JAGS** and scrUN. Table 18.1 shows summaries of 25,000 posterior draws, and suggests that while the true parameter values are easily covered by the 95% credible intervals, the intervals are rather wide. This low precision is not just a peculiarity of this particular data set—it will generally be

Table 18.1 Posterior summaries from the spatial count ("SC") model applied to simulated data using scrbook and **JAGS**. Twenty-five thousand samples were generated, but substantial Monte Carlo error is still evident. All parameters were given uniform priors.

Parameter	Mean	SD	2.5%	50%	97.5%
		scrUN (..., updateY=FALSE)			
$\sigma = 5$	4.718	0.922	3.239	4.615	6.833
$\lambda_0 = 0.4$	0.500	0.136	0.268	0.489	0.793
$N = 50$	60.653	31.067	21.000	54.000	137.000
		scrUN (..., updateY=TRUE)			
σ	4.554	0.784	3.216	4.486	6.264
λ_0	0.489	0.131	0.262	0.479	0.775
N	64.772	30.162	26.000	59.000	140.000
		***JAGS* (without latent encounter histories)**			
σ	4.70	0.88	3.24	4.66	6.63
λ_0	0.52	0.14	0.27	0.52	0.80
N	58.55	30.30	20.00	52.00	135.00

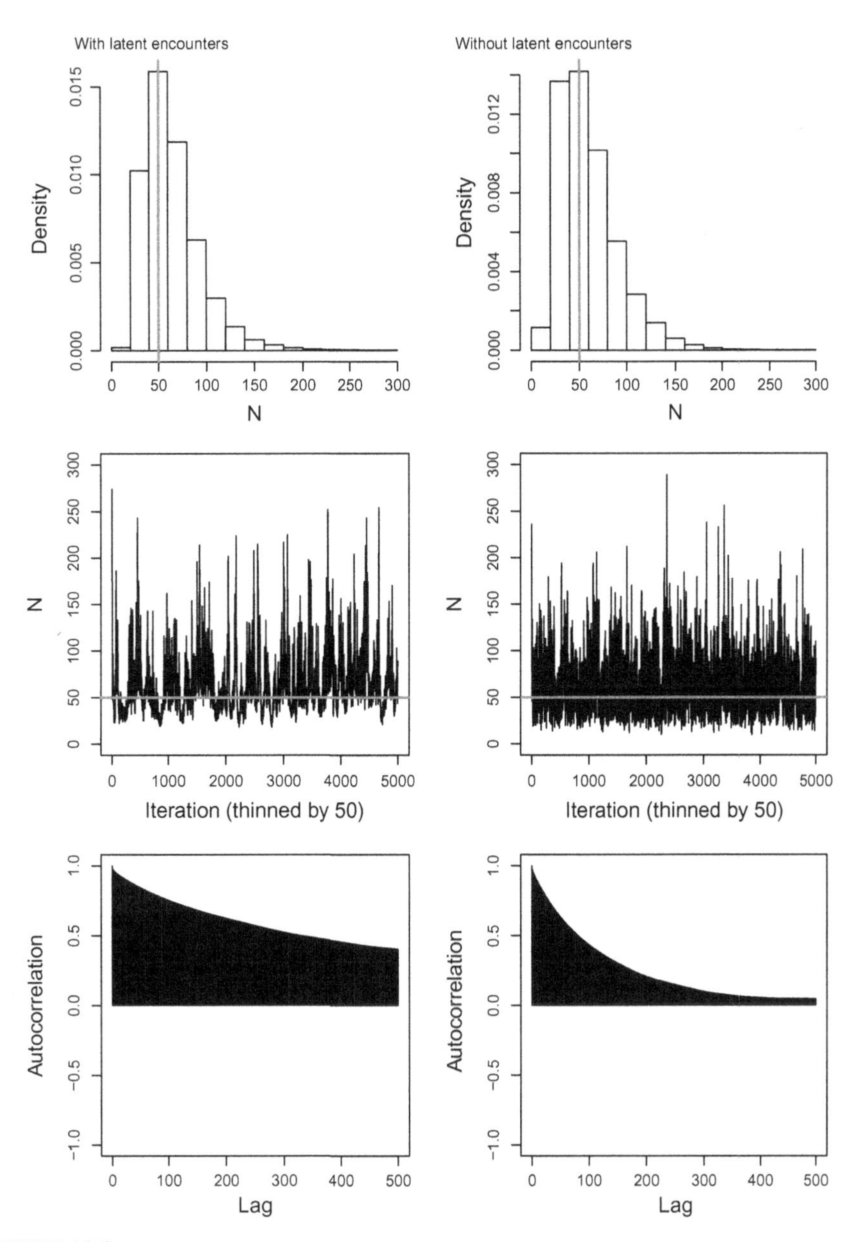

FIGURE 18.2

MCMC results for the parameter N from the two algorithms (with and without the latent encounter histories). The first row contains the histograms of the posterior distributions, the second row contains the history (or trace) plots, the third row shows the autocorrelation plots.

low unless the sample size is very large, as noted by Chandler and Royle (2013). Furthermore, autocorrelation of the samples will typically be high (Figure 18.2), and thus it may take many iterations to achieve convergence. The results shown in Figure 18.2 also indicate that the algorithm that includes the latent encounter histories seems to have a hard time exploring the region of the posterior in which N is low. Given these technical difficulties, we recommend using the **JAGS** implementation (based on Panel 18.2), and it is always a good idea to use MCMC diagnostic tools such as those available in the `coda` package (Chapter 17).

The take-home message is that, even with simulated data, the precision of the posterior distributions is low and mixing is poor. This should be expected given that we are asking so much from so little data. In essence, we are trying to fit a point process model while being twice removed from the actual point (activity center) locations. These difficulties may warrant the investigation of simpler models at the expense of the mechanistic description of the system. Another option is to figure out ways of improving model precision—options we discuss in Section 18.6. Before doing so, we re-analyze the northern parula (*Parula americana*) data described in Chandler and Royle (2013).

18.5.2 Northern parula in Maryland

The parula data are standard avian point count data, with one exception. Typically, when studying passerines, points are spaced by > 200m in order to maintain statistical independence. In contrast, the parula data were collected at 105 points located on a 50 m grid, which virtually ensures spatial correlation since the parula song can be

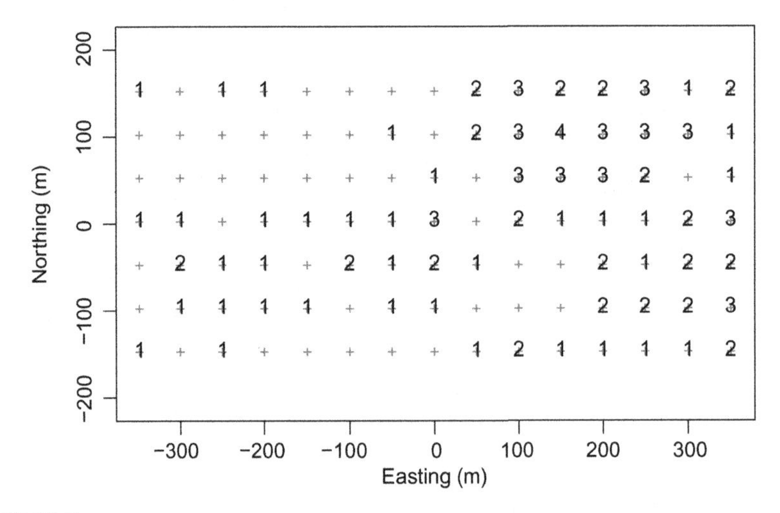

FIGURE 18.3

Spatially correlated counts of northern parula. Gray crosses are the locations of the 105 point count stations. Superimposed are the number of detections after three survey occasions.

heard from distances > 50m. Each point was surveyed three times during June 2006, and Figure 18.3 depicts the resulting spatially correlated counts ($n_{j.}$). A total of 226 detections were made with a maximum count of 4 during a single survey. At 38 points, no birds were detected. All but one of the detections were of singing males, and this one observation was not included in the analysis.

We fit the model using **JAGS** and the code from Panel 18.2, which does not include the latent encounter histories. For comparative purposes, we used two sets of priors: (1) the improper uniform priors considered by Chandler and Royle (2013) and (2) proper Gamma(0.001, 0.001) priors for σ and λ_0 as well as an approximate $N \propto 1/N$ prior, which is (almost) the Jeffreys prior (see Link (in press) and Link and Barker (2010) for details on Jeffreys prior). The state-space was created by buffering the grid of point count locations by 250 m. M was set to 300. To reduce computation time, we used the `parallel` package and distributed three chains to three separate cores. The entire example can be reproduced using the code on the help page for `nopa` in our **R** package `scrbook`. The following code illustrates the essential elements:

```
> library(scrbook)
> library(rjags)
> dat2 <- list(n = nopa$n, X = nopa$X, M=300, J=nrow(nopa$n), K=ncol(nopa$n),
+              xlim=c(-600, 600), ylim=c(-400, 400))
> init2 <- function() {
+   list(sigma=rnorm(1, 100), lam0=runif(1), z=rep(1, dat2$M))
+ }
> cl2 <- makeCluster(3) # Open 3 parallel R instances
> clusterExport(cl2, c("dat2", "init2", "pars1")) # send objects to 3 cores
> out2 <- clusterEvalQ(cl2, {# executes the folowing commands on each core
+   library(rjags)
+   jm <- jags.model("nopa2.jag", dat2, init2, n.chains=1, n.adapt=1000)
+   jc <- coda.samples(jm, pars1, n.iter=150000)
+   return(as.mcmc (jc))
+ })
> mc2 <- mcmc.list(out2) # put the 3 chains together
> plot(mc2)
> summary(mc2)
```

18.6 Extensions of the spatial count model

18.6.1 Improving precision

The results of the parula analysis are presented in (Table 18.2). Once again, we see wide credible intervals for N and high sensitivity to the priors. These limitations support the conclusions of Chandler and Royle (2013) that researchers should: (1) elicit prior information from the published literature and/or (2) mark a subset of individuals when applying the SC model. Both of these options should be readily accomplished in many studies, especially the first option because extensive information on home range size has been compiled for many species in diverse habitats (e.g., DeGraaf and Yamasaki, 2001), which can be embodied as a prior distribution for

Table 18.2 Posterior summary statistics for spatial count model applied to the northern parula data. Note the sensitivity of the posterior to the two different prior distributions.

Par	Prior	Mean	SD	2.5%	50%	97.5%
N	DUnif(0,300)	38.474	37.275	1.000	29.000	138.000
λ_0	Unif(0,∞)	0.310	0.183	0.082	0.269	0.817
σ	Unif(0,∞)	127.935	99.303	44.760	87.291	438.374
N	$\propto 1/N$	29.591	32.555	1.000	19.000	119.000
λ_0	Gamma(0.001,0.001)	0.309	0.194	0.078	0.261	0.843
σ	Gamma(0.001,0.001)	150.183	105.044	48.735	117.069	447.616

the encounter rate parameter σ in a Bayesian analysis (Chandler and Royle (2013), Chapter 5).

In some cases, it may not be possible to mark any individuals, and no prior information may exist about encounter parameters; however, it may be possible to improve precision by collecting auxiliary data, such as distance measurements. In fact, in the parula study, detections were classified as either within or beyond 150 m, and it seems sensible to expand the model to accommodate this rudimentary distance sampling data. But if auxiliary data such as distance measurements exist, why bother with the SC model at all since density can be estimated using the distance data alone? This is a good point, and in general, the simplest model that does the job should be preferred. The reasons why one might prefer an expanded SC model over a simple distance sampling model include the ability to model spatial correlation and the ability to model movement. But how exactly can the SC model be extended to accommodate such auxiliary data?

The basic extension that we consider here is to use a type of search-encounter model (Chapter 15) that includes the activity centers (**s**) and the actual locations of individuals (**u**). By including both activity centers and actual locations in the model, abundance in any region $\mathcal{B}$ is given by

$$N(\mathcal{B}) = \sum_i I(\mathbf{u}_i \in \mathcal{B}). \tag{18.6.1}$$

Thus, in the context of distance sampling studies in which the distance data are recorded in discrete intervals, the region $\mathcal{B}$ would be the area corresponding to a particular distance interval. The probability of detecting the individuals $N(\mathcal{B})$ would be the average detection probability $\bar{p}$, which is computed by integrating a distance-based detection function over the distance interval.

In other contexts, such as when conducting removal surveys, the region $\mathcal{B}$ could be a fixed-area plot, such as a stream segment. Again, Eq. (18.6.1) could be used to model local abundance ($N(\mathcal{B})$), and detection probability within the region could be modeled conditional on $N(\mathcal{B})$. A reasonably general description of this model is as follows:

$$\mathbf{s}_i \sim \text{Uniform}(\mathcal{S}),$$

$$\mathbf{u}_{ik} \sim \text{BVN}(\mathbf{s}_i, \mathbf{\Sigma}),$$

$$N(\mathcal{B}_{jk}) = \sum_{i=1}^{M} I(\mathbf{u}_{ik} \in \mathcal{B}_{jk}),$$

$$n_{jkl} \sim \text{Binomial}(N(\mathcal{B}_{jkl}), p),$$

where $\mathbf{\Sigma} = \begin{pmatrix} \tau & 0 \\ 0 & \tau \end{pmatrix}$, with τ governing the size of the bivariate normal home range. The interpretation of the parameter p will depend upon the survey protocol (Nichols et al., 2009).

When plots are far enough apart that individuals cannot move between them, the counts will be uncorrelated and the model can be approximated using a non-spatial N-mixture model allowing for temporary emigration (Chandler et al., 2011). In the next example, we consider data in which the plots are obviously not independent.

18.6.2 Dusky salamanders in Maryland

The independence assumption of the Chandler et al. (2011) model will not always hold. A prime example is in studies of aquatic species in stream networks. For example, consider the data depicted in Figure 18.4. The figure shows counts of northern dusky salamanders (*Desmognathus fuscus*) in 25 m stretches on a small stream in the Chesapeake and Ohio National Historic Park. The data were collected by E.H.C. Grant and colleagues with the objective of understanding the spatial and temporal dynamics of salamander populations in response to seasonal and annual variations in stream hydrology.

To sample the population, the stream networks were divided into 25 m stretches, and in each stretch, "temporary" removal sampling was conducted, which involved capturing and removing salamanders on three consecutive passes. The salamanders were placed in individual plastic baggies for the brief 30–60 min duration of sampling and then they were released at the location of capture. The entire process was repeated 3–4 times per season (May–August). In a subset of streams and years, individuals were marked, but in general it was too expensive to mark all of the captured individuals, and the data considered here consist entirely of unmarked individuals.

The sampling protocol may be thought of as a "robust design" (Pollock, 1982), with "occasions" (typically 1 day) being the primary period, and secondary samples being the removal passes within the primary periods. An obvious feature of these data is that the neighboring counts are spatially correlated. In this case, we have reason to believe that this correlation is the result of habitat preferences, with individuals actively selecting habitat in the upper reaches of the streams. This could be modeled as a function of a covariate describing the distance from the mouth of the stream. Another obvious feature of this data is that the pattern of spatial correlation remains consistent between occasions, but the overall counts decline markedly over the course of the season. These phenomena can be explained by the fact that the salamanders have

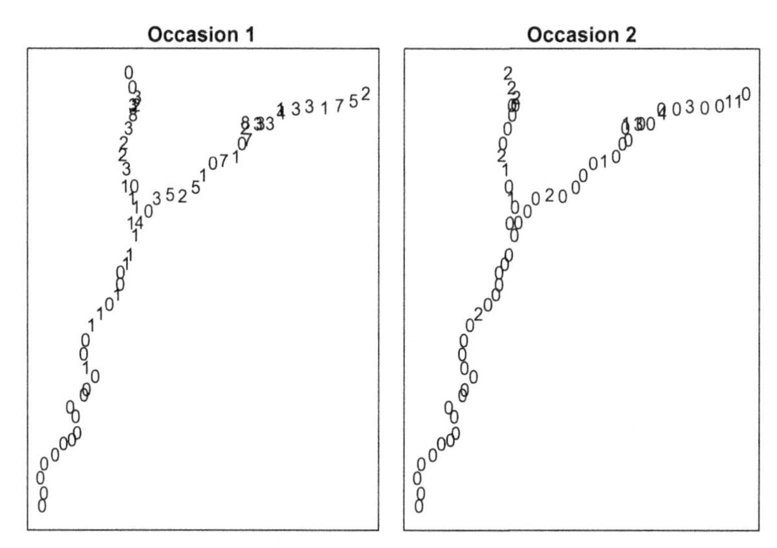

FIGURE 18.4

Stream segment counts of northern dusky salamanders in the Chesapeake and Ohio National Historic Park, VA/MD. Each number is the count associated with a 25 m stretch in which three removal passes were made on three occasions each summer (only two occasions are shown here). Notice the consistency of the spatial correlation between occasions and the temporal decline in the counts.

relatively small home ranges, and this results in the consistent pattern of correlation among occasions. Furthermore, as the season progresses, the streams dry out, and many individuals move underground.

Given the importance of movement within home ranges, which determines the correlation among occasions, and movement underground, which results in a decreasing number of individuals being available for sampling, it would be helpful to have a model that describes both processes and allows for evaluation of hypotheses regarding the effects of environmental variables. For example, one might ask how stream flow is related to the probability that an individual remains active, i.e., aboveground. A model describing this process could be used to predict activity levels under future conditions. Although we do not investigate covariate effects in this section, we do present a general model allowing for movement among occasions, and for decreasing availability over the season.

This expanded model is founded on the one described in the previous section, but it also includes a removal model for the observation process, and it includes a basic "open" population model to allow for a decline in abundance over time (Chapter 16). Actually, the population is not thought to decline substantially during the season, but rather, the number of individuals *available* for detection declines because many

```
model {
phi ~ dbeta(1,1)       # "availability" parameter
tau ~ dunif(0, 1000)   # "movement parameter" of Gaussian kernel model
p ~ dbeta(1,1)         # detection prob
psi ~ dbeta(1,1)       # data augmentation parameter
for(i in 1:M) {
  z[i,1] ~ dbern(psi)         # is the individual real?
  z[i,2] ~ dbern(z[i,1]*phi) # and still aboveground?
  z[i,3] ~ dbern(z[i,2]*phi) # ...
  s[i] ~ dcat(PrSeg[]) # location (stream segment) of activity center
  for(g in 1:G) {
    PrU[i,g] <- exp(-distmat[s[i],g]^2/(2*tau^2)) # Pr(u | s)
    }
  for(k in 1:K) {
    u[i,k] ~ dcat(PrU[i,]) # location of guy i at time k
    for(g in 1:G) {
      y[i,g,k] <- (u[i,k] == g)*z[i,k] # was individual at u==g?
      }
    }
  }
for(j in 1:J) {
  for(k in 1:K) {
    NB[j,k] <- sum(y[,seg[j],k]) # Number in seg j at time k
    # removal model:
    n[j,1,k] ~ dbin(p, NB[j,k])
    NB2[j,k] <- NB[j,k] - n[j,1,k]
    n[j,2,k] ~ dbin(p, NB2[j,k])
    NB3[j,k] <- NB2[j,k] - n[j,2,k]
    n[j,3,k] ~ dbin(p, NB3[j,k])
    }
  }
N[1] <- sum(z[,1]) # Total abundance, occasion 1
N[2] <- sum(z[,2]) # Total abundance, occasion 2
N[3] <- sum(z[,3]) # Total abundance, occasion 3
}
```

PANEL 18.3

BUGS description of model for the data shown in Figure 18.4. The model allows for spatially-explicit temporary emigration, and for a decrease in abundance as individuals move underground throughout the course of the season.

individuals move underground as the streams dry. Each of these components is included in the **BUGS** description of the model presented in Panel 18.3.

We fit this model to the data and obtained the posterior distributions summarized in Table 18.3. The results indicate that the population size available for detection did decrease rapidly during the season, the rate of which is determined by the ϕ parameter. Modeling this parameter as a function of water flow or volume would allow one to

Table 18.3 Posterior summarizes from removal model of salamander counts allowing for movement and decreasing population size over the course of a breeding season.

Parameter	Mean	SD	2.5%	50%	97.5%
N_1	178.393	16.346	151.000	177.000	214.000
N_2	62.322	6.884	51.000	62.000	77.000
N_3	21.202	3.695	15.000	21.000	29.000
ϕ	0.348	0.038	0.275	0.348	0.425
τ	27.427	3.200	21.293	27.173	33.706
p	0.396	0.053	0.294	0.394	0.502

predict salamander activity under future environmental conditions. Another result of the analysis is that the movement parameter, τ, was relatively low, indicating that adult salamanders rarely move more than 100 m from their home range center during a season. This explains why the distribution of individuals within the stream remains relatively constant over time.

18.7 Summary and outlook

Unlike traditional models of count data used in ecology, the SC model is parameterized in terms of *individuals*—individuals that just so happen not to be marked. Although developing the model in terms of latent encounters increases model complexity, several reasons exist for accommodating this latent structure. First, it allows for individual-level covariates, including the location of each individual in the population. Second, by including an underlying point process specific to individual activity centers (and possibly locations in time), the model allows for modeling continuous variation in density, which, if ignored, may result in bias in conventional models of count data. Third, accommodating the latent structure provides a more mechanistic description of ecological systems, for example by attaching a mechanism (movement) to the widely observed phenomenon of spatial correlation in count data.

The SC model is a conceptually simple extension of standard SCR models, but in terms of computational requirements and latent structure, it is perhaps at the extreme end of what is possible to do with count data. As is always true, the harder we try to mirror reality with our models, the harder it becomes to estimate the parameters of the system. In this chapter, we tried to emphasize that as conceptually appealing as the SC model may be, it is unlikely to produce satisfying results in the absence of additional information. However, additional information such as home range size estimates will often be available for many species, and if not, we have provided an alternative method of accommodating additional data in the form of distance measurements or removal counts. This can greatly increase precision of estimates from studies designed to make spatially explicit inferences about population processes.

Although we focused on the situation in which all individuals are unmarked, one of the reasons for developing this class of models was to handle the problem of estimating population size or density when only a subset of individuals are marked. This topic is the subject of the next chapter, which builds on a rich literature of combining data from marked and unmarked individuals for purposes of designing efficient studies Bartmann et al. (1987), Neal et al. (1993), Conroy et al. (2008) and McClintock and White (2009).

CHAPTER

Spatial Mark-Resight Models

19

Throughout most of this book we have dealt with the situation where all individuals are identifiable upon encounter because they carry some form of individual mark. In Chapter 18 we introduced and developed an SCR model for non-identifiable populations. These two extremes are common in the study of animal populations with non-invasive sampling methods. However, there is also an intermediate situation where part of the population is tagged or otherwise marked and can thus be identified upon recapture, while the unmarked portion remains unidentifiable. In this situation, so-called mark-resight models (Bartmann et al., 1987; Arnason et al., 1991; Neal et al., 1993; Bowden and Kufeld, 1995) can be used to estimate population size and density by combining data from both the marked and unmarked individuals.

Traditionally, capture-recapture studies involved physical capture and marking of individuals throughout the study. This methodology is still widely applied in the study of species that are relatively easy to capture, such as small mammals, but can be very costly, logistically challenging, and risky when dealing with larger species. In contrast, in mark-resight studies a sample of individuals is captured and tagged (or otherwise marked) during a single marking event (in some cases, part of the population may have recognizable natural marks). Marking is followed by resighting surveys, in which both the detection of marked and unmarked animals is recorded. Resighting surveys are usually non-invasive (hence the name "resighting"), so that they don't involve handling of animals. As such, mark-resight models have a major advantage over traditional capture-recapture models in that they only require individuals to be captured and handled once, during the initial marking. This reduces field costs and risks for the animals (and potentially the researchers).

Mark-resight models have a set of underlying assumptions, most of which are identical to those of capture-recapture models; e.g., demographic population closure (violation of geographic population closure can be accommodated by some models) and no loss or misidentification of marks (see also Chapter 5). Just like standard capture-recapture models, there are means to incorporate heterogeneity in capture probability. An essential assumption of mark-resight models is that the marked individuals are a representative sample of the study population, so that inference about detection can be made for the whole population from the marked sample. While this is also an implicit assumption of capture-recapture models, in mark-resight models this means that the process of marking individuals requires careful consideration in order

Spatial Capture-Recapture. http://dx.doi.org/10.1016/B978-0-12-405939-9.00019-0

to produce a random sample. This assumption is usually addressed by employing a different method for marking than for resighting.

Owing to the advantages of mark-resight over capture-recapture, especially when dealing with hard-to-trap species, mark-resight is a popular tool in wildlife population studies. The method has been applied for decades to a suite of species and survey techniques, ranging from banding and resighting Canada geese (Hestbeck and Malecki, 1989) to ear-tagging and camera trapping grizzly bears (Mace et al., 1994) to paintball marking and areal resightings of large ungulates (Skalski et al., 2005).

In this chapter we consider mark-resight within a spatial context and develop a spatial mark-resight (SMR) model. To motivate this model development, imagine you conduct a live-trapping study during which you capture and mark a number of animals with individually recognizable tags. Subsequently, you go back out to the field and conduct resighting surveys on an array of locations, and during these resighting surveys you see some of your marked individuals, as well as new, unmarked ones. Then, for the marked animals you obtain the same type of spatially explicit individual encounter histories as you would in a standard SCR study. In addition, you obtain site (and occasion) specific counts of individuals you did not mark. SMR models make use of both the encounter history data from the marked individuals and the counts of unmarked individuals to estimate density and detection parameters.

In the following sections we first provide some background information on mark-resight and the types of data such surveys can provide. We will further explore the implications of the assumption of the marked individuals being a random subset of the population, which, in the context of SMR models refers to not only the *demographic composition*, but also to the *spatial distribution* of the marked individuals in the state-space $\mathcal{S}$. In many real-life sampling situations, this assumption will not hold—animals will most often be marked in some region that does not represent the entire state-space. As a result, the distribution of marked individuals will generally *not* follow a homogeneous point process, but their activity centers will be concentrated in the vicinity of where marking took place. For the sake of model development, however, throughout the central part of this chapter we will make the assumption that marked animals are a random sample from the population in $\mathcal{S}$. We will show that SMR models are hybrids of standard SCR models and the models presented in Chapter 18 for data where individuals cannot be uniquely identified. We explore models for both known and unknown numbers of marked individuals, and for imperfect individual identification of marks, and approaches to incorporate telemetry location data. In the spatial framework, most of the information on model parameters comes from the marked individuals. But in Section 19.5 we will see that, analogous to the models we developed previously in Chapter 18, the spatial correlation in counts of unmarked individuals also contributes information about detection and movement. We conclude the chapter by presenting some general strategies for addressing a situation where marked individuals are not a random sample from $\mathcal{S}$.

19.1 Background

Before we start exploring spatial mark-resight approaches in more detail, we need to establish some terminology and gain a clear understanding of what types of mark-resight data we can have, in order to appreciate and understand the different flavors of mark-resight models.

19.1.1 Resighting techniques

As with capture-recapture surveys, there are numerous methods suitable for obtaining resightings. Possible methods are visits to a set of points for resightings by an observer, or camera trapping. In many mark resight applications resightings are not restricted to a particular set of locations. Rather, they may resemble search-encounter kind of methods, where a certain area is searched, systematically or opportunistically, for marked animals (see Chapter 15), or methods in which the accurate location of an individual may be difficult to ascertain (consider, for example, resightings at large distances). In this chapter we will only deal with fixed location resighting surveys assuming that resighting locations adequately mirror locations of individuals, and we will refer to the set of resighting locations as the resighting array. In some instances we will also be concerned with the locations of marking events and we refer to these locations as the marking locations.

19.1.2 Types of mark-resighting data

In general, we have (at least) two sets of data: encounter histories for marked, and thus, identifiable individuals i at resighting location j and occasion k, y_{ijk}, and counts of unmarked records, n_{jk}, for each resighting location j and occasion k. Depending on the sampling technique, we can conceive of three slightly different types of partial ID data:

(1) *Known number of marked individuals:* If you implement a resighting survey shortly after the marking session, you may be confident that none of the marked individuals have died or lost their marks. Under these circumstances you know that the number of marked individuals available for resighting, m, is equal to the number of individuals you marked. Alternatively, the marking technique might involve radio-transmitters, allowing you to confirm the presence or absence of marked individuals in the resighting survey area using radio-telemetry (White and Shenk, 2001). In both cases, you know the number of marked individuals in the surveyed population. In this situation, even though you may fail to resight some of the marked individuals, you know how many there are, and so you can simply assign all-zero encounter histories to the marked individuals not encountered—in other words, contrary to regular capture-recapture models, in mark-resight models with a known number of marked individuals, we can observe all-zero encounter histories. Under these circumstances, estimating N reduces to estimating the number of unmarked individuals, U.

(2) *Unknown number of marked individuals:* If m is not known, for example because we suspect that some of the marks may have been lost between tagging and conducting the resighting surveys, we obtain a slightly different type of mark-resight data. Here, we do not accurately know the number of marked individuals available for resighting. As a consequence, individuals have to be resighted at least once for us to know they are still marked and alive and thus available for resighting. So, contrary to the situation where we know m and analogous to regular capture-recapture models, we cannot observe all-zero encounter histories of marked individual. In this situation, estimating N involves estimating both m and U.

A special case of this kind of data can arise from camera trapping. Even when dealing with a species that has no spots or stripes, some individuals in the study population can have natural marks that make them identifiable on pictures, such as scars or a distinct coloration. In this scenario an individual has to be photographed at least once to be known. Here, the fact that both the "marking" method and the subsequent resighting method are the same (although marking in this case does not involve any actual physical marking) can be cause for concern: our sample of "marked" individuals may not be a random sample of the population but consist of individuals that for some reason are more likely to be photographed. In that case, a basic assumption of the mark-resight model is violated.

(3) *Unknown marked status:* Finally, consider a situation where we cannot always ascertain the marking status of an individual upon resighting. The part of the individual that holds the mark (e.g., the neck in case of a collar, or the leg in case of a leg band, etc.) may be hidden from our view or not exposed on a camera trap picture. In this scenario, n_{jk} can contain both completely new individuals that are not represented at all in the set of encounter histories of marked animals, $\mathbf{Y}$, but it can also contain records of individuals that we previously identified. This type of data violates one of the basic assumptions of mark-resight models, namely, that marked individuals are always correctly identified as such.

To our knowledge there are currently no mark-resight models available that account for possible misidentification of the marking status of individuals (although some literature is available on misidentification of individuals in capture-recapture studies, e.g., Lukacs and Burnham, 2005; Yoshizaki et al., 2009; Link et al., 2010). In this chapter we will ignore this kind of data and focus instead on types (1) and (2).

For both types of data a slightly different situation arises when we can only tell that an individual is marked, but not who it is. You may be able to see that an individual is marked but the identifying feature of the tag (a number or coloration) may have become unreadable, or may be hidden from view. In this case, in addition to the observed y_{ijk} and n_{jk}, you also observe a number of sightings of marked but unidentified individuals, say r_{jk}.

19.1.3 A short history of mark-resight models

Initially, mark-resight methods focused on radio-tagged individuals to estimate population size (White and Shenk, 2001). Radio-collars provide a means of determining which of the animals is in the study area and available for sampling, allowing us to identify the number of marked individuals in the population. Knowing this number was a prerequisite for most earlier mark-resight approaches (White, 1996). The oldest mark-resight model is the good old Lincoln-Petersen estimator, where individuals are marked and a single resight/recapture occasion is carried out (Krebs, 1999). We need not identify individuals, but only tell apart marked from unmarked individuals. Let m be the number of marked individuals in the population, $m_{(R)}$ the number of marked individuals seen on the resighting occasion, and $n_{(R)}$ the total number of marked and unmarked individuals observed during resighting. Population size N is then estimated as

$$N = m \times n_{(R)}/m_{(R)}.$$

Mark-resight models using individual capture histories over several resighting occasions were developed in the 1980s and 1990s and compiled into the program **NOREMARK** (White, 1996). Apart from the basic model with known number of marked individuals and no individual variation in resighting probabilities (joint hypergeometric maximum likelihood estimator) (Bartmann et al., 1987; Neal, 1990; White and Garrot, 1990; Neal et al., 1993), **NOREMARK** contains models that account for lack of geographic population closure (Neal et al., 1993), individual heterogeneity in resighting rates, and sampling with replacement (i.e., individuals can be seen more than once on any occasion, (Minta and Mangel, 1989; Bowden and Kufeld, 1995)). A first mark-resight model allowing for an unknown number of marked individuals was developed by Arnason et al. (1991).

While many of these models perform well under certain situations, they are somewhat limited in that they do not allow for combining data across several surveys (McClintock et al., 2006) and not all of them are likelihood based or allow for different parameterizations (e.g., including a time effect on detection), so that selection of the most appropriate model cannot be based on standard approaches such as AIC, but is largely left up to educated guesswork (McClintock et al., 2006). Recently, more flexible and generalized likelihood-based mark-resight models have been developed. These models can account for individual heterogeneity in detection, unknown number of marked individuals and lack of geographic closure, as well as imperfect individual identification of marked individuals; they can be applied to sampling with and without replacement and can combine data across several primary sampling occasions in a robust design type of analysis (McClintock et al., 2009a,b; McClintock and White, 2009). Since they are all likelihood based, model selection among different parameterizations and model averaging based on AIC is possible. Most of these models have also been incorporated into the program **MARK** (McClintock and White, 2012).

For a detailed treatment of these different non-spatial mark-resight models, we refer you to White and Shenk (2001) and McClintock et al. (2012). In short, these models are based on the joint likelihood of two model components: one describing the resighting process of marked individuals and one describing the number of unmarked individuals observed. The resighting process of marked individuals can use either a Poisson or a Bernoulli observation model, depending on whether sampling is with or without replacement, and the resighting probabilities can have both fixed effects to model individual and environmental covariates, and a random-effect component to accommodate variation in detection due to individual heterogeneity. The process describing the number of unmarked individuals observed (or, under a Poisson observation model, the number of times unmarked individuals are observed), n_t (t here and in the following description denotes a primary sampling occasion, for example, a year or a season) is approximated as a normal distribution (McClintock et al., 2006), or a normal distribution left-truncated at 0 (McClintock et al., 2009a):

$$n_t \sim \text{Normal}(\mathbb{E}(n_t), \text{Var}(n_t)).$$

For a single-season study, the t subscript does not need to be included. Although this is a simplification of the actual sampling process, McClintock et al. (2006) found this normal distribution to be a satisfactory approximation, which allows N to enter the model likelihood via $\mathbb{E}(n_t)$ and $\text{Var}(n_t)$.

In the simplest model without any variation in detection, the expected number of resightings of unmarked individuals, $\mathbb{E}(n_t)$, can be written as the number of unmarked individuals times the expected number of detections of a single individual. This is the mean or expected value of the underlying observation model:

$$\mathbb{E}(n_t) = (N - m) \times \theta, \tag{19.1.1}$$

where $\theta = K \times p$ for a Binomial observation model with K replicates and individual encounter probability p, or θ = expected individual encounter rate λ for a Poisson observation model. Similarly, $\text{Var}(n_t)$ depends on the underlying observation model and is based on the parameters that determine the individual encounter probability or rate. Combining these two components, N is directly incorporated into the joint likelihood of the model.

While these mark-resight models are very flexible, they share the shortcomings of traditional capture-recapture models when it comes to estimating population density (e.g., Chapters 1 and 4). As long as resightings are collected across a number of locations, however, they come with the same spatial information as (re)captures in a standard SCR study. In the following sections we will consider mark-resight sampling in the framework of spatial capture-recapture.

19.1.4 The random sample assumption

In mark-resight studies it is a prerequisite that the marked portion of the studied population is a random sample of that population, so that encounter probability for the population can be adequately estimated from the marked subset. If, for example, there is some latent group structure in the population where one group has a higher encounter probability than the other, the marked portion of the population should have the same composition with regard to this group structure as the study population. Intuitively, people think of this as a demographic problem. But this assumption also has spatial implications. In a non-spatial mark-resight survey, the study area is defined by the area exposed to marking efforts, and we need to mark a random sample of individuals from the population inhabiting that area. As in non-spatial capture-recapture, the difficulties with this approach lie in defining the area exposed to marking. We have claimed repeatedly that, at least for capture-recapture, the answer to this problem is to explicitly include space into the model, i.e., move to spatial capture-recapture. In spatial mark-resight, however, this turns out not to be as straight forward. The assumption that marked individuals are a random subset, demographically and spatially, from the study population, manifests itself in a peculiar manner in SMR models, for two reasons: (1) we define the spatial context of the population by setting a state-space; (2) we assume a certain distribution or point process for all individuals within that state-space, in most cases a uniform distribution or homogeneous point process (but see Chapter 11 for models with inhomogeneous spatial point processes). For the marked individuals to follow a homogeneous point process (i.e., be a random spatial subset of the population) in $\mathcal{S}$, marking must be done uniformly throughout the state-space. When we study a species where some individuals can be identified based on natural marks, while others do not have unique marks (for example, regular colored versus melanistic leopards), and we can assume that the distribution of these two groups of individuals across $\mathcal{S}$ is identical, then we can frame the estimation problem in terms of estimating the density of two homogeneous point processes, one for the marked and one for the unmarked population. But what if we actively need to mark individuals in order to distinguish them? Then, we have two options: (1) if we want to meet the random sample assumption, then the definition of $\mathcal{S}$ becomes part of our study design (contrary to SCR models, where $\mathcal{S}$ is set after data collection for analysis purposes); (2) if we don't want to, or cannot, meet the random sample assumption, we have to specify an alternative model that adequately describes the distribution of marked and unmarked individuals in $\mathcal{S}$.

Here is another way to think about this: In SCR models, once the state-space is chosen large enough, estimates of density are no longer sensitive to the size of $\mathcal{S}$, because N scales with the area of $\mathcal{S}$. In spatial mark-resight, however, our population of individuals consists of two groups, marked and unmarked. Consider the case where we have a known number of marks. Because we fix the size of the marked part of our population, total population size N no longer scales with the area of the state-space. While the number of unmarked individuals can go up as $\mathcal{S}$ increases in size, m is fixed by design, and thus, as $\mathcal{S}$ increases, overall density will decrease.

If we want to make sure, by design, that marked individuals are a random sample from $\mathcal{S}$, then, in practical terms, we need to define the state-space, which includes the resighting array plus a sufficient buffer to include all animals potentially exposed to this array, and uniformly mark individuals throughout $\mathcal{S}$. This does not mean that we necessarily have to achieve complete coverage of $\mathcal{S}$ with our marking effort; alternatively, we could also randomly distribute traps in $\mathcal{S}$ in order to randomly distribute marks throughout $\mathcal{S}$. We can see some sampling situations in which such a scenario might be reasonable, or at least reasonably approximated. For example, later on in this chapter we present a study where raccoons were caught and marked throughout an island, the boundaries of which are a natural limit for the state-space of this particular system.

For many studies, however, this might not be the case. Often, marking is the more difficult and logistically challenging part of a mark-resight study—think about the difficulty in physically capturing and tagging large carnivores. Especially for rare and cryptic species, areas over which resighting is conducted might have to be large to accumulate sufficient data, and marking over an even larger area—$\mathcal{S}$—would be logistically impossible. So what happens if we capture and mark individuals in a subset of the state-space? Then, whereas we may well have an overall constant density across $\mathcal{S}$, we will have a higher density of marked individuals in the vicinity of the marking locations—live traps, mist nets, whatever is used to catch animals—and the density of marks will generally go down as we get further away from the marking locations. As with all methods discussed in this book, the marking process of mark-resight studies also has a spatial component and induces a certain spatial distribution of marked individuals in the study area. We have to account for that when developing an SMR model. Thus, if we want to relax the assumption that marked animals are a random sample from $\mathcal{S}$, we need to describe the distribution of marked individuals' activity centers using an adequate spatial point process model. Developing a suitable point process model is one of the primary challenges when fitting SMR models, and one that at this point in time still requires substantial model development efforts. We provide some ideas on how to approach this problem in the last section of this chapter.

Although it might not be a reasonable assumption for many real-life survey situations, for now, we will continue by showing the development of SMR models assuming that the marked animals are, indeed, a random sample of N, following a homogeneous point process in $\mathcal{S}$. This simplifies the modeling problem substantially, thereby allowing us to focus on the underlying principles and possible useful extensions of SMR models.

19.2 Known number of marked individuals

We begin the model development with the simplest situation. Here, a known number of individuals constituting a random sample from the population within $\mathcal{S}$ are marked, and a series of resight surveys are conducted following marking. No marks (or marked animals) are lost between marking and resighting, all individuals are correctly identified as marked or unmarked, and marked individuals are 100% correctly identified to individual level.

Recall from Chapter 18 that without any individual identity, the observed counts at resighting location j and occasion k, n_{jk}, represent the sum of all latent individual detections at j and k, $\sum_{i=1}^{N} y_{ijk}$, where y_{ijk} is the latent encounter history of individual i at j and k. We can model these counts as

$$n_{jk} \sim \text{Poisson}(\Lambda_j),$$

where

$$\Lambda_j = \sum_{i=1}^{N} \lambda_{ij}.$$

Under this formulation, in order to carry out MCMC, we do not need to update the individual y_{ijk} in our model, which is more efficient in terms of computing. However, we can also formulate the model as conditional on the latent y_{ijk}. This is useful because if we have m marked animals in our study population, then y_{ijk} for those m individuals are no longer latent, but fully observed, and can easily be included in the analysis to provide information on detection parameters.

The formulation conditional on y_{ijk} basically brings us back to the original SCR model, where individual site and occasion-specific counts, y_{ijk}, are modeled as

$$y_{ijk} \sim \text{Poisson}(\lambda_{ij})$$

and

$$\lambda_{ij} = \lambda_0 \exp(-d_{ij}^2/(2\sigma^2)),$$

in the case of a Poisson encounter model. Unobserved y_{ijk} are treated as missing data and have to be updated as part of the MCMC procedure. We can do that by using their full conditional distribution, which is multinomial with sample size n_{jk}. For analysis of SMR models we once again make use of data augmentation, i.e., we add all-zero encounter histories to the observed y_{ijk} so that the augmented data set has size M. Let $\mathbf{u}$ be an index vector of the $M - m$ hypothetical unmarked individuals, $u = m+1, m+2, \ldots, M$, and let $\mathbf{y}_{ujk}$ be the vector of observations of *all* individuals in $\mathbf{u}$ at j and k. Then

$$\mathbf{y}_{ujk} \sim \text{Multinomial}(n_{jk}, \boldsymbol{\pi}_{uj}),$$

where $\boldsymbol{\pi}_{uj}$ is a vector of the expected encounter rates λ_{ij} (times the data augmentation variable z_i) of all unmarked individuals $\mathbf{u}$ at resighting location j, scaled to 1.

In non-spatial mark-resight models, marked individuals provide information about individual encounter probability (or rate); in a spatial framework they also inform σ, as described in Chapter 18. Including marked individuals into the analysis helps estimate model parameters more accurately and precisely. We will address the relationship between the number of marked individuals and accuracy of the estimated parameters in Section 19.5.

19.2.1 Implementing spatial mark-resight models

Implementing a spatial mark-resight model in **JAGS** is not straightforward, since the program does not accept partially observed multivariate nodes (in this case the partially observed individual encounter histories, which we model as coming from a multinomial distribution). We can, however, work around that by separating the marked from the unmarked data. The **JAGS** code for the model with a known number of marked individuals is shown in Panel 19.1. You see that data augmentation is only applied to the unmarked part of the population, and N is the sum of the estimated number of unmarked individuals (`sum(z[])`) and the number of marked individuals, which is known. Also, to reduce run time, we summed observations of marked individuals across occasions and account for that by multiplying λ_{ij} with K. Although the two data sets are separated, both parts of the population, marked and unmarked, have the same prior uniform distribution of activity centers. A last noteworthy detail in this code is the `dsum` distribution. This distribution is specific to **JAGS** (i.e., you cannot run this model in **BUGS**), and allows you to impose a sum constraint on observations. In other words, it allows you to model data—here the counts of unmarked individuals, n_{jk}—as the sum of a number of latent variables, which in this case are the latent encounter histories of unmarked individuals. While it can be a pain writing out all the arguments of dsum, it is this function that allows us to implement SMR models in **JAGS**.

Alternatively, we can use the technical concepts presented in Chapter 17 and derive our own MCMC algorithm. To do so, we only have to make relatively simple modifications to the MCMC code developed for regular SCR models in Chapter 17. Essentially, since we observe individual detections for the marked part of the population, we have to update only the unobserved part of the full—augmented—set of encounter histories, **Y**, and modify the updating steps for z_i and ψ, the parameters introduced by data augmentation, to reflect that these only apply to the unmarked part of the population, in other words, to the $M - m$ individuals in our data. You can find the full MCMC code in the accompanying **R** package scrbook by invoking scrPID. The **R** code below shows how to simulate SMR data using the scrbook function sim.pID.data, and running an SMR model on the data, both in **JAGS** and using scrPID. The model file mknown.jag in the jags.model call should contain the code from Panel 19.1:

```
> set.seed(2013)
> N = 80 # pop. size
> m <- 45 # No. marked
> sigma = 0.5
> lam0 = 0.5
> K = 5
# Make resighting array
> gx <- gy <- seq(0,6,1)
> X <- as.matrix(expand.grid(gx, gy))
> J = dim(X)[1]
> # Limits of S
> xlims <- ylims<-c(-1.5, 7.5)
# Simulate data
> dat = sim.pID.data(N = N, K = K, sigma = sigma, lam0 = lam0, knownID = m,
  X = X, xlims = xlims, ylims=ylims, obsmod='pois',nmarked='known')

### Prep data for analysis in JAGS
> n <- dat$n - apply(dat$Yknown,2:3,sum)
> y <- apply(dat$Yknown,1:2,sum)

> M <- 80 # Augmentation only for unmarked

# Initial values for latent y
> yin <- array(0, c(M,J,K))
> for(j in 1:J){
+   for(k in 1:K){
+     yin[1:M,j,k] <- rmultinom(1, n[j,k], rep(1/M, M))
+   }}

> data <- list(y=y, nU=n, m=m, M=M, J=J, X=X, xlim=xlims, ylim=ylims, K=K)
> inits <- function(){list(sigma=runif(1), lam0=runif(1),
  sm=cbind(runif(m, xlims[1], xlims[2]), runif(m, ylims[1], ylims[2])),
  s=cbind(runif(M, xlims[1], xlims[2]), runif(M, ylims[1], ylims[2])),
  z=rep(1, M),yu=yin)}
> params <- c('lam0', 'sigma', 'N', 'psi')
```

```
model{

#priors
psi ~ dbeta(1,1)
lam0 ~ dunif(0, 5)
sigma ~ dunif(0, 5)

#marked part
for(i in 1:m) {
  sm[i,1] ~ dunif(xlim[1], xlim[2])
  sm[i,2] ~ dunif(ylim[1], ylim[2])
  for(j in 1:J) {
    distm[i,j] <- sqrt((sm[i,1]-X[j,1])^2 + (sm[i,2]-X[j,2])^2)
    lambdam[i,j] <- lam0*exp(-distm[i,j]^2/(2*sigma^2))
    y[i,j]~dpois(lambdam[i,j]*K)
    }
  }

##unmarked part
for(i in 1:M) {
  z[i] ~ dbern(psi)
  s[i,1] ~ dunif(xlim[1], xlim[2])
  s[i,2] ~ dunif(ylim[1], ylim[2])
  for(j in 1:J) {
    dist[i,j] <- sqrt((s[i,1]-X[j,1])^2 + (s[i,2]-X[j,2])^2)
    lambda[i,j] <- lam0*exp(-dist[i,j]^2/(2*sigma^2))
    for(k in 1:K) {
      yu[i,j,k] ~ dpois(lambda[i,j]*z[i])
      }
    }
  }

for(j in 1:J) {
  for(k in 1:K) {nU[j,k] ~ dsum(yu[1,j,k],yu[2,j,k],yu[3,j,k],
[...code shortened...],
yu[79,j,k],yu[80,j,k])
}
  }

N <- sum(z[])+m

}
```

PANEL 19.1

JAGS model specification for SMR model with known number of marked individuals. In this example, M, the size of the augmented unmarked data set, is 80. Note that the arguments yu[4,j,k] to yu[78,j,k] of the dsum function are omitted from the code.

Table 19.1 Posterior summaries of the parameters of a spatial mark-resight model with known number of marks, analyzed in JAGS and using `scrPID`.

Implementation	Parameter	Mean	SD	2.5%	50%	97.5%
JAGS	N	88.72	6.75	77.00	88.00	103.00
	λ_0	0.53	0.08	0.39	0.53	0.70
	σ	1.29	0.02	1.26	1.30	1.32
	ψ	0.47	0.03	0.45	0.47	0.53
`scrPID`	N	86.01	7.58	73	85	102
	λ_0	0.54	0.08	0.39	0.53	0.72
	σ	0.48	0.03	0.42	0.48	0.53
	ψ	0.51	0.11	0.32	0.51	0.73

```
# Analysis in JAGS
> library(rjags)
> mod <- jags.model('mknown.jag',data, inits, n.chains = 1, n.adapt = 800)
> out <- coda.samples(mod,params, n.iter = 5000)

> # Analysis with scrbook MCMC code
> library(scrbook)
> library(coda)
> inits2 <- function(){list(psi=runif(1), sigma=0.5, lam0=0.5,
  S=cbind(runif(M+m, xlims[1],xlims[2]), runif(M+m, ylims[1],ylims[2])))}
> out2 <- scrPID(n=n, X=X, y=dat$Yknown, M=M+m, obsmod = "pois", niters=5800,
  xlims=xlims, ylims=ylims,inits=inits2(),delta=c(0.1,0.1,0.5))
```

You can look at the two sets of output by invoking `summary(out)` for the **JAGS** analysis and `summary(window(mcmc(out2),start = 801))` for the custom MCMC algorithm, excluding the first 800 iterations as burn-in. We summarized the results in Table 19.1. The posterior mean of N is slightly higher than the data-generating value of $N = 80$, but it falls comfortably within the credible intervals. As expected, estimates from both implementations are very similar; slight differences are probably the result of Monte Carlo error due to the relatively low number of iterations. You will find that sometimes, **JAGS** produces an error message upon trying to compile the model, saying that some of the observed y are inconsistent with parent nodes at initialization. We have mentioned before that **JAGS** cannot always auto-generate acceptable initial values, and we believe this is what is happening here. If this error occurs, just repeat the `jags.model` command, usually, model compilation is successful on a second attempt (assuming, of course, that you followed the code above correctly). We further find that the custom MCMC algorithm tends to be faster than **JAGS**, which is why the examples and simulation studies shown in the following sections were run solely in **R**.

19.3 Unknown number of marked individuals

Now let us consider the case where we do *not* know the exact number of marked individuals available for resighting so that we have to observe an individual at least once to be sure that it is available. Unless we have a direct means of confirming the number of marked animals available for resighting, treating this number as unknown is probably more realistic in most circumstances. As a consequence of not knowing the exact number of marked individuals, we cannot observe all-zero encounter histories. When using maximum likelihood inference, this situation requires a model where detection rates of known individuals are modeled using a zero-truncated distribution (McClintock et al., 2009a). If we did not account for the fact that zeros are unobservable, estimates of detection rates would be artificially inflated and estimates of population size would be negatively biased.

Working with zero-truncated distributions in a spatial mark-resight setting is less straightforward than for non-spatial mark-resight. A marked individual only has to show up once, anywhere on the resighting array, for us to know that it is there. When resightings are pooled across the entire sampling grid, then the total individual counts $\sum_j y_{ijk}$ have to be > 0 for all resighted individuals and a zero-truncated distribution can be used to model these counts. However, we are concerned with trap-specific encounters, y_{ijk}, which can easily be 0 for a resighted individual, as long as a single y_{ijk} is > 0. Thus, the zero truncation does not apply to the individual- and trap-specific counts we observe, but only to the sum of these counts over all traps.

As an alternative to a zero-truncated distribution, in a Bayesian framework, we can make use of data augmentation to estimate the number of marked individuals (McClintock and Hoeting, 2010). In the SMR framework that means that we create two augmented data sets, one for the marked individuals and one for the unmarked, and estimate their numbers separately, having them share the parameters of the detection model. Sometimes we may know the maximum number that were ever marked before a resighting survey, in which case we can use that number as the data augmentation limit for the marked data set. Panel 19.2 shows the **JAGS** code for the SMR model with unknown number of marks, which is identical to the one in Panel 19.1, but for the augmentation of the marked data set. This introduces both a data augmentation parameter, `psim`, and an auxiliary "alive state" variable, `zm[i]`, into the description of the marked data model. Again, we provide an alternative, **R**-only MCMC algorithm within `scrbook`, namely `scrPID.um`.

Note that we could look at the problem of not knowing the number of marked individuals in the study population as a manifestation of a lack of population closure. Marked individuals may have emigrated, died, or lost their marks in the time between marking and resighting. If we have information on the rates of these events, or a series of resighting surveys, we could develop an open population model for the marks in our population and estimate their number at a given resighting survey in this fashion. This kind of SMR model remains to be explored.

19.3.1 Canada geese in North Carolina

We applied the spatial mark-resight model with an unknown number of marks and a binomial encounter process (i.e., an encounter model where individuals can only be observed at most once during a resighting occasion at a given location; see Section 5.2) to a dataset of Canada goose resightings (Rutledge, 2013). During the molt of

```
model{

# Prior distributions
psim ~ dbeta(1,1)
psi ~ dbeta(1,1)
lam0 ~ dunif(0, 5)
sigma ~ dunif(0, 5)

# Marked part of the model
for(i in 1:max) {
  zm[i]~dbern(psim)
  sm[i,1] ~ dunif(xlim[1], xlim[2])
  sm[i,2] ~ dunif(ylim[1], ylim[2])
  for(j in 1:J) {
    distm[i,j] <- sqrt((sm[i,1]-X[j,1])^2 + (sm[i,2]-X[j,2])^2)
    lambdam[i,j] <- lam0*exp(-distm[i,j]^2/(2*sigma^2))*zm[i]
    y[i,j]~dpois(lambdam[i,j]*K*zm[i])
    }
  }

# Unmarked part of the model
for(i in 1:M) {
  z[i] ~ dbern(psi)
  s[i,1] ~ dunif(xlim[1], xlim[2])
  s[i,2] ~ dunif(ylim[1], ylim[2])
  for(j in 1:J) {
    dist[i,j] <- sqrt((s[i,1]-X[j,1])^2 + (s[i,2]-X[j,2])^2)
    lambda[i,j] <- lam0*exp(-dist[i,j]^2/(2*sigma^2))
    for(k in 1:K) {
      yu[i,j,k] ~ dpois(lambda[i,j]*z[i])
      }
    }
  }

for(j in 1:J) {
  for(k in 1:K) {nU[j,k] ~ dsum(yu[1,j,k],yu[2,j,k],yu[3,j,k],
[...code shortened...],
yu[79,j,k],yu[80,j,k])
}
  }

Nu <- sum(z[])
Nm<-sum(zm[])
N<-Nu+Nm

}
```

PANEL 19.2

JAGS model specification for SMR model with unknown number of marked individuals. In this example, M, the size of the augmented unmarked data set, is 80. Note that the arguments yu[4,j,k] to yu[78,j,k] of the `dsum` function are omitted from the code for space reasons.

FIGURE 19.1

Banded and unbanded Canada geese in a parking lot in Greensboro, North Carolina (*Photo credit: M.E. Rutledge, NCSU Canada Goose Project*).

2008, 751 individual geese were captured and marked with neck and leg bands in Greensboro, North Carolina (Figure 19.1). Geese were resighted at 87 locations on 81 resighting events over a period of 18 months. In addition to the banded geese, the number of unmarked geese was recorded during each resighting event. Here, we only looked at a subset of the data, from mid-July to the end of October 2008, which corresponds to the first part of the post-molt season, before migratory Canada geese arrive in North Carolina. We treated this population as closed over this period. During this part of the study, 57 of the resighting sites were visited and $n = 654$ marked geese were resighted 3,994 times at 40 different sites. In addition, 7,944 sightings of unmarked geese were recorded at 48 sites.

In the model, we allowed σ to vary between males and females. We set the size of the augmented unmarked data set to 7,000. We used the total number of marked geese (751) as the upper limit for the augmented marked data set. We ran 50,000 MCMC iterations and removed a burn-in of 5,000 iterations. To describe the state-space, we buffered the resighting locations by 4.5 km. We assumed that marked geese

Table 19.2 Posterior summaries of parameters of the spatial mark-resight model for Canada geese in North Carolina. N is the total population size of marked and unmarked individuals; m is the number of marked individuals.

	Mean	SD	2.5%	50%	97.5%
m	739.77	3.24	733.00	740.00	746.00
N	5,756.10	90.68	5,577.00	5,757.00	5,932.00
D	13.76	0.19	13.38	13.76	14.14
λ_0	0.19	<0.01	0.18	0.19	0.19
σ, females	1.29	0.02	1.26	1.30	1.32
σ, males	1.06	0.02	1.02	1.06	1.11
ψ, marked	0.99	<0.01	0.98	0.99	0.99
ψ, unmarked	0.72	0.01	0.69	0.72	0.74
ϕ	0.36	0.02	0.32	0.36	0.39

were a random sample from the state-space, which seems reasonable for this particular study because (a) marking took place across most of the extent of the resighting array; and (b) marking was done during the molting period, when geese are fairly immobile, and it seems reasonable to assume that, once the molt is complete, the marked geese redistributed themselves. Note that under this model formulation, estimates of density will be sensitive to the choice of the state-space (19.1.4), so that we face a similar problem as in non-spatial mark-resight, where we have to define a sampled area in order to obtain density estimates. Development of models that do not suffer from these shortcomings is ongoing (e.g., see Section 19.7). We provide all the data (`data('geesedata')`) and functions (`geeseSMR`) for you to repeat this analysis, but be aware that given the large data set it will take days to do so. The **R** code to set up the data and run 5,000 iterations of the model for the geese data is given as an example on the help page for `geeseSMR`. The model results, including the derived parameter density (D) in individuals per km^2, are shown in Table 19.2.

We see that credible intervals of estimates are pretty narrow, surely an effect of the large data set. Estimates of m indicate that most of the 751 geese originally banded were still alive and marked, which is not surprising, given that not much time passed between marking and this first resighting session. The parameter ϕ in this model is the probability of being a male, a measure of the sex ratio of the population, which is slightly biased in favor of females.

19.4 Imperfect identification of marked individuals

Often during resighting, it may be possible to see that an individual is marked but impossible to determine its individual identity. In this situation, in addition to y_{ijk} and n_{jk}, we also have site and occasion-specific counts of marked but unidentified individuals, r_{jk}. Here, the individual encounter histories of marked animals are incomplete, and if we used these incomplete data to inform the detection parameters of the model, we would run the risk of underestimating encounter rate and overestimating abundance. Some non-spatial mark-resight models do not require that marked animals be identified individually, as long as the marking status can be observed unambiguously, but ignoring individual-level information means that we cannot accommodate heterogeneity in detection (McClintock and White, 2012). In a spatial framework we could ignore marked and unmarked status completely and apply the model by Chandler and Royle (2013) discussed in Chapter 18. But, that would mean losing important information on individual detection and movement. Therefore, being able to retain the individual identity of records that can be identified while at the same time accounting for imperfect identification of marked individuals is extremely useful.

McClintock et al. (2009a, 2009b) suggest an ad hoc but intuitive means of correcting for this bias in a non-spatial model framework when dealing with a Poisson encounter model (a plausible model when sampling with replacement). When marked but unknown resightings are part of the data, the expected number of unmarked records, n, changes from Eq. (19.1.1) to:

$$\mathbb{E}(n) = (N - m)\lambda + \eta/m,$$

where λ is the individual encounter rate estimated from the known resighted individuals and η is the number of records of marked but unidentified individuals. So, because the observed λ is known to be too low, the average number of unidentified pictures per known individual is added as a correction factor. This procedure assumes that the inability to identify a marked individual occurs at random throughout the population, which seems to be a reasonable assumption under most circumstances. While this approach is straight forward, it has been shown to perform less than nominal when identification rate is low (i.e., <0.9) and when there are non-negligible levels of individual heterogeneity in resighting probability (White and Shenk, 2001).

We can translate this same concept to the spatial mark-resight models. In the spatial framework we are interested in the individual and trap-specific encounter rate, λ_{ij}. Further, we do not look at the sum of all records of unmarked individuals, but formulate the model conditional on the latent individual encounter histories. Thus, instead of using η/m as a correction factor, we need something that applies at the individual and trap level. If we take the sum of all correctly identified records of marked individuals, $\sum y_c$ and divide it by the total number of records of marked individuals, $\sum y_m$, we get the average rate of correct individual identification for marked individuals, say, c:

$$c = \sum y_c / \sum y_m.$$

We can then apply c as a correction factor for λ_0 for the marked individuals.

A more formal, model-based way to specify c is by assuming that

$$\sum y_c \sim \text{Binomial}\left(\sum y_m, c\right)$$

and estimating c as another model parameter, so that we account for the uncertainty about it. For the marked individuals we can then multiply λ_0 by c to account for the fact that we observe incomplete individual encounter histories. Since we don't have this identification issue for unmarked individuals, their baseline trap encounter rate remains as before simply λ_0 (or in other words, c for unmarked individuals equals 1).

Incomplete individual identification of marked individuals is easily incorporated into our **JAGS** model, no matter whether m known or unknown, by adding the following two lines of code:

```
c ~ dbeta(1,1) #prior for c
npics[1] ~ dbin(c, npics[2]) #model for c
```

and modifying the marked observation model description to

```
y[i,j] ~ dpois(lambdam[i,j]*c*K)
```

Here, the data object `npics` is a vector with the number of correctly identified records of marked individuals and the total number of marked records. Accounting for imperfect identification of marks is also included as an option in the `scrPID` and `scrPID.um` functions. Choosing an uninformative (and conjugate) Beta(1, 1) prior for c, within the `scrPID` algorithm, we can update c directly from its full conditional distribution, which is $\text{Beta}(1 + \sum y_c, 1 + (\sum y_m - \sum y_c))$. We show an example of using c in an analysis in Section 19.6.

Observe that now, in addition to assuming that failure to identify marked individuals occurs at random throughout the population, we also assume that it occurs at random throughout space, i.e., our success of identifying a marked individual does not depend on the trap we encounter it in. As long as individuals are identified based on the same type of tags this assumption should be valid. Identification of marked individuals could under certain circumstances not be spatially random, for example when varying habitat conditions influence the ability to recognize individual tags, or when an observer effect influences individual identification rates. While we haven't experimented with it, we believe that the approach described above could readily be extended to account for these differences. For example, identification rates could be calculated separately for different observers, or be modeled as functions of habitat covariates. As an alternative to the approach we present here, model development could explore assigning records of marked but unidentified individuals to marked individuals in a fashion similar to how unmarked records are assigned to hypothetical individuals in this model, namely, based on the location of the record and the estimates of activity centers of marked individuals. While this is computationally more advanced it would make full use of the spatial information of the unmarked records.

19.5 How much information do marked and unmarked individuals contribute?

It is intuitive that having marked individuals in the study population should lead to more accurate and precise parameter estimates than when no individuals are identifiable. To evaluate how strongly adding marked individuals to a population improves

Table 19.3 Posterior mean, mode, and associated relative RMSE for simulations in which m of $N = 75$ individuals were marked. One hundred simulations of each case were conducted. Table taken from Chandler and Royle (2013).

	Parameter	Mean	rRMSE	Mode	rRMSE	BCI
$m = 0$	N	85.866	0.259	77.720	0.242	0.950
	λ_0	0.506	0.180	0.488	0.182	0.960
	σ	0.495	0.115	0.486	0.113	0.960
$m = 5$	N	80.898	0.184	76.360	0.182	0.970
	λ_0	0.510	0.178	0.494	0.180	0.950
	σ	0.496	0.089	0.488	0.086	0.970
$m = 15$	N	79.028	0.148	76.250	0.147	0.950
	λ_0	0.508	0.163	0.494	0.164	0.950
	σ	0.496	0.073	0.492	0.071	0.970
$m = 25$	N	77.765	0.114	75.810	0.113	0.950
	λ_0	0.511	0.153	0.498	0.157	0.950
	σ	0.496	0.067	0.493	0.065	0.940
$m = 35$	N	76.446	0.085	74.900	0.085	1.000
	λ_0	0.513	0.142	0.501	0.144	0.950
	σ	0.497	0.056	0.493	0.057	0.940

parameter estimates, Chandler and Royle (2013) performed a simulation study. They used a 15×15 resighting grid and simulated detection data of $N = 75$ individuals in a 20×20 units state-space over $K = 5$ occasions with $\sigma = 0.5$ and $\lambda_0 = 0.5$. They generated 100 data sets each for $m = (0, 5, 15, 25, 35)$, where m is the known number of marked individuals randomly sampled from the population.

Without any marked individuals in the population, the posterior distribution of N turned out to be highly skewed, but the mode was still an approximately (frequentist) unbiased point estimator of N. As anticipated, posterior precision increased substantially with the proportion of marked individuals (Table 19.3 and Figure 19.2). The relative root-mean squared error decreased from 0.246 when no individuals were marked to 0.085 when 35 individuals were marked (Table 19.3). Coverage was nominal for all values of m and posterior skew greatly diminished with increasing m (Table 19.3).

As we saw in the previous chapter, the spatial correlation in unmarked counts can be sufficient to obtain estimates of movement and detection parameters. However, only marked and thus identifiable individuals provide us with direct information about these parameters and may well dominate estimates. To single out the contribution of marked and unmarked individuals to parameter estimates, we reran the same simulations but let σ and λ_0 be updated based solely on the data of marked individuals. Results are summarized in Table 19.4. We see that if we update λ_0 and σ based on marked individuals only, estimates of these parameters are more biased and less precise. For estimates of N, especially for $m = 5$ and $m = 15$, we observe a stronger positive bias, lower

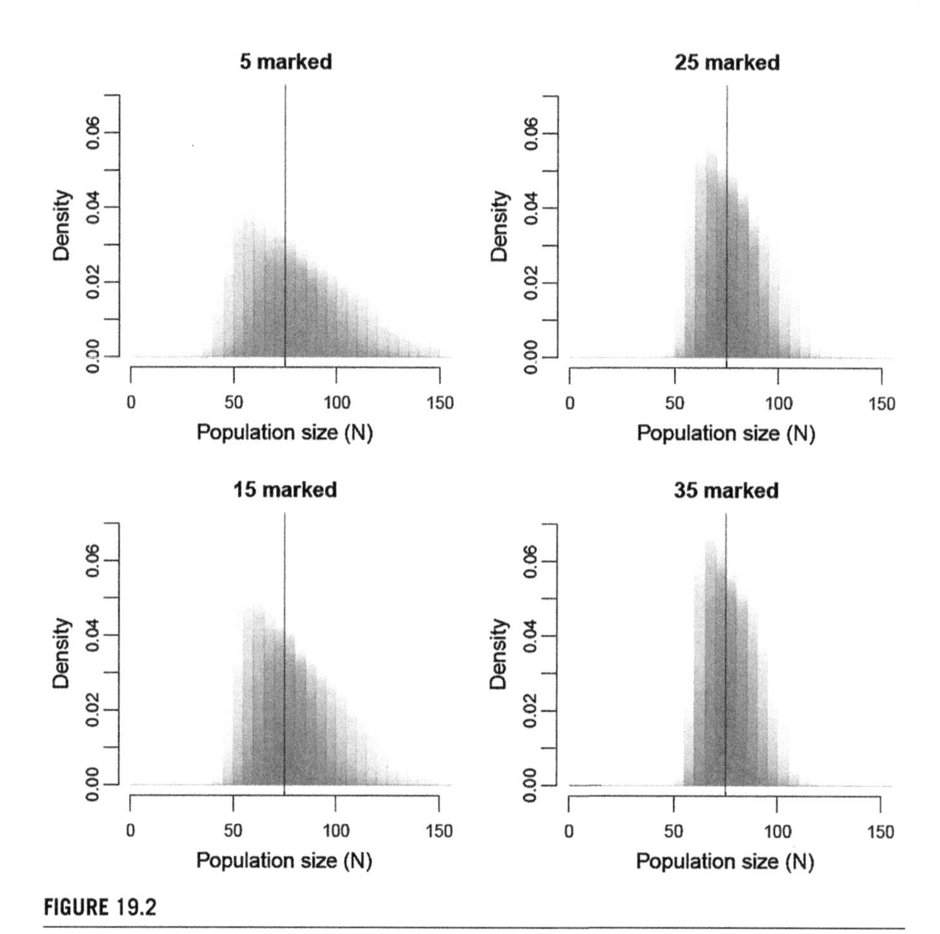

FIGURE 19.2

Overlaid posterior distributions of N from 100 simulations for four levels of marked individuals.

accuracy, and considerably lower BCI coverage as compared to when both marked and unmarked individuals contribute to parameter estimates (Table 19.4). Thus, unmarked individuals do actually contribute noticeably to estimating model parameters.

19.6 Incorporating telemetry data

As expected, parameter estimates of spatial mark-resight models get better the more marked individuals we have in our study population. While this is great advice in theory, it may not be very helpful in practice, especially when dealing with animals that are hard or somewhat dangerous to capture, such as large carnivores. Oftentimes, studies involving the physical capture of such animals will employ telemetry tags in order to learn about the study species' spatial ecology and behavior. In the context of spatial mark-resight models, the actual locations collected by telemetry tags can

Table 19.4 Posterior mean, mode, and associated relative RMSE for simulations in which m of $N = 75$ individuals were marked and unmarked individuals did not contribute to estimating λ_0 and σ. One hundred simulations of each case were conducted.

	Parameter	Mean	RMSE	Mode	RMSE	BCI
$m = 5$	N	88.621	0.369	83.139	0.421	0.810
	λ_0	1.255	1.247	0.606	1.148	0.950
	σ	0.472	0.252	0.426	0.333	0.910
$m = 15$	N	81.031	0.192	78.361	0.175	0.820
	λ_0	0.535	0.281	0.476	0.284	0.970
	σ	0.503	0.109	0.490	0.107	0.940
$m = 25$	N	78.206	0.129	76.594	0.123	0.920
	λ_0	0.531	0.204	0.496	0.202	0.960
	σ	0.497	0.081	0.489	0.084	0.950
$m = 35$	N	76.833	0.099	75.422	0.096	0.940
	λ_0	0.528	0.192	0.505	0.186	0.940
	σ	0.499	0.069	0.493	0.070	0.960

provide detailed information on individual location and movement, and being able to incorporate this information directly into the SMR model should improve estimates of these parameters, especially when resighting information is sparse.

So how could we combine resighting data and telemetry data in a unified mark-resight model? Recall that the basic SCR model underlying all the SMR models we discuss here uses a Gaussian kernel to describe the trap encounter model. By using this function, we can relate the parameters σ and $\mathbf{s}_i$ directly to those from a bivariate normal model of space usage, with mean $= \mathbf{s}_i$, variance in both dimensions $= \sigma^2$, and covariance $= 0$. Ordinarily, these parameters are estimated directly from the spatial distribution of individual captures/resightings. Telemetry data, however, provide more detailed information on individual location and movement, since the resolution and extent of the data are not limited by the trapping grid and potentially more locations can be accumulated through telemetry than resighting (depending on the monitoring frequency and resighting rates of individuals).

By assuming that the R_i locations of individual i, $\mathbf{l}_i$ (each consisting of a pair of x and y coordinates, l_{irx} and l_{iry}), are a bivariate normal (BVN) random variable:

$$\mathbf{l}_i \sim \text{BVN}(\mathbf{s}_i, \boldsymbol{\sigma}^2\mathbf{I}),$$

we can estimate σ as well as $\mathbf{s}_i$ for the collared individuals directly from telemetry locations, using their full conditional distributions:

$$[\sigma|\mathbf{l}, \mathbf{s}] \propto \left\{ \prod_{i=1}^{m} \prod_{r=1}^{R_i} \frac{1}{2\pi\sigma^2} \exp\left(-1/2 \left[\frac{(l_{irx} - s_{ix})^2}{\sigma^2} + \frac{(l_{iry} - s_{iy})^2}{\sigma^2} \right]\right) \right\} \times [\sigma]$$

and

$$[\mathbf{s}_i|\mathbf{l}, \sigma] \propto \left\{ \prod_{r=1}^{R_i} \frac{1}{2\pi\sigma^2} \exp\left(-1/2\left[\frac{(l_{irx} - s_{ix})^2}{\sigma^2} + \frac{(l_{iry} - s_{iy})^2}{\sigma^2}\right]\right) \right\} \times [\mathbf{s}_i].$$

For the unmarked individuals $\mathbf{s}_i$ are estimated as described before, conditional on their latent encounter histories. Note that the bivariate normal model assumes that locations are independent of each other. If you have frequent telemetry fixes, for example from GPS collars that report animal locations every few hours or more, this assumption seems unrealistic and it might be advisable to thin your telemetry data (maybe to daily fixes) in order to approximate independence. Alternatively, movement models could be used that acknowledge the temporal correlation in location data. We suggested some possible movement models in Chapter 15. Not all marked individuals need to be telemetry-tagged, but telemetry data should correspond to the period over which resighting surveys were conducted (as we discussed in Chapter 5, both $\mathbf{s}_i$ and σ should only be interpreted against the specific sampling period). Further, telemetry data need to be independent of the resighting data.

Implementation of this model extension is straightforward, both in **JAGS** and **R**. Take the SMR model description for the case where m is known (Panel 19.1). Then, all we have to do is add a description of the bivariate normal model for the telemetry locations, here `locs`, into the loop over the m marked individuals:

```
[...parts of model code omitted...]

for(i in 1:m) {

 sm[i,1] ~ dunif(xlim[1], xlim[2])
 sm[i,2] ~ dunif(ylim[1], ylim[2])

 #telemetry model
 for (r in off1[i]:off2[i]){
 locs[r,1]~dnorm(sm[i,1], 1/(sigma^2))
 locs[r,2]~dnorm(sm[i,2], 1/(sigma^2))
 }

 for(j in 1:J) {
  distm[i,j] <- sqrt((sm[i,1]-X[j,1])^2 + (sm[i,2]-X[j,2])^2)
  lambdam[i,j] <- lam0*exp(-distm[i,j]^2/(2*sigma^2))
  y[i,j]~dpois(lambdam[i,j]*K)
  }
 }

[...parts of model code omitted...]
```

The data object `locs` is a table with all $\sum_{i=1}^{m} R_i$ telemetry locations. The vectors `off1` and `off2` describe which subset of this matrix belongs to individual i. So if, say, the locations for individual 1 are contained in the first 10 rows of `locs`, `off1` and `off2` would be 1 and 10 for $i = 1$; and if the locations of individual 2 are in the following 15 rows, `off1` and `off2` for $i = 2$ would be 11 and 25, and so on. For the implementation of this SMR model with telemetry data in **R**, see the `scrPID.tel` function in `scrbook`. In a nutshell, in the MCMC algorithm we replaced the Metropolis-Hastings updating steps for σ and activity centers of marked individuals, which were originally conditional on the resighting data, with updating steps conditional on the telemetry data. This is not quite what the above **JAGS** code does; rather, **JAGS** will update these parameters conditional on both the telemetry *and* the resighting data. We could easily re-write `scrPID.tel` to do that, but believe that for most applications, the information on location and movement contained in the telemetry data will outweigh that in the resighting data, so that the resulting loss of information should be minimal.

19.6.1 Raccoons on the Outer Banks of North Carolina

Sollmann et al. (2013a) applied a spatial mark-resight model with telemetry data to a camera trap and radio-telemetry data set from the raccoon population on South Core Banks, a barrier island within Cape Lookout National Seashore, North Carolina. Between May and September 2007, 131 raccoons were marked with dog collars and large individually numbered cattle tags. Individuals were marked throughout the island, so that (a) we do not have to deal with sensitivity to choice of the state-space, because it is clearly defined by nature; and (b) it is reasonable to assume that marked raccoons are a random sample of individuals from this state-space. Of the 131 tagged individuals, 44 were also equipped with radio-collars. Collared individuals were located using a VHF receiver and antenna, and their locations were estimated approximately weekly. Twenty camera traps were set up along the length of South Core Banks, and camera trapping data collected between October 1, 2007 to January 22, 2008 constituted the resighting data used in this analysis. During this period, 104 marked individuals, 38 radio-collared, were alive and available for resighting with camera traps.

The state-space $\mathcal{S}$ was the entire area of South Core Banks island. A change in the number of photocaptures over the course of the study suggested a variation of encounter rate with time. Since date recording in cameras malfunctioned, photographic records could only be assigned to the time interval between subsequent trap checks, and these intervals between checks are referred to as sampling occasions. These occasions ranged from 2 to 43 days; λ_0 was standardized to 7-day intervals and allowed to change with sampling occasion. Since not all pictures of marked raccoons could be identified to the individual level, the authors applied the correction factor c as described in Section 19.4, estimated separately for each occasion.

Camera traps recorded 117 pictures of unmarked raccoons, 33 pictures of 18 marked and identifiable raccoons, and 49 records of marked but not individually identifiable individuals (Figure 19.3). An average of 16.32 telemetry locations (SD 4.91) were collected for each of the 38 radio-collared individuals. Raccoon abundance on the island was estimated at 186.71 (SE 14.81) individuals, which translated to a density of 8.29 (SE 0.66) individuals per km^2. Parameter estimates are listed in Table 19.5.

FIGURE 19.3

Camera trap picture of a raccoon marked with a cattle tag that cannot be read to determine individual identity. Taken on South Core Banks, North Carolina (*Photo credit: Arielle Parsons*).

Table 19.5 Posterior summary statistics from spatial mark-resight model for raccoon camera trapping and telemetry data, taken from Sollmann et al. (2013a). Baseline trap encounter rate λ_0 was standardized to 7-day intervals; λ_0 and the probability of identifying a picture of a marked individual, c, were allowed to vary among the 6 sampling occasions (t); σ is estimated from telemetry data of 38 radio-collared individuals.

Parameter	Mean (SE)	2.5%	50%	97.5%
N	186.71 (14.81)	162.00	185.00	220.00
D	8.29 (0.66)	7.19	8.22	9.77
λ_0 ($t=1$)	0.24 (0.05)	0.16	0.23	0.34
λ_0 ($t=2$)	0.40 (0.08)	0.26	0.39	0.57
λ_0 ($t=3$)	0.11 (0.03)	0.06	0.11	0.17
λ_0 ($t=4$)	0.30 (0.07)	0.17	0.29	0.46
λ_0 ($t=5$)	0.03 (0.01)	0.02	0.03	0.06
λ_0 ($t=6$)	0.03 (0.01)	0.02	0.03	0.05
σ	0.49 (0.01)	0.47	0.49	0.51
c ($t=1$)	0.55 (0.09)	0.38	0.55	0.71
c ($t=2$)	0.39 (0.11)	0.18	0.39	0.62
c ($t=3$)	0.29 (0.11)	0.11	0.29	0.52
c ($t=4$)	0.38 (0.16)	0.10	0.36	0.71
c ($t=5$)	0.38 (0.16)	0.10	0.36	0.71
c ($t=6$)	0.30 (0.14)	0.08	0.29	0.60

In this study, although a large number of raccoons were tagged, photographic data of these tagged individuals were surprisingly sparse. Analysis of the photographic data set without the telemetry data did not render usable estimates as parallel Markov chains did not converge. One reason for the relatively sparse data was the camera trap study design: traps were spaced on average 1.77 km apart, which is about 3.5 times σ. Consequently, very few individual raccoons were photographed at more than one trap. Under these circumstances, the telemetry data provide the necessary spatial information to estimate σ and the activity centers of individual animals and thus make other model parameters estimable. Similarly, in a camera trapping study on Florida panthers (*Puma concolor coryi*), Sollmann et al. (2013c), including telemetry data from the three individuals that were collared and known to use the study area, resulted in density estimates with considerably higher precision as compared to preliminary estimates *without* telemetry location data, reducing the width of the 95% BCI by about 60%. Such improvements in precision of estimates are especially important when we are interested in changes in the population over time.

19.7 Point process models for marked individuals

As discussed in Section 19.1.4, existing SMR models assume that marked individuals are a random sample, both spatially and demographically, from the population of the state-space. For many studies it may not be feasible to strive to meet or approximate the assumption of spatial randomness, and it is thus important to generalize SMR models to situations where marking does not take place throughout $\mathcal{S}$. We already stated that in this situation, we generally cannot assume that activity centers of marked individuals (and unmarked, for that matter) follow a homogeneous point process. In this final section, we will describe two possible approaches to formulating such an inhomogeneous point process model. We will only provide conceptual descriptions, not a full-blown model development, because, at the time of writing this book, these approaches are still somewhat experimental.

19.7.1 Homogeneous point process in a subset of $\mathcal{S}$

Imagine we perform an area search in a square, $\mathcal{X}$, for some species we want to study, maybe a reptile, and we mark all individuals we encounter. We conduct our sampling in a way that we can assume that marked individuals are randomly sampled within $\mathcal{X}$, and that there are no marked individuals with activity centers outside of $\mathcal{X}$. This design entails the assumption that $\mathcal{X}$ can be clearly defined. We will come back to these assumptions in a minute. We then perform resighting surveys of some sort in an area that overlaps $\mathcal{X}$, so that, when we set a state-space around the resighting locations, $\mathcal{X}$ is completely contained within $\mathcal{S}$ (Figure 19.4). We further assume that individuals that were marked in $\mathcal{X}$ continue to live within $\mathcal{X}$ when resighting surveys are conducted, i.e., their activity centers do not shift during the complete mark-resight study. This implies population closure across both the marking and the resighting part of the

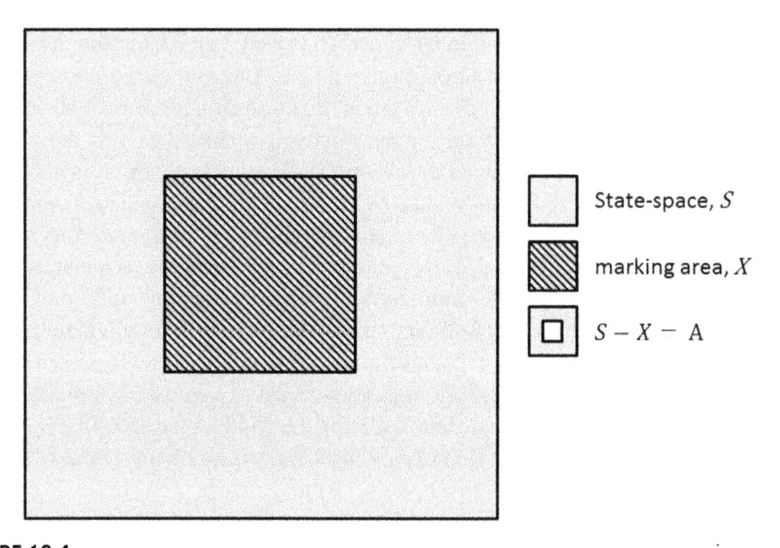

FIGURE 19.4

Relationship between marking area $\mathcal{X}$ and state-space, $\mathcal{S}$.

study, and in this situation we can treat the number of marked animals, m, as known. Let the total population of $\mathcal{X}$ be N_X. Under the conditions specified above, the number of marked animals m can be described as the outcome of a binomial random variable

$$m \sim \text{Binomial}(N_X, \theta),$$

where θ is the probability that an animal living in $\mathcal{X}$ is marked. Under these conditions we can describe the point process for marked individuals as uniform across $\mathcal{X}$, with zero probability of a marked activity center being located outside of $\mathcal{X}$. If the combined (or marginal) point process of marked and unmarked individuals is homogeneous across $\mathcal{S}$, i.e., overall density is constant, then, colloquially speaking, the point process for unmarked individuals is the complement of the marked process: outside of $\mathcal{X}$, unmarked animals occur at the average density of $\mathcal{S}$, D, while inside $\mathcal{X}$ they occur at $D \times (1 - \theta)$.

The above model is an approach to specifying a spatial reference frame for marked individuals that is independent of $\mathcal{S}$. Some of the assumptions of the model, however, are reminiscent of traditional capture-recapture and thus, suffer from the same shortcomings. $\mathcal{X}$ needs to be clearly defined as the area the marked individuals live in, but how do we define it? Imagine again that $\mathcal{X}$ is a square search plot. Surely, we could capture an individual at the edge of the plot, whose activity center is located *off* that plot. Not accounting for this effect would overestimate density in $\mathcal{X}$. This is the equivalent of having to define an effective area sampled in traditional capture-recapture in order to estimate density. Further, we assume that θ, the probability of an individual within the plot being marked, is the same for all individuals in $\mathcal{X}$. But we discussed early on in this book that this is unlikely to be true, because exposure to sampling depends on an individual's home range overlap with the sampled area. So individuals near the edge of $\mathcal{X}$ are less likely to be marked than those in the center, assuming we dispense marking effort uniformly across $\mathcal{X}$. In spite of the shortcomings of this approach, we believe it could serve as a reasonable approximation of some study systems. Moreover, it serves as a conceptual device because it presents a relatively simple way of thinking about two overlapping point processes, in the context of SMR the point processes describing the distribution of marked and unmarked individuals in $\mathcal{S}$.

19.7.2 Inhomogeneous point processes

An alternative, and more realistic, point process for marked individuals is one that describes a decline in the density of marks with increasing distance to the marking location(s). We would expect this kind of spatial pattern in marks to arise, because animals living in the center of the marking grid have a higher probability to be marked than those living on the edges, which in turn have a higher marking probability than those living beyond the marking grid. As a consequence, the density of marks is higher in the center of the marking grid and decreases as we move a away from it. Imagine that marking of animals takes place across some area or grid, and let C_m be the centroid of that marking area. Then, a plausible model for the distribution of marked animals is a bivariate normal model with mean C_m, where the variance in both dimensions is σ_C^2 and the covariance is 0 (Figure 19.5). Of course, there are alternative models to describe a decrease in density, such as a negative-exponential or hazard function. Once the distribution of marked animals conditional on C_m is adequately described, the inhomogeneous point process for the unmarked animals

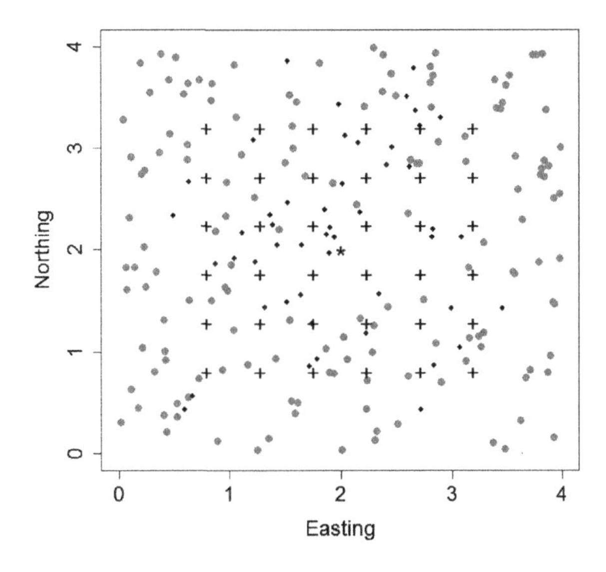

FIGURE 19.5

Plot of activity centers of marked (black dots) and unmarked (gray dots) individuals, with marked individuals following a bivariate normal distribution around the centroid (asterisk) of the marking locations (+); the marginal point process of both marked and unmarked individuals combined is homogeneous.

can again be modeled by assuming that the marginal density across $\mathcal{S}$ is constant, similar to the example above. The parameters of the inhomogeneous point process could also be modeled using the methods outlined in Chapter 11. Density under such a model should be invariant to the size of $\mathcal{S}$, as is the case with standard SCR models, because the marked individuals are probabilistically constrained to live in the vicinity of the marking locations, no matter how large we choose $\mathcal{S}$.

While this model formulation is more realistic and general, it has some minor drawbacks. Based on very limited experience, the variance parameter, σ_C^2, is barely estimable under realistic sample sizes. This is not surprising, given that this parameter describes the distribution of the activity centers, which are themselves latent. Further, the approach does entail some design constraints for marking. We implicitly assume that the centroid of the marking array is the area of highest probability of being marked, which means that marking effort has to be somewhat constant across the marking area. The assumption seems reasonable for systematic or random marking arrays (Figure 19.5), but, for example, for a "hollow grid" the centroid would not be an adequate representation of the point of highest marking probability. The take-home message here is that an adequate point process model for the distribution of marks depends on the spatial context of the marking process.

Dynamic activity centers: One issue we have not addressed explicitly in this section is what to do when enough time passes between marking and resighting so that animals may have rearranged themselves spatially. In example 1 this could mean that marked animals have shifted their activity center outside of the marking plot $\mathcal{X}$. In the second example, marked animals may no longer follow that initial distribution around the centroid of the marking array. Irrespectively, we will have to make some assumption about how marked individuals are distributed in space. We saw such a case in the Canada geese example in Section 19.3.1. In that example, we defined a reasonable state-space and assumed that animals redistributed themselves randomly across that state-space between marking and resighting, because they were captured while molting. This seems like a sensible solution for that particular study system, but it leads to a model formulation where estimates of density are sensitive to choice of $\mathcal{S}$, thus creating a situation that is similar to determining the area sampled in non-spatial mark-resight and capture-recapture studies. Alternatively, we could again use a bivariate normal distribution around the centroid of the original marking array—if movement of activity centers is random with respect to C_m it seems plausible that the overall underlying distribution of $\mathbf{s}$ would still be adequately described with a bivariate normal model. It should be clear that choice of a model for the inhomogeneous point process of marked individuals depends both on the spatial context of marking and what we know (or believe) about the biology of the study species.

19.8 Summary and outlook

In this chapter we combined SCR models (for marked individuals) with a model for unmarked individuals, to derive a spatial mark-resight (SMR) model, which accommodates that only part of the population is individually identifiable, often through artificial tags. Under the assumption that marked individuals are a random sample, both demographically and spatially, from the state-space, the basic model with known number of marked individuals and perfect individual identification of marks is easily modified for situations where the number of marked individuals is unknown, or where marked animals can sometimes not be identified to individual level. As expected, having marked individuals in the study population improved accuracy and precision of parameter estimates when compared to fully unmarked populations. Still, we also saw that the spatial counts of unmarked individuals noticeably contribute information to parameter estimates even when marked individuals are present. We further presented an approach to incorporate telemetry location data into the spatial mark-resight model to inform estimates of σ and activity centers. Just as in SCR, the spatial mark-resight models can account for a variety of factors that may influence individual movement and detection, as well as survey-related parameters, and we saw one example for the Canada geese, where σ was sex-specific, and another for raccoons, where λ_0 was time dependent.

Many details of SMR models remain to be explored and we noted a few of those topics throughout this chapter. For example, we mentioned the assignment of marked but unidentified records to actual marked individuals based on their spatial location, which provides some (though imperfect) information about their identity (Section 19.4). Similarly, records where the marked status cannot be determined could potentially be included in the model as some form of overall correction factor on detection. GPS telemetry devices and their ability to collect location data with much higher frequency offer the opportunity to assign records of collared animals to individuals based on how close to a given resighting location the collared individuals were, both in space and time. In this scenario, individual identity itself could be expressed probabilistically, leading to an SMR model accounting for potential misidentification. All these possible extensions can tailor SMR models to specific survey techniques.

A fundamental assumption of the SMR models developed in this chapter was that marked animals are a random sample from $\mathcal{S}$. This simplifies the model as we can assume a homogeneous point process for both the marked and the unmarked part of the population. While this is a convenient situation, it is neither likely to arise often in real life, nor strictly necessary. If marked animals are not a random sample from $\mathcal{S}$, we need to describe their distribution in the state-space using an adequate point process that will almost always be inhomogeneous across $\mathcal{S}$. We mention two possible approaches—a uniform (homogeneous) point process over a smaller area within $\mathcal{S}$, and an inhomogeneous point process where the intensity decreases as the distance to the centroid of the marking locations increases. Both formulations effectively attempt to describe the distribution of marks in space as a consequence of the spatial nature of the marking process. We believe that another way to approach this problem is to combine spatially explicit models for the marking process and the resighting process. Where marked animals were captured carries information on their spatial distribution, and it should be possible to make use of this information by formulating an integrated spatial capture-mark-resighting model. Such approaches have been developed in a non-spatial CR framework (Pledger et al., 2009; Matechou et al., 2013), but to our knowledge have not yet been addressed in the context of SCR.

Spatial mark-resight models are a recent development, and work on how to relax the spatial component of the random sample assumption and formulate adequate point process models for the distribution of marks is ongoing. While there is still a lot of work to be done, we believe that SMR modeling holds the potential to address a wide range of population estimation problems when dealing with animals that cannot be identified based on natural marks, a situation that is typical of a majority of animal population studies.

CHAPTER

2012: A Spatial Capture-Recapture Odyssey

20

You've finally made it to the last chapter and we realize it's been a long journey to get here. Congratulations! (and thank you!) We hope this book has provided you with many ideas on how to conduct ecological studies and address specific questions that were previously thought difficult or impossible to answer, and given you a solid foundation for carrying out SCR analyses using either Bayesian or classical methods of statistical inference. However, we believe this journey is only just beginning, and we leave you now with a few thoughts on what we see as the future of SCR methods.

Let us first briefly consider how we got here. Over a century ago, around 1786 in France, Pierre-Simon Laplace and others first developed capture-recapture methods and introduced the study of populations. This was of course regarding human population demography, but still, the foundation of how we would go on to study animal populations was being laid out then and there. The Lincoln-Petersen method had been described by the 1930s and development of capture-recapture models began to grow rapidly starting in the 1950s. Soon, capture-recapture methods had become a cornerstone of ecological and wildlife modeling and analysis. Today, spurred on by the advent and rapid development of non-invasive technologies like DNA sampling, camera trapping, acoustic sampling, and other methods, capture-recapture is more relevant and widely used than ever before. These new survey methods allow researchers to use capture-recapture models for species that could not be studied efficiently even a few years ago, especially those that are difficult to capture or handle including bears, mustelids such as fishers (*Martes pennanti*, Figure 20.1) or weasels (e.g., long-tailed weasel *Mustela frenata*, Figure 20.2), most felids (Figure 20.3), and many other species.

With these new sampling techniques, like many commonly used capture-recapture sampling methods, spatial information about location of capture is collected. Classical capture-recapture models ignore this information, and in doing so fail to provide a formal method for modeling spatial variation in density and encounter probability. It was these deficiencies that motivated the development of SCR models, starting around 2003–2004.

We have seen a great increase in the number of papers that use or cite SCR models, and to articulate and quantify this growth, we did a Google Scholar search on March 6, 2013 using the terms:

```
"spatial capture recapture" OR "spatially explicit capture recapture".
```

Spatial Capture-Recapture. http://dx.doi.org/10.1016/B978-0-12-405939-9.00020-7

FIGURE 20.1

Fisher assaulting tree # 8–12, outfitted with a baited hair snare.

Photo credit: NYSDEC (New York State Department of Environmental Conservation), A Fuller/NYSDEC camera trap and hair snare study of fishers in southern NY.

FIGURE 20.2

A long-tailed weasel taking bait on a hair snare, southern NY fisher study.

Photo credit: Marty DeLong.

FIGURE 20.3

Canada Lynx, ear-tagged and radio collared, producing high-quality data in the name of science.

Photo credit: A Fuller, Cornell University.

The results from this literature search are shown in Table 20.1. We see a rapid growth in citation counts after 2004 fueled by publication of Efford (2004) and the release of the software DENSITY (Efford et al., 2004). In 2012, there were 84 articles published, and 27 through the first 9 weeks of 2013. Most (but not all) of these papers are about the type of SCR models discussed in this book, although a handful had to do with other types of spatial analyses as related to capture-recapture models. The results, we think, suggest a bright future for the development and application of spatial capture-recapture models.

We believe that use and growth of SCR modeling in conservation biology, management, wildlife, fisheries, and many other disciplines that we place under the general umbrella of ecology will only continue. This prediction is based on the fact that SCR provides a flexible framework for studying spatial and temporal variation in ecological processes while acknowledging the fact that these processes are almost always observed imperfectly. The "big idea" of SCR, if you could distill the whole thing into one idea, is based on extending closed population models by augmenting them with a point process model that describes the distribution of individuals (Efford, 2004) in space. In a sense, that is really all there is to it. It seems like a little thing, a minor addition to a model, some incremental advance or "ϵ-improvement" of existing technology. But the relevance is much bigger and more profound because, once we have made space explicit in the model, we can think about building population models

Table 20.1 Google Scholar citations by year based on a search of "spatial capture recapture" OR "spatially explicit capture recapture" conducted on March 6, 2013. The estimated growth rate of this population of papers was 33.4%.

Time Period	Cumulative Cites	Cites in Year Previous
Since 2002	274 cites	
Since 2003	274 cites	0 articles published in 2002
Since 2004	271 cites	3 articles published in 2003
Since 2005	269 cites	2 articles published in 2004
Since 2006	264 cites	5 articles published in 2005
Since 2007	261 cites	3 articles published in 2006
Since 2008	253 cites	8 articles published in 2007
Since 2009	242 cites	11 articles published in 2008
Since 2010	222 cites	20 articles published in 2009
Since 2011	176 cites	46 articles published in 2010
Since 2012	111 cites	65 articles published in 2011
Since 2013	27 cites	84 articles published in 2012
		27 published so far in 2013, since March 6

that embody explicit spatial processes and using those models to improve our understanding of population biology and ecology, as well as test explicit hypotheses about mechanisms that govern populations.

We covered many ecological processes that can be studied using SCR, such as landscape connectivity, resource selection, and spatial variation in density. These are all by themselves profound extensions of the basic capture-recapture method, and they broaden and expand the relevance and utility of capture-recapture for studying animal populations. Although we filled almost 600 book pages (mostly) with SCR methods, there remains much to be done in the continued development of SCR models. In the following section, we highlight some emerging topics that show promise or might be in need of further development. Finally, we end with a few remaining thoughts on the use of SCR models in the future.

20.1 Emerging topics

In this book, we provided an overview and synthesis of capture-recapture methods as known to us around the end of 2012. There are many emerging topics that we have not covered either because of lack of technical knowledge, lack of time for satisfactory development, or lack of a good framework for implementation. Here, we present some of those topics. This is not a complete list by any means, just a subset of topics that we or our colleagues are currently working on, or that we think might make good PhD, Masters, or other research projects.

20.1.1 Modeling territoriality

While we discussed inhomogeneous point processes in Chapter 11, there is still a need for more general point process models to account for biological and behavioral processes such as territoriality and interactions among individuals. Very little work has been done on this topic and there are many useful models that could be borrowed from the point process literature to improve the biological reality of SCR models. In ongoing work, Reich et al. (in review) propose a model that accounts for spatial variation in density and potential interactions between individuals' territory centers. Under this model, the activity centers follow an inhomogeneous Strauss process (Strauss, 1975), which includes a parameter that determines the strength of repulsion between territory centers. The idea is based on the notion that territorial species would have well-defined (and defended) territories and thus activity centers may be more regularly distributed on the landscape than predicted by a homogeneous point process. A simulation study demonstrated that properly accounting for interactions between individuals can substantially improve population size estimates in terms of bias and precision relative to the usual independence model.

While the Strauss model is intuitive and shows great potential, it presents computational challenges. The first challenge is that the likelihood includes a high-dimensional integral that has no closed form. To address this issue, Reich et al. (in review) developed an approximation to the Strauss likelihood that allows for posterior sampling, extending related work for categorical Markov random fields (Green and Richardson, 2002; Smith and Smith, 2006). The second challenge is that N is treated as an unknown parameter to be updated and hence N varies and so does the dimension of the posterior distribution. In this case, the dimension-changing problem can be overcome by using data augmentation, as we have done in many situations in this book.

20.1.2 Combining data from different surveys

In some instances, researchers apply different survey techniques to the population of interest, because they yield complementary information. For example, camera trapping is the prime tool for estimating population size/density and other demographic parameters for uniquely marked species, while genetic surveys can yield additional information on the genetic diversity and health of a population that cannot be acquired using camera traps. At the same time, genetic surveys, when samples are analyzed to the individual level, also yield spatial capture-recapture data (see Chapter 15). In this situation, we have two data sets in hand that carry information on animal density, and we should be able to get more precise estimates of density if we combine these two data sets into a single SCR model.

Gopalaswamy et al. (2012a) developed two approaches to combining data from different survey types. In the first case, both surveys are carried out at the same time, so that we can assume that they both sample the same closed population, i.e., there are no possible changes in population density between the two surveys. For camera trapping and genetic surveys, we cannot match records of individuals between the two data sets.

However, models for the distinct sample methods may share some parameters (e.g., σ of the encounter probability model) and, if the studies were conducted simultaneously, they share a common population size N. A second approach of using information from one survey in the analysis of a second survey (that maybe does not yield quite as much data as the primary survey) is by analyzing the primary data set alone, then taking the posterior distribution of a parameter both surveys share and using it as an informative prior distribution in the analysis of the second data set. Gopalaswamy et al. (2012a) refer to this as the stepwise approach.

In summary, no matter which approach is chosen, combining data across surveys can help researchers obtain more precise population size or density estimates, which is especially valuable when dealing with rare and elusive species like big cats that almost always will produce sparse individual data sets. The paper by Gopalaswamy et al. (2012a) considers the situation where we have two SCR data sets, but we can imagine combining SCR data with other sources of information, such as telemetry data (see Chapters 13 and 19 for examples), and possibly opportunistic observations, although to our knowledge this latter issue has not yet been tackled in the context of SCR.

20.1.3 Misidentification

Imperfect identification of individuals can happen in a variety of ways. In genetic surveys there is usually some probability of misidentification due to genotyping error (e.g., Lukacs and Burnham, 2005). In camera trap surveys, a different type of imperfect identification can occur when only the only one flank of an animal is recorded in a detection event and cannot be matched to any of the individuals identified by both flanks. In that case, we can match single-flank pictures with the same side flank pictures, but not with opposite side flank pictures and thus cannot construct definite encounter histories for these single-flank individuals (a right flank and a left flank picture could be the same individual, or could be from two distinct individuals). Finally, in Chapter 19, in the context of mark-resight models, we discussed the case where individuals can not definitely be identified as marked —a violation of a basic mark-resight assumption. Further, we developed an approach to dealing with the situation where we can tell if an animal is marked, but we are not always able to ascertain its individual identity.

In non-spatial capture-recapture some efforts have been made to formally deal with misidentification. Stevick et al. (2001) addressed this problem by double-sampling to derive an error rate for genetic identification, and then including this error rate as a known constant into a Lincoln-Petersen estimator of abundance. Lukacs and Burnham (2005) developed an approach that includes an additional parameter in the model—the probability of a genotype being identified correctly, which is estimated as part of the model likelihood. Link et al. (2010) developed an approach toward solving the same problem implemented in a Bayesian framework that relaxes some of the assumptions of the initial approach. Yoshizaki et al. (2009) dealt with misidentification from camera trap pictures due to evolving marks (i.e., natural marks that change over time, such as scars). This situation is different from the genotyping error one. Here, a change in marks creates a supposedly "new" individual that can be recaptured several times, while the original individual is never captured again (its mark is no longer in the population). In contrast, in genotyping error it is assumed that misidentification creates a "new" individual that is never observed again, because each error leads to a new unique genotype. Yoshizaki et al. (2009) approach this situation similarly, by including a parameter describing the probability of correctly identifying an individual upon recapture (the parameter can also be interpreted as the probability that a mark does not change between capture occasions). Because of the dependencies between true and false detection histories (when a "new" individual is created, the "real" one can no longer be recaptured), the standard multinomial approach to coming up with a model likelihood does not work and implementing the model in a maximum likelihood framework is difficult. The authors instead demonstrate an implementation of the model based on minimizing a function of the squared differences between the observed and expected frequencies of the observed capture histories. Recently, McClintock et al. (in press) developed a Bayesian approach to dealing with single flank photographs in camera trapping studies.

To our knowledge no attempts have been made to deal with misidentification in an SCR framework. While all of the mis-ID cases described above require distinct approaches, we believe that there is one unifying theme to all of them: the capture locations of the potentially misidentified records should be informative about identity. For example, a right flank and a left flank camera trap picture that are taken at two neighboring camera traps should be more likely to belong to the same individual than a right and a left flank picture taken at cameras located at opposing ends of the trap array, especially if animal movement is smaller than the extent of the trap array. SCR models provide a natural way of using this additional information to reduce the uncertainty arising from misidentification.

20.1.4 Gregarious species

One of the key assumptions of the SCR models that we described throughout this book is that the activity centers are independent of one another, but this assumption will be violated for species that associate in pairs, family groups, or any other type of aggregation. However, we believe that general models can be developed for use in studies of gregarious species.

The two issues that must be addressed are that (1) detections are not independent—a trap that catches one individual of given group is likely to capture others in the same group, and (2) the activity centers $\mathbf{s}_i$ should appear clustered or, in fact, completely redundant in some cases. A possible way to account for this is to change our definition of $\mathbf{s}_i$ from the location of an individual's activity center, to the location of a group's activity center (Russell et al., 2012). Ideally, to accommodate unknown group size, the SCR model would be expanded to include a model component for group size, so that formal estimation of both group density and group size would be possible.

20.1.5 Single-catch traps

In Chapter 9 we covered multinomial models in which an individual's probability of being captured in a trap is independent of all other individuals. This is the multi-catch type of device in which traps never fill up, but an individual can only be caught in one trap in any given occasion. We suggested (following Efford et al., (2009a)) that the multi-catch independent multinomial model could be used for "single-catch" traps (traps that hold a single individual or "fill up"), and that bias associated with misspecifying the model would be low under certain conditions (i.e., when the proportion of occupied traps is low).

As discussed in Section 9.3, we recognize that the *time*, or order, of capture of an individual in any trapping interval will affect the encounter probabilities of subsequently captured individuals. Thus, if the order of capture was known, then this information could be used to write the likelihood of the detection model exactly. In practice, the order of capture is almost never known, but it should be possible to regard capture order as a latent variable and update the ordering based on the data. In ongoing work, we find in simulations that our updating process for the order of trap closures works well for estimating abundance, particularly when compared to models that allow multiple individuals to be captured in traps. We expect that this work will lead to a formal model for the single-catch trap problem in the near future.

20.1.6 Model fit and selection

Evaluation of model adequacy or "fit" is an important part of any applied analysis. In Chapter 8, we offered up a number of ideas based on standard considerations and adapted and applied them to SCR models. However, these ideas have not been widely applied, or evaluated, and much work needs to be done. In particular, some basic analysis of their power under meaningful alternatives would increase their relevance and possibly lead to insights for devising better methods. This applies to both Bayesian and likelihood-based methods, for which there are even fewer published applications of goodness-of-fit assessment.

Similarly, we discussed model selection strategies using more-or-less conventional ideas based on AIC/DIC, and model indicator variables using the Kuo and Mallick (1998) method. Calibration of these methods under alternatives is needed, along with some analysis of sensitivity of density estimates to misspecification of certain model components. Bayesian model selection remains an active area of research and development, with much work yet to be done to make existing methods accessible and practical to use (Millar, 2009; Tenan et al., 2013).

20.1.7 Explicit movement models

We briefly discussed the topics of dispersal, transience, and migration in Chapter 16 and sketched out a few ideas that allow for dynamics related to movement or migration. Temporary emigration and transience are two topics where a significant amount of work has been accomplished in non-spatial closed and open capture-recapture models

(Kendall et al., 1997; Pradel et al., 1997; Hines et al., 2003; Clavel et al., 2008; Chandler et al., 2011; Gilroy et al., 2012). Additionally, models for dispersal (e.g., Clobert et al., 2001; Ovaskainen, 2004; Ovaskainen et al., 2008) and other forms of movement (e.g., Jonsen et al., 2005; Johnson et al., 2008a; McClintock et al., 2012) have received quite a bit of attention and development in ecology.

With the recent development of SCR models, the framework is in place to provide a formal integration of the movement dynamics governing the processes of dispersal, emigration, and transiency. Further, the availability of SCR models that allow for explicit population dynamics (survival, recruitment) (Gardner et al., 2010a) now sets the stage to integrate models of movement dynamics directly with models of population demography, and parameterize interactions among population processes. What remains as an area of fruitful research is the development of realistic models of movement dynamics, dispersal, temporary emigration, and transience that can be effectively fitted given typical sparse individual encounter history data generated from capture-recapture studies. Dispersal and emigration can also be related to the life stage of an individual in a certain population. Ultimately, combining multi-state models, where the states are age classes or breeding status categories, with open population SCR models and explicitly modeling patterns of movement and dispersal as a function of state (e.g., age or size class) seems like an important area of development.

20.2 Final remarks

Everything in ecology is spatial, and now so too are capture-recapture models, models that have been the cornerstone of ecological research on populations for decades. Historically, the main use of capture-recapture was to obtain population size estimates, but SCR models move the focus from one of estimation to one of formalizing hypotheses about spatial and temporal variation in ecological processes. These processes include resource selection, landscape connectivity, movement, and how individuals organize themselves in space. SCR models allow for this formalization by borrowing methods from spatial statistics, but unlike many spatial models, SCR models include key demographic parameters such as density and survival and thus allow for mechanistic rather than just phenomenological descriptions of natural variation.

However, much work remains to be done to improve computational feasibility, to address many technical or methodological holes in the literature, and to make these methods more accessible to practitioners. We look forward to these developments and hope that this book will help catalyze further development and application of SCR models.

PART

Appendix

APPENDIX

Appendix I—Useful Software and R Packages

Throughout this book we have used a suite of software and **R** packages, all of which are freely available online. To make life a little easier for you, here we provide you with a list of all software and **R** packages, download links, and some (hopefully) helpful tips regarding their installation.

20.3 WinBUGS

Although **WinBUGS** (Gilks et al., 1994) is becoming increasingly obsolete with the faster and more flexible **OpenBUGS** (Lunn et al., 2009) and **JAGS** (Plummer, 2003), there are still situations in which the program comes in handy. The .exe file can be downloaded from `http://www.mrc-bsu.cam.ac.uk/bugs/winbugs/contents.shtml`. On 32-bit machines you can just go ahead and double-click on the .exe file and follow the installation instructions on the screen. On 64-bit machines, according to the **BUGS** project you should download a zip file (from the same page) and unzip it into a folder of your choice. There are a couple of additional steps to make **BUGS** run. First, you need to obtain a key (which is free and valid for life) here: `http://www.mrc-bsu.cam.ac.uk/bugs/winbugs/WinBUGS14_immortality_key.txt`. The key comes with instructions on how to activate it. Second, you need to update the basic **WinBUGS** version to the most current one (which is from August 2007) following the instructions given here: `http://www.mrc-bsu.cam.ac.uk/bugs/winbugs/WinBUGS14_cumulative_patch_No3_06_08_07_RELEASE.txt`. **WinBUGS** is ready to use after quitting and reopening it. Remember that **WinBUGS** only runs on Windows machines. Also, there appears to be a problem installing the program in Vista, although we have no personal experience with this.

20.3.1 WinBUGS through R

While you can run **WinBUGS** as a standalone application, we recommend you access it from within **R** using the package `R2WinBUGS` (Sturtz et al., 2005), so you can conveniently process your output, make graphs, etc. `R2WinBUGS` also allows you to run models in **OpenBUGS** (see below). You can install the package from within **R** directly from a cran mirror. In addition to the usual package help

Spatial Capture-Recapture. http://dx.doi.org/10.1016/B978-0-12-405939-9.00028-1

document(http://cran.r-project.org/web/packages/R2WinBUGS/R2WinBUGS.pdf), you can also download a short manual with some examples (http://voteview.com/bayes_beach/R2WinBUGS.pdf).

20.4 OpenBUGS

OpenBUGS is the up-to-date version of **WinBUGS** and can be downloaded here: http://www.openbugs.info/w/Downloads (Windows, Mac, and Linux versions are available). The name "**OpenBUGS**" refers to the software being open source, so users do not need to download a license key, like they have to for **WinBUGS** (although the license key for **WinBUGS** is free and valid for life). For Windows, install by double-clicking on the .exe file and following the instructions on the installer screen. Compared to **WinBUGS**, **OpenBUGS** has more built-in functions. The method of how to determine the right updater for each model parameter has changed and the user can manually control the MCMC algorithm used to update model parameters. Several other changes have been implemented in **OpenBUGS** and a detailed list of differences between the two **BUGS** versions can be found at http://www.openbugs.info/w/OpenVsWin. We have encountered convergence problems with simple SCR models in this program. There is an extensive help archive for both **WinBUGS** and **OpenBUGS** and you can subscribe to a mailing list, where people pose and answer questions of how to use these programs at http://www.mrc-bsu.cam.ac.uk/bugs/overview/list.shtml.

20.4.1 OpenBUGS through R

Like **WinBUGS**, **OpenBUGS** can be used as a standalone application or through **R**. There are several packages that allow **R** to interface with **OpenBUGS**, all of which can be installed directly from a cran mirror:

`R2WinBUGS`: One of the arguments in the `bugs` call is `program`, which lets you specify either **WinBUGS** or **OpenBUGS**. This is a convenient option, because after having worked through some of this book you will likely be familiar with the format of `bugs` output and other functions of the `R2WinBUGS` package.

`R2OpenBUGS` (Sturtz et al., 2005) is very similar to, and actually based on, `R2WinBUGS` and it is unclear to us what can be gained by using one over the other. Arguments of the `bugs` call differ slightly between the two packages, and given that `R2WinBUGS` allows for the use of both **OpenBUGS** and **WinBUGS**, it is probably easiest to stick with it.

`BRugs` (Thomas et al., 2006) has the convenient feature that all pieces of a **BUGS** analysis can be run from within **R**, including checking the model syntax, something that requires opening the **BUGS** GUI with other packages. In addition to the help document at http://www.biostat.umn.edu/brad/software/~BRugs/BRugs_9_21_07.pdf there is a **WinBUGS** style manual you can access at http://www.rni.helsinki.fi/openbugs/OpenBUGS/Docu/BRugs%20Manual.html.

20.5 JAGS

JAGS (Just Another Gibbs Sampler) (Plummer, 2003) runs SCR models considerably faster than **WinBUGS**, does not have the convergence problem with simple SCR models we have encountered in **OpenBUGS**, but similar to the latter program, is flexible and constantly updated. Writing a **JAGS** model is virtually identical to writing a **WinBUGS** model. However, some functions may have slightly different names and you can look up available functions and their use in the **JAGS** manual (`http://iweb.dl.sourceforge.net/project/mcmc-jags/Manuals/3.x/jags_user_manual.pdf`). One potential downside is that **JAGS** can be very particular when it comes to initial values. These may have to be set as close to truth as possible for the model to start. Although **JAGS** lets you run several parallel Markov chains, this characteristic interferes with the idea of using overdispersed initial values for the different chains. Also, we have found that, when running models, sometimes **JAGS** crashes for unclear reasons, taking **R** down with it. Oftentimes, in order to make the program run again, you'll have to go through downloading and installing it again (remove the non-functioning version first).

JAGS has a variety of functions that are not available in **WinBUGS**. For example, **JAGS** allows you to supply observed data for some deterministic functions of unobserved variables. In **BUGS**, we cannot supply data to logical nodes. Another useful feature is, that the adaptive phase of the model is run separately from the sampling from the stationary Markov chains. This allows you to easily add more iterations to the model run if necessary, without the need to start from 0. There are other, more subtle differences, and there is an entire manual section on differences between **JAGS** and **OpenBUGS**.

JAGS is available for download at `http://sourceforge.net/projects/mcmc-jags/files/`, together with the **R** package `rjags` (Plummer, 2011), which allows running **JAGS** through **R**, user and installation manuals and examples. Under the above address **JAGS** is available for Windows and Mac; Linux binaries are distributed separately, and you can find links to various sources here: `http://mcmc-jags.sourceforge.net/`. **JAGS** comes with a 32-bit and a 64-bit version and can be installed by double-clicking on the .exe file and following the instructions on the installer screen. For questions and problems concerning **JAGS**, there is a forum online at `http://sourceforge.net/projects/mcmc-jags/forums/forum/610037`.

20.5.1 JAGS through R

Unlike the two **BUGS** programs, **JAGS** does not have a GUI interface, but a command line interface that can be used to run the program as a standalone application. **JAGS** will solely perform the MCMC simulation; analyzing and summarizing the output has to be done outside of **JAGS**. To run **JAGS** through **R**, you have two options.

`rjags`: As mentioned above, `rjags` can be downloaded from the same site as **JAGS** and is being maintained by the creator of **JAGS**, which means it is guaranteed to stay up to date as **JAGS** changes. The package can also be installed from a cran mirror and the help document can be accessed at `http://cran.r-project.org/web/packages/rjags/rjags.pdf`.

`R2jags`: Alternatively, the package `R2jags` (Su and Yajima, 2011) provides a means of accessing **JAGS** through **R**. We prefer `rjags` for the reason named above, as well as because it stores data in a more memory-efficient way and has better `plot` and `summary` methods.

20.6 R

At the time of the preparation of this list, **R** for Windows is at version 2.15.0, which can be downloaded at `http://cran.r-project.org/bin/windows/base/`. This site also contains helpful tips on how to install **R** in Windows Vista, how to update **R** packages, etc. Installation of **R** in Windows is straightforward: download the .exe file, double-click on it, and follow the instructions of the Windows installer. The later versions of **R** come with versions for both 32-bit and 64-bit machines. The **R** site (`http://mirrors.softliste.de/cran/`) has an extensive FAQ section (Hornik, 2011), which includes instructions on how to install **R** on Unix and Mac computers.

20.6.1 R packages

This section provides an alphabetical list of useful **R** packages. There is a huge number of **R** packages and by no means is this list intended to be complete in terms of what is useful. Rather, we list packages that we are familiar with and that we employ at one point or the other in this book. Unless explicitly stated otherwise, all packages can be installed directly from within **R** trough a cran mirror.

`adapt` (Genz et al., 2007): A package for multi-dimensional numerical integration. The package has been removed from the CRAN repository but can be obtained from `http://cran.r-project.org/src/contrib/Archive/adapt/`.

`coda` (Plummer et al., 2006): Lets you summarize and perform diagnostics on MCMC output. For a list and description of functions, see the manual at `http://cran.r-project.org/web/packages/coda/coda.pdf`.

`gdistance` (van Etten, 2011): A package for calculating distances and routes on geographical grids that can be used to calculate least cost path surfaces. Manual at `http://cran.r-project.org/web/packages/gdistance/gdistance.pdf`.

`igraph` (Csardi and Nepusz, 2006): Provides routines for graphs and network analysis. Manual at `http://cran.r-project.org/web/packages/igraph/igraph.pdf`.

`inline` (Sklyar et al., 2010): Allows the user to define **R** functions with in-lined **C**, **C++** or **Fortran** code. Manual at `http://cran.r-project.org/web/packages/inline/inline.pdf`.

maps (Becker et al., 2012): A package for the display of maps. Manual at http://cran.r-project.org/web/packages/maps/index.html.

maptools (Lewin-Koh et al., 2011): Provides a set of tools for reading and manipulating spatial data, especially ESRI shapefiles. Manual at http://cran.r-project.org/web/packages/maptools/maptools.pdf.

mvtnorm (Genz et al., 2012): Computes multivariate normal and t probabilities, quantiles, random deviates and densities. Manual at http://cran.r-project.org/web/packages/mvtnorm/mvtnorm.pdf.

parallel: Contains a suite of functions for parallel computing on multiple computer cores and comes with **R** versions 2.14.0 or higher. More information about the package can be found at http://stat.ethz.ch/R-manual/R-devel/library/parallel/doc/parallel.pdf.

R2cuba (Hahn et al., 2010): Another package for multi-dimensional integration. Manual at http://cran.r-project.org/web/packages/R2Cuba/R2Cuba.pdf.

raster (Hijmans and van Etten, 2012): Provides functions for geographic analysis and modeling with raster data. Manual at http://cran.r-project.org/web/packages/raster/raster.pdf.

Rcpp (Eddelbuettel and François, 2011): Provides **R** functions as well as a **C++** library that facilitates the integration of **R** and **C++**. Manual at http://cran.r-project.org/web/packages/Rcpp/Rcpp.pdf.

RcppArmadillo (François et al., 2011): A templated **C++** linear algebra library, integrating the Armadillo library and **R**. Manual at http://cran.r-project.org/web/packages/RcppArmadillo/RcppArmadillo.pdf.

reshape (Wickham and Hadley, 2007): Allows you to easily manipulate, summarize and reshape data. Manual at http://cran.r-project.org/web/packages/reshape/reshape.pdf.

rgeos (Bivand and Rundel, 2011): Provides many useful functions for spatial operations, such as intersecting or buffering spatial features. Manual at http://cran.r-project.org/web/packages/rgeos/rgeos.pdf.

SCRbayes (Russell et al., 2012): Provides a Bayesian implementation of certain kinds of SCR search-encounter models. This package is not available on CRAN but can be downloaded at http://www.mbr-pwrc.usgs.gov/software/SCRbayes.shtml.

secr (Efford et al., 2009a): An allround package for fitting a wide array of SCR models in a frequentist framework. Manual at http://cran.r-project.org/web/packages/secr/secr.pdf.

shapefiles (Stabler, 2006): Allows you to read and write ESRI shapefiles (i.e., shapefiles you would use in **ArcGIS**). Manual at http://cran.r-project.org/web/packages/shapefiles/shapefiles.pdf.

snow (Tierney et al., 2011), snowfall (Knaus, 2010): Provide functionality for parallel computing. The latter is a more user-friendly wrapper around the former. Manuals at http://cran.r-project.org/web/packages/snowfall/

snowfall.pdf and http://cran.r-project.org/web/packages/snow/snow.pdf.

sp (Pebesma and Bivand, 2011): A package for plotting, selecting, subsetting, etc. spatial data. sp and spatstat (see below) are complementary in may ways and data formats can be easily converted between the two packages. Manual at http://cran.r-project.org/web/packages/sp/sp.pdf.

SPACECAP (Gopalaswamy et al., 2012a): Provides a user friendly GUI interface to fit SCR models with a binomial observation model in a Bayesian framework. Manual at http://www.icesi.edu.co/CRAN/web/packages/SPACECAP/SPACECAP.pdf.

spatstat (Baddeley and Turner, 2005): An extensive package for analyzing spatial data. We use it, for example, to generate random points within a state-space that cannot be described as a rectangle but consists of a (or several) arbitrary polygon (s). Manual at http://cran.r-project.org/web/packages/spatstat/spatstat.pdf.

unmarked (Fiske and Chandler, 2011): Fits hierarchical models of animal abundance and occurrence to data collected using a range of predominantly direct observation based methods. Manual at http://cran.r-project.org/web/packages/unmarked/unmarked.pdf.

Bibliography

Adriaensen, F., Chardon, J.P., De Blust, G., Swinnen, E., Villalba, S., Gulinck, H., Matthysen, E., 2003. The application of least-cost modelling as a functional landscape model. Landscape and Urban Planning 64, 233–247.

Agresti, A., 2002. Categorical Data Analysis. John Wiley and Sons.

Aitkin, M., 1991. Posterior bayes factors. Journal of the Royal Statistical Society, Series B (Methodological) 53, 111–142.

Alho, J.M., 1990. Logistic regression in capture-recapture models. Biometrics 46, 623–635.

Alpízar-Jara, R., Pollock, K.H., 1996. A combination line transect and capture-recapture sampling model for multiple observers in aerial surveys. Environmental and Ecological Statistics 3, 311–327.

Amstrup, S.C., McDonald, T.L., Manly, B.F.J., 2005. Handbook of Capture-Recapture Analysis. Princeton University Press.

Anderson, D.R., Burnham, K.P., White, G.C., Otis, D.L., 1983. Density estimation of small-mammal populations using a trapping web and distance sampling methods. Ecology 64, 674–680.

Arnason, N.A., 1972. Parameter estimates from mark-recapture experiments on two populations subject to migration and death. Researches on Population Ecology 13, 97–113.

Arnason, N.A., 1973. The estimation of population size, migration rates and survival in a stratified population. Researches on Population Ecology 15, 1–8.

Arnason, A., Schwarz, C., Gerrard, J., 1991. Estimating closed population size and number of marked animals from sighting data. Journal of Wildlife Management 55, 716–730.

Baddeley, A., Turner, R., 2005. Spatstat: an R package for analyzing spatial point patterns. Journal of Statistical Software 12, 1–42.

Bales, S., Hellgren, E., Leslie Jr., D., Hemphill Jr., J., 2005. Dynamics of a recolonizing population of black bears in the Ouachita Mountains of Oklahoma. Wildlife Society Bulletin 33, 1342–1351.

Balme, G.A., Slotow, R., Hunter, L.T.B., 2010. Edge effects and the impact of non-protected areas in carnivore conservation: leopards in the Phinda-Mkhuze complex, South Africa. Animal Conservation 13, 315–323.

Spatial Capture-Recapture. http://dx.doi.org/10.1016/B978-0-12-405939-9.00029-3

Bartmann, R.M., White, G.C., Carpenter, L.H., Garrott, R.A., 1987. Aerial mark-recapture estimates of confined mule deer in Pinyon-Juniper woodland. The Journal of Wildlife Management 51, 41–46.

Bayne, E., Hobson, K., 2002. Annual survival of adult American Redstarts and Ovenbirds in the southern boreal forest. The Wilson Bulletin 114, 358–367.

Becker, R.A., Wilks, A.R., Brownrigg, R., Minka, T.P., 2012. maps: Draw Geographical Maps. R package version 2.2-5. <http://CRAN.R-project.org/package=maps>.

Beier, P., Majka, D.R., Spencer, W.D., 2008. Forks in the road: choices in procedures for designing wildland linkages. Conservation Biology 22, 836–851.

Belant, J., Van Stappen, J., Paetkau, D., 2005. American black bear population size and genetic diversity at Apostle Islands National Lakeshore. Ursus 16, 85–92.

Berger, J., Liseo, B., Wolpert, R., 1999. Integrated likelihood methods for eliminating nuisance parameters. Statistical Science 14, 1–28.

Besag, J., Kooperberg, C., 1995. On conditional and intrinsic autoregressions. Biometrika 82, 733–746.

Best, N.G., Ickstadt, K., Wolpert, R.L., 2000. Spatial poisson regression for health and exposure data measured at disparate resolutions. Journal of the American Statistical Association 95, 1076–1088.

Bibby, C.J., Buckland, S.T., 1987. Bias of bird census results due to detectability varying with habitat. Acta Ecologica 8, 103–112.

Bibby, C.J., Burgess, N.D., Hill, D.A., 1992. Bird Census Techniques. Academic Press, London, UK.

Bivand, R., Rundel, C. 2011. Rgeos: interface to geometry engine—open source (GEOS). R package version 0.1-8.

Blair, W., 1940. A study of prairie deer-mouse populations in southern Michigan. American Midland Naturalist 24, 273–305.

Bolker, B., 2008. Ecological Models and Data in R. Princeton University Press.

Bondrup-Nielsen, S., 1983. Density estimation as a function of live-trapping grid and home range size. Canadian Journal of Zoology 61, 2361–2365.

Borchers, D.L., 2012. A non-technical overview of spatially explicit capture-recapture models. Journal of Ornithology 152, 1–10.

Borchers, D.L., Buckland, S.T., Zucchini, W., 2002. Estimating Animal Abundance: Closed Populations. Springer Verlag.

Borchers, D.L., Efford, M.G., 2008. Spatially explicit maximum likelihood methods for capture-recapture studies. Biometrics 64, 377–385.

Borchers, D.L., Zucchini, W., Fewster, R.M., 1998. Mark-recapture models for line transect surveys. Biometrics 54, 1207–1220.

Boulanger, J., McLellan, B., 2001. Closure violation in DNA-based mark-recapture estimation of grizzly bear populations. Canadian Journal of Zoology 79, 642–651.

Bowden, D.C. 1993. A simple technique for estimating population size. Technical Report 93/12, Department of Statistics, Colorado State University, Fort Collins, Colorado, USA.

Bowden, D.C., Kufeld, R.C., 1995. Generalized mark-sight population size estimation applied to Colorado moose. Journal of Wildlife Management 59, 840–851.

Box, G.E.P., Draper, N.R., 1959. A basis for the selection of a response surface design. Journal of the American Statistical Association 54, 622–654.

Box, G.E.P., Draper, N.R., 1987. Empirical Model-Building with Response Surfaces. Wiley, New York.

Boyce, M.S., McDonald, L.L., 1999. Relating populations to habitats using resource selection functions. Trends in Ecology and Evolution 14, 268–272.

Brooks, S.P., Catchpole, E.A., Morgan, B.J.T., 2000. Bayesian animal survival estimation. Statistical Science 15, 357–376.

Brownie, C., Hines, J.E., Nichols, J.D., Pollock, K.H., Hestbeck, J.B., 1993. Capture-recapture studies for multiple strata including non-Markovian transitions. Biometrics 49, 1173–1187.

Buckland, S.T., Anderson, D.R., Burnham, K.P., Laake, J.L., Borchers, D.L., Thomas, L. (Eds.), 2004. Advanced Distance Sampling: Estimating Abundance of Biological Populations. Oxford University Press, USA.

Buckland, S.T., Anderison, D.R., Burnham, K.P., Laake, J.L., Borcher, D.L., 2001. Introduction to distance sampling: estimating abundance of biological populations. Oxford University Press, Oxford, UK.

Burnham, K.P., 1997. Distributional results for special cases of the Jolly-Seber model. Communications in Statistics 26, 1395–1409.

Burnham, K.P., Anderson, D.R., 2002. Model Selection and Multimodel Inference: A Practical Information-Theoretic Approach. Springer Verlag.

Burnham, K.P., Anderson, D.R., Laake, J.L., 1980. Estimation of density from line transect sampling of biological populations. Wildlife Monographs 72, 3–202.

Burnham, K.P., Anderson, D.R., White, G.C., Brownie, C., Pollock, K.H., 1987. Design and Analysis Methods for Fish Survival Experiments Based on Release-Recapture. American Fisheries Society Monograph, vol. 5. American Fisheries Society, Bethesda Maryland.

Burnham, K.P., Overton, W.S., 1978. Estimation of the size of a closed population when capture probabilities vary among animals. Biometrika 65, 625–633.

Burt, W., 1943. Territoriality and home range concepts as applied to mammals. Journal of Mammalogy 24, 346–352.

Carlin, B.P., Chib, S., 1995. Bayesian model choice via Markov chain Monte Carlo methods. Journal of the Royal Statistical Society, Series B 57, 473–484.

Casella, G., Berger, R.L., 2002. Statistical Inference. Duxbury Press.

Casella, G., George, E.I., 1992. Explaining the Gibbs sampler. American Statistician 46, 167–174.

Caswell, H., 1989. Matrix Population Models. Sinauer Association, Sunderland.

Caswell, H., Werner, P.A., 1978. Transient behavior and life history analysis of teasel (*Dipsacus sylvestris* Huds.). Ecology 59, 53–66.

Chandler, R.B., Royle, J.A., 2013. Spatially-explicit models for inference about density in unmarked or partially marked populations. Annals of Applied Statistics 7, 936–954.

Chandler, R.B., Royle, J.A., King, D., 2011. Inference about density and temporary emigration in unmarked populations. Ecology 92, 1429–1435.

Clavel, J., Robert, A., Devictor, V., Julliard, R., 2008. Abundance estimation with a transient model under the robust design. Journal of Wildlife Management 72, 1203–1210.

Clobert, J., Danchin, E., Dhondt, A., Nichols, J.D., 2001. Dispersal. Oxford.

Cochran, W., 2007. Sampling Techniques. John Wiley & Sons, USA.

Compton, B.W., McGarigal, K., Cushman, S.A., Gamble, L.R., 2007. A resistant-kernel model of connectivity for amphibians that breed in vernal pools. Conservation Biology 21, 788–799.

Conde, D., Colchero, F., Zarza, H., Christensen, N., Sexton, J., Manterola, C., Chávez, C., Rivera, A., Azuara, D., Ceballos, G., 2010. Sex matters: modeling male and female habitat differences for jaguar conservation. Biological Conservation 143, 1980–1988.

Conn, P.B., Cooch, E.G., 2009. Multistate capture-recapture analysis under imperfect state observation: an application to disease models. Journal of Applied Ecology 46, 486–492.

Conroy, M.J., Carroll, J.P., 2009. Quantitative Conservation of Vertebrates. Wiley-Blackwell.

Conroy, M.J., Runge, J.P., Barker, R.J., Schofield, M.R., Fonnesbeck, C.J., 2008. Efficient estimation of abundance for patchily distributed populations via two-phase, adaptive sampling. Ecology 89, 3362–3370.

Converse, S.J., Royle, J.A., 2012. Dealing with incomplete and variable detectability in multi-year, multi-site monitoring of ecological populations. In: Gitzen, R.R., Millspaugh, J.J., Cooper, A.B., Licht, D.S. (Eds.), Design and Analysis of Long-term Ecological Monitoring Studies. Cambridge University Press, pp. 426–442.

Converse, S.J., White, G.c., Block, W., 2006a. Small mammal responses to thinning and wildfire in ponderosa pine-dominated forests of the southwestern United States. Journal of Wildlife Management 70, 1711–1722.

Converse, S.J., White, G.C., Farris, K., Zack, S., 2006b. Small mammals and forest fuel reduction: national-scale responses to fire and fire surrogates. Ecological Applications 16, 1717–1729.

Cooch, E.G., White, G.C., 2006. Program MARK: A Gentle Introduction. Available online with the MARK programme.

Cormack, R.M., 1964. Estimates of survival from the sighting of marked animals. Biometrika 51, 429–438.

Coull, B.A., Agresti, A., 1999. The use of mixed logit models to reflect heterogeneity in capture-recapture studies. Biometrics 55, 294–301.

Cox, D., 1955. Some statistical methods connected with series of events. Journal of the Royal Statistical Society, Series B (Methodological) 17, 129–164.

Cressie, N.A.C., 1991. Statistics for Spatial Data. Wiley Series in Probability and Mathematical Statistics.

Csardi, G., Nepusz, T., 2006. The igraph software package for complex network research. International Journal of Complex Systems 1695, 38.

Cushman, S.A., Compton, B.W., McGarigal, K., 2010. Habitat fragmentation effects depend on complex interactions between population size and dispersal ability: modeling influences of roads, agriculture and residential development across a range of life-history characteristics. In: Cushman, S.A., Huettmann, F. (Eds.), Spatial Complexity, Informatics, and Wildlife Conservation, Springer, New York, pp. 369–385.

Cushman, S.A., McKelvey, K.S., Hayden, J., Schwartz, M.K., 2006. Gene flow in complex landscapes: testing multiple hypotheses with causal modeling. The American Naturalist 168, 486–499.

Dawson, D.K., Efford, M.G., 2009. Bird population density estimated from acoustic signals. Journal of Applied Ecology 46, 1201–1209.

DeGraaf, R.M., Yamasaki, M., 2001. New England Wildlife: Habitat, Natural History, and Distribution. University Press of New England.

Dellaportas, P., Forster, J., Ntzoufras, I., 2000. Bayesian variable selection using the Gibbs sampler. In: Dey, D., Ghosh, S., Mallick, B. (Eds.), Generalized Linear Models: A Bayesian Perspective. CRC Press, pp. 273–286.

DeSante, D.F., Burton, K.M., Saracco, J.F., Walker, B.L., 1995. Productivity indices and survival rate estimates from MAPS, a continent-wide programme of constant-effort mist-netting in North America. Journal of Applied Statistics 22, 935–948.

Dice, L.R., 1938. Some census methods for mammals. Journal of Wildlife Management 2, 119–130.

Dice, L.R., 1941. Methods for estimating populations of mammals. Journal of Wildlife Management 5, 398–407.

Diggle, P.J. 2003. Statistical Analysis of Spatial Point Processes, second ed. Arnold, London.

Dijkstra, E.W., 1959. A note on two problems in connexion with graphs. Numerische Mathematik 1, 269–271.

Dillon, A., Kelly, M., 2007. Ocelot Leopardus pardalis in Belize: the impact of trap spacing and distance moved on density estimates. Oryx 41, 469–477.

Dixon, P.M., 2002. Bootstrap resampling. In: Encyclopedia of Environmetrics. John Wiley & Sons.

Dorazio, R.M., 2007. On the choice of statistical models for estimating occurrence and extinction from animal surveys. Ecology 88, 2773–2782.

Dorazio, R.M., Royle, J.A., 2003. Mixture models for estimating the size of a closed population when capture rates vary among individuals. Biometrics 59, 351–364.

Dormann, C.F., McPherson, J.M., Araújo, M.B., Bivand, R., Bolliger, J., Carl, G., Davies, G.R., Hirzel, A., Jetz, W., Kissling, W.D., et al., 2007. Methods to account for spatial autocorrelation in the analysis of species distributional data: a review. Ecography 30, 609–628.

Durban, J.W., Elston, D.A., 2005. Mark-recapture with occasion and individual effects: abundance estimation through Bayesian model selection in a fixed dimensional parameter space. Journal of Agricultural, Biological, and Environmental Statistics 10, 291–305.

Eddelbuettel, D., François, R., 2011. Rcpp seamless R and C++ integration. Journal of Statistical Software 40, 1–18.

Efford, M.G., 2011a. SECR: spatially explicit capture-recapture models. R package version 2.3-1.

Efford, M.G., 2004. Density estimation in live-trapping studies. Oikos 106, 598–610.

Efford, M.G., 2011b. Estimation of population density by spatially explicit capture-recapture analysis of data from area searches. Ecology 92, 2202–2207.

Efford, M.G., 2012. Spatially explicit capture-recapture for bear researchers and managers, Western Black Bear Workshop. Coeur dAlene, ID.

Efford, M.G., Dawson, D.K., 2009. Effect of distance-related heterogeneity on population size estimates from point counts. The Auk 126, 100–111.

Efford, M.G., Dawson, D.K. 2010. SECR for acoustic data, <http://www.otago.ac.nz/density/pdfs/secr-sound.pdf> (updated May 14 2012).

Efford, M.G., Borchers, D.L., Byrom, A.E. 2009a. Density estimation by spatially explicit capture-recapture: likelihood-based methods. In: Thomson, D.L., Cooch, E., Conroy, M.J. (Eds.), Modeling Demographic Processes in Marked Populations. Springer Verlag, US, pp. 255–269.

Efford, M.G., Dawson, D.K., Borchers, D.L., 2009b. Population density estimated from locations of individuals on a passive detector array. Ecology 90, 2676–2682.

Efford, M.G., Dawson, D.K., Robbins, C.S., 2004. DENSITY: software for analysing capture-recapture data from passive detector arrays. Animal Biodiversity and Conservation 27, 217–228.

Efford, M.G., Fewster, R.M., 2012. Estimating population size by spatially explicit capture-recapture. Oikos 122, 918–928.

Efford, M.G., Warburton, B., Coleman, M.C., Barker, R.J., 2005. A field test of two methods for density estimation. Wildlife Society Bulletin 33, 731–738.

Epps, C.W., Palsbøll, P.J., Wehausen, J.D., Roderick, G.K., Ramey, R.R. I.I., McCullough, D.R., 2005. Highways block gene flow and cause a rapid decline in genetic diversity of desert bighorn sheep. Ecology Letters 8, 1029–1038.

Epps, C.W., Wehausen, J.D., Bleich, V.C., Torres, S.G., Brashares, J.S., 2007. Optimizing dispersal and corridor models using landscape genetics. Journal of Applied Ecology 44, 714–724.

Farnsworth, G.L., Pollock, K.H., Nichols, J.D., Simons, T.R., Hines, J.E., Sauer, J.R., 2002. A removal model for estimating detection probabilities from point-count surveys. Auk 119, 414–425.

Fedorov, V.V., 1972. Theroy of Optimal Experiments. Academic Press, New York

Fedorov, V.V., Hackl, P, 1997. Model-Oriented Design of Experiments. Springer.

Fienberg, S.E., Johnson, M.S., Junker, B.W., 1999. Classical multilevel and Bayesian approaches to population size estimation using multiple lists. Journal of the Royal Statistical Society of London A 163, 383–405.

Fiske, I.J., Chandler, R.B., 2011. Unmarked: an R package for fitting hierarchical models of wildlife occurrence and abundance. Journal of Statistical Software 43, 1–23.

Forester, J.D., Im, H.K., Rathouz, P.J., 2009. Accounting for animal movement in estimation of resource selection functions: sampling and data analysis. Ecology 90, 3554–3565.

Forester, J.D., Ives, A.R., Turner, M.G., Anderson, D.P., Fortin, D., Beyer, H.L., Smith, D.W., Boyce, M.S., 2007. State-space models link elk movement patterns to landscape characteristics in Yellowstone National Park. Ecological Monographs 77, 285–299.

Fortin, D., Beyer, H.L., Boyce, M.S., Smith, D.W., Duchesne, T., Mao, J.S., 2005. Wolves influence elk movements: behavior shapes a trophic cascade in Yellowstone National Park. Ecology 86, 1320–1330.

Foster, R.J., Harmsen, B.J., 2012. A critique of density estimation from camera-trap data. Journal of Wildlife Management 76, 224–236.

Fowler, C., 1981. Density dependence as related to life history strategy. Ecology 62, 602–610.

François, R., Eddelbuettel, D., Bates, D., 2011. RcppArmadillo: Rcpp integration for Armadillo templated linear algebra library. R package version 0.2-25.

Frary, V., Duchamp, J., Maehr, D., Larkin, J., 2011. Density and distribution of a colonizing front of the American black bear Ursus americanus. Wildlife Biology 17, 404–416.

Fujiwara, M., Anderson, K., Neubert, M., Caswell, H., 2006. On the estimation of dispersal kernels from individual mark-recapture data. Environmental and Ecological Statistics 13, 183–197.

García-Alaníz, N., Naranjo, E.J., Mallory, F.F., 2010. Hair-snares: a non-invasive method for monitoring felid populations in the Selva Lacandona. Mexico, Tropical Conservation Science 3, 403–411.

Gardner, B., Reppucci, J., Lucherini, M., Royle, J.A., 2010a. Spatially explicit inference for open populations: estimating demographic parameters from camera-trap studies. Ecology 91, 3376–3383.

Gardner, B., Royle, J.A., Wegan, M.T., 2009. Hierarchical models for estimating density from DNA mark-recapture studies. Ecology 90, 1106–1115.

Gardner, B., Royle, J.A., Wegan, M.T., Rainbolt, R.E., Curtis, P.D., 2010b. Estimating black bear density using DNA data from hair snares. Journal of Wildlife Management 74, 318–325.

Garshelis, D.L., Hristienko, H., 2006. State and provincial estimates of American black bear numbers versus assessments of population trend. Ursus 17, 1–7.

Gelfand, A., Smith, A., 1990. Sampling-based approaches to calculating marginal densities. Journal of the American Statistical Association 85, 398–409.

Gelman, A., 2006. Prior distributions for variance parameters in hierarchical models. Bayesian Analysis 1, 515–533.

Gelman, A., Carlin, J.B., Stern, H.S., Rubin, D.B., 2004. Bayesian Data Analysis, second ed. CRC/Chapman & Hall, Bocan Raton, Florida, USA.

Gelman, A., Meng, X.L., Stern, H., 1996. Posterior predictive assessment of model fitness via realized discrepancies. Statistica Sinica 6, 733–759.

Geman, S., Geman, D., 1984. Stochastic relaxation, Gibbs distributions, and the Bayesian restoration of images. IEEE Transactions on Pattern Analysis and Machine Intelligence PAMI-6, 721–741.

Genz, A., Bretz, F., Miwa, T., Mi, X., Leisch, F., Scheipl, F., Hothorn, T., 2012. mvtnorm: Multivariate Normal and t Distributions. R package version 0.9-9992. <http://CRAN.R-project.org/package=mvtnorm>.

Genz, A.S., Meyer, M.R., Lumley, T., Maechler, M., 2007. The adapt package. R package version 1.0-4.

Gerlach, G., Musolf, K., 2000. Fragmentation of landscape as a cause for genetic subdivision in bank voles. Conservation Biology 14, 1066–1074.

Gilks, W.R., Wild, P., 1992. Adaptive rejection sampling for Gibbs sampling. Applied Statistics 41, 337–348.

Gilks, W.R., Thomas, A., Spiegelhalter, D.J., 1994. A Language and Program for Complex Bayesian Modelling. Journal of the Royal Statistical Society. Series D (The Statistician) 43, 169–177.

Gilroy, J., Virzi, T., Boulton, R.L., Lockwood, J., 2012. A new approach to the apparent survival problem: estimating true survival rates from mark-recapture studies. Ecology 93, 1509–1516.

Gimenez, O., Rossi, V., Choquet, R., Dehais, C., Doris, B., Varella, H., Vila, J.P., Pradel, R., 2007. State-space modelling of data on marked individuals. Ecological Modelling 206, 431–438.

Gopalaswamy, A.M., 2012. Capture-recapture models, spatially explicit. In: El-Shaarawi, A.H., Piegorsch, W. (Eds.), Encyclopedia of Environmentrics, second ed. John Wiley & Sons Ltd, Chichester, UK.

Gopalaswamy, A.M., Royle, J.A., Hines, J.E., Singh, P., Jathanna, D., Kumar, N., Karanth, K.U., 2012a. Program SPACECAP: Software for estimating animal density using spatially explicit capture–recapture models. Methods in Ecology and Evolution 3, 1067–1072.

Gopalaswamy, A.M., Royle, J.A., Delampady, M., Nichols, J.D., Karanth, K.U., Macdonald, D.W., 2012b. Density estimation in tiger populations: combining information for strong inference. Ecology 93, 1741–1751.

Grant, E.H.C., Nichols, J.D., Lowe, W.H., Fagan, W.F., 2010. Use of multiple dispersal pathways facilitates amphibian persistence in stream networks. Proceedings of the National Academy of Sciences 107, 6936–6940.

Graves, T.A., Chandler, R.B., Royle, J.A., Beier. P., 2013. Estimating landscape resistance to dispersal (unpublished report).

Green, P., Richardson, S., 2002. Hidden Markov models and disease mapping. Journal of the American Statistical Association 97, 1055–1070.

Greig-Smith, P., 1964. Quantitative, Plant Ecology. Butterworths (Washington).

Grimm, V., Revilla, E., Berger, U., Jeltsch, F., Mooij, W., Railsback, S., Thulke, H.-H., Weiner, J., Wiegand, T., DeAngelis, D., 2005. Pattern-oriented modeling of agent-based complex systems: lessons from ecology. Science 310, 987–991.

Hahn, T., Bouvier, A., Kiêu, K., 2010. R2Cuba: multidimensional numerical integration. R package version 1.0-6.

Hall, R.J., Henry, P.F.P., Bunck, C.M., 1999. Fifty-year trends in a box turtle population in Maryland. Biological Conservation 88, 165–172.

Hanks, E.M., Hooten, M.B., 2013. Circuit theory and model-based inference for landscape connectivity. Journal of the American Statistical Association 108, 22–33.

Hanski, I.A., 1999. Metapopulation Ecology. Oxford University Press.

Hardin, R.H., Sloane, N.J.A., 1993. A new approach to the construction of optimal designs. Journal of Statistical Planning and Inference 37, 339–369.

Hastings, W., 1970. Monte Carlo sampling methods using Markov chains and their applications. Biometrika 57, 97–109.
Hawkins, C.E., Racey, P.A., 2005. Low population density of a tropical forest carnivore, cryptoprocta ferox: implications for protected area management, Oryx 39, 35–43.
Hayes, R.J., Buckland, S.T., 1983. Radial distance models for the line transect method. Biometrics 39, 29–42.
Hayne, D.W., 1950. Apparent home range of Microtus in relation to distance between traps. Journal of Mammalogy 31, 26–39.
Hayne, D.W., 1949. An examination of the strip census method for estimating animal populations. Journal of Wildlife Management 13, 145–157.
Hedley, S.L., Buckland, S.T., Borchers, D.L., 1999. Spatial modelling from line transect data. Journal of Cetacean Research and Management 1, 255–264.
Hendriks, I., Tenan, S., Tavecchia, G., Marbá, N., Jordá, G., Deudero, S., Álvarez, E., Duarte, C.M., 2013. Boat anchoring impacts coastal populations of the pen shell, the largest bivalve in the Mediterranean. Biological Conservation 160, 105–113.
Hestbeck, J.B., Malecki, R.A., 1989. Mark-resight estimate of Canada Goose midwinter numbers. Journal of Wildlife Management 53, 749–752.
Hijmans, R.J., van Etten, J., 2012. Raster: geographic analysis and modeling with raster data. R package version 1.9-67.
Hines, J.E., Kendall, W.L., Nichols, J.D., 2003. On the use of the robust design with transient capture-recapture models. Auk 120, 1151–1158.
Holdenried, R., 1940. A population study of the long-eared chipmunk (Eutamias quadrimaculatus) in the central Sierra Nevada. Journal of Mammalogy 21, 405–411.
Hooten, M.B., Johnson, D.S., Hanks, E.M., Lowry, J.H., 2010. Agent-based inference for animal movement and selection. Journal of Agricultural, Biological, and Environmental Statistics 15, 523–538.
Hooten, M.B., Wikle, C.K., 2010. Statistical agent-based models for discrete spatio-temporal systems. Journal of the American Statistical Association 105, 236–248.
Hornik, K., 2011. The R FAQ. ISBN 3-900051-08-9.
Huggins, R.M., 1989. On the statistical analysis of capture experiments. Biometrika 76, 133–140.
Hurlbert, S.H., 1984. Pseudoreplication and the design of ecological field experiments. Ecological monographs 54, 187–211.
Illian, J., Penttinen, A., Stoyan, H., Stoyan, D., 2008. Statistical Analysis and Modelling of Spatial Point Patterns. Wiley.
Ivan, J.S., 2012. Density, demography, and seasonal movements of snowshoe hares in central Colorado, PhD Thesis. Colorado State University.
Ivan, J.S., White, G.C., Shenk, T.M., 2013a. Using auxiliary telemetry information to estimate animal density from capture-recapture data. Ecology 94, 809–816.
Ivan, J.S., White, G.C., Shenk, T.M., 2013b. Using simulation to compare methods for estimating density from capture-recapture data. Ecology 94, 817–826.

Jackson, R.M., Roe, J.D., Wangchuk, R., Hunter, D.O., 2006. Estimating snow leopard population abundance using photography and capture-recapture techniques. Wildlife Society Bulletin 34, 772–781.

Jennelle, C.S., Runge, M.C., MacKenzie, D.I., 2002. The use of photographic rates to estimate densities of tigers and other cryptic mammals: a comment on misleading conclusions. Animal Conservation 5, 119–120.

Jett, D.A., Nichols, J.D., 1987. A field comparison of nested grid and trapping web density estimators. Journal of Mammalogy 68, 888–892.

Jhala, Y.V., Qureshi, Q., Gopal, R., 2011. Can the abundance of tigers be assessed from their signs? Journal of Applied Ecology 48, 14–24.

Johnson, D.S., Laake, J.L., Ver Hoef, J.M., 2010. A model-based approach for making ecological inference from distance sampling data. Biometrics 66, 310–318.

Johnson, D., 1980. The comparison of usage and availability measurements for evaluating resource preference. Ecology 61, 65–71.

Johnson, D.H., 1999. The insignificance of statistical significance testing. Journal of Wildlife Management 63, 763–772.

Johnson, D.S., London, J.M., Lea, M.A., Durban, J., 2008a. Continuous-time correlated random walk model for animal telemetry data. Ecology 89, 1208–1215.

Johnson, D.S., Thomas, D.L., Ver Hoef, J.M., Christ, A., 2008b. A general framework for the analysis of animal resource selection from telemetry data. Biometrics 64, 968–976.

Jolly, G.M., 1965. Explicit estimates from capture-recapture data with both death and dilution-stochastic model. Biometrika 52, 225–247.

Jonsen, I.D., Flemming, J.M., Myers, R.A., 2005. Robust state-space modeling of animal movement data. Ecology 86, 2874–2880.

Kadane, J.B., Lazar, N.A., 2004. Methods and criteria for model selection. Journal of the American Statistical Association 99, 279–290.

Karanth, K.U., 1995. Estimating tiger Panthera tigris populations from camera-trap data using capture-recapture models. Biological Conservation 71, 333–338.

Karanth, K.U., Nichols, J.D., 1998. Estimation of tiger densities in India using photographic captures and recaptures. Ecology 79, 2852–2862.

Karanth, K.U., Nichols, J.D., 2000. Ecological status and conservation of tigers in India, WCS, US Fish and Wildlife Service. Centre for Wildlife Studies, Bangalore, India.

Karanth, K.U., Nichols, J.D., 2002. Monitoring tigers and their prey: a manual for researchers, managers and conservationists in tropical Asia. Centre for Wildlife Studies, Bangalore, India.

Kass, R.E., Wasserman, L., 1996. The selection of prior distributions by formal rules. Journal of the American Statistical Association 91, 1343–1370.

Kays, R.W., Slauson, K.M., 2008. Remote cameras. In: Long, R.A., MacKay, P., Zielinski, W.J., Ray, J.C. (Eds.), Noninvasive Survey Methods for Carnivores. Island Press, Washington, DC, pp. 110–140.

Kelly, M.J., Noss, A.J., Di Bitetti, M.S., Maffei, L., Arispe, R.L., Paviolo, A., De Angelo, C.D., Di Blanco, Y.E., 2008. Estimating puma densities from camera trapping across three study sites: Bolivia. Argentina, and Belize. Journal of Mammalogy 89, 408–418.

Kendall, K.C., Stetz, J.B., Boulanger, J., Macleod, A.C., Paetkau, D., White, G.C., 2009. Demography and genetic structure of a recovering grizzly bear population. Journal of Wildlife Management 73, 3–16.

Kendall, W.L., 1999. Robustness of closed capture-recapture methods to violations of the closure assumption. Ecology 80, 2517–2525.

Kendall, W.L., Nichols, J.D., Hines, J.E., 1997. Estimating temporary emigration using capture-recapture data with Pollock's Robust design. Ecology 78, 563–578.

Kéry, M., 2008. Estimating abundance from bird counts: binomial mixture models uncover complex covariate relationships. Auk 125, 336–345.

Kéry, M., 2010. Introduction to WinBUGS for Ecologists: A Bayesian Approach to Regression, ANOVA and Related Analyses. Academic Press.

Kéry, M., 2011. Towards the modelling of true species distributions. Journal of Biogeography 38, 617–618.

Kéry, M., Gardner, B., Monnerat, C., 2010. Predicting species distributions from checklist data using site-occupancy models. Journal of Biogeography 37, 1851–1862.

Kéry, M., Gardner, B., Stoeckle, T., Weber, D., Royle, J.A., 2011. Use of spatial capture-recapture modeling and DNA data to estimate densities of elusive animals. Conservation Biology 25, 356–364.

Kéry, M., Royle, J.A., Schmid, H., 2005. Modeling avian abundance from replicated counts using binomial mixture models. Ecological Applications 15, 1450–1461.

Kéry, M., Schaub, M., 2012. Bayesian Population Analysis Using WinBugs. Academic Press.

Kiefer, J., 1959. Optimal experimental designs (with discussion). Journal of the Royal Statistical Society, Series B 21, 272–319.

King, R., Brooks, S.P., 2001. On the Bayesian analysis of population size. Biometrika 88, 317–336.

King, R., Brooks, S.P., Coulson, T., 2008. Analyzing complex capture-recapture data in the presence of individual and temporal covariates and model uncertainty. Biometrics 64, 1187–1195.

Knape, J., de Valpine, P., 2012. Are patterns of density dependence in the Global Population Dynamics Database driven by uncertainty about population abundance? Ecology Letters 15, 17–23.

Knaus, J., 2010. Snowfall: easier cluster computing (based on snow). R package version 1.8-4.

Koehler, G.M., Pierce, D.J., 2003. Black bear home-range sizes in Washington: climatic, vegetative, and social influences. Journal of Mammalogy 84, 81–91.

Kohn, M., York, E., Kamradt, D., Haught, G., Sauvajot, R., Wayne, R., 1999. Estimating population size by genotyping faeces. Proceedings of the Royal Society of London. Series B: Biological Sciences 266, 657–663.

Krebs, C.J., 1999. Ecological Methodology. Benjamin/Cummings, Menlo Park, CA.

Kucera, T.E., Barrett, R.H., 2011. A history of camera trapping. In: O'Connell, A.F., Nichols, J.D., Karanth, K.U. (Eds.), Camera Traps in Animal Ecology Methods and Analyses. Springer, Toyoko, Japan, pp. 9–26.

Kuo, L., Mallick, B., 1998. Variable selection for regression models, Sankhyā: The Indian Journal of Statistics, Series B 60, 65–81.
Laird, N.M., Ware, J.H., 1982. Random-effects models for longitudinal data. Biometrics 38, 963–974.
Langtimm, C.A., Dorazio, R.M., Stith, B.M., Doyle, T.J., 2011. New aerial survey and hierarchical model to estimate manatee abundance. Journal of Wildlife Management 75, 399–412.
Le Cam, L., 1990. Maximum likelihood: an introduction. International Statistical Review/Revue Internationale de Statistique 58, 153–171.
Lebreton, J.D., Burnham, K.P., Clobert, J., Anderson, D.R., 1992. Modeling survival and testing biological hypotheses using marked animals: a unified approach with case studies. Ecological Monographs 62, 67–118.
Leggett, W.C., Carscadden, J.E., 1978. Latitudinal variation in reproductive characteristics of American shad (Alosa sapidissima): evidence for population specific life history strategies in fish. Journal of the Fisheries Research Board of Canada 35, 1469–1477.
Lele, S.R., Keim, J.L., 2006. Weighted distributions and estimation of resource selection probability functions. Ecology 87, 3021–3028.
Lele, S.R., Moreno, M., Bayne, E., 2012. Dealing with detection error in site occupancy surveys: what can we do with a single survey? Journal of Plant Ecology 5, 22–31.
Lele, S.R., Nadeem, K., Schmuland, B., 2010. Estimability and likelihood inference for generalized linear mixed models using data cloning. Journal of the American Statistical Association 105, 1617–1625.
Leonard, J.B.K., McCormick, S.D., 1999. Effects of migration distance on whole-body and tissue-specific energy use in American shad (Alosa sapidissima). Canadian Journal of Fisheries and Aquatic Sciences 56, 1159–1171.
Lewin-Koh, N.J., Bivand, R., contributions by Edzer J. Pebesma, Archer, E., Baddeley, A., Bibiko, H.-J., Dray, S., Forrest, D., Friendly, M., Giraudoux, P., Golicher, D., Rubio, V.G., Hausmann, P., Hufthammer, K.O., Jagger, T., Luque, S.P., MacQueen, D., Niccolai, A., Short, T., Stabler, B., Turner, R., 2011. Maptools: Tools for reading and handling spatial objects. R package version 0.8-10.
Lichstein, J.W., Simons, T.R., Shriner, S.A., Franzreb, K.E., 2002. Spatial autocorrelation and autoregressive models in ecology. Ecological Monographs 72, 445–463.
Link, W.A., 2003. Nonidentifiability of population size from capture-recapture data with heterogeneous detection probabilities. Biometrics 59, 1123–1130.
Link, W.A., in press. A cautionary note on the discrete uniform prior for the binomial N. Ecology (in press).
Link, W.A., Barker, R.J., 1994. Density estimation using the trapping web design: a geometric analysis. Biometrics 50, 733–745.
Link, W.A., Barker, R.J., 2005. Modeling association among demographic parameters in analysis of open population capture-recapture data. Biometrics 61, 46–54.
Link, W.A., Barker, R.J., 2006. Model weights and the foundations of multimodel inference. Ecology 87, 2626–2635.

Link, W.A., Barker, R.J., 2010. Bayesian Inference: With Ecological Applications. Academic Press, London, UK.
Link, W.A., Eaton, M.J., 2011. On thinning of chains in MCMC. Methods in Ecology and Evolution 3, 112–115.
Link, W.A., Yoshizaki, J., Bailey, L.L., Pollock, K.H., 2010. Uncovering a latent multinomial: analysis of mark-recapture data with misidentification. Biometrics 66, 178–185.
Liu, Wu, 1999. Parameter expansion for data augmentation. Journal of American Statistical Association 94, 1264–1274.
Lukacs, P.M., Burnham, K.P., 2005. Estimating population size from DNA-based closed capture-recapture data incorporating genotyping error. Journal of Wildlife Management 69, 396–403.
Lunn, D.J., Spiegelhalter, D., Thomas, A., Best, N., 2009. The BUGS project: evolution, critique, and future directions. Statistics in Medicine 28, 3049–3067.
Lunn, D.J., Thomas, A., Best, N., Spiegelhalter, D., 2000. WinBUGS-a Bayesian modelling framework: concepts, structure, and extensibility. Statistics and Computing 10, 325–337.
Mace, R., Minta, S., Manley, T., Aune, K., 1994. Estimating grizzly bear population size using camera sightings. Wildlife Society Bulletin 22, 74–83.
MacEachern, S.N., Berliner, L.M., 1994. Subsampling the Gibbs sampler. American Statistician 48, 188–190.
MacKay, P., Smith, D.A., Long, R.A., Parker, M., 2008. Scat detection dogs. In: Long, R.A., MacKay, P., Zielinski, W.J., Ray, J.C. (Eds.), Noninvasive Survey Methods for Carnivores. Island Press, Washington DC, pp. 183–222.
MacKenzie, D.I., Nichols, J.D., Lachman, G.B., Droege, S., Royle, J.A., Langtimm, C.A., 2002. Estimating site occupancy rates when detection probabilities are less than one. Ecology 83, 2248–2255.
MacKenzie, D.I., Nichols, J.D., Royle, J.A., Pollock, K.H., Bailey, L.L., Hines, J.E., 2006. Occupancy Estimation and Modeling: Inferring Patterns and Dynamics of Species Occurrence. Academic Press.
Maffei, L., Noss, A.J., 2008. How small is too small? Camera trap survey areas and density estimates for ocelots in the Bolivian Chaco. Biotropica 40, 71–75.
Magoun, A.J., Long, C.D., Schwartz, M.K., Pilgrim, K.L., Lowell, R.E., Valkenburg, P., 2011. Integrating motion-detection cameras and hair snags for wolverine identification. Journal of Wildlife Management 75, 731–739.
Manel, S., Schwartz, M.K., Luikart, G., Taberlet, P., 2003. Landscape genetics: combining landscape ecology and population genetics. Trends in Ecology and Evolution 18, 189–197.
Manly, B.F.J., McDonald, L., Thomas, D.L., McDonald, T., Erickson, W., 2002. Resource Selection by Animals: Statistical Design and Analysis for Field Studies, second ed. Springer.
Marques, T.A., Buckland, S.T., Borchers, D.L., Tosh, D., McDonald, R.A., 2010. Point transect sampling along linear features. Biometrics 66, 1247–1255.

Marques, T.A., Thomas, L., Ward, J., DiMarzio, N., Tyack, P.L., 2009. Estimating cetacean population density using fixed passive acoustic sensors: an example with Blainville's beaked whales. Journal of the Acoustical Society of America 125, 1982–1994.

Marques, T.A., Thomas, L., Royle, J.A., 2011. A hierarchical model for spatial capture-recapture data: comment. Ecology 92, 526–528.

Matechou, E., Morgan, B.J.T., Pledger, S., Collazo, J.A., Lyons, J.E., 2013. Integrated analysis of capture-recapture-resighting data and counts of unmarked birds at stop-over sites. Journal of Agricultural, Biological, and Environmental Statistics 18, 120–135.

Matthysen, E., 2005. Density-dependent dispersal in birds and mammals. Ecography 28, 403–416.

McCarthy, M.A., 2007. Bayesian Methods for Ecology. Cambridge University Press, Cambridge.

McClintock, B.T., Hoeting, J.A., 2010. Bayesian analysis of abundance for binomial sighting data with unknown number of marked individuals. Environmental and Ecological Statistics 17, 317–332.

McClintock, B.T., King,R., Thomas,L., Matthiopoulos,J., McConnell,B.J., Morales,J.M., 2012. A general discrete-time modeling framework for animal movement using multi-state random walks. Ecological Monographs 82, 335–349.

McClintock, B.T., White, G.C., 2012. From NOREMARK to MARK: software for estimating demographic parameters using mark-resight methodology. Journal of Ornithology 152, 641–650.

McClintock, B.T., White, G.C., Antolin, M.F., Tripp, D.W., 2009a. Estimating abundance using mark-resight when sampling is with replacement or the number of marked individuals is unknown. Biometrics 65, 237–246.

McClintock, B.T., White, G.C., Burnham, K.P., 2006. A robust design mark-resight abundance estimator allowing heterogeneity in resighting probabilities. Journal of Agricultural, Biological, and Environmental Statistics 11, 231–248.

McClintock, B.T., Conn, P., Alonso, R., Crooks, K.R., 2013. Integrated modeling of bilateral photo-identification data in mark-recapture analyses. Ecology. http://dx.doi.org/10.1890/12-1613.1.

McClintock, B.T., White, G.C., 2009. A less field-intensive robust design for estimating demographic parameters with markresight data. Ecology 90, 313–320.

McClintock, B.T., White, G.C., Burnham, K.P., Pryde, M.A., 2009b. A generalized mixed effects model of abundance for mark-resight data when sampling is without replacement. In: Thomson, D., Cooch, E.G., Conroy, M.J. (Eds.), Modeling Demographic Processes in Marked Populations. Springer, New York, pp. 271–289.

McCullagh, P., Nelder, J.A., 1989. Generalized Linear Models. Chapman & Hall/CRC.

McRae, B.H., Beier, P., 2007. Circuit theory predicts gene flow in plant and animal populations. Proceedings of the National Academy of Sciences 104, 19885–19890.

McRae, B.H., Dickson, B.G., Keitt, T.H., Shah, V.B., 2008. Using circuit theory to model connectivity in ecology, evolution, and conservation. Ecology 89, 2712–2724.

Metropolis, N., Rosenbluth, A.W., Rosenbluth, M.N., Teller, A.H., Teller, E., et al., 1953. Equation of state calculations by fast computing machines. Journal of Chemical Physics 21, 1087–1092.

Metropolis, N., Ulam, S., 1949. The Monte Carlo method. Journal of the American Statistical Association 44, 335–341.

Meyer, R.K., Nachtsheim, C.J., 1995. The coordinate-exchange algorithm for construction exact optimal experiemental designs. Technometrics 37, 60–69.

Millar, R.B., 2009. Comparison of hierarchical Bayesian models for overdispersed count data using DIC and Bayes' factors. Biometrics 65, 962–969.

Mills, L.S., Citta, J.J., Lair, K.P., Schwartz, M.K., Tallmon, D.A., 2000. Estimating animal abundance using noninvasive DNA sampling: promise and pitfalls. Ecological Applications 10, 283–294.

Minta, S., Mangel, M., 1989. A simple population estimate based on simulation for capture-recapture and capture-resight data. Ecology 70, 1738–1751.

Mitchell, T.J., 1974. An algorithm for the construction of D-optimal experimental designs. Techometrics 16, 203–210.

Mohr, C., 1947. Table of equivalent populations of North American small mammals. American Midland Naturalist 37, 223–249.

Molinari-Jobin, A., Kéry, M., Marboutin, E., Marucco, F., Zimmermann, F., Molinari, P., Frick, H., Wölfl, S., Bled, F., Breitenmoser-Würsten, C., Fuxjäger, C., Huber, T., Koren, I., Kos, I., Manfred Wölfl, M., Breitenmoser, U., 2013. Mapping range dynamics from opportunistic data: spatio-temporal distribution modeling of lynx Lynx lynx L. in the Alps. (unpublished manuscript).

Mollet, P., Kéry, M., Gardner, B., Pasinelli, G., Royle, J.A., in review. Population size estimation for capercaillie (Tetrao urogallus L.) using DNA-based individual recognition and spatial capture-recapture models.

Morrison, M.L., Strickland, M.D., Block, W.M., Collier, B.A., Peterson, M.J., 2008. Wildlife Study Design. Springer.

Müller, W.G., 2007. Collecting Spatial Data: Optimum Design of Experiments for Random Fields. Springer.

Murdoch, W.W., 1994. Population redulation in theory and practice. Ecology 75, 271–287.

Neal, A.K., White, G.C., Gill, R., Reed, D., Olterman, J., 1993. Evaluation of mark-resight model assumptions for estimating mountain sheep numbers. Journal of Wildlife Management 57, 436–450.

Neal, A.K. 1990. Evaluation of Mark-Resight Population Estimates using Simulations and Field Data from Mountain Sheep. MS Thesis, Colorado State University, Fort Collins, Colorado, USA.

Neal, R., 2003. Slice sampling. Annals of Statistics 31, 705–741.

Nelder, J.A., Wedderburn, R.W.M., 1972. Generalized linear models, Journal of the Royal Statistical Society. Series A (General) 135, 370–384.

Nichols, J.D., Pollock, K.H., 1990. Estimation of recruitment from immigration versus in situ reproduction using Pollock's robust design. Ecology 71, 21–26.

Nichols, J.D., Thomas, L., Conn, P.B., 2009. Inferences about landbird abundance from count data: recent advances and future directions. In: Thomson, D., Cooch, E.G., Conroy, M.J. (Eds.), Modeling Demographic Processes in Marked Populations. Springer, pp. 201–235.

Nichols, J.D., Hines, J.E., Lebreton, J.-D., Pradel, R., 2000a. Estimation of contributions to population growth: a reverse-time capture-recapture approach. Ecology 81, 3362–3376.

Nichols, J.D., Hines, J.E., Pollock, K.H., Hinz, R.L., Link, W.A., 1994. Estimating breeding proportions and testing hypotheses about costs of reproduction with capture-recapture data. Ecology 75, 2052–2065.

Nichols, J.D., Hines, J.E., Sauer, J.R., Fallon, F.W., Fallon, J.E., Heglund, P.J., 2000b. A double-observer approach for estimating detection probability and abundance from point counts. Auk 117, 393–408.

Nichols, J.D., Karanth, K.U., 2002. Statistical concepts: assessing spatial distributions. In: Karanth, K.U., Nichols, J.D., (Eds.), Monitoring tigers and their prey: a manual for researchers, managers and conservationists in Tropical Asia. Centre for Wildlife Studies, Bangalore, India, pp. 29–38.

Niemi, A., Fernández, C., 2010. Bayesian spatial point process modeling of line transect data. Journal of Agricultural, Biological, and Environmental Statistics 15, 327–345.

Norris, J.L., Pollock, K.H., 1996. Nonparametric MLE under two closed capture-recapture models with heterogeneity. Biometrics 52, 639–649.

Nowak, R.M., 1999. Walker's Mammals of the World, vol 1, sixth ed. John's Hopkins University Press, Baltimore.

Nychka, D., Yang, Q., Royle, J.A., 1997. Constructing spatial designs for monitoring air pollution using regression subset selection. In: Barnett, V., Turkman, K.F. (Eds.), Statistics for the Environment: Pollution Assesment and Control, vol. 3. Springer Verlag, New York, NY, pp. 131–154.

O'Brien, T.G., 2011. Abundance, density and relative abundance: a conceptual framework. In: O'Connell, A.F., Nichols, J.D., Karanth, K.U. (Eds.), Camera Traps in Animal Ecology: Methods and Analyses. Springer, Toyoko, Japan, pp. 71–96.

O'Connell, A.F., Nichols, J.D., Karanth, K.U. (Eds.), 2011. Camera Traps in Animal Ecology: Methods and Analyses. Springer, Tokoyo, Japan.

O'Hara, R., Sillanpää, M., 2009. A review of Bayesian variable selection methods: what, how and which. Bayesian Analysis 4, 85–118.

Otis, D.L., Burnham, K.P., White, G.C., Anderson, D.R., 1978. Statistical inference from capture data on closed animal populations. Wildlife Monographs, 3–135.

Ovaskainen, O., 2004. Habitat-specific movement parameters estimated using mark-recapture data and a diffusion model. Ecology 85, 242–257.

Ovaskainen, O., Rekola, H., Meyke, E., Arjas, E., 2008. Bayesian methods for analyzing movements in heterogeneous landscapes from mark-recapture data. Ecology 89, 542–554.

Parmenter, R.R., MacMahon, J.A., 1989. Animal density estiamtion using a trapping web design: field validation experiments. Ecology 70, 169–179.

Parmenter, R.R., Yates, T.L., Anderson, D.R., Burnham, K.P., Dunnum, J.L., Franklin, A.B., Friggens, M.T., Lubow, B.C., Miller, M., Olson, G.S., et al., 2003. Small-mammal density estimation: a field comparison of grid-based vs. web-based density estimators. Ecological Monographs 73, 1–26.

Patterson, T.A., Thomas, L., Wilcox, C., Ovaskainen, O., Matthiopoulos, J., 2008. State-space models of individual animal movement. Trends in Ecology and Evolution 23, 87–94.

Paviolo, A., De Angelo, C.D., Di Blanco, Y.E., Di Bitetti, M.S., 2008. Jaguar Panthera onca population decline in the Upper Paraná Atlantic Forest of Argentina and Brazil. Oryx 42, 554–561.

Paviolo, A., Di Blanco, Y.E., De Angelo, C.D., Di Bitetti, M.S., 2009. Protection affects the abundance and activity patterns of pumas in the Atlantic Forest. Journal of Mammalogy 90, 926–934.

Pebesma, E., Bivand, R., 2011. Package sp. R package version 0.9-91.

Pledger, S., 2004. Unified maximum likelihood estimates for closed capture-recapture models using mixtures. Biometrics 56, 434–442.

Pledger, S., Efford, M.G., Pollock, K.H., Collazo, J., Lyons, J.E., 2009. Stopover duration analysis with departure probability dependent on unknown time since arrival. Modeling Demographic Processes in Marked populations 3, 349–363.

Plummer, M., 2003. JAGS: A program for analysis of Bayesian graphical models using Gibbs sampling. In: Proceedings of the 3rd International Workshop on Distributed Statistical Computing (DSC 2003). March, pp. 20–22.

Plummer, M., 2011. rjags: Bayesian graphical models using mcmc. R package version 3-5.

Plummer, M., Best, N., Cowles, K., Vines, K., 2006. CODA: convergence diagnosis and output analysis for MCMC. R News 6, 7–11.

Pollock, K.H., 1982. A capture-recapture design robust to unequal probability of capture. Journal of Wildlife Management 46, 752–757.

Pollock, K.H., Nichols, J.D., Brownie, C., Hines, J.E. 1990. Statistical inference for capture-recapture experiments. Wildlife Monographs 107, 3–97.

Porneluzi, P.A., Faaborg, J., 1999. Season long fecundity, survival, and viability of Ovenbirds in fragmented and unfrangmented landscapes. Conservation Biology 13, 1151–1161.

Pradel, R., 1996. Utilization of capture-mark-recapture for the study of recruitment and population growth rate. Biometrics 52, 703–709.

Pradel, R., Hines, J.E., Lebreton, J.D., Nichols, J.D., 1997. Capture-Recapture Survival Models Taking Account of Transients. Biometrics 53, 60–72.

Raabe, J. 2012. Factors Influencing Distribution and Survival of Migratory Fishes following Multiple Low-Head Dam Removals on a North Carolina River. Ph.D dissertation. Ph.D Thesis, North Carolina State University, Raleigh, NC.

Rathbun, S.L., 1996. Estimation of Poisson intensity using partially observed concomitant variables. Biometrics 52, 226–242.

Rathbun, S.L., Cressie, N.A.C., 1994. A space-time survival point process for a longleaf pine forest in southern Georgia. Journal of the American Statistical Association 89, 1164–1174.
Rathbun, S.L., Shiffman, S., Gwaltney, C., 2007. Modelling the effects of partially observed covariates on Poisson process intensity. Biometrika 94, 153–165.
Reich, B.J., Gardner, B., Wilting, A., in review. A spatial capture-recapture model for territorial species.
Robert, C.P., Casella, G., 2004. Monte Carlo Statistical Methods, Springer, New York, USA.
Robert, C.P., Casella, G., 2010. Introducing Monte Carlo Methods with R, Springer, New York, USA.
Roberts, G.O., Rosenthal, J.S., 1998. Optimal scaling of discrete approximations to Langevin diffusions. Journal of the Royal Statistical Society: Series B (Statistical Methodology) 60, 255–268.
Rowcliffe, M.J., Carbone, C., Jansen, P.A., Kays, R., Kranstauber, B., 2011. Quantifying the sensitivity of camera traps: an adapted distance sampling approach. Methods in Ecology and Evolution 2, 464–476.
Rowcliffe, M.J., Field, J., Turvey, S.T., Carbone, C., 2008. Estimating animal density using camera traps without the need for individual recognition. Journal of Applied Ecology 45, 1228–1236.
Royle, J.A., 2004a. Generalized estimators of avian abundance from count survey data. Animal Biodiversity and Conservation 27, 375–386.
Royle, J.A., 2004b. N-mixture models for estimating population size from spatially replicated counts. Biometrics 60, 108–115.
Royle, J.A., 2006. Site occupancy models with heterogeneous detection probabilities. Biometrics 62, 97–102.
Royle, J.A., 2008. Modeling individual effects in the Cormack-Jolly-Seber model: a state-space formulation. Biometrics 64, 364–370.
Royle, J.A., 2009. Analysis of capture-recapture models with individual covariates using data augmentation. Biometrics 65, 267–274.
Royle, J.A., Chandler, R.B., 2012. Integrating Resource Selection Information with Spatial Capture-Recapture. arXiv, preprint arXiv:1207.3288.
Royle, J.A., Chandler, R.B., Gazenski, K.D., Graves, T.A. 2013a. Spatial capture-recapture for jointly estimating population density and landscape connectivity. Ecology 94, 287–294.
Royle, J.A., Chandler, R.B., Sun, C.C., Fuller, A.K., 2013b. Integrating resource selection information with spatial capture-recapture. Methods in Ecology and Evolution 4, 520–530.
Royle, J.A., Chandler, R.B., Yackulic, C., Nichols, J.D., 2012a. Likelihood analysis of species occurrence probability from presence-only data for modelling species distributions. Methods in Ecology and Evolution 3, 545–554.
Royle, J.A., Converse, S.J., in review. Hierarchical spatial capture-recapture models: Modeling population density in stratified populations.
Royle, J.A., Converse, S.J., Link, W.A., 2012b. Data Augmentation for hierarchical capture-recapture models. arXiv, preprint arXiv:1211.5706.

Royle, J.A., Dawson, D.K., Bates, S., 2004. Modeling abundance effects in distance sampling. Ecology 85, 1591–1597.
Royle, J.A., Dorazio, R.M., 2006. Hierarchical models of animal abundance and occurrence. Journal of Agricultural, Biological, and Environmental Statistics 11, 249–263.
Royle, J.A., Dorazio, R.M., 2008. Hierarchical Modeling and Inference in Ecology: the Analysis of Data from Populations, Metapopulations and Communities. Academic Press.
Royle, J.A., Dorazio, R.M., 2012. Parameter-expanded data augmentation for Bayesian analysis of capture-recapture models. Journal of Ornithology 152, S521–S537.
Royle, J.A., Dorazio, R.M., Link, W.A., 2007. Analysis of multinomial models with unknown index using data augmentation. Journal of Computational and Graphical Statistics 16, 67–85.
Royle, J.A., Dubovsky, J.A., 2001. Modeling spatial variation in waterfowl band-recovery data. Journal of Wildlife Management 65, 726–737.
Royle, J.A., Gardner, B. 2011. Hierarchical models for estimating density from trapping arrays. In: O'Connel, A.F.J., Nichols, J.D., Karanth, U., (Eds.), Camera Traps in Animal Ecology: Methods and Analyses. Springer Verlag, Toyoko, Japan, pp. 163–190.
Royle, J.A., Karanth, K.U., Gopalaswamy, A.M., Kumar, N.S., 2009a. Bayesian inference in camera trapping studies for a class of spatial capture-recapture models. Ecology 90, 3233–3244.
Royle, J.A., Kéry, M., 2007. A Bayesian state-space formulation of dynamic occupancy models. Ecology 88, 1813–1823.
Royle, J.A., Kéry, M., Guélat, J., 2011a. Spatial capture-recapture models for search-encounter data. Methods in Ecology and Evolution 2, 602–611.
Royle, J.A., Link, W.A., 2006. Generalized site occupancy models allowing for false positive and false negative errors. Ecology 87, 835–841.
Royle, J.A., Magoun, A.J., Gardner, B., Valkenburg, P., Lowell, R.E., 2011b. Density estimation in a wolverine population using spatial capture-recapture models. Journal of Wildlife Management 75, 604–611.
Royle, J.A., Nichols, J.D., 2003. Estimating abundance from repeated presence-absence data or point counts. Ecology 84, 777–790.
Royle, J.A., Nichols, J.D., Karanth, K.U., Gopalaswamy, A.M., 2009b. A hierarchical model for estimating density in camera-trap studies. Journal of Applied Ecology 46, 118–127.
Royle, J.A., Nychka, D., 1998. An algorithm for the construction of spatial coverage designs with implementation in SPLUS. Computers and Geosciences 24, 479–488.
Royle, J.A., Young, K.V., 2008. A hierarchical model for spatial capture-recapture data. Ecology 89, 2281–2289.
Russell, R.E., Royle, J.A., Desimone, R., Schwartz, M.K., Edwards, V.L., Pilgrim, K.P., McKelvey, K.S., 2012. Estimating abundance of mountain lions from unstructured spatial samples. Journal of Wildlife Management 76, 1551–1561.

Rutledge, M., 2013. Impacts of resident Canada geese in a suburban environment, PhD Thesis. North Carolina State University.

Sacks, J., Welch, W.J., Mitchell, T.P., Wynn, H., 1989. Design and analysis of computer experiments. Statistical Science 4, 409–435.

Sæther, B.-E., Bakke, 2000. Avian life history variation and contribution of demographic traits to the population growth rate. Ecology 81, 642–653.

Sæther, B.E., Engen, S., Matthysen, E., 2002. Demographic characteristics and population dynamical patterns of solitary birds. Science 295, 2070–2073.

Saïd, S., Servanty, S., 2005. The influence of landscape structure on female roe deer home range size. Landscape Ecology 20, 1003–1012.

Salom-Pérez, R., Carrillo, E., Sáenz, J., Mora, J., 2007. Critical condition of the jaguar Panthera onca population in Corcovado National Park, Costa Rica. Oryx 41, 51–56.

Sanathanan, L., 1972. Estimating the size of a multinomial population. Annals of Mathematical Statistics 43, 142–152.

Sauer, J.R., Link, W.A., 2002. Hierarchical modeling of population stability and species group attributes from survey data. Ecology 83, 1743–1751.

Schaub, M., Royle, J.A., 2013. Estimating true instead of apparent survival using spatial Cormack-Jolly-Seber models (in review).

Schofield, M.R., Barker, R.J., 2008. A unified capture-recapture framework. Journal of Agricultural, Biological, and Environmental Statistics 13, 458–477.

Schwartz, M.K., Copeland, J.P., Anderson, N.J., Squires, J.R., Inman, R.M., McKelvey, K.S., Pilgrim, K.L., Waits, L.P., Cushman, S.A., 2009. Wolverine gene flow across a narrow climatic niche. Ecology 90, 3222–3232.

Schwartz, M.K., Monfort, S.L., 2008. Genetic and endocrine tools for carnivore surveys. In: Long, R., MacKay, P., Ray, J., Zielinski, W. (Eds.), Noninvasive Survey Methods for Carnivores. Island Press Washington, DC, USA, pp. 228–250.

Schwarz, C.J., Arnason, A.N., 1996. A general methodology for the analysis of Capture-recapture experiments in open populations. Biometrics 52, 860–873.

Schwarz, C.J., Arnason, A.N., 2005. Jolly-Seber models in MARK. In: Cooch, E.G., White, G. (Eds.), Program MARK: A Gentle Introduction, fifth ed. (book accessed online <http://www.phidot.org/software/mark/docs/book/>).

Schwarz, C.J., Bailey, R.E., Irvine, J.R., Dalziel, F.C., 1993. Estimating salmon spawning escapement using capture-recapture methods. Canadian Journal of Fisheries and Aquatic Sciences 50, 1181–1191.

Seber, G.A.F., 1965. A note on the multiple-recapture census. Biometrika 52, 249–259.

Seber, G.A.F., 1982. The Estimation of Animal Abundance and Related Parameters. Macmillan Publishing Company.

Sepúlveda, M.A., Bartheld, J.L., Monsalve, R., Gómez, V., Medina-Vogel, G., 2007. Habitat use and spatial behaviour of the endangered Southern river otter (Lontra provocax) in riparian habitats of Chile: conservation implications. Biological Conservation 140, 329–338.

Shirk, A.J., Wallin, D.O., Cushman, S.A., Rice, C.G., Warheit, K.I., 2010. Inferring landscape effects on gene flow: a new model selection framework. Molecular Ecology 19, 3603–3619.

Sillett, T.S., Chandler, R.B., Royle, J.A., Kéry, M., Morrison, S.A., 2012. Hierarchical distance-sampling models to estimate population size and habitat-specific abundance of an island endemic. Ecological Applications 22, 1997–2006.
Sillett, T.S., Rodenhouse, N.L., Holmes, R.T., 2004. Experimentally reducing neighbor density affects reproduction and behavior of a migratory songbird. Ecology 85, 2467–2477.
Skalski, J.R., Millspaugh, J.J., Spencer, R.D., 2005. Population estimation and biases in paintball, mark-resight surveys of elk. Journal of Wildlife Management 69, 1043–1052.
Skaug, H.J., Schweder, T., 1999. Hazard models for line transect surveys with independent observers. Biometrics 55, 29–36.
Sklyar, O., Murdoch, D., Smith, M., Eddelbuettel, D., François, R., 2010. Inline C, C++. Fortran function calls from R. R package version 0.3.8.
Smith, D., Smith, M.S., 2006. Estimation of binary Markov random Fields using Markov chain Monte Carlo. Journal of Computational and Graphical Statistics 15, 1–21.
Smith, M.H., Blessing, R., Chelton, J.G., Gentry, J.B., Golley, F.B., McGinnis, J.T., 1971. Determining density for small mammal populations using a grid and assessment lines. Acta Theriologica 16, 105–125.
Soisalo, M.K., Cavalcanti, S.M.C., 2006. Close-up space in radio-telemetry. Biological Conservation 129, 487–496.
Sollmann, R., Furtado, M.M., Gardner, B., Hofer, H., Jacomo, A.T.A., Trres, N.M., Silveira, L., 2011. Improving density estimates for elusive carnivores: accounting for sex-specific detection and movements using spatial capture-recapture models for jaguars in central Brazil. Biological Conservation 144, 1017–1024.
Sollmann, R., Gardner, B., Belant, J.L., 2012. How does spatial study design influence density estimates from spatial capture-recapture models? PLoS One 7, e34575.
Sollmann, R., Mohamed, A., Samejima, H., Wilting, A., 2013a. Risky business or simple solution-Relative abundance indices from camera-trapping. Biological Conservation 159, 405–412.
Sollmann, R., Gardner, B., Parsons, A., Stocking, J., McClintock, B., Simons, T., Pollock, K.H., O'Connell, A., 2013b. A spatial mark-resight model augmented with telemetry data. Ecology 94, 553–559.
Sollmann, R., Gardner, B., Chandler, R.B., Shindle, D., Onorato, D.P., Royle, J.A., O'Connell, A.F., 2013c. Using multiple data sources provides density estimates for endangered Florida panther. Journal of Applied Ecology 50, 961–968.
Sólymos, P., Lele, S.R., Bayne, E., 2012. Conditional likelihood approach for analyzing single visit abundance survey data in the presence of zero inflation and detection error. Environmetrics 23, 197–205.
Spiegelhalter, D.J., Best, N.G., Carlin, B.P., Van Der Linde, A., 2002. Bayesian measures of model complexity and fit. Journal of the Royal Statistical Society: Series B (Statistical Methodology) 64, 583–639.
Stabler, B., 2006. Shapefiles: read and write ESRI shapefiles. R package version 0.6.
Stanley, T.R., Burnham, K.P., 1999. A closure test for time-specific capture-recapture data. Environmental and Ecological Statistics 6, 197–209.

Stanley, T.R., Richards, J.D., 2013. Software review: a program for testing capture-recapture data for closure. Wildlife Society Bulletin 33, 782–785.

Stevens, Jr. D.L., Olsen, A.R., 2004. Spatially balanced sampling of natural resources. Journal of the American Statistical Association 99, 262–278.

Stevick, P.T., Palsbøll, P.J., Smith, T.D., Bravington, M.V., Hammond, P.S., 2001. Errors in identification using natural markings: rates, sources, and effects on capture-recapture estimates of abundance. Canadian Journal of Fisheries and Aquatic Sciences 58, 1861–1870.

Stoyan, D., Penttinen, A., 2000. Recent applications of point process methods in forestry statistics. Statistical Science 15, 61–78.

Strauss, D., 1975. A model for clustering. Biometrika 63, 467–475.

Sturtz, S., Ligges, U., Gelman, A., 2005. R2WinBUGS: A package for running WinBUGS from R. Journal of Statistical Software 12, 1–16.

Su, Y.-S., Yajima, M., 2011. R2jags: A Package for running jags from R. R package version 0.02-17.

Sun, C.C., 2013. Population Estimation, Genetic Diversity, and Structure of Black bears in Southwestern New York, Master's Thesis. Cornell University.

Taberlet, P., Bouvet, J., 1992. Bear conservation genetics. Nature 358, 197–197.

Takemura, A., 1999. Some superpopulation models for estimating the number of population uniques. In: Proceedings of the International Conference on Statistical Data Protection SDP. Citeseer 98, 45–58.

Tanner, M.A., Wong, W.H., 1987. The calculation of posterior distributions by data augmentation. Journal of the American Statistical Association 82, 528–540.

Tenan, S., O'Hara, R.B., Hendriks, I., Tavecchia, G., 2013. Bayesian model and variable selection in ecological studies (in review).

Thomas, A., O'Hara, R.B., Ligges, U., Sturtz, S., 2006. Making BUGS Open. R News 6, 12–17.

Thompson, C.M., Royle, J.A., Garner, J.D., 2012. A framework for inference about carnivore density from unstructured spatial sampling of scat using detector dogs. Journal of Wildlife Management 76, 863–871.

Thompson, S.K., 2002. Sampling. Wiley, New York.

Tierney, L., Rossini, A.J., Li, N., Sevcikova, H., 2011. Snow: Simple Network of Workstations. R package version 0.3.7.

Tilman, D., Kareiva, P., 1997. Spatial Ecology: the Role of Space in Population Dynamics and Interspecific Interactions. Princeton University Press.

Tischendorf, L., Fahrig, L., 2000. On the usage and measurement of landscape connectivity. Oikos 90, 7–19.

Tischendorf, L., Grez, A., Zaviezo, T., Fahrig, L., 2005. Mechanisms affecting population density in fragmented habitat. Ecology and Society 10, 7.

Tobler, M.W., Carrillo-Percastegui, S.E., Leite Pitman, R., Mares, R., Powell, G., 2008. An evaluation of camera traps for inventorying large-and medium-sized terrestrial rainforest mammals. Animal Conservation 11, 169–178.

Tobler, M.W., Hibert, F., Debeir, L., Hansen, C., 2013. Density and sustainable harvest estimates for the lowland tapir in the Amazon of French Guiana using a spatial capture-recapture model. Oryx (in press).

Tracy, J.A. 2006., Individual-based movement modeling as a tool for conserving connectivity. In: Crooks, K., Sanjayan, M., (Eds.), Connectivity Conservation. Cambridge University Press, pp. 343–368.

Trolle, M., Kéry, M., 2003. Estimation of ocelot density in the Pantanal using capture-recapture analysis of camera-trapping data. Journal of Mammalogy 84, 607–614.

Trolle, M., Kéry, M., 2005. Camera-trap study of ocelot and other secretive mammals in the northern Pantanal. Mammalia 69, 409–416.

Tufto, J., Andersen, R., Linnell, J., 1996. Habitat use and ecological correlates of home range size in a small cervid: the roe deer. Journal of Animal Ecology 65, 715–724.

Tyre, A.J., Tenhumberg, B., Field, S.A., Niejalke, D., Parris, K., Possingham, H.P., 2003. Improving precision and reducing bias in biological surveys: estimating false-negative error rates. Ecological Applications 13, 1790–1801.

Valiere, N., Taberlet, P., 2000. Urine collected in the field as a source of DNA for species and individual identification. Molecular Ecology 9, 2150–2152.

van Etten, J. 2011. Package gdistance. R package version 1.1-2.

Venables, W.N., Ripley, B.D., 2002. Modern Applied Statistics with S, Springer Verlag, New York, NY.

Venables, W.N., Smith, D., Team, R.D.C., 2012. An Introduction to R.

Ver Hoef, J.M., Boveng, P., 2007. Quasi-poisson vs. negative binomial regression: How should we model overdispersed count data? Ecology 88, 2766–2772.

Ver Hoef, J.M., 2012. Who Invented the Delta Method? American Statistician 66, 124–127.

Wallace, R.B., Gomez, H., Ayala, G., Espinoza, F., 2003. Camera trapping for jaguar (Panthera onca) in the Tuichi Valley, Bolivia. Journal of Neotropical Mammalogy 10, 133–139.

Wegan, M.T., Curtis, P., Rainbolt, R., Gardner, B., 2012. Temporal sampling frame selection in DNA-based capture mark-recapture investigations. Ursus 23, 42–51.

Wegan, M.T., 2008. Aversive conditioning, population estimation, and habitat preference of black bears (Ursus Americanus) on fort drum military installation in Northern New York. Master's Thesis. Cornell University, January.

Wegge, P., Pokheral, C.P., Jnawali, S.R., 2004. Effects of trapping effort and trap shyness on estimates of tiger abundance from camera trap studies. Animal Conservation 7, 251–256.

White, G.c., 1996. NOREMARK: population estimation from mark-resighting surveys. Wildlife Society Bulletin 24, 50–52.

White, G.C., Bennetts, R., 1996. Analysis of frequency count data using the negative binomial distribution. Ecology 77, 2549–2557.

White, C.G., Shenk, T.M. 2001. Population estimation with radio-marked inividuals. In: Millspaugh, J., Marzluff, J., (Eds.), Radio Tracking and animal populations. Academic Press, San Diego, USA, pp. 329–350.

White, G.C., Anderson, D.R., Burnham, K.P., Otis, D., 1982. Capture-Recapture and Removal Methods for Sampling Closed Populations. Los Alamos National Laboratory, Los Alamos.

White, G.C., Garrot, R., 1990. Analysis of Wildlife Radiolocation Data, Academic Press, New York, USA.
White, G.C., Shenk, T.M., 2000. Population Estimation with Radio-Marked Animals. Academic Press, San Diego, California.
Whitman, J.S., Ballard, W.B., Gardner, C.L., 1986. Home range and habitat use by wolverines in southcentral Alaska. Journal of Wildlife Management 50, 460–463.
Wickham, H., 2007. Reshaping data with the reshape package. Journal of Statistical Software 21.
Wikle, C.K., 2010. Hierarchical Modeling with spatial data. In: Gelfand, A., Diggle, P., Fuentes, M., Guttorp, P., (Eds.), Handbook of Spatial Statistics. Chapman and Hall, pp. 89–106.
Williams, B.K., Nichols, J.D., Conroy, M.J., 2002. Analysis and Management of Animal Populations: Modeling, Estimation, and Decision Making. Academic Press.
Wilson, K.R., Anderson, D.R., 1985a. Evaluation of a density estimator based on a trapping web and distance sampling theory. Ecology 66, 1185–1194.
Wilson, K.R., Anderson, D.R., 1985b. Evaluation of two density estimators of small mammal population size. Journal of Mammalogy 66, 13–21.
With, K., Crist, T., 1995. Critical thresholds in species' responses to landscape structure. Ecology 76, 2446–2459.
Wolpert, R.L., Ickstadt, K., 1998. Poisson/gamma random field models for spatial statistics. Biometrika 85, 251–267.
Woods, J.G., Paetkau, D., Lewis, D., McLellan, B.N., Proctor, M., Strobeck, C., 1999. Genetic tagging of free-ranging black and brown bears. Wildlife Society Bulletin 27, 616–627.
Wright, J.A., Barker, R.J., Schofield, M.R., Frantz, A.C., Byrom, A.E., Gleeson, D.M., 2009. Incorporating genotype uncertainty into mark-recapture-type models for estimating abundance using DNA samples. Biometrics 65, 833–840.
Wynn, H.P., 1970. The sequential generation of D-optimum experimental designs. Annals of Mathematical Statistics 41, 1655–1664.
Yang, H.C., Chao, A., 2005. Modeling animals' behavioral response by Markov Chain models for capture-recapture experiments. Biometrics 61, 1010–1017.
Yoshizaki, J., Pollock, K.H., Brownie, C., Webster, R.A., 2009. Modeling misidentification errors in capture-recapture studies using photographic identification of evolving marks. Ecology 90, 3–9.
Zeller, K.A., McGarigal, K., Whiteley, A.R., 2012. Estimating landscape resistance to movement: a review. Landscape Ecology 27, 777–797.
Zuur, A.F., Ieno, E.N., Walker, N.J., Saveliev, A.A., Smith, G.M., 2009. Mixed Effects Models and Extensions in Ecology with R. Springer Verlag.
Zylstra, E.R., Steidl, R.J., Swann, D.E., 2010. Evaluating survey methods for monitoring a rare vertebrate, the Sonoran desert tortoise. Journal of Wildlife Management 74, 1311–1318.

Index

A

B

C

T

U

W

Z

Printed in the United States
By Bookmasters